Advances in
Cross-Cultural
Decision Making

Advances in Human Factors and Ergonomics Series

Series Editors

Gavriel Salvendy
Professor Emeritus
Purdue University
West Lafayette, Indiana

Chair Professor & Head
Tsinghua University
Beijing, People's Republic of China

Waldemar Karwowski
Professor & Chair
University of Central Florida
Orlando, Florida, U.S.A.

Advances in
Cross-Cultural
Decision Making

Edited by
Dylan Schmorrow
Denise Nicholson

CRC Press
Taylor & Francis Group
Boca Raton London New York

CRC Press is an imprint of the
Taylor & Francis Group, an **informa** business

CRC Press
Taylor & Francis Group
6000 Broken Sound Parkway NW, Suite 300
Boca Raton, FL 33487-2742

ISBN-13: 978-1-4398-3495-4 (hbk)
ISBN-13: 978-1-138-11674-0 (pbk)

Visit the Taylor & Francis Web site at
http://www.taylorandfrancis.com

and the CRC Press Web site at
http://www.crcpress.com

Table of Contents

Section III: Civilizational Change: Ideological, Economic, and Historical Change

Section IV: Cross Cultural Decision Making: Implications for Individual and Team Training

Section VII: Hybrid & Multi-Model Computational Techniques for HSCB Applications

Section VIII: Sense Making in Other Cultures: Dynamics of Interaction

Section XI: Understanding and Mitigating the Impact of Culture on Collaboration and Negotiation

Section XII: Use Cases of Cross Cultural Decision Making

Preface

This book is concerned with how decisions are made within a specific culture and across different cultures. The primary focus of the Cross Cultural Decision Making field is specifically on the intersections between psychosocial theory provided from the social sciences and methods of computational modeling provided from computer science and mathematics. While the majority of research challenges that arise out of such an intersection fall quite reasonably under the rubric of "human factors", although these topics are broad in nature, this book is designed to focus on crucial questions regarding data acquisition as well as reconciliation of mathematical and psychosocial modeling methodologies. The utility of this area of research is to aid the design of products and services which are utilized across the globe in the variety of cultures and aid in increasing the effectiveness of cross-cultural group collaboration.

Each of the chapters of the book were either reviewed by the members of Editorial Board or germinated by them. For these our sincere thanks and appreciation goes to the Board members listed below.

Explicitly, the book contains the following subject matters. Applications and Use Cases and of CCDM can be found in sections to aid a researcher in defining the requirements and metrics for this complex topic:

I. Applications of Human, Social, Culture Behavioral Modeling Technology

IV. Cross Cultural Decision Making: Implications for Individual and Team Training

X. Tactical Culture Training: Narrative, Personality, and Decision-Making

XII. Use Cases of Cross Cultural Decision Making

Theories and Techniques for understanding, capturing, and modeling the components of Culture are covered in Sections:

II. Assessing and Developing Cross-Cultural Competence
III. Civilizational Change: Ideological, Economic, and Historical Change
V. Cultural Models for Decision Making
VI. Extracting Understanding from Diverse Data Sources
VII. Hybrid & Multi-Model Computational Techniques for HSCB Applications
IX. Socio-cultural Models and Decision-Making
VIII. Sense Making in Other Cultures: Dynamics of Interaction
XI. Understanding and Mitigating the Impact of Culture on Collaboration and Negotiation

The science and technology provided in this book represents the latest available from the international community. It is hoped that this content can be used to tackle two of the biggest challenges in this area: 1) Unification and standardization of data being collected for CCDM applications/research so these data can support as many different thrusts under the CCDM umbrella as possible; and 2) Validation and verification with respect to utility and underlying psychosocial theory. Solutions for both of these must be in the context of, and will require, sound methods of integrating a complex array of quite different behavioral models and modeling techniques.

This book would of special value to researchers and practitioners in involved in the design of products and services which are marketed and utilized in a variety of different countries

April 2010

Dylan Schmorrow
USA

Denise Nicholson
USA

Editors

Chapter 1

Cultural Decision Making Through Aggregate Models of Human Behavior

Julie A. Rosen, Anne V. Russell, Mark A. Clark, Wayne L. Smith

Science Applications International Corporation
8301 Greensboro Drive, m/s E-4-5
McLean, Virginia 22102-3600

ABSTRACT

Computational social scientists who support culturally aware decision making have investigated how scientifically based human, social, culture behavioral (HSCB) knowledge can explain human-driven events. In today's state of the art, most approaches have proven difficult to translate to operational, decision-making processes. In this paper, we extend the existing state of HSCB modeling and analysis to enable an understanding of the evidence underlying the HSCB model and the analysis output. The authors recognize that "understanding" extends to the "visualization" of findings of HSCB modeling. To address this, the authors' approach toward such HSCB forecasts, especially the uncertainty of these estimates, is naturally presented as contours on an interactive geospatial map.

Keywords: HSCB model, reasoning under uncertainty, aggregation of influences, geospatial visualization

INTRODUCTION

Computational social scientists are researching how observations of human behavior might be used to develop scientifically based models of human, social, culture, and behavioral (HSCB) events. Today's state-of-the-art efforts, many of which have been funded by the U.S. Department of Defense (DoD), have proven difficult to translate to operational, decision-making processes. For example, the majority of processes used to implement HSCB models and their data remain cumbersome and time-consuming, demanding substantial processing power and prohibitive subject matter expertise with limited evidence of operational applicability [1]. Some modeling approaches have attempted to apply mathematical precision to assess human events, but have sacrificed usability by requiring data that are unavailable in real-world situations (e.g., volume and type). Other approaches have focused on engineering HSCB platforms and models that present analysts with an overwhelming burden to collect data sets, manually input data, and/or manipulate variables to operate in environments that do not have such luxuries of time. As such, most approaches fail to accommodate workflows, availability of data, and the transparency required to explain results in terms relevant to decision makers who must trust the output.

Robust modeling of human behavior requires methods that accommodate rapidly changing situations and uncertainty in the quality and timeliness of the information. To achieve the confidence of the decision maker responsible for culturally aware strategies and operations, models of human behavior must incorporate the physical, societal, and behavioral context of the operation's environment. Moreover, to make HSCB modeling more operationally useful, the discipline must embrace the operation's environmental constraints without sacrificing the models' theoretical foundations. These foundations must accrue (over time) and aggregate (across disciplines) evidence from a variety of sources.

AN HSCB REASONING FRAMEWORK

Several HSCB modeling attributes impart a less-than-certain confidence in both the underlying evidence and the relationships (existence, correlated, chronological, or causal) among the factors described in the evidence. Subjective judgments, which are valuable especially when "hard" data are scarce, come with human biases that must be understood and bounded, qualitatively and quantitatively. Moreover, academic disciplines that contribute information about past and evolving cultural trends employ terminology with varying levels of granularity, which in turn generates uncertainty when describing "truth" or "strength of influence" for the decision maker. These two examples highlight the need and potential for developing approaches that incorporate reasoning under uncertainty. The benefit of these approaches lies in their ability to forecast the likelihood of undesirable outcomes and to reduce the impact of such outcomes in the future.

Reasoning under uncertainty [2] begins with the assumption that human decision making is based on uncertain premises, which are interpreted by measurable and perceived evidence that is sparse, ambiguous, or conflicting. There are many approaches to managing uncertainty in decision making, ranging from the cognitively accessible, rule-based systems to the more robust but less-generally recognized algorithms based on probability theory. The framework illustrated in Figure 1 emphasizes probabilistic approaches including Bayesian inference networks. The emphasis reflects the need for algorithms that support the human decision-making process under rapidly evolving and complex situations.

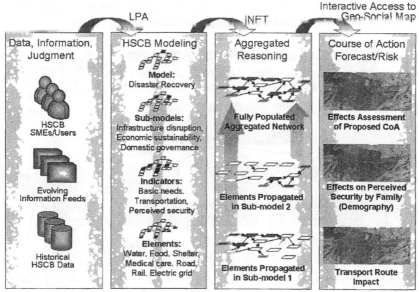

FIGURE 1. A framework for HSCB reasoning, from evidence through visualization, must address variation of granularity and terminology of the knowledge producers and decision-making consumers.

Data describing human behavior, including identity of factors and relationships among them, must be interpreted in the context of the evolving *field* in which social interactions take place and the effects they create in that space. Specifically, Lewin [3] speaks of *field theory* as being "characterized as a method of analyzing causal relations and of building scientific constructs." This seminal work in psychology is relevant to today's evolving work in HSCB modeling. For instance, traditional, physics-based modeling of medical epidemics after a natural disaster employs mathematical models of causality that incorporate influencing factors such as time, distance between population centers, estimates of medical resources in population centers, and volume and destination of emigration trends. When making culturally aware decisions, these factors must be expanded to include factors of human behavior for which sufficient empirical data do not exist.

In addition, robust reasoning of human behavior should address the notion of actor aggregates (e.g., groups of humans, nation-states, multinational organizations), the composition of which changes depending on the roles played by the atomic actors in the specific, but evolving, situation. Using the notional disaster model depicted in Figure 2, an actor might be an international non-governmental organization (NGO) providing medical care to earthquake victims. Elements under

4

the NGO's influence include providing basic necessities of shelter, water and food, in addition to the medical care that is its first priority. The NGO's role and influence with aid organizations that are providing medical supplies can be quite different than its role and influence with security forces responsible for protecting medical supply convoys. In this context, the leader/actor is an aggregate of the various factors associated with its role and the context-based set of relationships that derive from both the individual actors and their dynamic roles.

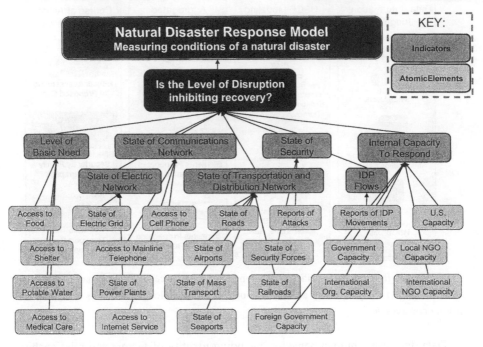

FIGURE 2. The SAIC Natural Disaster Response Model measures the level of disruption using data collected at the street level

Even with all these challenges, there is a natural framework that lies at the foundation of traditional inference modeling and that can be expanded to address the needs for cross-cultural forecasting. This framework is the mathematical concept of a multiply-connected network of entities (nodes or vertices) and relationships (links or edges). This structure is especially useful for HSCB situations when we use graphs to represent "local dependencies among conceptually related propositions" [4] While traditional HSCB modeling emphasized the nodes of such graphs, the authors focus on the links joining the nodes that model the perspective-dependent role of the actor that lies at the heart of our framework.

The remaining sections of this paper describe the authors' approach to searching available information (Section 3), using this information to reason about alternate, possibly conflicting, perspectives (Section 4), and presenting the findings to the culturally aware decision maker (Section 5). To illustrate our approach, we consider a notional disaster response model developed from evolving data collection and dissemination methods. In this real-world example, collecting, analyzing, and visualizing crisis data for operational planning has been a minute-by-minute endeavor where plans must be updated frequently.[1] Automated HSCB modeling and reasoning analytics can help streamline the information collection and exploitation to scale to the rapidity of the decision-making needs of the situation.

HSCB MODELING

In recent government-funded experiments[2] Science Applications International Corporation's (SAIC's) Linguistic Pattern Analyzer (LPA) has emerged as one of the more effective methods for instantiating and formalizing expert "paper" HSCB models. These include models that measure state stability, rebel activity, nuclear proliferation, pandemic flu outbreaks and the notional street-level disaster model of Figure 2. The LPA enables the expert model's use in automated environments and the rapid comprehension of current conditions. The model instantiation process depicted in Figure 3 applies to any evidence-based HSCB model that interprets text, has indicators that can be observed, and has a logical link among the indicators.

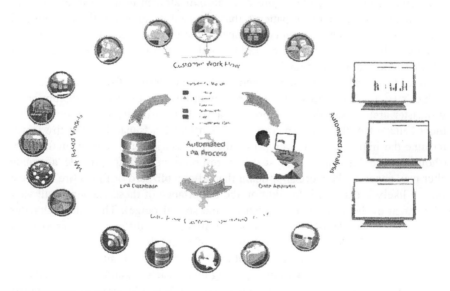

---

[1] Organizations such as Ushahidi have developed manual processes that take advantage of open source platforms to rapidly update and map crisis data online. The process currently involves manual manipulation to ensure proper data mapping. During a conversation with Ushahidi in late January 2010, the authors were told that a maximum of 50 percent of the data were mapped In a timely fashion during the first two weeks after the Haiti earthquake.
[2] For the Defense Advanced Research Projects Agency's (DARPA's) Pre Conflict Assessment and Shaping (PCAS) program, LPA was used to formalize and populate SAIC's Rebel Activity Model and other instability indicators to probabilistically based forecast engines. As a whole, the PCAS program successfully forecasted levels of instability in Thailand and Indonesia via computational models.

FIGURE 3. SAIC's Linguistic Pattern Analyzer's (LPA) process for formalizing and automating expert HSCB models

For each HSCB model, subject matter experts (SMEs) provide insight into methods for finding evidence and related linguistic patterns. Within the LPA, indicators are decomposed into atomic elements with related intensity scales, weights, and decision algorithms to mimic the intuitive decision-making process. Once the model has been instantiated, the LPA implements the expert's analytic process by identifying evidence for the indicator, calculating the severity of the occurrence and assessing the events of interest. The LPA outputs timelines that depict the event dynamics, provide inputs to other analytical engines, and facilitate scientific confirmation and measurement of the models' performance.

To illustrate the HSCB modeling and reasoning framework, the authors developed a conceptual disaster response model, formalizing Ushahidi's categories for organizing disaster data. This conceptual model measures the level of disruption to normal life that occurs in the wake of a natural disaster. We hypothesize that the disaster response model would be most effectively used in the four-month window immediately after a disaster. SAIC's formalized disaster response model concept uses seven key indicators to measure and aggregate the level of disruption to daily life within a disaster area, see Figure 2. Each indicator, in turn, is associated with atomic elements and linguistic patterns that are used to automatically capture and assess incoming granular data on conditions on the ground. Each atomic element is associated with a discrete-valued[3] rating scale that measures the intensity of conditions reported on the ground.

While these atomic elements and indicators derive from categories on Ushahidi's Haiti crisis map web site, they might be used to measure disruption in any disaster area. These elements provide a credible structure for reasoning. The authors believe that the type scenario – e.g., earthquake, tsunami, forest fire – will determine the importance (or weight) of the various elements. For example, after earthquakes and hurricanes, the scale of response needed to provide adequate shelter may generally be more significant than after a blizzard or forest fire, where a populace likely is displaced. In addition the importance of these elements will vary by the extent of the disaster (e.g., intensity and spatial range). The natural disaster response model can identify the nuances that affected the difference in response to the Haiti earthquake 2010, Hurricane Katrina 2005, and the Asian Tsunami in 2004.

In a notional concept of an operational disaster relief situation, the LPA platform would run in the background, polling both structured and unstructured data sources for evidence of atomic elements. The LPA would assess the evidence, aggregate the assessments, and provide to the reasoner frequent snapshots of current conditions on the ground.

---

[3] For purposes of this paper, we assume the typical 5-point scale ranging from 1 (manageable conditions or no damage to a staging area for response activities) to 5 (destruction is total and disaster response is completely lacking). The discrete values, in turn, are mapped to intervals for probabilistic reasoning.

REASONING UNDER UNCERTAINTY

At the heart of our inquiry into the modeling of cross-cultural human behavior situations is the desire to "exploit unique characteristics of sub-populations of data to make strong inferences about cause-and-effect dependencies." [5] Models constructed from data gathered from human observations to infer causal dependencies among the factors from these data are called causal models.[4] Whether causal models are fully automated[5] or SME-created to include human interaction, the goal of such modeling is to support decisions about which actions should be made in a given (and evolving) situation. For example, causal models can be applied to inform decisions to allocate limited resources to seemingly low probability but catastrophic outcomes, which reflect the nonlinear, asymmetric social value of undesirable behaviors from competing cultural perspectives.

SAIC colleagues developed a technique for analyzing the causal dependencies of complex and evolving situations. This technique, known as influence net (INET) modeling, is a marriage of two established methods of decision analysis: Bayesian inference net modeling, originally employed by the mathematical community and influence diagramming techniques originally employed by operations researchers [8, 9]. Our principal goal in creating the INET modeling process and associated decision support applications, Causeway™ and SIAM™, was the combination of an intuitive, graphical method for model construction, with a foundation in robust mathematical technique for the rigorous analysis of such models. Specifically, the reasoning engine performs forecasting with uncertain evidence and model elements of varying levels of aggregation.

As Figure 1 illustrates, HSCB model variables can be aggregated through an analytical representation that embodies the variables as nodes in a probabilistic reasoning framework. In contrast to the traditional rule-based, tree-like HSCB hierarchy, the variables are joined along multiply-connected chains within a Bayesian inference network. The uncertainty in these variables naturally transforms into the marginal and conditional probabilities of the inference network. These uncertainties are propagated along the multiple, possibly conflicting, network fragments reflecting the overall influence of an actor, whether individual or an aggregate of factors, in order to forecast the confidence in achieving the decision maker's goal. This forecasting technique applies to the full range of culturally aware disaster planning, such as the unintended effects of a long-view policy statement, the slowly evolving shift in cultural patterns of the populace, or the direct impact of proposed plans.

---

[4] We make the distinction between statistical models of association, which employ correlation associations derived from data and probabilistic models that identify causal dependencies. It is the latter that is considered in this paper.

[5] Recent research findings indicate great promise for automating the discovery of node and link structures of causal networks from relational data, see [6, 7]. Future research is recommended to explore the benefit of such methods to HSCB modeling and multi-discipline sources of evidence.

We can examine the results of the propagation of uncertainty along multiple paths using distinct perspectives, called *excursions* in our INET modeling, see Figure 4. For example, excursions might address a temporally changing influence of a given actor, say relative to a given culturally accepted period of mourning after a humanitarian disaster. Excursions also can be used to investigate the influence of factors that differ over neighborhoods or locations, e.g., rural and urban area disaster response. And for the decision maker responsible for planning culturally aware policies and operations, excursions encourage investigation of the effect of influence enacted with alternate hypothetical futures.

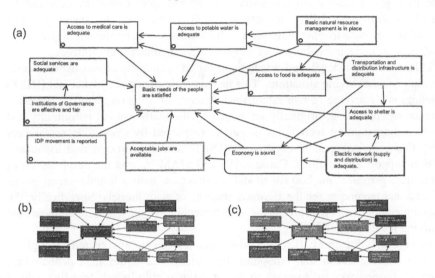

FIGURE 4. (a) Notional inference network model with individual (green highlighting) and aggregate (purple highlighting) model elements for Haiti 2010. Excursions capture the on-the-ground perspective of (b) city populace and (c) rural communities, which indicates the city center has lower likelihood of recovery.

VISUALIZATION

Even the most robust reasoning algorithms are only as good as the confidence they enjoy from the culturally aware policy maker and operations planner. In short, the visualization of findings from modeling efforts must address the human's subject matter background, talent and skills, and the task/charter at hand.

The authors recognize that the term "visualization" implies a cognitive understanding of the situation within which decision makers must plan and execute policies and operations. This visualization goes beyond traditional forecasting and planning tools that simply render data on a screen; it allows view and edit access to the elements of the reasoning model to incorporate real-world factors as knowledge of the situation evolves. Interactive graphical and tabular presentation offers a transparency that allows users to drill down to evidence supporting the reasoner's findings and incorporate evolving real-world events that traditional forecast methods do not address. This interaction is promoted through the "jargon-free" rendering of the reasoner's findings as geospatial overlays associated with the perspective of an excursion's timeline and location. In addition to the end-state (or goal) of the HSCB scenario, this presentation facilitates a "deeper" investigation of the second- and third-order effects of real-world events, see Figure 5.

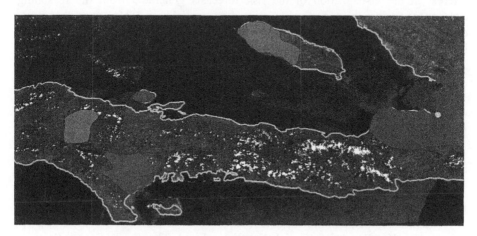

Figure 5. When excursions represent locational "grids," a geospatial map can be used to present the belief that a given event is true. Here, we analyze the likelihood of access to basic needs (Figure 4) for Haiti. Red color saturation indicates that recovery is highly unlikely near the city center and becomes a bit more likely (less saturated red) as we move away from the capital to the rural villages. Analogously, we use blue hues to indicate a greater than 50 percent (true) status for an event.

Moreover, belief in a given event can be compared across excursions to determine if all perspectives are influenced in similar fashion. For example, suppose the national government of our hypothetical country plans to deploy a limited number of medical personnel and supplies. Should they be sent to locations with the greatest number of civilians, say the city center where the populace works in the service economy? This statistics-based approach would be optimal if all medical issues held the same impact. However, we must account for the "Black Swan Events" [10] that have low probability, but catastrophic impact. By mapping the HSCB-based perspectives to a location grid, the traditional geographic map is converted to a geo-social map on which we can present the need for medical care.

SUMMARY

In this paper, the authors present a modeling approach to decision making in HSCB situations. We have presented a conceptual framework that includes probabilistic reasoning of model elements from both the physical and societal disciplines of study. We focus on the human user—whether policy maker or operational planner—as we perform the aggregated reasoning, with particular interest in the visualization of the risk forecasts generated from the reasoning component against the cognitively accepted geospatial map. This paper documents our concept formulation using extant modeling platforms, but further research into implementation of vetted HSCB models with the reasoning framework is required.

REFERENCES

[1] Russell, A. and Clark, M., Modeling Human, Social, Cultural or Behavioral Events for Real World Applications: Results and Implications from the State Stability Project. In Proceedings of 2009 IEEE Computer Society International Conference on Computational Science and Engineering, http://www.computer.org/portal/web/csdl/doi/10.1109/CSE.2009.410.

[2] Shafer, G. and Pearl, J., Readings in Uncertain Reasoning, Morgan Kaufmann Publishers, Inc., San Mateo, CA1990.

[3] Lewin, K., Resolving Social Conflicts: Field Theory in Social Science, American Psychological Association, reissued 1997, p 201.

[4] Pearl, J., Probabilistic Reasoning in Intelligent Systems: Networks of Plausible Inference, Morgan Kaufmann Publishers, San Mateo, CA, 1988.

[5] Jensen, D., Beyond Prediction: Directions for Probabilistic and Relational Learning In Inductive Logic Programming, Lecture Notes in Computer Science 4894 (H. Blockeel J. Ramon, J. Shavlik and P. Tadepalli, eds.). Berlin: Springer, 2008.

[6] Rattigan, M. and Jensen, D., The Case For Anomalous Link Discovery. ACM SIGKDD Explorations vol. 7 issue 2, December, 2005.

[7] Getoor, L., Koller, D., Taskar, B., and Friedman, N., "Learning Probabilistic Relational Models with Structural Uncertainty," Proceedings of the AAAI-2000 Workshop on Learning Statistical Models from Relational Data , Austin, TX (July 2000).

[8] Rosen, J. and Smith, W.L., "Influence Net Modeling: A Collaborative Approach to Forecasting," Proceedings of AIPA97 Symposium.

[9] Rosen, J. and Smith, W.L., "Influence Net Modeling for Strategic Planning: A Structured Approach to Information Operations," Phalanx Journal, special issue on Information Operations, December 2000.

[10] Taleb, N., The Black Swan: The Impact of the Highly Improbable, Random House, 2007.

Chapter 2

Information Channels in MMOGs: Implementation and Effects

Michael Zyda, Marc Spraragen, Balakrishnan Ranganathan,
Bjarni Arnason, Huahang Liu

GamePipe Laboratory, University of Southern California
zyda@usc.edu; marc.spraragen@gmail.com;
brangana@usc.edu; barnason@usc.edu; huahang.liu@usc.edu

ABSTRACT

We have integrated information channels into Cosmopolis, a Massively Multiplayer Online Game (MMOG), providing news from the real world, commercial advertisements, in-game announcements, and chat capability. The means for display of these information channels may include player UI elements as well as in-game objects like bulletin boards and news outlets. The technologies employed for this integration and display include Natural Language Processing (NLP), artificial intelligence-driven software agent non-player characters (NPCs), knowledge / information dissemination algorithms, and player-based news generation tools. There are several predicted effects of this information channel integration on MMOG elements (the game world, its economy, NPCs, and player statistics).

INTRODUCTION AND MOTIVATION

A 2008 study by the National Research Council entitled "Behavioral Modeling and Simulation – from Individuals to Societies" [NRC 2008] discusses the need to expand research in modeling and simulation of individual and societal behaviors. In that study, it is pointed out that a technological infrastructure needs to be developed for behavioral modeling such that the researchers can properly develop, test and then deploy such models. That study, in fact, suggests the development of a massively multiplayer online game (MMOG) for that infrastructure. Such an MMOG can be utilized as a testbed for models of individual and group phenomena. We have developed an MMOG with an eye towards applying that MMOG for the understanding of issues of peace maintenance and globalization.

Cosmopolis, the MMOG we are developing [Zyda et al. 2009], doubles as a test bed for social and behavior modeling. As such, Cosmopolis contains a new approach to incorporating information channels into a game environment. Previous virtual environments do not have the ability to analyze information feeds from real-world events. Creating a more realistic in-game experience based on analysis of real-world news headlines is a new and, given the escapist nature of many on-line games and virtual worlds, a counter-intuitive proposition. However, such integration is an important factor for effective simulation of virtualized social and organizational environments, as human reaction to various local and world news is a hallmark of global civilization.

RELATED WORK

In-game information channels, providing player chat and administrative announcements, have long been included in MMOGs and their Multi-User Domain (MUD) ancestors [Bartle 1990, 2007]. A more recent development is the inclusion of real-world advertisements in virtual world applications such as Second Life [Linden Research 2009]. There has also been some work on integration of real-world weather forecasts to simulate weather in sports games, such as the Madden football games enlisting data feeds from the Weather Channel [Bush 2008].

Principles strongly corresponding to real world economies were empirically shown to exist in virtual world economies [Castronova et al. 2009]. This mapping provides a background for experimentation with economic feeds from the real world to the game world, and by extrapolation studies that integrate other real world newsfeeds into the game world. One caveat found [Wikström et al., 2009] is that in order to engage players, real-world information integration needs to fit seamlessly into the game context (e.g., no car ads in a fantasy-roleplaying virtual environment).

APPROACH

Implementation of Information Channel Integration

In Cosmopolis, we structure our information system as a collection of channels through which messages flow. Channels may display news feeds from the real world or commercial advertisements; channels may publish in-game announcements publicly or regionally; channels may be configured as special chat lines between players. One of the goals of our project is to present messages efficiently and effectively to players, without breaking the immersion of the players' gaming experience.

Fig. 1 Cosmopolis: Newsfeed scrollbar (at bottom) and chat window (upper left)

For this purpose, we utilize some 2D UI elements (Fig.1). A scrolling line of text on the bottom of the screen displays important real-world and in-game headlines. The content of that line is generated automatically with the help of an open-source natural language processor. The intention is to capture users' attention without overly interfering with the 3D game world. The users can choose to see more details by clicking on the text. Elsewhere on the screen, there is a small scrollable window for in-depth observation of specific information channels. Users can also type into that window to broadcast messages to a channel, to chat with friends or NPCs, or to spread other kinds of news.

Another method of channel presentation in Cosmopolis is integrating information feeds within the 3D game world, using bulletin boards and message cubes. A bulletin board can be placed prominently as in the real world, to attract the attention of players and NPCs. For example, a bulletin board publishing general game announcements may be at the busy center of a city. A message cube is similar to a billboard, but more player-controlled and interactive. When players approach or click on a message cube, an in-game interface allows players to interact with the information channels they choose.

Effects of Information Channel Integration

Effects on the game environment

We are building an in-game virtual economy system that has a commodity market and currency exchange market. Commodity prices and currency exchange rates are synchronized periodically with incoming real-world rates. We will maintain a database of real-world commodities we are tracking, and updates to this database 1) will be available when in-game commodities are updating their prices and 2) may trigger a news broadcast or alerts to configured devices/applications.

Weather is similarly synchronized. Different weather conditions in the real world trigger different graphics renderings in our game. For example, different skyboxes, environment maps, lighting, and particle effects are used to simulate different weather conditions. Similar to a. above, our database contains regional mappings of real-world weather conditions, for 1) lookup when in-game weather changes, and 2) broadcast of extraordinary in-game weather conditions.

Effects on NPCs

People in the real world may behave differently according to the information they receive. In our game, by using data extracted from information channels, we change the behaviors of NPCs (implemented as artificial intelligence-driven software agents). An NPC exposed to news broadcasts may be configured to change state (i.e., take on a new action/belief/goal). News processors, one for each NPC and more generally, one for each different region or group, determine which news events will trigger responses from particular NPCs. For instance, a news warning of "earthquake imminent in region 1" published via UI or in-game objects may cause flight or other defensive behavior for NPCs in that region. Rumors of "Unrest" may coincide with NPCs behaving in a less friendly manner towards players or each other. A more mundane example: weather forecasts may cause changes in attire, behavior and mood. Stock market gradients can also change the aspects of NPCs, e.g. increasing stock prices makes certain NPCs happier.

Also, knowledge as a transferable commodity may spread through communities. Besides spreading rumors, NPCs and players may teach each other particular in-game skills that can only be learned through channeled knowledge-exchange transactions with other characters.

Effects on players

As our news channel system is relatively untested, it is an open question as to all the ways players will react to it. However, some speculation is possible. First, our system provides an efficient way for players to get real-world news while playing the game. Players can also be influenced by the game world's or NPCs' responses to certain events as described above. For example, NPCs are programmed to respond to earthquake predictions properly, and the players can learn from NPCs' actions.   Another example is that players may choose to move to a region where commodities are becoming relatively more valuable.

Furthermore, we provide a mechanism for players to spread news of their own. Each player will have a dedicated channel to send news messages. Their friends' news updates may be merged into an optional incoming stream or streams in a player's UI. As a player can spread any kind of message, it is important to encourage good use of the system. To achieve that effect, we borrow the ratings idea of "digging" and "burying": messages that get more "digs" will be displayed more frequently/prominently, and players will get reputation boosts or decrements as their messages get dug or buried, respectively.

CONCLUSION

Information flow in the real world is fairly complex. Because we are building an MMOG test bed for behavioral models, we are trying to address that complexity by bringing together the real world and the MMOG world with an information channel system. We intend to show via our own and third-party experimentation that our system will create a realistic simulated environment for behavior models.

REFERENCES

Bartle, R. Early MUD History. Linnaean.org, 1990. http:// www. linnaean. org/ ~lpb/ muddex/ bartle.txt

Bartle,   R.   ...On   the   State   of   Virtual   Worlds.   The   Guardian,   2007. http://www.guardian.co.uk/technology/gamesblog/2007/jul/17/idcloseworld

Bush, E. Real-Life Weather Effects come to Madden NFL 10 Games. Planet Xbox360, 2008.   http://   www.   planetxbox360.   com/   article_7414/   Real-Life_Weather_Effects_Come_to_Madden_NFL_10_Games

Castronova, E., Williams, D., Shen, C., Ratan, R., Xiong, L., Huang, Y., and Keegan, B. As real as real? Macroeconomic behavior in a large-scale virtual world. *New Media Society* 2009; 11; 685-707.

Linden Research, Inc. Advertising in Second Life. http:// wiki. secondlife. com/ wiki/ Advertising_in_Second_Life, 2009.

National Research Council. Behavioral Modeling and Simulation: from Individuals to Societies, Committee on Human Factors, Division of Behavioral and Social Sciences and Education, National Research Council, National Academies Press, Washington, DC, 2008, ISBN 0-309-11862-X.

Wikström, P., Comas, J., and Tschang, T. Once Upon A Social Web: Social Media And Firms' Learning Behavior In Two Worlds. Copenhagen Business School, 2009.

Zyda, M., Spraragen, M., and Ranganathan, B. Testing behavioral models with an online game. *IEEE Computer 42*, 4 (Apr. 2009), 103–105.

Chapter 3

Modeling Social Conflict: Theory, Data and Integration Across Multiple Levels

Michael Salwen, Elisa Jayne Bienenstock, Benton McCune, Ashley Arana

NSI, Inc.
Arlington, VA, USA

ABSTRACT

U.S. engagement around the world involving ethnic conflict, genocide and other humanitarian challenges raises a number of key challenges for military and civilian planners. Social crises are riddled with highly complex micro- and macro-level interactions. The Ethnic conflict, Repression, Insurgency, and Social strife (ERIS) system is a multi-paradigm model of ethnic conflict at multiple levels of analysis and implementation. ERIS aims to model the complexity of micro- and macro-level interactions within a society and provide insight into the range of possible social outcomes given varying sets of initial conditions. Social science theories such as relative deprivation, social capital and electoral incentives, among others, inform the system design. So that the project can scale, generalize and apply to a variety of contexts, it is built upon flexible theoretical drivers that draw together methods and ideas from empirical social science and computer science.

Keywords: Human Social Culture Behavior (HSCB) Modeling Program, agent-based model, system dynamics model, ethnic conflict, social strife, relative deprivation theory, electoral incentives theory, social capital theory, hybrid model

INTRODUCTION

Asymmetric warfare and social conflict involving non-state entities will be a persistent threat to U.S. national security in the foreseeable future. Not only will the U.S. need to defend against asymmetric attacks upon its national interests, but given its prominent role in world politics, U.S. forces will advise allied governments against insurgencies, mitigate ethnic conflicts and genocides and manage humanitarian and a range of counterinsurgency (COIN) and stability, security, transition, and reconstruction (SSTR) operations. This poses new challenges to military operational planning and raises the following key questions for analysts: What are the drivers of conflict and social strife? Which actions can be used to mitigate problems? What are the immediate and second- and third-order effects U.S. actions may have in conflict situations?

To address these questions, NSI is developing the Ethnic conflict, Repression, Insurgency and Social strife (ERIS) system. The initial ERIS design and development focuses on four states in northern India: Jammu and Kashmir, Himachal Pradesh, Punjab, and Haryana, which together comprise 62 districts and 306 sub-districts.

Ethnic conflict, repression, insurgency, social strife and a host of other well recognized forms of social crisis are examples of problems riddled with highly complex interactions between micro-level actors (e.g. individuals), between macro-level actors (e.g. government, media and communal leadership) and across these levels. The ERIS system attempts to address this complexity by systematically, methodologically and objectively simulating and integrating both micro-level phenomena via an agent-based model and macro-level phenomena via a system dynamics model.

The U.S. Department of Defense, under the aegis of the Human Social Culture Behavior (HSCB) Modeling Program, is supporting the development of several models akin to ERIS. In this paper, we use the ERIS project to discuss several important factors germane to the success of this overall effort.

BASIC PRINCIPLES OF ERIS SYSTEM DESIGN

The premise guiding ERIS design and development is that the system must be comprehensive, grounded in social science theory, based upon empirical data, yet flexible and generalizable beyond the case of India. We take comprehensive to mean accounting for contributing factors to intercommunal hostilities both percolating up from the population and moving out from centers of power or influence. A project like ERIS that is informed by theory and based upon data is built on general principles of social behavior that are instantiated in a particular model through the data available for that case. These general theories are derived from a large body of empirical work on societies spanning the globe at different points in history. The related concept of flexibility implies a system that can be readily modified or tailored to other areas of interest, or scaled to various levels of geographical or temporal resolution.

On the micro level, ERIS simulates the behavior of individuals with an agent-based model. The agent decision calculus in this model is governed by social science theory explaining individual choices with respect to fleeing or joining civil unrest, spatial segregation in response to intercommunal tension or violence and the role of interpersonal communication in escalating or dampening strife.

For substantive and practical reasons, agents in the model represent an aggregate of individuals. Substantively, a one-to-one ratio between persons and agents implies a level of precision in no way justified by the data available to construct the model and in no way relevant for the types of social forces we attempt to capture. On a practical level, reducing the many millions of individuals simulated in the model is necessary for execution efficiency. The representation used at the conclusion of the ERIS prototyping phase was 1000 persons per agent, uniform with respect to religious affiliation and heterogeneous with respect to age and sex.

On the macro level, ERIS includes a system dynamics component designed in relation to theories describing state actions in response to conflict or population migration and the effects of mass information dissemination from state, media or communal actors.

The ERIS system is intended to apply to conflict scenarios in areas beyond India and to scale in spatial or temporal resolution provided the availability of the data required to instantiate the simulation. In its development phase, the system operates spatially from the level of an Indian state down to the lowest geographic level for which data are available. The temporal resolution is not fixed, but presumed in the range of two to five years at a time step of one week. Many of these choices are determined by the availability of data at resolutions supporting these levels of analysis. The goal of ERIS is a system that depends upon broadly applicable and flexible theoretical drivers, with model parameters and scale determined by the specifics of the data in hand for a particular instance.

FROM OBSERVABLES TO CONCEPTS: BUILDING FLEXIBILITY AND IMPLEMENTING THEORY THROUGH ONTOLOGY

The ERIS model is being designed with a focus on India; however, the objective is to produce a general model that can apply across boundaries and cultures. To achieve this requires a blending of approaches from two fields: empirical social science and computer science. Empirical social scientists, while focused on the details and context of the social group of interest, rely on general principles of human behavior and interaction to frame their research. This provides both a basis for understanding and organizing the idiosyncratic details of their particular area, and also provides a means for communicating and relating these details to the work of others. It is as a result of comparing and contrasting the work of many scholars across many contexts that the regularities of human behavior have become known and can be refined and refuted (see Babbie 1990). In order to ensure standards across cultures and context, social science research methods have developed to facilitate relating the particulars observed or collected by one researcher to the work of others through a mapping to general theory. This mapping is known as operationalizing, and it requires the researcher to clearly define the general theory, which relates high-level social science concepts or latent variables, such as inequality and civic unrest, explicitly to measurable indicators of these concepts (see Babbie 1990). The connection between each high-level concept and its indicators must make sense theoretically and be empirically verifiable. Statements of the relationships between different latent variables define what most people think of as social theory. Testing and generalizing the theory to new contexts requires flexibility.

The challenge for ERIS is to develop a way to automate this flexibility. The hybrid ERIS model links micro- and macro-level concepts at the conceptual level rather than directly linking observables specific to India. With different data inputs the model could be adapted to other contexts and cultures if a reliable method is created to translate and map available indicators into the specified latent variables. To achieve this, ERIS is tied to an ontology, an explicit formal articulation of the domain knowledge of the social science required to operationalize ERIS latent variables (see Bollen 1989). This ontology has multiple levels. At the lowest level are potential observables that might be input as raw data. At the highest level are the latent variables included in the ERIS model. The ontology provides middle level concepts that build candidate constructs mapping to ERIS concepts. These intervening constructs are required to capture the complexity and allow the flexibility required for ERIS to accept and translate a range of data inputs so that the model can be applied to varied cultural contexts. They also provide an additional benefit: by making explicit the mappings of observable to higher-order concepts, the system will allow a non-expert to guide the operationalization of ERIS concepts with unanticipated observables as they become available (see Noy and McGuiness 2001).

SOCIAL THEORY INFORMING ERIS DESIGN

A wide range of theoretical traditions informs ERIS. Social science literatures on state breakdown, social stratification, social movements, ethnic conflict, and migration are all utilized (see, for example, Goldstone 1991, Grusky 2008, Horowitz 1985, Mann 2005, Skocpol 1979, Tarrow 1998 and Tilly 2003). Like many models of conflict, ERIS builds on the well known theory of relative deprivation most prominently developed by Ted Robert Gurr. More recent developments in social theory, though perhaps less familiar to modelers of social conflict, are also proving useful for the ERIS model. The ideas captured in social capital and electoral incentives theory, while of general applicability, have recently been profitably applied to the study of ethnic conflict in India. We briefly describe these theories below and discuss their relevance to the ongoing development of the ERIS model.

SOCIAL CAPITAL

Many social theorists, including Pierre Bourdieu and James Coleman, developed and popularized the concept of social capital. There is some variation in the usage of the term (for a discussion, see Glanville and Bienenstock 2009), but generally, social capital refers to the value inherent in social ties for both individuals and groups.

In his study of regional governments in Italy, Robert Putnam showed that areas with high levels of social capital and civic participation produce better outcomes in the form of stronger, more responsive democratic institutions. Putnam would go on to distinguish between *bridging* and *bonding* social capital. Bridging capital exists when there are ties between individuals across salient social identities such as race or religion. Bonding social capital is created by ties among people in one social group and can increase group solidarity, helping to achieve individual and group ends for members of that group. Bonding social capital also can also increase insularity and lead to conflict with people outside of the group.

For the ERIS project, we turn to Varshney, who applied social capital concepts to the Indian context and found that higher levels of bridging social capital corresponded to lower levels of intercommunal violence. Varshney argues that there is a key distinction between everyday cross-cultural civic engagement and associational civic engagement. Everyday engagement between Hindus and Muslims, while helpful, is not found to be as robust in preventing the outbreak of violence. A business group with Hindu and Muslim members is an example of associational civic engagement where ongoing contacts between ethnic groups can serve to quell rumors and defuse tensions.

Varshney compares Aligarh and Calicut, cities with very similar compositions of Hindus and Muslims, but differing levels of cross-cultural contact. Aligarh has a long history of communal conflict while Calicut has been largely peaceful. It has been argued that caste or class conflict has been more central historically in Calicut, thus reducing the saliency of religious identity, but Varshney argues that the most immediate cause of relative peace in Calicut is its rich intercommunal civic life. The social networks engendered by these organizations work to form an institutionalized peace system that renders Calicut resistant to provocations that would spark violence in cities such as Aligarh.

ELECTORAL INCENTIVES THEORY

The initial ERIS implementation utilized a wide array of social and demographic variables. One factor that was missing was state level politics. Donald Horowitz (2001), the noted authority on ethnic conflict, argues that the anticipated response of state actors to the initiation of an ethnic riot is a crucial factor in determining whether agents ultimately decide to initiate one. There are many local-level incentives promoting or constraining violence, but most important are the incentives at the level of government that controls the police or the army. Most actors initiate or continue violence when such acts can be conducted with relative impunity. In many cases, the promoters of violence are local elites pursuing their own ends (see Brass 2003). Horowitz went so far as to say "[t]here is not a single riot... in which rioters miscalculated their own tactics and power, the intentions of the police, or the response of their targets, such that the rioters suffered more casualties than their targets did" (Horowitz 2001, page 527).

In India, the prime responsibility for suppressing ethnic violence is at the state rather than the local or national level. Wilkinson shows that an Indian state government's willingness to do so largely depends on its electoral incentives. Wilkinson found that a key factor in whether the state would intervene to prevent Hindu on Muslim rioting was the extent to which the elected government was dependent on Muslim votes. One factor that made dependence on Muslim votes more likely was party fractionalization, a proliferation of parties at the state level that make coalitions necessary and where a (usually Muslim) minority can swing the election. Local Hindu leaders, aware of the political situation, use this knowledge in their calculation of whether to riot and at what level of violence. Wilkinson found these results to hold even when controlling for a variety of social and economic variables such as urban inequality, literacy, and the number of recent conflict events. These findings also appear to generalize outside the Indian context, as it is shown how groups in Africa, Ireland, Malaysia, Romania, and the United States utilized violence to increase ethnic saliency and solidify majority support in an intense electoral competition.

HYBRID MODELING AND EMPIRICAL VALIDATION

HYBRID MODELING

There has long been a macro–micro divide in social science. In economics, it is reflected in the very language used to describe the fundamental areas of the discipline. In sociology, there have been lengthy debates on structure versus agency. One can even see analogs in physics with the difficulties in unifying general relativity with quantum mechanics.

In ERIS, we have the micro level represented by an agent-based model and the macro represented by a system dynamics model. These are not to be thought of as simply two separate models running side by side, each residing in their own separate spheres, but intricately linked, often in quite natural and intuitive ways. For example, a conflict event generated at the macro level might drive a Muslim minority out of one location (an agent-based decision) and those people would decide where to emigrate based on a variety of variables including macro-level demographic and state political variables of the candidate regions. This then alters the macro-level demographic variables of the region left behind, which could serve to reduce tension due to less crowding and competition over scarce resources. A schematic of this type of interaction is depicted in Figure 1. The new region's changed demographics could result in new electoral incentives and more resource competition, which would combine with the fact that forced migration hardens ethnic attitudes of the immigrants (see, for example, Benard 1986). This would then feed back into the system dynamics model and alter the probability of future conflict events.

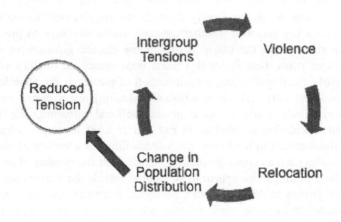

FIGURE 1. A schematic of the type of multi-level interaction modeled by ERIS.

The social theories discussed earlier quite naturally interlink across the micro and macro levels in ERIS. For example, the willingness of authorities to intervene against the instigators of an ethnic riot is a macro-level variable (in the Indian case, it lies primarily at the state level), but it must enter into the micro level when agents decide whether to engage in violent activity or not. Similarly, agent hostility towards other groups at the micro level is in part based on relative deprivation, which must be inferred from aggregated data on social inequality at the macro level. The social theory is not stretched to fit a hybrid model, but finds its natural expression there.

EMPIRICAL VALIDATION

ERIS will undergo rigorous internal validity testing to determine whether the simulation, in fact, captures the social theory we intend to implement. There will be a battery of tests that will modulate variables that would not normally be changed within the lifetime of a single ERIS instantiation. For example, simulation driven events are unlikely to alter a state government's willingness to suppress violence during the time frames under consideration by U.S. planners; however, we will test that the model reacts to changes in those variables appropriately. For among other reasons, this internal validity testing must be done to ensure that ERIS will behave appropriately when ported to other scenarios outside the Indian context.

One important aspect of the ERIS project is that not only will the model design be driven by the appropriate social science at both the macro and micro level, but the model will also be tested and externally validated against empirical data to the fullest extent possible. Data from multiple sources, including Indian census statistics and, where available, ethnic conflict event reports, will serve to validate the model and allow probabilistic predictions about the effects of proposed U.S. Diplomatic, Informational, Military or Economic (DIME) actions.

ERIS is not designed to predict the future, but to help analysts understand social context and its probable impact on events. Any computational model is founded on certain assumptions or data that may be wrong. Classical validation would ensure that the model correctly instantiates assumptions. This is often sufficient when the system modeled is well understood. Social systems are not understood well enough conceptually for this to be sufficient. While there are general regularities in the social world, at best these theories address the form of the equations: specific parameters and constants remain unknown. As a result, social science models require external validation. The challenge here is that the events of interest are rare and true experiments cannot be run. External validation, comparing model predictions to the unfolding of real events, is, therefore, not a reliable indicator of the value of the model. An excellent model may be wrong in the one instance that presents itself, or a bad model may successfully make a correct prediction. Without a sufficient number of cases and controls, it is impossible to determine the value of the real world event in validating the model.

The ERIS approach is to replicate the model rather than the reality in order to understand how errors in assumptions or in the data collected affect or change outcomes. The objective of ERIS is not to make the right prediction about what will occur, but to provide the user with the tools to understand the likelihood of a particular outcome given different sets of initial conditions. The ERIS model is not run once, but is iterated with key assumptions modified to demonstrate how robust or vulnerable outcomes are to both real perturbations in the system or misspecifications of the model. The output presented to the user is not a specific future, but a range of futures, and explanations of what differentiates these futures.

CONCLUSION

The ERIS system design incorporates explicit operationalization of high-order latent social variables by constructing micro-level (agent-based) and macro-level (system dynamics) model components that act upon these variables. This approach gives ERIS the robustness of the well developed social theory from which these latent variables are constructed and a flexible structure that can adapt to contexts outside the one in which the project is being developed.

The implementation challenge posed by this design is the creation of a mid-level ontology to map observable data of various types into the latent variables that the system processes. However well this may be done, ERIS, like any other model of its type, will always depend upon the data available to instantiate the simulation.

ERIS delivers to the end user a suite of possible future scenarios that can result from alternative applications of DIME actions. With this knowledge, the user will develop a richer understanding of both the country of interest and the strengths and weaknesses of the ERIS modeling environment.

Going forward, the ERIS team faces the critical challenge of transitioning this theoretically driven, empirically validated modeling system into a tool for the military planner or analyst. The ERIS design places the operational goal upon a solid theoretical and methodological foundation because it is the best way to deliver a tool that is functional, relevant and, above all, reliable.

ACKNOWLEDGEMENTS

The authors wish to thank ERIS team members Kari Kelton, Pamela Toman, Dominic Giordano, Robert Popp and Tom Allen for their invaluable insights and support in the drafting of this paper. Special thanks go to Bruce Bullock and Stacy Lovell Pfautz for their efforts on behalf of the ERIS project during its initial prototype development phase and to Stacy additionally for her work on the initial conception of this paper.

This work was sponsored by the Human Social Culture Behavior (HSCB) Modeling Program through the Combating Terrorism Technical Support Office (CTTSO).

REFERENCES

Babbie, Earl. (1990), *Survey Research Methods*. 2nd ed. Wadsworth. pp. 4 and 20.

Benard, Cheryl. (1986), *Politics and the Refugee Experience*. Political Science Quarterly, Vol. 101, No. 4. pp. 617-636.

Bollen, Kenneth A. (1989), *Structural Equations with Latent Variables*. John Wiley & Sons, Inc.

Bourdieu, P. (1986), *The Forms of Capital*. J. G. Richardson (Ed.), Handbook of Theory and Research for the Sociology of Education. New York: Greenwood. pp. 241-258.

Brass, Paul R. (2003), *The Production of Hindu-Muslim Violence in Contemporary India*. University of Washington.

Coleman, J. S. (1988), *Social Capital in the Creation of Human Capital*. American Journal of Sociology, 94(Suppl.), S95-S120.

Glanville, Jennifer. and Bienenstock, Elisa. (2009), "A Typology for Understanding the Connections among Different Forms of Social Capital". American Behavioral Scientist , Vol. 52, No. 11, pp. 1507-1530.

Goldstone, Jack A. (1991), *Revolution and Rebellion in the Early Modern World*. University of California Press.

Grusky, David. (2008), *Social Stratification: Class, Race, and Gender in Sociological Perspective*, Westview Press.

Gurr, Ted Robert. (1970), *Why Men Rebel*. Princeton University Press.

Horowitz, Donald L. (1985), *Ethnic Groups in Conflict*. University of California Press.

Horowitz, Donald L. (2001), *The Deadly Ethnic Riot*. University of California Press.

Mann, Michael. (2005), *The Dark Side of Democracy: Explaining Ethnic Cleansing*. Cambridge University Press.

Noy, Natalya F., and Deborah L. McGuinness. (2001), "Ontology Development 101: A Guide to Creating Your First Ontology". Stanford Knowledge Systems Laboratory Technical Report No. KSL-01-05. Available online at: http://www.ksl.stanford.edu/KSL_Abstracts/KSL-01-05.html.

Putnam, Robert. (1993), *Making Democracy Work: Civic Traditions in Modern Italy*/Robert D. Putnam with Robert Leonardi and Rafaella Y Nanetti. Princeton University Press.

Skocpol, Theda. (1979), *States and Social Revolutions: A Comparative Analysis of France, Russia, and China*. Cambridge University Press.

Tarrow, Sidney. (1998), *Power in Movement: Social Movements and Contentious Politics*. Cambridge University Press.

Tilly, Charles. (2003), *The Politics of Collective Violence*. Cambridge University Press.

Varshney, Ashutosh. (2002), *Ethnic Conflict and Civil Society: Hindus and Muslims in India*. Yale University Press.

Wilkinson, Steven I. (2004), *Votes and Violence: Electoral Competition and Ethnic Riots in India*. Cambridge University Press.

CHAPTER 4

Social Radar
for Smart Power

Mark Maybury

The MITRE Corporation
202 Burlington Road
Bedford, MA 01730
maybury@mitre.org

ABSTRACT

The center of gravity in modern warfare not only includes military targets such as tanks, ships, planes, command and control facilities, and military forces but equally important the perceptions, intentions and behaviors of citizens and leaders. This vision document develops the metaphor for a *Social Radar*, describes a framework for a novel set of capabilities, and identifies the need to create a community of interest that can advance the concept, development, and deployment of a Social Radar capability.

ORIGINS

With the global scourge of the second world war fresh in his mind and foreign air power a clear and present danger to dense US metropolitan populations, President Eisenhower set into motion a series of events to create a US air defense system that would ultimately lead to the founding of The MITRE Corporation in 1958 and the creation of the U.S. Air Force Semi Automatic Ground Environment (SAGE).

While SAGE contained many innovations such as high resolution display, photonic pointing device, database, time sharing, and track management, a core component of the system was radar (radio detection and ranging), which was invented in 1941. Radar provided a superhuman ability to see objects at a distance through the air.

FIGURE 1.1 . SAGE Operator Seeing Through the Air

While radar enables stand-off tracking of airborne objects, sonar (sound navigation and ranging) was invented to provide an ability to see through water to detect and locate submerged objects. Additional invention enabled humans to see non visible spectra, infrared, providing an ability to see through the dark.

Each of these inventions provided unprecedented improvements in situational awareness by increasing our "vision" through air, water, and night. Sensor and processing advances have provided increasing levels of fidelity, distance, and spectra (e.g., multispectral ability to remotely see chemical gases and liquids or vegetation types and conditions). Methods (e.g., foliage penetration, ground penetration, etc.) have even been developed to overcome the use of cover (foliage), concealment (underground), camouflage, and deception (CCCD) to deter/mitigate detection by earlier systems. In spite of these advances, however, radar, sonar, and infrared are blind to human adversary attitudes and intentions and often even behaviors toward our messages and activities. Our failure to track and affect instability, poverty, disease, corruption, conflict and natural disasters is done our own expense.

PURPOSE

The objective of this vision document is to develop the metaphor for a Social Radar (defined below), to envision a framework for a novel set of capabilities (since a social radar will require a multi-modal solution), and to create a community that can define and refine the requirements for and solutions to identified social radar needs.

NEED

The center of gravity in modern warfare not only includes military targets such as tanks, ships, planes, command and control facilities, and military forces but equally important the perceptions, intentions and behaviors of citizens and leaders. While radar, sonar, and infrared vision serve our military forces well, they provide limited insight into the social, cultural and behavioral and activities of populations. While hard power will always play a key role in warfare, increasingly soft power (Nye 2004), the ability to not coerce but to encourage or motivate behavior, will be necessary in the future of our increasingly connected and concentrated global village.

A *Social Radar* needs to sense perceptions, attitudes, beliefs and behaviors (via indicators and correlation with other factors) and geographically and/or socially localize and track these to support the smart engagement of foreign populations and the assessment and replanning of efforts based on indicator progression. As a modern center of gravity, the perceptions, cognitions, emotions, and behaviors of populations encompass the hopes, fears, and dreams of many publics. Accordingly, a social radar needs to be not only sensitive to private and public cognitions and the amplifying effect of human emotions but also sensitive to cultural values as they can drive or shape behavior. Conventional radar requires signatures for different kinds of objects and events: it needs to be tuned to different environmental conditions to provide accurate and reliable information. Analogously, a social radar needs signatures, calibration, and correlation to sense, if not forecast, a broad spectrum of phenomena (e.g., political, economic, social, environmental, health) and potentially forecast changing trends in population perceptions and behaviors. For example, radar or sonar enable some degree of forecasting by tracking spatial and temporal patterns (e.g. they track and display how military objects or weather phenomena move in what clusters, in which direction(s) and at what speed.) A user can thus project where and when objects will be in the future. Similarly, a social radar should enable us to forecast who will cluster with whom in a network, where, and when in what kinds of relationships.

One long term lesson from counter insurgency operations (COIN) is that while certain individuals or groups will always remain hard liners committed to their cause and yielding only to hard power, the only known successful exit strategy from an insurgency is re-integration of the disaffected into the political process. With respect to COIN, a successful social radar (and underlying models) should be able to sense and assess the trends of social engagement and provide indictors (based on or feeding models) of the positive and negative effects of engagement actions and messages on desired outcomes such as reintegration.

KEY SYSTEM PROPERTIES

For social radar to be as revolutionary as radar or sonar, it must exhibit some fundamental properties, including:

Global Access – Worldwide, real time capture, processing and analysis to include even areas with limited connectivity, denied access, or active censorship.

Multilingual and Multicultural – Ability to transcribe, summarize, translate, and interpret across languages, cultures, and societies.

Multimodal – Ability to process multiple media (e.g., radio, television, newspapers, websites, blogs, wikis) and multiple modalities (e.g., text, audio, imagery, action) which present challenges such as text understanding, speech recognition, and image and video understanding.

Persistent – Conventional access to foreign public beliefs and opinions is via polling or focus groups which are expensive, episodic (in anticipation of or in response to events), manually intensive, and subject to interviewer bias and interpretation error. Automated, large scale, continuous analysis of communications is required to provide wide area, multidimensional, long term dwell.

Real-Time, Geolocated – Social media (e.g., YouTube, blogs, wikis, twitter, Facebook, Flickr) can be analyzed in real time to provide sometimes attributed and localized to regional/group foreign public beliefs, opinions and behaviors.

Social – Detecting and tracking interactions among humans (individuals, groups, tribes, societies) using direct and indirect indicators to sense perceptions, attitudes, beliefs, opinions, and behaviors as well as the ability to infer roles and relationships, support social network analysis, and enable social network psychology (e.g., differentiating personal roles such as instigator or peacemaker and structural roles such as maven vs. connector).

Multispectrum – Ability to capture and correlate perceptions, beliefs, attitudes, and behavior in multiple domains including politics (e.g., governance), economics, military/law enforcement (including crime and corruption), society, healthcare, education, and the environment.

Passive and Anonymous – Preserving the anonymity and safety of the sensing activity (e.g., deterring traceback to the origin of the sensor) often by relying upon sources and methods that do not require active polling or engagement (e.g., using human to human typically public communications to assess perceptions, attitudes, beliefs and desires). Passivity is also important because anonymous collection helps mitigate bias that is inevitable when the person/population is aware of the data collection.

Security and Privacy Preserving – Although there are legitimate needs to track activities in economic, political, social, health and other spheres, methods and technology are needed to preserve individual security and privacy.

SOURCES

To detect, model, and forecast a broad range of phenomena, social radar will rely upon a rich set of sources including but by no means limited to:

Broadcast Media – Global, regional, and local broadcast services in print, radio, and television.

Social Media – User created content that is captured and shared via services such as in wikis, blogs, flickr, twitter, and YouTube, as well as social networking sites (e.g., Facebook, Linked-in).

Domain Specific Sources: Specialized sources can enable the detection of signatures in various domains such as health (e.g., ProMed, WHO medical reports), economics (e.g. World Bank reports, SEC filings), governance (e.g., UN corruption reports), or security (e.g., IAEA safeguards inspections).

METHODS

Like radar, sonar, or LIDAR (Light Detection and Ranging) sensors, a social radar will need to be calibrated and have signatures developed to detect and track various phenomena. A fully functional social radar will require elements including:

Calibration – Ability to baseline, benchmark (e.g., compare with traditional media), focus and/or refine indicators to enhance fidelity, accuracy, signal to noise ratio.

Signatures – identification of particular individuals (biometrics) or groups (sociometrics), sentiments (e.g., lexical, acoustic, or visual signatures expressing various shades of positive or negative reactions), and/or behaviors (e.g., economic, political, social, cultural).

Foreground/Background – The ability to provide foundational and/or baseline data such as the geography, demography, socioeconomic, political, and/or cultural environment which can provide a background for the integration and interpretation of foreground sensed events.

Noise Mitigation – Algorithms and methods are needed to ensure high signal to noise ratios by minimizing noise arising from variations in the signal or from the background environment (e.g., an individual's true attitudes or behaviors toward a message can be masked by their reactions to the messenger, local environment, or a global situation.)

De-cluttering - Algorithms and methods are needed to remove signals from irrelevant or duplicative signals from people, organizations, networks, topics, or events of non interest that interfere or obfuscate even a clean signal and thus impede sensemaking. Modeling of the various "terrains" (e.g., economic, political, environmental, social) promises to assist in developing countermeasures to clutter so that extraneous "returns" (i.e., passive interference) can be eliminated.

Jamming/Counter Denial – Just as radar needs to overcome interference, camouflage, spoofing and other occlusion, so too social radar needs to overcome denied access, censorship, and deception. Active interference against a social radar could mask targets, create false targets, or change how targets are sensed.

Correlation/Integration – Social "signatures" need to be correlated with indicators from other domains, such as demographic, economic, governance and health indicators. A particular challenge will be not only integrating across these domains but also integrating various levels of granularity within domains (e.g., micro versus macro economics; individual vs. group vs. tribal vs. national political models).

Spatio-Temporal Event Tracking – Density in time and space of phenomena intensity and progression as well as event correlation if not causation. Trend analysis is particularly challenging, as there may be a significant offset between stimulus and response.

Analytics – Sources and methods need to be developed to model, understand and forecast sociological events. For example, economic indicators might be microeconomic or macroeconomic. In health care, pandemic disease monitoring might use direct indicators (e.g., blood samples) or indirect ones (e.g., school closings, prescription supply and demand). Models across societies might help detect leading or lagging indicators as well as tipping points to forecast opinion or behavior change.

ARCHITECTURE

Figure 1 illustrates a high level systems architecture of a social radar including key sources, processing components and work flow. As detailed above, heterogeneous information sources include traditional news media (radio, television, print), polls, and surveillance sources as well as user generated, social media such as wikis, blogs, myspace, facebook, twitter, etc. These are processed using a variety of technologies and methods to support processes including media analysis, detection and tracking of signatures, and ultimately social indicator analysis. Indicators may be of group or individuals to include measuring perceptions, attitudes, sentiments, and intentions. Ultimately, these support collaborative analyses of military, law enforcement, religious, political, economic and health dimensions to support a range of missions including strategic communication, counter insurgency or counter radicalism, and humanitarian relief.

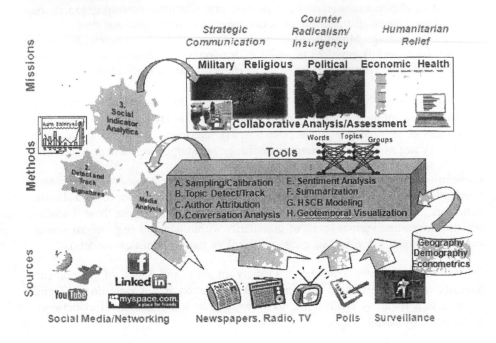

Figure 1: Social Radar Architecture

OPERATIONS

The creation of a social radar imposes the need for the development of sensors, signatures, and methods to collect, extract, process, correlate, and visualize social and behavioral phenomena. Tactics, techniques, and procedures will need to be developed to overcome sensor limitations in fidelity and coverage as well as denial and deception but also to discover how best to employ this new class of sensor or sensors. For example, sophisticated adversaries will employ viral communications that both infect (rapidly distribute) and affect vulnerable populations. Countering violent and viral communications require an ability to anticipate and counter message. Adversaries will actively "jam" social interaction, ranging from censorship to disruption of social media to physical intimidation.

A social radar should feed decision support tools and visualizations that can provide tailored support to defense, diplomatic, and development users performing a variety of functions ranging from security to stability to prosperity in roles encompassing policy formulation, public affairs, public diplomacy, intelligence, strategic communication, information operations, disaster relief, and military engagements. Results might be visualized and explored by time, geography, demography, and topic, depending upon the task at hand. Also, there likely will need to be strategic and tactical systems, e.g., a mobile, hand held, local area and task oriented version will be necessary for individual, focused use.

SOURCES, INDICATORS AND SIGNATURES

Table 1 (see Appendix) illustrates the range of some of the indicators, signatures and sources for various dimensions of a social radar such as military, political, and economic. This provides a sense of the range of open sources that can inform a social radar. For the political dimension, signatures that measure electoral fraud, trafficking (arms, drugs or human), laundering, public trust and the degree of freedom of the press can provide input to indicators such as the quality of governance, corruption, and balance of powers. These can be gleaned from reports from the United Nations, World Bank or Human Rights Watch, or journalistic reports and/or polling.

MEASURING AND GUIDING ENGAGEMENT

MITRE's research program has taken the first steps toward realizing the social radar vision. We have initiated several projects specifically focused on analysis of traditional and social media to understand perceptions and sentiments. These include:

- **Sentiment Analysis for Strategic Communication Assessment (SASCA):** Sentiment analysis using natural language processing to monitor attitude and behavior trends in key strategic areas of operation. Principal Investigators: Dr. David Day and Dr. John Boiney

- **Forum and Blog Threaded Comment Analysis (FABTAC).** Analysis of forum comment threads (e.g., YouTube comments, blogs) to provide interactive visualizations and thread summaries for intelligence and operations. Principal Investigator: Dr. Christy Doran

- **Public Opinion Polling by Proxy (POP/P).** An exploration of the ability of social media (e.g., Twitter) to serve as a proxy for traditional opinion polling methods to overcome their latency, expense, and invasiveness. Principal Investigator: Dr. John Henderson

- **Exploring Soft Power in Weblogistan.** Developing foundational Farsi and Dari language processing tools to enable analysis of large volumes of foreign

language   blog   and   social   media   content.
Principal Investigator: Dr. Karine Megerdoomian

- **Assessing Health Cognitions.** Integrating nationally-representative survey data into agent-based models of H1N1 virus transmission to assess the combined effects of individual health-protective attitudes (e.g., vaccination), behavioral intentions, and federal and local public health guidance. Principal Investigators: Dr. Jill Egeth and Dr. Jennifer Mathieu

TOWARD A "SOCIAL RADAR" COMMUNITY

Our grand vision of social sensing and tracking and the creation of a social radar will require a broad set of participants from a range of institutions including government, academic, industrial, non-governmental organizations (NGOs), and FFRDCs. Stakeholders must include users, developers, system integrators, and evaluators. This will by nature require an interdisciplinary set of skills from technical to operational to social sciences from communities spanning defense, development and diplomacy. Existing programs and partnerships (including coalition and international agencies) will need to be leveraged to address the many legal, data, processing, privacy/security and political/social impediments. Subject matter experts will be needed from many disciplines including but not limited to economics, political science, military science, history, psychology, anthropology, sociology, medicine, and environmental studies. Major steps will include establishing requirements, formulating a concept of operations, and assessing impact on various stakeholder communities. While daunting, the result could be nothing less than revolutionary.

CONCLUSION

Social radar is a long range vision for a capability that is essential to address a new center of gravity in modern affairs: public perceptions and behaviors. Realizing this vision will require the development of new methods of sensing, collecting, (socio-cultural-behavioral) modeling, processing, interpreting and acting on this new class of sensor. Successfully created, social radar would provide critical situational awareness and guidance in order to engage all of government power smartly in modern defense, diplomacy, and development.

REFERENCES

Damianos, Laurie E., Zarrella, Guido and Hirschman, Lynette. 2004. The MiTAP System for Monitoring Reports of Disease Outbreak.

Nye, Joseph. 2004. *Soft Power: The Means to Success in World Politics*. Public Affairs.

Final Report of the Defense Science Board Task Force on Strategic Communication. January 2008.

Report of the Defense Science *Board Task Force on Understanding Human Dynamics*. March 2009.

Stavridis, James G. "Strategic Communication and National Security," *Joint Force Quarterly*, 3rd Quarter, 2007.

ACKNOWLEDGEMENTS

Appreciation to Lynette Hirschman, Marty Ryan, Bill Neugent, John Boiney, Barry Costa, Gary Shaeff, Lisa Costa, Bob Nesbit, Danny Tromp, and Greg Crawford for valuable comments and feedback on earlier drafts.

The social radar vision originated from MITRE's Corporate Initiative in Smart Power (www.mitre.org/smartpower). Contact maybury@mitre.org for more information.

Table 1. Social Radar Example Sources, Indicators and Signatures

Elements	Military & Law	Political	Economic	Social	Health	Environment
Indicators	Violent/Border Conflict	Quality of governance	GDP	Displacement	Medical Access	Water/Air/Soil pollution
	Criminal Activity	Corruption	Employment, Poverty	Education Quality	Medical Outcomes	Climate
	Human rights	Balance of powers	Infrastructure	Hunger, Dissatisfaction	Mortality/Disability	Natural disaster
Signatures	Violent incidents	Electoral fraud	Currency stability	% homeless, % refugees	Care Access, % Insured	CO2, Smog, water quality
	Public safety	Trafficking, laundering	Consumer prices	% Graduates, Literacy rates	Absenteeism	Temperature, precipitation
	Grievances	Public Trust, Free media	Land Rights grievances	Grievances	AIDS, Birth/Mortality rates	Emergency Preparedness
Sources	UN reports	UN reports	World Bank	UN, NGOs	UN, NGOs	satellites
	Polls	World Bank, Human RightsWatch	SARs, DEA Reports	Newswire	World Health Organization	World Health Organization
	Newswire	Pew/Gallup	Bloomberg	Social Media	ProMED	Environmental NGOs

Chapter 5

Enabling a Comprehensive Approach to Operations: the Value of Human Social Culture Behavior Modeling

John A. Boiney[1], Dylan Schmorrow[2]

[1]MITRE Corporation
Bedford, MA 01730, U.S.A

[2] Office of the Director, Defense Research and Engineering
Office of the Secretary of Defense
Rosslyn, VA 22209, U.S.A

ABSTRACT

In the current global security environment, the success of operations increasingly depends on leveraging all instruments of national and international power in a coherent fashion. Such a "comprehensive approach to operations" (CA) involves coordinated and coherent action by multiple operational entities that may include national/international government agencies, militaries, non-governmental organizations, corporations, and other actors. We review the history of the CA and related concepts, identify core technical challenges associated with effective implementation of a CA, discuss ways that computational modeling may be leveraged to address those challenges, review selected initiatives and tools that are helping develop models and tools that may be useful in supporting the CA, and identify key science and technology gaps.

Keywords: Comprehensive approach, complex operations, modeling, social, cultural.

INTRODUCTION

Today, the U.S. military rarely acts on its own. Rather, the challenges to our national security are such that the services and other national defense elements are routinely required to coordinate or collaborate not only with other agencies of the U.S. government, but with other governments and non-governmental actors. What's more, the military's operational role has expanded considerably, not so much from "conventional" to "irregular" warfare, as is often written (When has war ever been "regular"?). Rather, where historically the military has focused on leading operations involving declared, kinetic-oriented conflict with a state-sponsored adversary, now it also routinely participates in operations with objectives focused on impacting non-combatant (green and grey) populations, and that may well involve facilitating the post-conflict recovery, reconstruction, and transition of a region.

With this reality, military operations are more likely to succeed if all instruments of national and international power (military, diplomatic, developmental, and economic) are employed in a coherent fashion. The multi-disciplinary concept of a "Comprehensive Approach to Operations" (CA) encourages coherent, coordinated, and constructive engagement amongst operational partners, such as national/international organizations, non-governmental organizations, businesses, international corporations and other local actors.

A number of key features characterize the CA to operations and have implications for its effective implementation. One, alluded to above, is the highly diverse array of actors who may be involved. The CA is a means to leverage the combined capabilities of these actors, which could include personnel from the military, public security, intelligence, diplomatic and development staff, host and local governments, allies, international organizations, non-governmental organizations (NGO), and private sector interests. Each type of actor brings its own resources, organizational structure, situation awareness, constituencies, priorities and politics. Yet, ideally, all will function as part of an effectively coherent whole for a given operation.

A related feature is the range of resources that are to be effectively applied in a given operation. The CA is a framework concept that enables the resources of the entirety of government, international organizations, NGOs, business, private actors, and local resources to be applied effectively in response to a crisis situation. Among the likely options are kinetic military action, disaster relief, humanitarian assistance, diplomacy, development, reconstruction, security sector reform, judicial reform, and social sector reform. These options vary enormously in a number of ways, not the least of which is their intended effects. All may be employed in stages as a situation evolves, or perhaps simultaneously. In any event, the ideal is for all to be coordinated, coherent, cooperative, and collaborative.

Historically, while elements of the CA to operations have been a part of effective counter-insurgency (COIN) operations, the concept is not restricted to or defined by COIN. Rather, the CA is defined in part by the fact that its elements should be applicable across the full spectrum of potential operations--domestic, expeditionary, or humanitarian.

A final feature worth noting is that, to be successful, implementation of a comprehensive approach to operations must rest on a shared understanding of the situation, the strategy and objectives. Despite the diversity and even the possible conflict among goals of the actors, there must be some common picture of the context and de-conflicted understanding of goals and objectives. That said, it is important also to stress that the CA does not (necessarily) entail an integrated or centrally controlled response.

There has been recent international attention to the CA, most notably by NATO which has moved to establish policy and recommended practices for its implementation (NATO, 2010). In the U.S., there is no national definition of CA, but there is widespread support for developing one. Concepts such as Comprehensive Approach, Whole of Government, Complex Operations, DIME (diplomatic, information, military, and economic) and Smart Power are widely used and discussed in policy, and in doctrinal, strategic, and operational documents. Much of the U.S. policy and doctrine that now governs CA can be traced to the aftermath of the 1994 Operation Restore Democracy, in Haiti.

> "...senior policymakers observed that agencies had not sufficiently coordinated their planning efforts. More specifically, they found gaps in civil-military planning, disconnects in synchronization of agency efforts, and shortfalls in resources needed to support mission accomplishment." (National Defense University, 2003, page 2)

On the diplomatic side of the government, National Security Presidential Directive 44 (NSPD-44) is the key document, specifying the Department of State's leading role in interagency efforts for stability and reconstruction. On the defense side, Department of Defense (DoD) Directive 3000.05(2005) has a comparable significance, providing guidance on military support to stability, security, transition and reconstruction operations. Much of U.S. joint doctrine addresses establishing, operating, and evaluating combined and joint task forces. Joint Publication 3-08 (2006) discusses interagency, intergovernmental, and non-governmental environments and provides fundamental principles and guidance to facilitate coordination between the DoD and other agencies and organizations. U.S. Army

Field Manual 100-7 (1995) specifies how theatre campaign plans are to be developed, a process that accounts for the roles played by domestic and international actors likely to be involved in a CA to operations.

IMPLEMENTATION CHALLENGES

Based on the preceding discussion, it may be self-evident that successful implementation of a comprehensive approach to operations is difficult. The types of challenges that may be encountered range widely, with roots in organizational, cultural, and communication issues. For some of the challenges discussed below, science and technology initiatives may be appropriate and helpful.

A starting point challenge is the simple lack of a common terminology and understanding within and between individual governments. It may literally be the case that potential CA partners don't speak the same language. Beyond that baseline challenge, it is likely that those partners don't use the same terms to characterize core operational elements, let alone what it means to take a "comprehensive" approach to the operations. There is a need at minimum for a CA taxonomy. Absent that, the lack of common understanding will lead to unclear direction across the set of actors, weakening communication and coordination.

The mix of military and non-military actors that may define the application of a CA to operations brings a set of issues. Many NGOs are circumspect, at best, about entering into any explicit coordinating relationship with the military. Typically, they stress non-kinetic means to their ends, and have constituencies and stakeholders that would react negatively to even indirect support of such means. There is also significant risk to consider. It is possible that opposing forces may target non-military elements, supposing them to be comparatively vulnerable, in an effort to disrupt whatever coordination may exist.

Another core challenge is the reality of highly stove-piped policy making, management, and implementing bureaucracies both within and across government actors. Demands associated with executing a CA to operations could dilute strong singular action that is at times necessary. Actors embedded in structures like this may well be unable to effectively respond to demands that comprehensive planning products place on them. In addition, information sharing will be compromised, a major problem as the management of information across government units and with other international participants is essential to effective implementation of a CA to operations. The integration challenge alone is enormous. At present, it is unlikely that any nation or organization is capable of integrating their management practices and information to fully support field operations. This challenge will be strengthened where organizational and cultural features compel actors to resist sharing information in a timely fashion.

Developing metrics and measuring effects following implementation of a CA to operations is a major challenge area. The problem is not simply developing reliable ways to detect and track the impact that such operations have on populations and conditions on the ground. It is also a question of anticipating the interdependent effects of different powers that are utilized as part of a CA to operations. For instance, how does the effect of a diplomatic action impact a military strike (if at all)? How might a "say-do" gap open, what will its consequences be, and how can its effects be mitigated? These are the kinds of questions that concern the ways that constituent elements of a CA to operations interact.

There is also the "cat herding" challenge. Orchestrating a CA to operations is a highly complex logistical matter. It is difficult to align and control such operations given that the CA is defined in part by its inclusion of NGOs and other actors who are not controlled or directed by participating governments. Different goals can conflict at times within government departments and agencies –not insurmountable, but potentially a significant obstacle to progress. These can lead to different views on priorities, principles, and mandates. Different planning and execution time horizons between actors also contribute to the problem, especially when contrasting military and development agencies. A common planning process or tool, which could be used by various government departments and agencies, as well as NGOs, would help—especially with the foundational task of establishing a common vision of the desired end state. However, it must be noted that not all participating entities will have the resources or expertise to engage in the required level of planning and engagement to be fully integrated into an operation.

A final challenge to note here is emergent behaviour. Even relatively simple systems exhibit behaviour that emerges more or less predictably from the interaction of their components. However, especially in more complex systems, behaviour may emerge that cannot readily be predicted or, indeed, accounted for by examining the system components. That is, the whole is greater than the sum of the parts. Such behaviour is highly likely in a system as complex as any operations that attempts to coordinate a set of actors as diverse as those involved in the CA to operations.

THE ROLE OF SCIENCE AND TECHNOLOGY

In the United States there is a robust and increasingly coherent set of S&T programs that, while not defined by a focus on the CA to operations, nonetheless are contributing to the evolution of the concept, exploration of its use, and development of tools to enable its effective application. Research and development programs span the full range of technical capability, including basic research, applied research, testing and evaluation, and transition to operational use.

Within the Department of Defense, programs are sponsored and executed by the Armed Services, the Office of the Secretary of Defense (OSD), Joint Improvised Explosive Devices Defeat Organization (JIEDDO), Defense Advanced Research Projects Agency (DARPA), Defense Threat Reduction Agency (DTRA) among others. The national laboratories (e.g. Sandia, Los Alamos, Livermore) conduct basic research that enables CA at least indirectly. DoD also coordinates and at times partners on research and development (R&D) with other departments, including the Department of Homeland Security, Department of State, and the national intelligence community. The DoD Command and Control Research Program sponsors and publishes a variety of material related to the C2 aspects of CA.

A recent report from the Defense Science Board Task Force on Understanding Human Dynamics (2009) includes results of a DoD-wide call for data on relevant research efforts. While that call was not specifically focused on CA, the work reported back to the Task Force spans most of the science and technology areas that enable a CA. Categories of R&D (and technology evaluation) identified in the report include:

- tools and techniques for collection and shared use of data, and for validation of socio-cultural models;
- research, modeling and analysis of adversaries, insurgents, and terrorists;
- influence operations and strategic communication;
- geospatial framework and services to enable integration of spatial, temporal, and socio-cultural information;
- simulation, training and mission rehearsal applications;
- research and analysis to build understanding and capability regarding groups, populations;
- modeling and research on operations, including tools for forecasting first to third order effects to support intelligence, course of action (COA) development, and decision making;
- infrastructure and applications to enable visualization of model outputs, multiple forms of information, and uncertainty; and
- socio-culture based indications and warnings and threat analysis.

In 2001, the Director, Defense Research and Engineering (DDR&E) established a Strategic Multi-Layer Assessment program that provides support to COCOMs and warfighters, and coordinates with the Joint Staff and U.S. Strategic Command (USSTRATCOM) to support global mission analysis. The program also integrates human, social, cultural, and behavioral factors, producing focused multi-disciplined strategic and technical assessments, and provides training and education regarding the development and application of new analytic tools. National Security Council exercises serve the same purpose – determining how inter-agency contingencies can be better reviewed at various levels. At much lower levels, ongoing lessons learned may be utilized in shaping doctrine and training by examining recent experiences.

Another effort, the Coalition Warfare Program, provides funding to projects that conduct collaborative research, development, testing, and evaluation with foreign government partners. The program (administered by the Director of Planning and Analysis, for the Under Secretary of Defense of Acquisition, Technology and Logistics) exists to assist Combatant Commanders, Services, and DoD Agencies in integrating coalition enabling solutions into existing and planned programs. It focuses on short term interoperability solutions, along with the early identification of coalition solutions to long term interoperability issues, such as architectures and major systems acquisitions.

In addition to R&D, joint experimentation, wargaming, and other exercises are critical elements of overall U.S. capability-building re the CA to operations. Joint Chiefs of Staff exercises, experiments, and concepts evolve, test, and evaluate new joint doctrine and concepts. These in turn influence the U.S. Combatant Command (COCOM) leaders and their staff to better evaluate war and contingency plans.

Supporting the DoD-related institutions and programs are a wide-ranging set of research-oriented institutions, both public and private sector. These include think tanks and foundations (e.g. The Markle Foundation), academia, and industry. Federally Funded Research and Development Corporations (FFRDC), including MITRE, RAND, and the Institute for Defense Analyses, provide technical support to DoD and other government sponsors. MITRE emphasizes engineering of complex systems across DoD and other U.S. government communities; RAND specializes in strategic and policy-level studies; the Institute for Defense Analysis (IDA) focuses on technology systems assessment and strategic planning.

Finally, there is a significant S&T role played by training and education programs that are sponsored by or in other ways support the DoD. For example, the National Defense University conducts quarterly symposia to create a cadre of professionals familiar with interagency processes and initiatives. These programs may be mechanisms for experimenting with or prototyping new technology and tools. And, of course, there is great interest in developing tools and technology specifically for the purpose of delivering better training (e.g. leveraging gaming technology). Another program, the Interagency Counterinsurgency Initiative conducts outreach through conferences, workshops, consultation, and collaboration. Included in this is the Consortium for Complex Operations, which networks training, education, research, and lessons learned programs underway across the U.S. government.

This survey of DoD-related programs and initiatives should not be considered comprehensive. Rather, it is intended to illustrate the fact that there is a good deal of research, development, and training work going on across the DoD and interagency that is aimed at supporting and better enabling a comprehensive approach to operations.

LEVERAGING COMPUTATIONAL MODELING

In this section, we describe a few of the U.S. programs that are supporting work on computational modeling that will benefit effective execution of a CA to operations. We then provide a list of some particular tools and, based on the earlier discussion of challenges, characterize how each tool can support implementation of a CA to operations. That discussion also indicates some of the areas where further work—research, development, and testing—would be helpful.

PROGRAMS

A number of U.S. programs support research, development, testing, and transition of models and model-based tools that support effective implementation of a CA to operations. The OSD Human Social Culture Behavior (HSCB) Modeling program is a vertically integrated effort to research, develop, and transition technologies, tools, and systems to Programs of Record (POR) and users in need. Administered by the Director of Defense Research and Engineering, the HSCB program is funded via three Program Elements, one focused on conducting applied research, one on maturing and demonstrating the tools and software outputs of that research, and another on testing and transition of tools and systems to formal acquisition programs and users. Rooted firmly in social science theory and methodology, the program's overarching goal is to provide DoD and the U.S. government with the ability to understand and effectively operate in human social culture terrains inherent to non-conventional missions. The program exists to support development of capabilities/tools for use in intelligence analysis, operations analysis and decision-making, training, and joint experimentation activities.

The U.S. Joint Forces Command (USJFCOM) sponsors the Joint Concept Development and Experimentation (JCD&E) program, which includes a multinational experimentation series. That series provides opportunities to explore new concepts and capabilities for multinational and interagency operations. "These capabilities include a 'whole of government' comprehensive approach to harmonize civilian and military efforts on a multinational basis." (http://www.jfcom.mil/about/experiments/multinational.htm). Modeling and simulation are critical elements of multinational experimentation, helping to generate recommendations to leadership, and to deliver validated innovations to practitioners.

The Air Force Research Lab (AFRL) supports research that addresses the fundamental long term challenge of predictive adversary behavior, with the objective of providing a detailed understanding of probable intent and future strategy in order to identify the set of potential courses of action that both adversaries and other entities' commanders have to consider while taking action. One major product of the AFRL research is the National Operational Environment Model (NOEM), a holistic modeling environment that supports baseline forecasts, analysis of pressure points for resolving instabilities, and what-if analysis. For more information on AFRL, go to http://www.wpafb.af.mil/AFRL/.

TOOLS

A number of methodologies and tools are in use and emerging that enable a CA. At the strategic level, a number of analytical tools enable modeling to anticipate and understand instability, and may be used to inform programmatic, operational, and tactical level plans. The U.S. State Department Office of the Coordinator for Reconstruction and Stabilization has developed the Interagency Conflict Assessment Framework (ICAF). The ICAF is designed to help agencies develop a common picture of the drivers of violent conflict in a given country, and to facilitate establishing a baseline against which to evaluate the impacts of U.S. involvement. A similar tool is the Global Forecasting Model of Political Instability, a product of the Political Instability Task Force, a government-sponsored grouping of researchers and scholars from a number of U.S. universities (http://globalpolicy.gmu.edu/pitf/).

At something closer to the operational level, there are several modeling tools either in development or already in use. An example is Senturion, a simulation capability that analyzes the political dynamics within local, domestic, and international contexts and predicts how the policy positions of competing interests will evolve over time. Developed by Sentia Group, Senturion relies on agent-based modeling to structure a simulation of the behavior of the individuals and groups that influence political outcomes. For an analysis of Senturion's application to a series of case studies, see Abdollahian, et al. (2006).

The Defense Advanced Research Projects Agency (DARPA) is overseeing development of the Integrated Crisis Early Warning system (ICEWS). The system will provide commanders with a capability to proactively manage and respond to security risks in their area of operations--spanning the entire spectrum of the crisis early warning and mitigation cycle. The system integrates social science models, theories, and data across multiple levels of analysis to systematically identify antecedents to a variety of destabilizing events. See the DARPA Website for more information (http://www.darpa.mil/ipto/Programs/icews/icews.asp.)

Social network analysis, game theory, systems dynamics, and red-teaming are methods and techniques with promise for improving our understanding of network behavior. Needed are analytical methods, models, and simulations that support analysis of emergent and directed behaviour in CA networks, and which increase understanding of trust-building and cohesion factors. The Office of Naval Research sponsors the program on Command Decision Making and Adaptive Architectures for Command and Control (A2C2). Using a combination of mathematical and computational models and empirical experiments with Navy officers, the A2C2 program has investigated the effectiveness of the alternative innovative organizational structures that are being enabled by the explosion in network connectivity. See the ONR Website for more information (http://www.onr.navy.mil/en/Media-Center/Fact-Sheets/Command-Decision-Making-Adaptive-Architectures.aspx).

Modeling-related R&D can also help advance capabilities for force synchronization and development of a common and coherent operational approach. COMPOEX is a first generation systems architecture for executing various computational models including systems dynamics and agent-based models (Waltz, 2008). COMPOEX has a suite of tools designed to help military commanders and their civilian counterparts to plan, analyze and conduct complex campaigns. It allows staff to explore sources of instability and centers of power in a conflict environment, visualize and manage a comprehensive campaign plan, and explore multiple courses of action in different environments to see the range of outcomes.

Visualization is an important mechanism for enabling better coordination in conditions like those characteristic of a CA to operations, where the actors are not only diverse but often distributed. Effective coordination will rest in part on a common operating picture, and planning will be facilitated by having the ability to display geospatial layers of social, cultural, and behavioral factors that define the human terrain. To help meet this need, the OSD HSCB Modeling program is supporting development of visualization tools and infrastructures that display hybrid data sources such as geospatial layers, between individual and group relationships, and related socio-cultural data in ways that are easy for the user to assimilate and that address how evidence is created using provided data and how uncertainty propagates throughout the system. Another initiative, the MAP HT Joint Capability Technology Demonstration, addresses the limited Joint, Service, and Interagency capability to collect, visualize, and understand the socio-cultural information necessary to assist Commanders in understanding the "human terrain" in which they operate (http://www.mapht.org/).

Models are also being used to support effective training to build cultural awareness and understanding. Modeling and simulation technologies are leveraged for serious games and other virtual interactive tools in the context of military training. One R&D area for the HSCB Modeling program is demonstration of distributed training technologies to speed the development of socio-cultural skills of coalitions in current military operations.

Measuring the impacts and effectiveness of complex operations is a major area of need and ongoing work. The U.S. Army Corps of Engineers has developed the Measuring Progress in Conflict Environments (MPICE) methodology for measuring outcomes in the transition from open conflict to stability and reconstruction operations. MPICE includes a comprehensive, generic metrics framework, procedures to tailor the metrics to the environment and mission, and a computer-based tool to archive, analyze, and visualize the collected data. Another effort to look at effects, sponsored by the HSCB Modeling program, will develop and validate software which models the outcomes of collaboration between U.S. military forces and NGOs.

One area where there is a significant need for further work is capability planning and force optimization. Determining the optimal set of multi-agency capabilities for a given operation is a significant challenge, and modeling can help inform decisions about those capabilities. In the United Kingdom, a number of models have been developed that can inform capability planning. These include the Peace Support Operations Model (PSOM) and its Stabilization Operations Analysis Tool (STOAT). In addition, the UK has developed DIAMOND (Diplomatic and Military Operations in a Non-Warfighting Domain), which is a model intended to help assess the effectiveness of variations in force mixes.

CONCLUSIONS

This paper is a modest attempt to highlight some of the current significant work that is helping to better define, understand, and execute a comprehensive approach to operations. Given the complexity of the topic, the paper would be incomplete under any circumstances. In addition, we have kept our attention almost exclusively on U.S.-based ideas, programs, and tools. There is a good deal of thoughtful, innovative work being done that is led by other nations. Fortunately, much of that work is being done in partnership with representatives of the U.S. military and government.

REFERENCES

Abdollahian, M., Baranick, M., Efird, B., and Kugler, J. (2006). "Senturion: A Predictive Political Simulation Model." Center for Technology and National Security Policy, National Defense University. Washington, D.C.

Bush, George W. (2005). National Security Presidential Directive 44. Washington, D.C.

Defense Science Board. (2009). Report of Task Force on Understanding Human Dynamics, Washington, D.C.

North Atlantic Treaty Organization. (February 2, 2010) Topic: Comprehensive Approach. NATO Website: http://www.nato.int/cps/en/natolive/topics_51633.htm

National Defense University. Interagency Management of Complex Crisis Operations Handbook. January 2003.

U.S. Under Secretary of Defense, Policy. (2005) "Military Support for Stability, Security, Transition, and Reconstruction (SSTR) Operations". Department of Defense Directive 3000.05, November 2005.

U.S. Joint Chiefs of Staff. (2006). "Joint Publication 3-08: Interagency, Intergovernmental Organization, and Nongovernmental Organization Coordination during Joint Operations. Volume II, The Comprehensive Approach." Washington, D.C.

48

U.S. Army, Headquarters. (1995). "Field Manual 100-7. Decisive Force: The Army in Theater Operations." Washington, D.C.

Waltz, E. (2008). Situation Analysis and Collaborative Planning for Complex Operations. Paper delivered at 13th International Command and Control Research Program. Bellevue, WA.

CHAPTER 6

Identifying and Assessing a Schema for Cultural Understanding

Joan R. Rentsch[1], Ioana R. Mot[1], Allison Abbe[2]

[1]Organizational Research Group, LLC
Knoxville, Tennessee 37919, USA

[2]U.S. Army Research Institute for the Behavioral and Social Sciences
Arlington, Virginia 22202-3926, USA

ABSTRACT

Cultural training typically focuses on understanding culture specific information. In contrast, the goal of the present research was to identify a generalizable understanding of culture that will improve individuals' functioning upon entering any novel culture. The overall objective was to build on previous research and identify the core elements of a schema for cultural understanding. The core content of a schema for cultural understanding relevant to Soldiers was identified as being pertinent to their mission context. However, it is multileveled and interconnected, thus exhibiting features of expert schemas. Items representing core culture schema content and the structural relationships between these items were identified .

Keywords: Schema for cultural understanding, cultural schema, expertise in cultural understanding

INTRODUCTION

Military and business organizations must increasingly operate effectively in multinational and, therefore, multicultural environments. Working in such settings is particularly challenging when collaboration across cultures is required. Teamwork within such settings requires the ability to see events as members of other cultures see them.

Traditional cultural awareness training often focuses on understanding another culture from a non-native's perspective, conveying facts and customs of a specific cultural group. In contrast, the goal of the present research was to identify a general schema for cultural understanding, a key multicultural perspective taking competency that will enable individuals to function effectively in multinational settings.

A primary objective for the overall research program is to identify the core content and structure of a schema for cultural understanding that can be used in future military and business training. The culture schema is relevant for individuals who will be working in a novel culture and/or interacting in multicultural settings.

THE NEED FOR CULTURE GENERAL UNDERSTANDING

Military and organizational leaders are urging training and education to address cultural awareness. Some of this might focus on knowledge of social structures (e.g., McFate, 2005) and multicultural perspective taking (Abbe, Rentsch, & Mot, 2009; Rentsch, Gundersen, Goodwin, & Abbe, 2007). Though cultural training is widely available for specific countries and cultures, it can be difficult to provide specific cultural content that is timely, accurate, and sufficient to address the needs of a deploying unit. Foundational culture-general education would provide service members with frameworks and skills to better learn about and function in unfamiliar cultural contexts.

In addition, military and business leaders would benefit from training, development, and other resources that enable them to understand many different cultures. Indicators of the need for additional cultural training include Special Forces Soldiers who must encounter a novel culture relying primarily on the internet for cultural information beyond basic regional information (e.g., economic, demographic) (McFate, 2005).

Unfortunately, evidence suggests that, with respect to abilities related to working effectively in other cultures, the United States seems to be trailing other nations (Stewart & Bennett, 1991). Furthermore, many Americans may not recognize their cultural understanding limitations or the implications of not understanding or misunderstanding cultural differences (Stewart & Bennett, 1991). Whereas individuals from other cultures tend to be aware that they do not know the

American culture, they may also take advantage of their awareness that Americans do not understand their culture.

To be effective, cultural training should certainly include language, history, and customs. However, traditional approaches to training cultural understanding tend to be focused on specific cultures rather than on developing generalizable cultural knowledge. A culture-specific approach to training and education has some limitations. Though knowledge about a specific culture will be useful in that particular culture, the training content must accurately represent the exact content of the particular culture of interest. Erroneous information can be potentially detrimental. A culture general approach is more versatile for circumstances when specific cultural information is unknown or varies widely in different subcultures or regions of a country.

A culture-general foundation may also help alleviate some of the anxiety commonly experienced when entering a novel culture. Typically, when entering a new culture, the individual may have only superficial knowledge of the region. A general cultural understanding may alleviate the anxiety by promoting rapid learning of important cultural elements.

A culture general approach is supported by a schema for cultural understanding (Rentsch, Mot, & Abbe, 2009). A schema represents general and abstract knowledge of a topic. Schemas influence actions, thoughts, and expectations. A schema for cultural understanding enables navigating in, interacting with, and learning novel cultures.

SCHEMA FOR CULTURE UNDERSTANDING

The purpose of the present study was to identify a schema for cultural understanding. Because schemas serve to develop meanings and interpretations in any content domain, they are useful for interpreting ambiguous stimuli (Rumelhart, 1980). Of particular interest in the present study were expert schemas that are useful in a military context. Expert schemas tend to help an individual learn and retain knowledge easily (Bereiter & Scardamalia, 1986) and are typically associated with adaptability. Furthermore, expert schemas contain information or solutions related to a variety of issues relevant to the schema content, making it easier for experts to identify solutions when encountering a novel problem (Van Lehn, 1989).

Because schemas are developed as a result of experience, the focus of the present study was on subject matter experts with operational experience in the Army. They possessed the experience that required them to develop a schema for cultural understanding relevant to the military mission domain of the Army. Rentsch et al. (2007) identified expert schemas for cultural understanding as a key multicultural perspective taking competency. Individuals who possess (or develop) expert schemas for cultural understanding will also possess increased ability to identify and adapt to novel cultural situations. Rentsch et al. (2007) articulated two major multicultural competencies associated with a schema for cultural understanding. They encompass the abilities to extract and interpret cultural influences in novel situations. These competencies in combination with an expert schema for cultural understanding will enable individuals to be more adaptable and effective as they navigate novel cultures. An illustration of a representation of a schema for cultural understanding is presented in Figure 1.

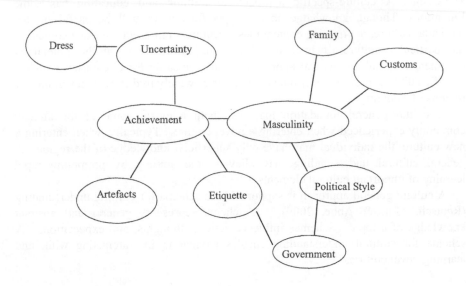

Figure 1. Hypothetical representation of a schema for cultural understanding.

The key, however, is to identify the schema for cultural understanding which enables individuals to develop general categories for interpreting cultural information. Multicultural perspective taking theory (Rentsch et al., 2007) argues that various types of knowledge will be included in expert schemas for cultural understanding. This type of knowledge includes, for example, the understanding that individuals exist simultaneously in multiple cultures. These cultures include national, regional, professional, organizational cultures, and so on. In addition, the schema for cultural understanding is theorized to include the knowledge that the multiple cultures have a significant impact on individuals' identity, thoughts, and behavior.

Furthermore, a schema for cultural understanding will include knowledge associated with understanding others' perspective on their own culture. An expert schema for cultural understanding is hypothesized to include the realization that variation exists within cultures, as well as the explicit recognition that similarities and differences exist across cultures (e.g., Hofstede, 2001; Pedersen, 2004). This includes an understanding of cultural barriers and identifiers (Rentsch et al., 2007). Cultural barriers make learning another culture difficult. For example, understanding that some cultural knowledge is tacit would be indicative of understanding a cultural barrier. Cultural identifiers are, for example, artifacts that are embedded with and convey cultural information.

An expert schema for cultural understanding might also include knowledge of general, macro-level features of national cultures. Many of these features have been articulated in scholarly research and include identification of cultural profiles that differentiate nations.   Hofstede (2001; 1980), Schwartz (1992), Trompenaars (1994), and House et al. (2004) have written about such profiles. These models provide frameworks for perceiving and organizing differences between cultures.

All of the components of a schema for cultural understanding described here are based on the literature in anthropology, psychology, sociology, and intercultural learning.   Although these components represent an expert schema for cultural understanding, they have been identified from the perspective of scholars.   A schema for cultural understanding that will be most applicable to a trainee is that which is developed based on cultural experts in the relevant context. Therefore, the present study was conducted to address this issue.

IDENTIFYING A SCHEMA FOR CULTURE GENERAL UNDERSTANDING

The present research was conducted using two samples of Soldiers who were interviewed about their understanding of culture. All Soldiers were returning from active duty in a foreign country and had been deployed for about a year. Therefore, they had some experience working successfully in another culture.

Based on the interview data, a set of cultural descriptors were identified as components of a schema for understanding culture. Another sample of Soldiers with similar experience then sorted and rated the descriptors. The ratings were analyzed and used to identify a core set of descriptors.

The core set of descriptors were used to develop an assessment of a schema for cultural understanding. Subject matter experts subsequently responded to the assessment and the results were used to identify the soldiers' schemas for cultural understanding. Thus, an assessment tool was developed and pilot tested as part of the study.

The study findings can inform training development and guide further research on the skills needed to function effectively in multicultural environments. Academics and Soldiers have highly different purposes for understanding culture. Therefore, experienced Soldiers were expected to have a unique and expert schemas for cultural understanding. Although the academic work is highly informative and useful to Soldiers, we expected that a schema for cultural understanding that is directly relevant to Soldiering would be the most informative to military trainees.

This initial research has pointed to the fact that Soldiers understand culture in terms that are relevant for their work context. For example, no Soldier described culture using terms such as "uncertainty avoidance" (which is a term used by Hofstede to describe the extent to which people in a culture tend to feel uncomfortable with and avoid ambiguity). Rather, Soldiers were more likely to mention mission goals and perceived danger in their descriptions of cultural features. Because the study is still in progress, the true nature of the Soldier expert schema continues to be articulated and represented. At this point, our expectation is that it will contain features of other expert schemas, namely that it will be multileveled and interconnected.

POTENTIAL APPLICATIONS

The ultimate goal of this research program and the research on multicultural perspective taking (Rentsch et al., 2007; Rentsch et al., 2009) is to contribute to the development of training designed to enable trainees to function effectively in novel cultures. Any type of training should be guided by sound training development principles such as those outlined by Goldstein and Ford (2002). Furthermore, it should adhere to principles and knowledge of adult learning processes (i.e., Kolb, 1984). Such components might include a multidisciplinary research team, best practices for adult and intercultural learning, and training.

The results of this study provide a foundation for the development of culture general training and have the potential to provide instructors with a better understanding of their training audience. The use of schemas in training is not unusual. It capitalizes on developing learners' cognitive structures used to organize cultural information. Future research will further inform cultural education and training. These findings demonstrate that the cultural expert of interest is that who operates in the context under study.

Cultural scholars (e.g., Kluckhohn, 1951; Kluckhohn and Strodtbeck, 1961; Hofstede, 1980; Hofstede and Bond, 1984; Schwartz 1992; Trompenaars, 1994) understand culture very differently than the subject matter experts who participated in the study. These subject matter experts approached and utilized the problem of cultural information in terms most relevant to their intercultural context. Therefore, the schema from the present research may be more effective when used to develop military training programs than frameworks drawn from the scholarly literature.

Thus, this research contributes to the development of cultural education and training. It will translate cultural knowledge into a cultural schema to guide training content with greater utility in the operational context of interest.

CONCLUSIONS

Identification of a schema for cultural understanding is critical for developing beneficial culture general training. The research described here is illustrates an innovative and useful approach. Future work will examine different levels of cultural expertise.

REFERENCES

Abbe, A., Rentsch, J. R., and Mot, I. (2009) "Cultural Schema: Mental Models Guiding Behavior in a Foreign Culture", paper presented at the Defense Equal Opportunity Management Institute Biennial Research Symposium, Patrick Air Force Base, Florida.

Bereiter, C., and Scardamalia, M. (1986). Educational Relevance of the Study of Expertise. INTERCHANGE Volume 17.

Goldstein, I. L., and Ford, J. K. (2002). Training in organizations (4th ed.). Belmont, CA: Wadsworth.

Hofstede, G. (2001). Culture's consequences. Thousand Oaks, CA: Sage.

Hofstede, G. H. (1980). Culture's consequences: International differences in work-related values. Beverly Hills, Calif.: Sage Publications.

Hofstede, G., and Bond, M. (1984). Hofstede's Culture Dimensions. JOURNAL OF CROSS CULTURAL PSYCHOLOGY Volume 15.

House, R. J., Hanges, P. J., Javidan, M., Dorfman, P. W., and Gupta, V. (Eds.) (2004). Culture, leadership, and organizations: The GLOBE study of 62 societies. Thousand Oaks, CA: Sage Publications.

Kluckhohn, C. (1951) "The study of culture", in: The policy sciences, D. Lerner & H.D. Haswell (Eds.), pp. 86-101

Kluckhohn, F.R., and Strodtbeck, F.L. (1961). Variations in value orientations. Oxford, England: Row, Peterson.

Kolb, D. A. (1984). Experiential learning: Experience as the source of learning and development. New Jersey: Prentice Hall.

McFate, M. (2005). The military utility of understanding adversary culture. JOINT FORCE QUARTERLY, 38, 42-48.

Pedersen, P. B. (2004). 110 experiences for multicultural learning. Washington, DC: APA.

Rentsch, J. R., Gundersen, A., Goodwin, G. F., and Abbe, A. (2007) "Conceptualizing Multicultural Perspective Taking Skills" (TR 1216). Arlington, VA: U.S. Army Research Institute for the Behavioral and Social Sciences.

Rentsch, J. R., Mot, I. , and Abbe, A. (2009) "Identifying the Core Content and Structure of a Schema for Cultural Understanding". Technical Report 1251. The U.S. Army Research Institute for the Behavioral and Social Sciences, 2511 Jefferson Davis Highway, Arlington, Virginia 22202-3926. Army Project Number: W91WAW-07-C-0053.

Rumelhart, D. E. (1980). On Evaluating Story Grammars, COGNITIVE SCIENCE Volume 4.

Schwartz, S. H. (1992). Universals in the Content and Structure of Values: Theoretical Advances and Empirical Tests in 20 Countries. ADVANCES IN EXPERIMENTAL SOCIAL PSYCHOLOGY Volume 25.

Stewart, E. C., & Bennett, M. J. (1991). AMERICAN CULTURAL PATTERNS: A CROSS-CULTURAL PERSPECTIVE (Rev. ed.). Yarmouth, ME: Intercultural Press.

Trompenaars, A. (1994). Riding the waves of culture: Understanding diversity in global business. Burr Ridge, Ill.: Irwin Professional Publishing.

Van Lehn, K. (1989). Problem solving and cognitive skill acquisition. In M. Posner (Ed.), FOUNDATIONS OF COGNITIVE SCIENCE (pp. 527-579). Cambridge, MA: MIT Press.

Chapter 7

Modeling and Assessing Cross-Cultural Competence in Operational Environments

Michael J McCloskey, Kyle J Behymer

361 Interactive, LLC
408 Sharts Drive, Suite 7
Springboro, OH 45066, USA

ABSTRACT

Contemporary operational environments are often characterized by ambiguous, multi-cultural contexts in which deployed military personnel must operate effectively without extensive prior knowledge of a region or its people. Ongoing training development efforts are addressing the need for general cross-cultural competence (3C), but this broad competence must be clearly understood in order to determine if our forces are being adequately prepared.

This chapter describes a cognitively-based approach to identify the key factors of 3C and related measures that assess that competence. A research-based model describing 3C development in small-unit leaders has been generated and it serves as the foundation for an assessment battery that is being developed. The battery includes self-report measures, situational judgment tests, cultural vignettes and a cultural attribute ranking task. These metrics are based on extensive cognitive and behavioral task analyses and surveying with over 300 Soldiers to date. An online tool incorporating these metrics will assess an individual's mission-centric 3C regardless of the operating environment and provide customized competence ratings and performance improvement guidance.

Keywords:
Cross-cultural competence, performance metrics, competence modeling, cultural development, cultural assessment, assessment tools, competence assessment

INTRODUCTION

Tactical military operations have always been and will always be complex and inherently uncertain endeavors. The fog of war that permeates small unit leader missions becomes more prevalent as operations become more culturally immersive. Identifying the true enemy, for instance, becomes much more complicated when the insurgent walks among the civilians that US forces are tasked to protect, support and win over. Broad mission goals of promoting stability, security and pro-US sentiment often conflict with more immediate tactical objectives. For example, the forcible removal of a hostile that would increase short-term security in a village might have long-lasting and far-reaching negative consequences on the "hearts and minds" of the residents. The broader impact of such actions is often difficult to envision, especially for those making life and death decisions on a daily basis.

As US Forces are being deployed to regions where US sentiment is mixed and mission sets require high levels of interactions with local populaces, the need for cross-cultural competence (3C) becomes more critical. For these interactions to be successful, US military personnel must be able to adapt to multicultural contexts despite limited experience with, or knowledge of particular regions.

UNDERSTANDING GENERAL CROSS-CULTURAL COMPETENCE

General 3C is the focus of several ongoing training development efforts across all branches of the military. However, even as these training interventions are being developed, the research community struggles to agree on a consistent definition of 3C (See Abbe, Gulick, & Herman for a discussion of this issue.). Varying operational definitions of 3C drive investments in training and assessment across a number of domains of practice, and the military is no exception. In order to accurately evaluate training initiatives and to focus research goals and investments, the military requires an operational definition of 3C that captures the actual field requirements of deployed military personnel in novel cultural environments. This research effort specifically focuses on addressing the following questions:

- What are the key attributes and skills that make military personnel effective in cross-cultural settings regardless of deployment location?
- How do these attributes and skills develop in military personnel over time?
- How can we assess this 3C in military personnel and predict their likelihood for success without being able to observe them in deployment settings?

Before measures could be developed to assess 3C, the research team first had to understand 3C *in operational settings*. Over 100 cognitive task analysis (CTA) data elicitations were conducted with Soldiers having experience operating in cross-cultural environments. The primary intent was to elicit critical incidents where 3C (or the lack thereof) impacted mission success in cross-cultural interactions. Many of the interviewed Soldiers had been deployed to at least five different countries, often providing them with a unique perspective on the nature of, and need for *generalizable* 3C. Using an additional CTA technique developed specifically for this effort, participants were prompted to rank team members from a recent deployment on a cultural competence continuum, and then provide descriptors of each clustering of team members in terms of their cultural characteristics, skills and abilities (See McCloskey, Grandjean, Behymer & Ross, in press, for more detail on data collection methodologies.).

Multiple rounds of data analysis yielded an initial set of KSAAs (Knowledge, Skills, Attitudes and Abilities) that were reduced to five key developmental factors that together comprise mission-centric 3C. Steps taken to reduce the CTA data into the five factors included multiple card sorts, incorporation of literature review findings, and pattern/thematic analyses within the data (McCloskey, Behymer & Ross, in preparation). Table 1 describes the developmental factors as well as the main KSAAs that comprise them.

Table 1 Developmental factors of cross-cultural competence

FACTOR	DESCRIPTION	KSAAS SUBSUMED BY FACTOR
Cultural Maturity	The ability to remain confident, calm, professional and dedicated in cross-cultural settings, and to further seek interactions to promote mission success	Emotional Self-Regulation Self-Efficacy Dedication Willingness to Engage Emotional Empathy
Cognitive Flexibility	The ability to withhold judgment in the face of limited information, remain open to alternative explanations and easily adjust perceptions based on new information	Openness Flexibility Uncertainty-Tolerance
Cultural Knowledge	The knowledge that cultural differences are deeper than customs, with an awareness of how they influence one's own behaviors and perceptions and those of others.	Awareness
Cultural Acuity	The ability to form accurate cross-cultural understandings and assessments of: situational dynamics, the perspectives and emotions of others, and the impact of cultural actions on the broader mission	Perspective Taking Sensemaking Big Picture Mentality
Interpersonal Skills	The ability to consistently present oneself in a manner that promotes positive short- and long-term relationships in order to achieve mission objectives	Self-Monitoring Rapport Building Relationship Building Manipulation/ Persuasion

Analysis of team member clusters from 68 team ranking tasks, along with a review of models of learning and expertise led to the identification of four stages of 3C development. As shown in Figure 1, at the initial stage of Pre-Competent, an individual has not developed enough cultural maturity or knowledge to move into a space where true learning and development can begin. This individual may be unwilling to interact for any number of reasons (fear, ignorance, and apathy, for example), so developmental opportunities will be extremely limited. Once the

60

individual advances to Novice/Advanced Beginner, he/she can begin to develop the cognitive and behavioral skills that promote successful assessments and interactions. Each further stage of development is characterized by varying degrees of advancement within the other factors. Note that as one becomes more competent he/she will advance to the Intermediate level and eventually to the Proficient/Expert level with enough positive learning experiences. However, within the Intermediate level, a series of negative cross-cultural encounters may generate frustration or hostility, resulting in decreased tolerance and a subsequent reversion to a lower level of competence. Interviewed personnel often refer to this phenomenon as mission "burnout".

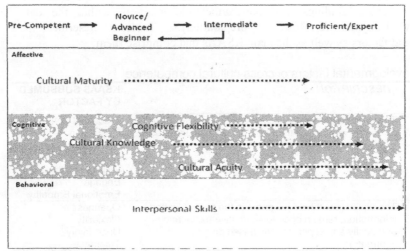

FIGURE 1 Developmental model of cross-cultural competence.

This model is the foundation for the development of a series of metrics that will assess general 3C. Ultimately, these metrics will be integrated into a computer-based tool that will assess the general 3C development of individual military personnel and provide customized feedback on how to leverage cross-cultural strengths and address identified competence gaps. The remainder of this chapter will discuss the development of these metrics and the ongoing evaluation process.

METHOD

While most existing 3C assessment batteries rely purely on self-report measures, we have expanded our measures to also include situational judgment tests (SJTs), cultural vignettes, and KSAA value judgment tasks. Self-report measures have been found to be effective in many cases in providing insight into an individual's affective attributes, but we have found that for this domain, they alone are not sufficient in providing a clear and rounded view of 3C. 3C in the military domain includes not only affective components, but also cognitive and behavioral elements as well. We cannot rely solely on the individual's own perceptions of their cognitive and behavioral abilities. Novices are, by definition, often unaware of their own developmental gaps and therefore tend to rate themselves inaccurately. Further, we have observed a social desirability bias in our self-report data collected to date. Thus, while we have developed self-report measures that assess affective, cognitive, and behavioral factors of 3C, we are also utilizing scenario/situational based measures that we believe will provide richer and more accurate insight into an individual's cognitive and behavioral abilities. Together, we believe that these measures will provide an innovative and realistic predictor of cross-cultural performance in deployment settings. The development of these predictor measures and our criterion measures is described below, and our overall validation framework is shown in Figure 2.

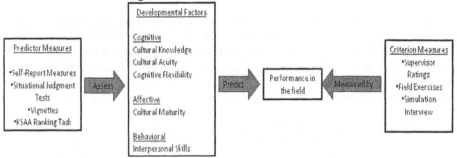

FIGURE 2 Metrics Validation Framework

PREDICTOR MEASURES

Multiple predictor measures have been developed to assess the five developmental factors of 3C and predict job performance as measured by our criterion metrics. Four metrics have been developed for this purpose:

- A Self-Report Measure,
- Three Situational Judgment Tests (SJTs),
- Three Cultural Vignettes,
- A KSAA Importance Ranking Task.

The development of each measure is described below.

Self-Report Measure

The critical incident and team ranking data collected over the course of this research effort provided the content for a self-report scale. Statements were developed by reviewing the raw data for instances of the earlier-identified KSAAs in cross-cultural descriptions as provided by the interview subjects. For example, in one team ranking task, a Soldier described a highly-ranked team member as "seek(ing) out opportunities to experience the local culture." This statement was converted to the self-report item: "During deployments, I would seek out opportunities to experience the local culture" with a Likert scale. The measure initially consisted of sixteen subscales with eight items in each subscale targeting each of the KSAAs identified during model development.

The self-report questions included a balance of deployment-specific and general questions to examine the impact of mission-relevant context. An eight-item social desirability scale was included as well. The overall instrument was pilot-tested and items were then eliminated if they yielded item-total correlation below .3 to optimize the reliability of each subscale. The revised 79-item self-report scale is currently being validated.

Situational Judgment Tests

According to McDaniel and Nguyen (2001), one of the best predictors of job performance is performance on a simulation of the job. This is the rationale for the development of three situational judgment tests (SJTs), in which the framework utilized by Motowidlo, Dunnette, and Carter (1990) was followed. To promote acceptance from the user community, the actual situations were based on critical incidents gathered from Soldiers in earlier cognitive task analysis interviews.

After the initial situations were developed, 80 Soldiers were asked to provide open-ended responses to the situations in order to develop realistic response options. An example of an open-ended question in a situation where a prized village goat is struck by a US vehicle is: *As the squad leader, what immediate courses of action (COAs) would you take in this situation and why?* For each question, the data gathered from the 80 participants were reduced to five to eight main response options that represented the range of response options. For example, one common response to the goat-strike query was along the following lines: *Through the interpreter, I would apologize to the group for hitting the goat and have the interpreter offer to pay for the animal.* Another frequent, but very different response was: *I would have my men get back in the trucks and leave to go straight to the village and secure it because our mission is to find that weapons cache, so we could deal with the goat problem later.*

Next, the SJTs were administered with the new response options to an additional 56 Soldiers. Participants were asked to rate both the *tactical* competence and the *cultural* competence of each response. In the analysis, any response item that had a tactical competence rating that was a standard deviation above or below the mean was eliminated. The intent here was to identify response options that varied significantly in cultural effectiveness, but did not vary significantly in tactical effectiveness. This procedure aimed to control for tactical considerations, ensuring that, with the remaining responses, any variation we find would not be a result of differences in tactical effectiveness of the responses.

To ensure the accuracy of the cultural effectiveness ratings of the remaining response options, we then vetted each response during additional pilot testing with Army cultural trainers.

Cultural Vignettes

The research team developed a series of cultural vignettes that differ from SJTs primarily in that they can include response options that go beyond multiple choice measures. For example, to assess an individual's observational skills (a subcomponent of cultural acuity), the participant may choose when to "stop the action" in a scene, and then select from a static scene view those perceptual cues that are most critical in the unfolding situation.

The initial versions of these vignettes, however, involve an unfolding scenario, interspersed with several series of questions that focus on specific developmental factors. For these vignettes, four responses to each question are presented to participants, who are instructed to rank each response option from the response they would most likely do to the response they would least likely do. The four response options were taken directly from Soldiers' responses to simulation interview questions that had been previously rated by the research team. One response option was selected to represent each of the four levels of competence (pre-competent, novice/advanced beginner, intermediate, and proficient/expert). An example is provided below:

How would you feel at that moment? (remember to rank the 4 options from most likely to do, feel, or think "1", to least likely to do, feel or think "4", using each number (1-4) only once).

___ A. I would be annoyed, but I would try to keep from letting it show for this first meeting, although I am not sure I could do that completely. (3)
___ B. I would be very angry and would be unable to keep it in. We just got done with a long plane ride and would be tired and he just shows up with no explanation. He would get yelled at. (1)
___ C. I would definitely be frustrated, but I would be able to suck it up and keep my cool. I would probably vent to one of my Soldiers later and cool off before I bring it up to the terp. (4)
___ D. I would be tired and frustrated and would probably raise my voice to him a little right there even though there would probably be a better way to handle it. (2)

KSAA Importance Ranking Task

Abbe et al. (2007) states that: "....the *perceived* importance of culture-specific knowledge and skills is sometimes higher than their *actual* importance." This has implications for providing competence feedback. Perhaps an individual will be less receptive to feedback on how to improve a particular KSAA if they do not consider that KSAA important. For this reason, we initially developed a KSAA importance ranking/rating scale. The developed survey therefore incorporates both normative (rating) and ispative (ranking) measures as each are known to offer unique benefits (Rankin & Grube, 1980; Baron, 1996). For this measure, participants rank the 16 KSAAs in order of importance and rate each KSAA's importance individually on a

Likert scale. Early indications suggest that an invidual's perceived value placed on certain KSAA's may correlate with their measured performance on the other predictor measures. This is currently being further explored and may result in the inclusion of this task within the final assessment battery.

CRITERION MEASURES

While this research effort is attempting to assess 3C, the main objective is to predict *performance* as it relates to 3C. Campbell (1983) defined performance as the degree to which an individual helps the organization reach its goals. Accepting that definition, we are primarily interested in assessing the KSAAs that improve an individual's ability to achieve mission objectives in a cross-cultural environment. As such, our criterion measures need to be performance based.

We have adopted the Motowidlo, Borman, and Schmit (1997) concept of performance which posits that the performance domain is composed of only those behavioral episodes that make a difference in accomplishing an organization's goals. These differences can range from an extremely positive effect to an extremely negative effect. With this definition of performance in mind, we are leveraging three measures of performance: a simulation interview, a supervisor/peer rating/ranking scale, and field training exercise ratings.

Simulation Interview

A simulation interview was developed using critical incidents collected earlier in the project. Within the interview, participants are first presented with situational background information, and then are presented with a series of six evolving situations. Each of the situations incorporates at least one specific challenge or decision point, with multiple probes addressing the challenge. The simulation is set in the Central African Republic of Burundi, which was chosen because of its current political and economic instability as well as its potential for future increased military presence in non-kinetic roles. Burundi is also a country that few participants were familiar with, reducing the prior regional knowledge that could be brought to bear on responses. Further, the realism of involving a genuine location has been found in our past research to support engagement by participants, versus fictional locations which often lead to participant frustration. The six segments of the interview involve cross-cultural situations such as how to prepare a team for a rapidly approaching deployment to an unfamiliar country, how to structure initial interactions with a new interpreter, and how to manage a village elder who may be supporting insurgents. At the end of each segment, a series of specific interview probes were used to elicit detail on the critical cognitive, affective, and behavioral KSAAs that could influence mission performance.

Supervisor/Peer Rating/Ranking Scales

Since supervisors and peers are more likely than researchers to be in positions to observe extended performance in the field, we developed a 3C rating scale to be used as an additional criterion measure. To reduce common errors with supervisor ratings such as ranking subordinates higher/lower than their actual performance (leniency/severity), avoiding extreme ratings even if they are warranted (central tendency), and having an overall impression lead to ranking an individual high on all dimensions (halo effect) (Kleiman, 2006), we followed the guidelines listed in Knapp et. al (2002). We also included instructions that provided: 1) an overview of the instrument and performance areas to be rated 2) directions on how to mark the ratings and 3) examples of common rating errors to avoid. Additionally, we are in the early stages of developing a peer rating/ranking scale to compliment the supervisor ratings. The scale has four categories that match the four developmental stages within our model, and it specifically addresses the model's developmental factors. Descriptive anchors were also developed for each developmental factor at each stage of development using data gathered from the simulation interview.

Field Training Exercises

In addition to the criterion measures that were developed in-house, the research team realized the importance of also having an independently evaluated measure. In the validation stage of the effort, we will administer the assessment battery (and eventually the computer-based tool) and conduct the simulation interview with Civil Affairs Soldiers who will be preparing to take part in a culmination training exercise prior to deployment. During this exercise, Soldiers will interact with role players in a simulated cross-cultural environment, and their performance will be evaluated by Army cultural trainers.

CONCLUSION & FUTURE DIRECTIONS

Unlike traditional performance domains where expertise develops along a somewhat predictable continuum, cross-cultural competence is inherently more complicated. Whereas a firefighter or medical professional is expected to begin as a novice and become more proficient as experiences accrue, expertise development in cross-cultural environments is heavily influenced by affective variables. Resistance to engage with local populaces, upfront hostility based on lack of awareness, and even refusal to accept that 3C is even real or of value all can inhibit development along the expertise continuum. This effort, therefore, represents only a beginning in understanding the complexities of 3C and how it develops in our military personnel. We have uncovered five key factors that influence 3C development as well as hypotheses about their interactions and relative importance at different stages of 3C development. We have developed candidate metrics to assess the levels of these factors present in an individual and are currently validating their predictive value. The resultant computer-based tool that incorporates these metrics will be field-tested in late 2010 with simulation interviews, supervisor ratings, and independently rated performance in field exercises all serving as criterion measures. It is our hope that the resultant tool will provide insight into an individual's likelihood for success in interpreting cross-cultural events and interacting in cross-cultural environments regardless of the deployment setting. We further envision that grouped results from the tool will be of use to commanders in assessing the cross-cultural readiness of their troops and to training developers to assess the effectiveness of their interventions. Further research is required on the nature, assessment and training of general 3C, but it is our hope that this current effort will be useful as a basis to generate further promising research directions in this emerging field.

REFERENCES

Abbe, A., Gulick, L. M.V., & Herman, J.L. (2007). *Cross-cultural competence in Army leaders: A conceptual and empirical foundation.* U.S. Army Research Institute for the Behavioral and Social Sciences, Study Report 2008-1. Arlington, VA: ARI.

Baron, H. (1996). Strengths and limitations of ipsative measurement. *Journal of Occupational and Organizational Psychology, 69,* 49-56.

Campbell, J. P. (1983). Some possible implications of "modeling" for the conceptualization of measurement. In F. Landy, S. Zedeck, & J. Cleveland (Eds.), *Performance measurement and theory* Hillsdale, NJ: Lawrence Erlbaum Associates, Inc.

Kleiman, L. (2006). *Human resource management: A managerial tool for competitive advantage.* Cengage Custom Publishing.

Knapp, D. J., Burnfield, J. L., Sager, C. E., Waugh, G. W., Campbell, J. P., Reeve, C. L. Campbell, R. C., White, L. A., Heffner, T. S. (2002). *Development of predictor and criterion measures for the NCO21 research program.* (Technical Report DASW01-98-D-0047, DO 0015). Alexander, VA: U.S. Army Research Institute.

McCloskey, M.J., Grandjean, A., Behymer, K.J. & Ross, K.G. (in press). *Assessing the development of cross-cultural competence in Soldiers* (Technical Report). Alexander, VA: U.S. Army Research Institute.

McCloskey, M.J., Behymer, K.J. & Ross, K.G. (in preparation). *Modeling cross-cultural competence to support the development of an assessment system* (Technical Report). Alexander, VA: U.S. Army Research Institute.

McDaniel, M.A. & Nguyen, N.T. (2001). Situational judgment tests: A review of practice and constructs assessed. *International Journal of Selection and Assessment, 9,* 103-113.

Motowidlo, S. J., Borman, W. C., & Schmit, M. J. (1997). A theory of individual differences in task and contextual performance. *Human Performance, 10,* 71-83.

Motowidlo, S.J., Dunnette, M.D., & Carter, G.W. (1990). An alternative selection procedure: The low-fidelity simulation. *Journal of Applied Psychology, 75,* 640-647.

Rankin, W. & Grube, J. (1980). A comparison of ranking and rating procedures for value system measurement. *European Journal of Social Psychology, 10,* 233-246.

CHAPTER 8

Using Cultural Models of Decision Making to Develop and Assess Cultural Sensemaking Competence

Louise J. Rasmussen[1], Winston R. Sieck[1], Joyce Osland[2]

[1]Applied Research Associates
Culture & Cognition Group
Fairborn, OH 45324-6362, USA

[2]Global Leadership Advancement Center
Department of Organization & Management
College of Business, San Jose State University
San Jose, CA 95192-0070, USA

ABSTRACT

In this chapter we outline a theoretical framework for cultural sensemaking that connects high level metacognitive skills to region-specific knowledge. We also describe a novel instructional analysis and design approach, specifically developed to identify learning objectives and content for cultural sensemaking training. This approach leverages cultural models of decision making in the development and assessment of cultural competence. The cultural sensemaking framework describes a possible avenue through which culture-specific learning can contribute to culture-general competence.

Keywords: Culture, Sensemaking, Competence, Training, Assessment

INTRODUCTION

> *"(The Afghan National Police officer) came in and shook everyone's hand. He came in and it was a quick walk around the room. And then he sat on the couch and blew up. Afghans will shake your hand and walk out the door and then have you killed. A handshake means nothing to them. ... It makes no sense."*
> (Marine Corps SGT, Special Operations Team Leader)

Understanding and having the ability to influence foreign decision makers within the cultural terrain is increasingly recognized as a core warfighter competency. Providing warfighters with region-specific knowledge needed for current missions, as well as the cross-cultural competence required for future missions presents a theoretical as well as practical challenge for culture programs across the services. Current efforts towards defining and scoping culture-general capabilities, or Cross-Cultural Competence (3C) include high level cognitive skills such as sensemaking and perspective taking (Abbe, Gulick, and Herman, 2008)—however, the field has yet to effectively characterize the cognitive processes that these skills entail.

In this paper we will present a theoretical framework that specifies the role high level cognitive and metacognitive processes, such as cultural sensemaking and perspective taking, play in 3C and connect these processes to specific content knowledge. We will also describe the role of cultural models of decision making within the cultural sensemaking process and explain how cultural models can be used within an instructional analysis and design process to assess competence and develop learning objectives. For purposes of illustration, we will describe a specific study where we employed cultural models of Afghan decision making to assess competence in a target learner population and to derive learning objectives.

CULTURAL SENSEMAKING

Sensemaking in general refers to the processes involved in understanding events and behaviors in a broad sense. *Cultural sensemaking* refers to the processes by which people make sense of and explain culturally different behaviors (Osland and Bird, 2000). When people try to make sense of events, they begin with some perspective, viewpoint, or framework (Klein, Phillips, Rall, and Peluso, 2004). Within the context of culturally different behaviors, this initial perspective is often grounded in expectations stemming from the normal situational behavior learned within one's own culture (Archer, 1986).

In the social and personality literature, perspective taking is described as the capacity to think about the world from other viewpoints that *"allows an individual to anticipate the behavior and reactions of others"* (Davis, 1983, p. 115). We propose that perspective taking is a component of cultural sensemaking in that it is an approach people can use to generate explanations for cultural behavior. We further propose that in order to use perspective taking to generate culturally appropriate explanations for a behavior, one needs insight into what people from the other culture think and care about, what motivates them. In the following section we describe a model that outlines the components of cultural sensemaking competence.

CULTURAL SENSEMAKING COMPETENCE

To talk about cultural sensemaking as a competency, we need to define its requisite knowledge, skill, and attitude components. Based on our previous research into the cognition that drives the decision making of US military servicemen within intercultural situations, we have developed a model of cultural sensemaking competence. This model, described below, defines the cognitive content (knowledge) as well as metacognitive processes (skills) involved in cultural sensemaking and relates these to attitudinal outcomes.

Cognition: Culture-Specific Content Knowledge

In our model, the knowledge component of cultural sensemaking competence is the knowledge that allows you to successfully explain and predict the behavior of people with different cultural backgrounds within specific situations. In order to effectively take the perspective of another within an intercultural situation, a person requires insight into what the other person thinks and cares about, what motivates them. That is, a *mental model of the factors that influence members of another culture's decision making within specific contexts* can assist a person in making sense of and anticipating their behaviors. Particularly, this knowledge should enable the person to make sense of cultural behaviors that appear paradoxical. Cultural scholars have long argued that culture is both paradoxical and context-specific (Kluckhohn and Strodtbeck, 1961), which means that an etic approach utilizing comparative bipolar cultural value dimensions is not sufficient to understand cultural complexity (Osland and Bird, 2000). Importantly, we are proposing an *emic* approach to defining the knowledge that is necessary for cultural competence.

Metacognition: Culture-General Question Asking Skills

The first sensemaking challenge within an intercultural situation is recognizing when the frames one would normally use for sensemaking no longer apply (Osland and Bird, forthcoming). Next, one must seek the information one needs in order to be able to develop culture-appropriate understanding. We therefore propose that the most important skill-component of cultural sensemaking is "skill in the process of knowledge-getting" (Bruner, 1966). That is, the ability to obtain and/or construct the knowledge required to successfully 'make sense of' and subsequently explain and predict behavior. Further, this overall skill is embedded within a framework of related metacognitive skills which allow the individual to obtain, apply, test, and refine their cultural knowledge. These metacognitive skills are culture-general in the sense they support attainment of culture-specific knowledge within any culture. Figure 1 illustrates the metacognitive skill components of cultural sensemaking.

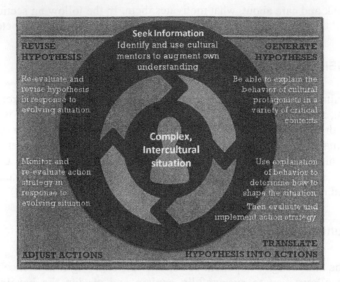

FIGURE 1. Metacognitive skill components of cultural sensemaking.

So, how does one 'get knowledge'? Educational research has shown that students who ask more questions during classroom sessions tend to acquire learning materials better (Ciardiello, 1998). Further, the students who ultimately end up developing the highest levels of competence (i.e. comprehend the learning materials more deeply) are students who tend to ask a certain kind of questions. These are questions that tap explanatory reasoning (Graesser, Baggett, and Williams, 1996). Similarly, cultural research has shown that cultural sensemaking experts ask explanation-based questions and, more specifically, they ask questions that can explicitly challenge the fundamental assumptions underlying their conception of a culture (Sieck, Smith, and Rasmussen, 2008). Question-asking is an indicator that a student is self-regulating their learning by (metacognitively) reasoning across their knowledge base, identifying knowledge deficits and asking questions to repair them. However, educational research has also shown that students often need training to improve these skills (Rosenshine, Meister, and Chapman, 1996). In the following we will describe an instructional analysis approach for building training that targets both the knowledge and metacognitive skill components of cultural sensemaking.

INSTRUCTIONAL ANALYSIS

Cultural learning presents a unique challenge to traditional approaches for instructional analysis and design. One reason is that the 'meaning' of cross-cultural situations is subjective—it depends on the person's cultural perspective. This includes the outcomes of actions and interventions. For instance, because a person believes he has successfully resolved an intercultural conflict does not mean that the person from a different culture on the other side of the conflict holds the same opinion. Cognitive Task Analysis methods, for example, are useful for identifying the knowledge and cognitive skills needed for complex tasks, but they need to be embedded within an empirical framework that allows examination of cross-cultural situations from different cultural perspectives. We have developed a suite of field research and analysis methods that meets this requirement, and can be employed to identify knowledge and skill requirements for cultural sensemaking training.

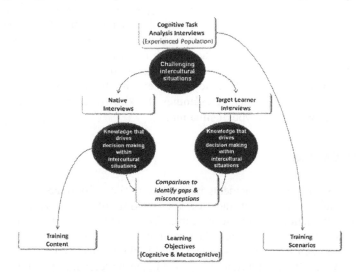

FIGURE 2. Instructional analysis methodology for cultural sensemaking.

This approach starts with Cognitive Task Analysis methods to identify the culturally and cognitively challenging intercultural interactions to include in a scenario-based training program (see figure 2). It then uses the *Cultural Network Analysis (CNA)* mental models based approach to cognitive modeling to characterize native decision making within these challenging interactions. Further, specific knowledge learning objectives result from a comparison between target learner models and native models that identifies gaps and misconceptions.

COGNITIVE TASK ANALYSIS

Cognitive Task Analysis is a set of methods for identifying and documenting the cognitive processes, cognitive challenges, and cognitive requirements for a task or work domain. We employed Critical Decision Method (CDM) interview techniques, adapted to focus on the cognitive nature of the challenges experienced by warfighters within intercultural interactions. We conducted 21 individual, face-to-face CTA interviews with Marines and soldiers who had returned from deployments in Afghanistan less than one month prior. The interviews were conducted by a pair of interviewers at Camp LeJeune, NC, and Fort Riley, KS. We asked the interviewees to *"tell us about a time, in Afghanistan, when you interacted with the local populace (civilians, tribal leaders, local officials, etc), and found the interaction challenging or puzzling."* A full description of the methodology and results can be found in Rasmussen, Grome, Crandall, and Sieck (2009).

The outcomes of a thematic analysis of the data included a typology of situations in which U.S. warfighters are experiencing sensemaking challenges as well as a typology of cultural sensemaking challenges. We selected six incidents that represented both typologies—a range of types of interactions and a significant proportion of the identified sensemaking challenges. An example incident described a situation in which a Mullah is working with Americans to provide humanitarian assistance supplies to villages near his own. Towards the end of the operation, the American team leader discovers that the Mullah has set aside one truck load of the supplies. At first, the Mullah denies the existence of left-over supplies; when pressured, however, he declares that they were set aside to be passed out to villages on the return trip. The next section examines the knowledge that drives decision making in natives within this type of situation.

NATIVE INTERVIEWS

We used the six challenging incidents indentified in the CTA stage as the basis for developing cultural models of the decision making of native Afghans within intercultural interactions with Americans. We conducted 14 scenario-based interviews with Afghan expatriates. The sample included 12 men and 2 women, with an average age of 27 years, who had lived in the United States for an average of 3.3 years. Within these interviews we probed their understanding of the situation, their understanding of the beliefs and values that would likely be driving the behavior of the Afghan characters within the scenarios, and their expectations about how the Afghans in the scenario might respond to various actions the American might take in the situation. The interviews were conducted individually and lasted 2 hours each. Each participant responded to two different scenarios. The interviews were recorded and transcribed.

In order to develop Afghan cultural models of the situations described in the scenarios, the following procedure was conducted separately for each of the scenarios. First, two independent coders read through all of the transcripts and identified excerpts from the interviews that contained causal beliefs. Next, the two analysts coded each excerpt by identifying for each causal belief the antecedent, the consequence, and the direction of the relationship between them (i.e. a certain antecedent increases or decreases the likelihood of a certain consequence). The inter-rater reliability across scenarios was .81. For the Mullah scenario, examples of common causal beliefs from the native interviews include:

1. Reciprocity: Mullah aims to help those who will help him
2. No theft: Mullah does not consider himself to be stealing
3. Status/Power: Mullah aims to increase his status among his own people
4. Privacy: Talk to Mullah in private; he loses face if confronted in public
5. No lie: Mullah does not think he's lying; ok to "tell stories" to save face

Importantly, the natives did not attribute to the Mullah the desire to increase his own wealth. Alternatively, they believed that the Mullah aimed to use the supplies to increase his status among his own people. A full description of this study can be found in Sieck, Smith, Simpkins, and Rasmussen (forthcoming). In the next section we will present the results of the target learner analysis, in which we evaluate the novice American perspective on these six challenging intercultural situations.

TARGET LEARNER ANALYSIS

In the learner analysis, content knowledge learning objectives are derived from a comparison between a native cultural model of the concepts, beliefs, and values that drive the decision making of natives within specific contexts and the target learners' understanding of their decision making. We employed the *Cultural Network Analysis* process for creating cultural models (Sieck, Rasmussen, and Smart, 2010) to represent native (in this case, the Afghan protagonist) thinking and decision making in the context of a specific scenario. For the purposes of deriving learning objectives for training, the cultural model provides the target concepts for training Americans to understand Afghan behavior—in other words, *the cultural model represents the hypothesized learning objectives.*

We conducted scenario-based target learner interviews with 20 newly recruited Marine Corps officers enrolled at the Marine Basic School in Quantico, Virginia. These officers of course had no prior deployments and overall had very little experience traveling overseas. We presented them with the same six scenarios we used in the native interviews. We probed their understanding of the situation, the cultural characters within the scenarios and their informational requirements (i.e. what would they like to know prior to making a decision), and their strategies for acting/interacting in this situation.

The interviews were taped, transcribed and then segmented into simple idea units. The idea units were coded using a coding scheme that consisted of five broad categories corresponding to the key cognitive and metacognitive aspects of cultural sensemaking probed in the interview guide: Understanding, Actions, Questions, Attitudes and an Other category. We then derived learning objectives by comparing the Understanding and Action ideas to the native cultural model. We evaluated the Questions against the expert sensemaking questions, and we coded the Attitudes idea units in terms of valence (positive/negative) and target (Afghan protagonist or Afghans as a group). In the following sections we will discuss the outcomes of this analysis and the resulting learning objectives (a complete description can be found in Rasmussen, Grome, Sieck, and Simpkins, 2009).

Cognitive Learning Objectives

The high-level knowledge oriented learning objectives for cultural sensemaking includes the ability to *identify likely belief-value drivers in critical situations for the target culture.* We used the previously described coding scheme to perform a quantitative assessment of the accuracy of the target learners' understanding of the cultural model and identify critical belief-value relationships that they either failed to perceive (gaps) or had misunderstood (misconceptions).

The analysis revealed a number of gaps and misconceptions listed below. For example, the Americans failed to apply the concepts of reciprocity, status/power, and privacy which constitute a gap. For example, none of the target learners considered the possibility that the Mullah would give the supplies to his own people in order to maintain his status to possibly increase his power. In terms of misconceptions, while a few target learners acknowledged that the Mullah may not consider himself to be 'stealing' or 'lying', these were common interpretations.

1. Reciprocity: *Missing concept* (gap)
2. Theft: Mullah is stealing (misconception)
3. Status/Power: *Missing concept* (gap)
4. Privacy: Discuss with Mullah in public (gap/misconception)
5. Lie: Mullah is lying; change in story means covering up lie (misconception)

Overall, an analysis of the target learners' errors indicated gaps between their conceptions and that of the Afghan natives related to facework, theft, lying, reciprocity, status and power, privacy, boldness, revenge, US status/reciprocity, and attention. Therefore, these are the knowledge-specific learning objectives for the Mullah scenario for the target learners. The next section addresses the metacognitive learning objectives that prepare trainees for any situation that requires cultural sensemaking.

Metacognitive Learning Objectives

The cognitive learning objectives are culture-specific, but effective intercultural training also includes culture-general learning or metacognitive objectives that build sensemaking competence. In the following we will focus on the metacognitive skills related to information seeking for illustrational purposes. For a full description of learning objectives see Rasmussen et al. (2009[b]). We derived learning objectives targeting the improvement of information seeking strategies by comparing the questions that the target learners asked to better understand the situation to the questions that expert cultural sensemakers tend to ask in order to create deep understanding. Generally, such questions take the form of "why," "why not," "how," "what if," or "what if not" (Graesser et al., 2003). Expert sensemaking questions provide deeper insight into the belief-value relationships driving behavior and as such support perspective taking (see Sieck, et al., 2008). For example, for the Mullah scenario, expert sensemakers could ask *"why did the Mullah take the goods?"* and *"how did the Mullah decide?"*

The analysis revealed that very few of the target learners' queries addressed the kind of information that expert cultural sensemakers would pose within a surprising intercultural situation. Instead their questions tended to focus on aspects of the situation that would allow them to determine the severity of the transgression, e.g. *"how much stuff did the Afghans set aside?"* or *"What kind of supplies were they?"* Many of the target learner's questions also directly illustrated their misdirected application of Western or U.S. belief-value systems to interpret the behavior of the Afghans. For example *"Why is the Mullah lying?"* and *"Why do they feel the need to short-change their fellow citizens and keep it for themselves?"*

This particular target learner population did not ask the questions that could provide insight into the native model. This raises the question of whether they are a likely target population for complex cognitive skills training. Interestingly, this population was rarely confident in the explanations they generated for the Afghans' behaviors. They often followed up an explanation with "I don't really know." Further, some of them would even hint towards cultural relativity and cite it as a source of uncertainty for them: *"Is lying looked upon in the same way in Afghan culture as it is in the U.S.?"* Their lack of confidence places them in Howell's (1982) "conscious incompetence" quadrant; i.e., "they know that they do not know." For these reasons, this population may constitute a favorable audience for cultural sensemaking training.

DISCUSSION

In this chapter we have outlined a theoretical framework for cultural sensemaking that connects high level metacognitive skills to region-specific knowledge. As such, the cultural sensemaking framework describes a possible avenue through which culture-specific learning can contribute to culture-general competence. The fundamental proposition put forth is the following: *if you provide people with baseline (cognitive) content knowledge of the factors that influence culturally different people's decision making within specific contexts* and *if you provide them with metacognitive skills needed to build upon that initial understanding* they will be able to use this initial knowledge and basic skill set to learn from the complex, real-life situations they encounter and thereby build and expand their database of experiences. In this way, learning programs designed to enhance cultural sensemaking competence can provide a springboard for life-long cultural learning.

In terms of performance outcomes, this chapter has mainly focused on the early stages of the cultural sensemaking process, information seeking and hypothesis/ explanation generation. To truly be effective within cross-cultural contexts it is paramount that people are also able to translate their understanding of a situation into appropriate behaviors and adjust their actions (see Figure 1). In the context of normative or highly rule-based behavior, this is sometimes referred to as code switching. Cross-cultural code-switching is *the act of purposefully modifying one's behavior in an interaction in a foreign setting in order to accommodate different cultural norms for appropriate behavior* (Molinsky, 2007, p.624). In very complex situations, learning culturally specific norms for behavior is not enough. In the Mullah scenario, a person could rely on norms relating to interpersonal interactions to decide whether he or she should confront the Mullah in public or private. But there is no 'norm' that can help a person decide whether to confront the Mullah at all. In situations, such as this, where there is no right or wrong decision, people must rely instead on their ability to make sense of the complexity in order to make better informed decisions.

CONCLUSION

This chapter has provided a theoretical framework and outlined a practical approach for assessing cultural sensemaking competence. The instructional analysis and design approach outlined in this chapter, i.e. comparison between novice understanding and native cultural models can meaningfully be applied as part of a pre-test, post-test paradigm to evaluate a trainee's increase in competence as a result of a training intervention. Within this paradigm, changes in the trainees' content knowledge as well as in their metacognitive skills can be assessed, seeing that clear learning objectives can be established for both areas. Conceivably, the same assessment process could be used to evaluate cultural sensemaking competence for the purposes of job placement and/or promotion.

REFERENCES

Abbe, A., Gulick, L. M. V., & Herman, J. L. (2007). Cross-Cultural Competence in Army leaders: A conceptual and Empirical Foundation. United States Army Research Institute for the Behavioral and Social Sciences. Final report.

Archer, C.M. 1986. Culture bump and beyond. In J. M. Valdes (Ed.), Culture bound: Bridging the cultural gap in language teaching. Cambridge University Press, 170-178.

Bruner, J. S. (1966). Toward a theory of instruction. Cambridge: Harvard University Press.

Ciardiello, A. V. (1998). Did you ask a good question today? Alternative cognitive and metacognitive strategies. Journal of Adolescent & Adult Literacy, 42, 210–219.

Graesser, A. C., Baggett, W., & Williams, K. (1996). Question-driven explanatory reasoning. Applied Cognitive Psychology, 10, S17-S32.

Klein, G., Phillips, J.K., Rall, E., & Peluso, D.A. (2004). A data/frame theory of sensemaking. In R.R. Hoffman (Ed.), Expertise out of context: Proceedings of the 6th International Conference on NDM. Mahwah, NJ: Erlbaum.

Kluckhohn, F.R., & Stroedtbeck, F.L. (1961). Variations in value orientations. New York: Harper Collins.

Molinsky, A. (2007). Cross-cultural code-switching: The psychological challenges of adapting behavior in foreign interactions. Academy of Management Review, 32, 622-40.

Osland, J.S., & Bird, A. (2000). Beyond sophisticated stereotyping: Cultural sensemaking in context. Academy of Management Executive, 14, 65-79.

Osland, J.S., & Bird, A. (forthcoming). Trigger events in Intercultural Sensemaking.

Rasmussen, L. J., Grome, A., Crandall, E., & Sieck, W. R. (2009[a]). The puzzling Afghan: Making sense of Afghan interactions. Technical Report No. W91CRB-09-C-0098.

Rasmussen, L. J., Grome, A., Sieck, W., & Simpkins, B. (2009[b]). Cultural Sensemaking: Competence assessment and learning objectives for new leaders in the Marine Corps. Technical Report No. W91CRB-09-C-0028.

Rosenshine, B., Meister, C., & Chapman, S. (1996). Teaching students to generate questions: A review of the intervention studies. Review of Educational Research, 66, 181–221.

Sieck, W. R., Rasmussen, L. J., & Smart, P. (2010). Cultural Network Analysis: A Cognitive Approach to Cultural Modeling. In D. Verma (Ed.), Network Science for Military Coalition Operations: Information Extraction and Interactions, IGI Global.

Sieck, W. R., Smith, J. L, & Rasmussen, L. J. (2008). Expertise in making sense of cultural surprises. I/ITSEC, December 2008, Orlando, FL.

Sieck, W. R., Smith, J., Simpkins, B., & Rasmussen, L. J. (forthcoming). Cultural models of Afghan decision making. Technical Report.

CHAPTER 9

Designing Games as Social-Process Simulation Crucible Experiences: Toward Developing and Assessing Intercultural Adaptability

Elaine M. Raybourn

Sandia National Laboratories*
Albuquerque, NM 87185, USA

ABSTRACT

The focus of the present chapter is intercultural social-process simulation design for crucible experiences in computer games and virtual worlds. Social-process simulation is a methodology used to create emotionally challenging multi-player experiences that can exercise intercultural adaptability and self-awareness. The present chapter introduces the social-process simulation and the Reflective Observer/Evaluator role employed to create engaging game-based or virtual world crucible experiences for teaching individuals and teams to interact and communicate more purposefully in intercultural contexts, and exercise metacognitive agility toward the development of intercultural adaptability.

Keywords: Crucible Experience, Reflection, Intercultural Adaptability, Games, Social-Process Simulations, Metacognitive Agility, Peer, Evaluations, Emotions

INTRODUCTION

In recent years, global demands have placed increased emphasis on the assessment and development of adaptive, self-aware individuals who are also competent intercultural communicators. Multi-national organizations have faced this challenge for some time and have tried a variety of techniques to both assess and foster intercultural adaptability in their employees. The field of intercultural communication is on the forefront of this research. A number of different approaches exist within the field of intercultural communication to familiarize learners with both cultural-general and culture-specific dynamics (paper exercises, multi-media critical incidents, role plays, simulations, games, etc.). A methodology used frequently by intercultural trainers to foster intercultural adaptability is social-process simulation (Fowler & Pusch, 2010). Social-process simulations may be instantiated as face-to-face live action simulations, tabletop exercises, card or board games, or computer games, etc. Often social-process simulations are used to distill complex phenomena into crucible experiences.

Often the best way to learn about and internalize the nuances of a challenging cultural dynamic is to experience it directly, or as directly as is safe or possible. A crucible experience is "…a defining moment that unleashes abilities, forces crucial choices, and sharpens focus. It teaches a person who he or she is." (Bennis & Thomas, 2002; p.16; Wong, 2004). Crucible experiences engage one's emotional intelligence and allow trainees to identify their strengths and weaknesses regarding specific situations. For example learning about the stages of culture shock by being told about them and later taking a test will not be as effective for internalizing the feelings associated with culture shock as participating in a social-process simulation. Arguably both approaches would be used in a training program but only the latter is a crucible experience. Many cognitive exercises (even though they may be presented in engaging 3D virtual environments) often do not teach a person *who* he or she *is*, but rather *what* he or she *knows*. Crucible experiences focus on unpacking the reality of authentic relationships (e.g. with self, others, environment). For example, using crucible experiences to develop a warrior-diplomat ethos is at the heart of military training (Wong, 2004).

Yet how does one design a crucible experience for intercultural adaptability in a virtual environment such as a game or virtual world? A working definition of intercultural adaptability in the present chapter is the ability to fully grasp the role of diverse mindsets or perspectives, be self-aware regarding the role of one's own cultural background in intercultural communication, and expertly change communicative behaviors in cross-cultural settings to facilitate understanding. This working definition incorporates Communication Adaptation Theory or CAT (Gallois et. al., 1988), Intercultural Adaptation Model or IAM (Cai & Rodriguez, 1996), Developmental Model of Intercultural Sensitivity or DMIS (Bennett, 1996) and the serious game Simulation Experience Design Method based on Intercultural

Communication (Raybourn, 2006, 2007a,b). An effective crucible experience for adaptability would allow learners to develop self-awareness, fully understand diverse emotions and perspectives while managing their own, and gain experience with changing communication behaviors to facilitate understanding. The emotional intelligence required for intercultural adaptability cannot be overemphasized—and yet there are very few simulations, games, or virtual environments aimed at tapping into this unique and powerful capability.

The author has employed social-process simulations and experience design methods to create computer game-based, intercultural crucible experiences that allow learners to explore and reflect on their emotions. A social-process simulation is defined as an environment that is used to replicate behavioral processes that usually employ a human in a role-playing situation (Gredler, 1992). Social-process simulations focus on human interactions and communication in the pursuit of social goals. Social-process simulations are crucible experiences in which trainees engage in safe, simulated situations that help them understand their strengths and weaknesses and how to develop new skills. Human-computer interaction experience design methods require that designers understand what makes a good experience first, and then translate these principles, as well as possible, into the desired medium without the technology dictating the form of the experience. Experience designers strive to create desired perceptions, cognitions, emotions, and behaviors among users, customers, visitors, or the audience.

Two instantiations of the social-process Simulation Experience Design method have resulted in deployed government game-based training systems focused on developing intercultural adaptability. One has been in use at Ft. Bragg, NC (Raybourn et. al., 2005) since 2003 and another module was transitioned to PEO-STRI in 2007 (Raybourn et. al., 2008). In each case, the Simulation Experience Design Method was employed to create an immersive intercultural communication practice environment specifically to develop and assess adaptability. Results from the author's doctoral dissertation and recent studies conducted using these two social-process simulation games that were designed using the Simulation Experience Design Method and Real-time In-Game Assessment, Evaluation and Feedback (Raybourn, 1998, 2009; Raybourn et. al., 2005) confirm that in all three cases learners understand more about their strengths and weaknesses by participating in the simulation game than they would if they had not participated. In other words, through the game-based, crucible experience designed using the principles discussed in subsequent sections, learners become more self-aware—a critical skill to develop and *assess* intercultural adaptability.

Intercultural adaptability is a lifelong process, not a destination. Therefore assessment of intercultural adaptability is also a lifelong process that each person can learn. The present chapter treats intercultural adaptability assessment as the *act of evaluation*. To assess one's own or others' intercultural adaptability requires metacognitive agility. Metacognitive agility is critical to developing adaptability and is defined in the present chapter as possessing the ability to analyze the way one or others think, discern different tasks or problems requiring different types of cognitive strategies, and employ those strategies to enhance learning and performance (Raybourn, 2007b). Discussion of measures or scales that have been developed to assess or predict one's intercultural adaptability (Bennett, 1996; Abbe et. al., 2007) are beyond the scope of the present chapter.

The focus of the present chapter is intercultural social-process simulation and experience design for computer games and virtual worlds. Some sections contain excerpts from the author's I/ITSEC conference paper on designing simulation games for intercultural adaptability (Raybourn, 2006). Social-process simulation is a methodology used to create multi-player crucible experiences that can exercise intercultural adaptability and self-awareness. The present chapter also introduces the Reflective Observer/Evaluator Role employed by the author to create engaging computer game-based crucible experiences for training individuals and teams to interact and communicate more purposefully in intercultural contexts, and exercise metacognitive agility toward the development of intercultural adaptability. This role utilizes the Real-time In-Game Assessment, Evaluation and Feedback System which promotes in-game reflection, observation, and evaluation (Raybourn, 2004; 2007a). These methods have been used to design multiplayer games, collaborative virtual environments, and other text-based social-process simulations that treat intercultural communication and cross-cultural discovery as core interaction goals.

DESIGNING INTERCULTURAL COMMUNICATION CRUCIBLE EXPERIENCES

Simulations are sophisticated, interactive, role-play exercises that are popular in education and training at various levels. Modern interest in social-process simulations and role-playing can be traced to the work of Lewin (1951) and Piaget (1972). According to Vincent and Shepherd (1998),

> Both [Lewin and Piaget] argued that effective learning occurred when there was a sustained interaction between the learner and the environment and when there was an opportunity via social interaction to reflect on the experiences in that environment. Piaget (1972) also stressed the importance of social interaction in providing stimulus for challenging existing beliefs, as a first step in changing those beliefs.

An important component of social-process simulation design is to challenge existing beliefs. The designer's task is to get learners to interact, take actions that affect others implicit assumptions and create cognitive dissonance or conflict among participants' goals, then guide the learners to develop skills in conflict negotiation, empathy and awareness, etc. Finally, learners successfully arrive at the learning outcome by monitoring their feedback and the feedback of others (see Table 1).

Table 1. Design Characteristics of Social-Process Simulations (Gredler, 1992)

Task	Focus	Problem	Actions	Feedback
Interact with others to address challenge	Effects of one's own assumption, goals, strategies on action	Arises from conflict in roles, goals or actions	Use of social interaction, i.e. negotiation, persuasion, mediation	Reactions of other participants and self-assessment evoke change

In general, learning through experience has been described as occurring either in a real situation, such as a workplace, or in role play. Key to social-process simulation is that as learners role-play they may experience intense feelings of frustration, rejection, pride, acceptance, conflict, cooperation, and a host of other emotions. The opportunities to experience these emotions are artfully designed into the simulation game. Social-process intercultural simulations are designed to allow role-players to develop self-awareness of their emotions and the impact of emotions on decisions, actions, and interpersonal communication. Raising the emotional capital associated with the problem and challenging the learner to rethink her assumptions as she negotiates with others through a conflict the structure of the crucible experience. As such one important component of social-process simulations is to explore the origins of emotional reactions and their relationships to the larger sphere of human experience and its impact on decision making. Exploring emotional reactions in a crucible experience such as a social-process simulation helps a person discern *who* he or she *is* and how their behaviors impact others and decisions.

Social-process simulations are also designed to challenge the assumptions that role-players bring to their roles. The designer's task is to get learners to interact, take actions that affect others implicit assumptions and create cognitive dissonance or conflict among participants, then guide the development of self-regulating skills such as monitoring their feedback and the feedback of others (Raybourn, 2006). According to Bennis & Thomas (2002; p.93):

> People with ample adaptive capacity may struggle in the crucibles they encounter, but they don't become stuck in or defined by them. They learn important lessons, including new skills that allow them to move on to new levels of achievement and new levels of learning. This ongoing process of challenge, adaptation, and learning prepares the individual for the next crucible, where the process is repeated. Whenever significant new problems are encountered and dealt with adaptively, new levels of competence are achieved, better preparing the individual for the next challenge.

As designers of educational environments supporting crucible experiences, our goal is to create dynamic, changing situations that challenge and prepare learners for the next crucible they will encounter, and so on. Crucible experiences are not just simulated; they are also real world experiences. Salen and Zimmerman (2004) describe the play experience outside of a game as the metagame. The time before, between, or after game play is ripe with metagame activities such as planning, reflecting on strategy, discussing in groups what happened previously during game training, sharing lessons learned, etc. These metagame activities are components of the total crucible system of experiences addressed by the Simulation Experience Design Method (Raybourn, 2004, 2006; 2007a,b).

The debriefing is the most vital element of successful intercultural simulation game design. During the debriefing, learners are usually guided by a facilitator to reflect on the lessons learned from the simulation game experience, by extending what was learned to "real" situations, or by identifying strategies that could have enhanced performance. Facilitators also may use the debriefing as an opportunity to ease learners out of their game roles, and all of the feelings associated with it, back to "reality" (Sisk, 1995). This becomes particularly important if some learners are playing roles aimed at honing cultural relativism, or empathy. Care should be taken to debrief their roles and the intense emotions that can be experienced during the training session.

Social-process simulation games should always be accompanied by other methods of instruction and in the context of a pedagogical curriculum when introducing *new* concepts to learners. Learners may come to the training experience with different backgrounds, perceptions, tolerances, or levels of domain knowledge. Social-process simulations are practice environments that require a certain level of willingness to explore oneself. It is not uncommon for participation in social process simulations to be the first time an individual confronts intimate details about her/himself. Used out of context of a sound instructional framework or in the absence of skilled facilitators, negative training could occur. For instance, the concepts explored in social-process simulation may trigger emotional responses that are deeply rooted and have remained unexplored by the trainee until surfaced by the training event (Sisk, 1995).

There are several benefits to using computer game-based or virtual world-based social-process simulations to facilitate intercultural learning through crucible social-process simulations. First, learners practice critical thinking skills that better prepare them to plan future strategies as well as spontaneously intuit the consequences of their decisions. Second, learners also learn to apply the theories and models explored in the simulated situation to real-world situations. Third, learners develop their emotional intelligence and come to better understand their own emotional strengths and weaknesses. The simulation gaming process also provides players an opportunity to practice real-world behaviors associated with competition, empathy, and communication in a simulated reality (Sisk, 1995;

Raybourn 1998). Finally perhaps one of the most valuable benefits is that a simulated reality is a safer arena for many people to confront their emotions toward cultural differences. Particularly when addressing some cross-cultural issues of potential controversy, simulation games provide a safe place to explore dangerous questions (Pedersen, 1995; Raybourn, 1998, 2001).

CRUCIBLE EXPERIENCE ROLE FOR REFLECTIVE OBSERVATION & EVALUATION

In developing social-process simulation game-based training systems that are deployed today (Raybourn et. al., 2005; Raybourn et. al., 2008), the author instantiated a reflective observer/evaluator role in software so learners could participate in crucible experiences through real-time, in-game reflection and evaluation of abstract concepts. The discussion below describes the approach to providing in-game opportunities for honing metacognitive agility toward intercultural communication competence and adaptability.

An approach to fostering metacogntive agility and intercultural adaptability through experiential learning is to give learners opportunities to practice concrete experience, reflective observation, abstract conceptualization and active experimentation through evaluating their own actions and those of others (Kolb, 1984). In order to truly foster the development of metacognitive skills, we must provide learners with opportunities to both internalize and use their knowledge in crucible experiences. We should provide opportunities to manage emotions and be able to recognize behaviors that can be changed to facilitate intercultural understanding. A goal of the Reflective Observer/Evaluator Role is to provide learners with the opportunity to reflect on actions and strategies that are enacted by themselves and others during a crucible exercise (Raybourn, 2007a). This role allows learners to practice the act of evaluation and self-assessment without extracting the emotional element that is critical to developing intercultural adaptability.

The Reflective Observer/Evaluator Role consists of providing an interface and role for in-game evaluations of learners' actions, communications, etc. as they occur in real-time and as they correspond to core competencies or specific training objectives. For example, students learn core competencies or criteria by which intercultural adaptability is evaluated before a training session begins. During the session Reflective Observer/Evaluators engage in the act of assessing (applying what they learned) role-players' adaptability. Their feedback is both quantitative and qualitative and corresponds to logged, time-stamped events. These evaluations that correspond to actual events are later aggregated and statistical analyses performed on the individual and group evaluations. Team and individual assessments are displayed either in realtime or during the debriefing. By switching roles that allow learners to *act* (role-playing self or person from another culture) and *reflect* (Reflective Observer/Evaluator Role) the learners perform different cognitive tasks and learn the *act of evaluation, or assessment.* The complete act of observation, reflection, and evaluation is a crucible experience because it occurs in real-time and with emotional consequences. The act of evaluation can be a conduit for an "aha" experience regarding one's own behavior.

This method is very flexible and has also been used with observer controllers, peers, subject matter or cultural experts, instructors, or training cadre. Those in the Reflective Observer/Evaluator Role introduce quantitative/qualitative in-game assessments of decisions made, actions taken, or strategies employed. In fact, a training goal of this role is to foster the skill development necessary for each learner to become his or her *own* Observer Controller, by *developing the habit of actively evaluating his or her own behaviors and identifying salient strengths and weaknesses.* By including Observer Controller or expert participation in training sessions, appropriate behaviors are also modeled for learners and serve as further learning reinforcement.

During the facilitated debriefing learners in *both* player and Reflective Observer/Evaluator roles can now participate in the discussion of the same phenomenon. Additionally, the Reflective Observer Controller/Evaluator role guides the group to discuss more than what went right, and what went wrong. The core competencies become a focal point of discussion as trainees are held accountable for their evaluations and values. A community-based debriefing in which each individual voice counts engenders a culture of thoughtful participation, increased risk taking (due to the creation of a safe learning environment), and sharing of novel solutions that expand each trainee's potential solution set outside of the simulated crucible experience to their reality.

CONCLUSION

The present chapter sought to address how to design games and virtual worlds as social-process simulation crucible experiences toward developing and assessing intercultural adaptability. Inspirations from intercultural communication, social-process simulation, and experience design have influenced this approach. Social-process simulation and the Reflective Observer/Evaluator role was introduced. The present chapter addressed how the instantiated role functionality and methods are used by observer controllers, peer learners, subject matter or cultural experts, instructors, etc. to provide quantitative feedback of actions taken, (including communications) as they occur in real-time crucible exercises. These methods are critical to developing the metacognitive strategies and self-monitoring skills necessary to assess and develop interculturally competent, adaptive, self-aware leaders. Statistically significant results from three quasi-experimental studies of crucible experiences designed using the methods discussed in the present chapter

(Raybourn, 1998, 2009; Raybourn et. al., 2005) confirm that trainees learn more about their strengths and weaknesses by participating in these crucible simulation games than they would if they had not participated. Many cognitive exercises (even though they may be presented in an elaborate 3D virtual environment) often do not teach a person who he or she is, but rather *what* he or she *knows*. The methods described in this chapter can be used to design crucible experiences that teach a person *who he or she is* in a safe, simulated environment. Being self-aware, managing one's emotions, and learning to honestly assess one's behaviors without fear are fundamental to intercultural adaptability.

Crucible experiences engage one's emotional intelligence and allow learners to identify their strengths and weaknesses regarding specific situations. Virtual environments such as multi-player games and virtual worlds provide promising setting for practicing skills associated with intercultural adaptability. Under certain circumstances virtual environments provide less threatening settings for social-process simulations and crucible experiences than face-to-face (Raybourn, 1998). Further research is needed to explore the full capability of using multi-player games and virtual worlds as crucible experience environments.

There is no explicit formula for designers on how to create crucible experiences, but by using the social-process Simulation Experience Design Method (Raybourn, 2007a) we begin to systematically unpack the concept of "crucible experiences" and identify structures that lead to the design of successful experientially and emotionally challenging situations. As designers, if we can achieve this, we will have truly contributed to preparing learners for intercultural adaptability needed in real world crucible experiences.

REFERENCES

Abbe, A., Gulick, L. M. V., & Herman, J. L. (2007). Cross-cultural competence in Army leaders: A conceptual and empirical foundation. (SR 2008-01). Arlington, Virginia: U.S. Army Research Institute for the Behavioral and Social Sciences.

Bennett, M. J. A developmental approach to training for intercultural sensitivity. International Journal of Intercultural Relations, 10, (1986), 179-96.

Bennis, W.G., & Thomas, R.J. (2002). Geeks & geezers: How era, values, and defining moments shape leaders. Harvard Business School Press, Boston, MA.

Cai, D. A., & Rodriguez, J. I. (1996). Adjusting to cultural differences: The intercultural adaptation model. Intercultural communication studies, V1:2.

Fowler, S.M., & Pusch, M.D. (2010). Intercultural simulation games: A review (of the United States and beyond). Simulation & Gaming, 41, 1, 94-115, Sage.

Gallois, C., Franklyn-Stokes, A., Giles, H., & Coupland, N. (1988). Communication accommodation in intercultural encounters. In Y.Y. Kim & W.B. Gudykunst (Eds.), Theories in intercultural communication. Newbury Park, C, Sage. 157-85.

Gredler, M. (1992). Designing and evaluating games and simulations: A Process Approach. Kogan Page, London.

Kolb, D. A. (1984). Experiential learning: Experience as the source of learning and development. New Jersey: Prentice-Hall.

Lewin, K. (1951) Field theory in social science, Harper and Row, New York.

Pedersen, P. (1995). Simulations: A safe place to take risks in discussing cultural differences. Simulation & Gaming, 26, 2, 201-6.

Piaget, J. (1972), The principles of genetic epsitemology, Basic Books, New York.

Raybourn, E. M., Deagle, E., Mendini, K., & Heneghan, J. (2005). Adaptive Thinking & Leadership Simulation Game Training for Special Forces Officers. I/ITSEC 2005 Proceedings, Interservice/ Industry Training, Simulation and Education Conference Proceedings, November 28-December 1, Orlando, Florida, USA.

Raybourn, E.M. (1998). An intercultural computer-based multi-user simulation supporting participant exploration of identity and power in a text-based networked virtual reality: DomeCity MOO. Unpublished doctoral dissertation, University of New Mexico.

Raybourn, E.M. (2007a). Applying simulation experience design methods to creating serious game-based adaptive training systems. Interacting with Computers 19, Elsevier.207-14.

Raybourn, E.M. (2009). Beyond game effectiveness part I: An empirical study of multi-role experiential learning. I/ITSEC 2009 Proceedings, Interservice/ Industry Training, Simulation and Education Conference Proceedings, November 30- December 04, Orlando, Florida, USA.

Raybourn, E. M. (2001) Designing an emergent culture of negotiation in collaborative virtual communities: The DomeCityMOO Simulation. In E. Churchill, D. Snowden, & A. Munro (Eds.) Collaborative Virtual Environments: Digital Places and Spaces for Interaction, Springer, London UK, 247-64.

Raybourn, E. M. (2004). Designing intercultural agents for multicultural interactions. In Sabine Payr & Robert Trappl (Eds.), Agent Culture: Human-Agent Interaction in a Multicultural World, Lawrence Erlbaum, 267-285.

Raybourn, E.M., Roberts, B., Diller, D. & Dubow, L. (2008). Honing intercultural engagement skills for stability operations with DARWARS Ambush! NK game-based training. In Proceedings of 26th Army Science Conference, Orlando, Fl, December 1-4.

Raybourn, E.M. (2006). Simulation experience design methods for training the forces to think adaptively. I/ITSEC 2006 Proceedings, Interservice/ Industry Training, Simulation and Education Conference Proceedings, December 4-7, Orlando, Florida, USA.

Raybourn, E.M. (2007b). Training approaches for honing junior leader adaptive thinking, cultural awareness and metacognitive agility. I/ITSEC 2007 Proceedings, Interservice/ Industry Training, Simulation and Education Conference Proceedings, November 26-29, Orlando, Florida, USA.

Salen, K. & Zimmerman, E. (2004). Rules of play. The MIT Press, Cambridge, MA.

Sisk, D. A. (1995). Simulation games as training tools. In Sandra M. Fowler and Monica G. Mumford (Eds.) Intercultural sourcebook: Cross-cultural training methods, vol. 1. Yarmouth, Maine: Intercultural Press, Inc. 81-92.

Vincent, A. & Shepherd, J. (1998). Teaching Middle East politics by interactive computer simulation. Retrieved June 23, 2006 from http://www-jime.open.ac.uk/98/11/vincent-98-11-01.html.

Wong, L. (2004). Developing adaptive leaders: the crucible experience of Operation Iraqi Freedom. Strategic Studies Institute, Carlisle Barracks, PA.

*Sandia is a multiprogram laboratory operated by Sandia Corporation, a Lockheed Martin Company, for the United States Department of Energy under Contract DE-AC04-94AL85000.

Chapter 10

Development of the Cross-Cultural Competence Inventory (3CI)

K.G. Ross[1], C.A. Thornson[1], D.P. McDonald[2], B.A. Fritzsche[3], H. Le[3]

[1]Cognitive Performance Group
3662 Avalon Park E. Blvd, Suite 2023
Orlando, FL 32828, USA

[2]Defense Equal Opportunity Management Institute
366 Tuskegee Airmen Drive
Patrick Air Force Base, FL 32925, USA

[3]Department of Psychology
University of Central Florida
Orlando, FL 32816-1390, USA

ABSTRACT

To date, there are no current validated instruments designed to assess cross-cultural competence (3C) in the military domain. To address this need, we developed the Cross-Cultural Competence Inventory (3CI), a 58-item self-report instrument to measure the six hypothesized dimensions of 3C. The purpose of this tool is to assist commanders in evaluating the readiness of their troops to interact effectively and appropriately with foreign nationals, multi-national coalition forces, and other individuals, agencies and organizations.

Keywords: Cross-cultural Competence, Military Readiness, Validation

INTRODUCTION

Peacekeeping and stability operations are central to today's military missions. Within such contexts, the need to establish and maintain relationships with local populations is essential. As Marine Corps General Charles C. Krulak (1999) noted, lower ranking personnel often represent American foreign policy across humanitarian assistance, peacekeeping, and traditional operations. As our efforts in Afghanistan and Iraq illustrate, the military is increasingly involved in advising and training roles as well (U.S. Department of Army, 2006, as cited in Zbylut et al., 2009). Enlisted personnel have been called up to engage in such diverse duties as serving as town mayor of an Iraqi village, negotiating with tribal leaders in Afghanistan, or training indigenous forces worldwide (Stringer, 2009). Therefore, it seems that no matter what the job, rank, or specific mission, working with foreign counterparts to create and maintain stability in fragile regions of the globe is critical, and the potential for cross-cultural conflict and international-level consequences of incompetence is high (Abbe, Gulick, & Herman, 2007). For these reasons, the Department of Defense has identified 3C, the capability to interact effectively and appropriately with others who are linguistically and culturally different from oneself, as a critical determinant of success in military missions today.

UNDERSTANDING CROSS-CULTURAL COMPETENCE

In order to understand and assess this multidimensional construct, a two-tiered approach was undertaken. An extensive review of the literature provided a theoretical domain upon which to base a deductive approach to item development. However, because of the lack of consensus among researchers and academicians, an inductive approach to item development was also employed. Following extensive literature review (Ross & Thornson, 2008), in-depth interviews were conducted with subject matter experts (SMEs). Qualitative data were collected from nine higher-ranking enlisted Army soldiers and Army officers who had been deployed to countries outside the United States (Ross, 2008). Thus, both inductive and deductive approaches to item generation were undertaken to enhance content validity, or the adequacy with which the measure assesses the domain of interest.

LITERATURE REVIEW

The efforts to identify individuals who possess the relevant characteristics associated with 3C in the military domain have not been fully explored to date. When describing 3C, a variety of constructs (e.g., knowledge, skills, attributes, cognitive dimensions, and attitudes) have been proposed and measured across different academic and scientific disciplines. Research into what types of people are likely to succeed in living and working outside their country of origin for extended periods of time (e.g., expatriate managers, study-abroad students, Doctors without Borders, and Peace Corps volunteers) has accumulated. We explored several of these validated scales, including The Big Five (Costa & McCrae, 1992); the Intercultural Adjustment Potential Scale (ICAPS; Matsumoto et al., 2001); the Multicultural Personality Questionnaire (van der Zee & Van Oudenhoven, 2000); the Scale of Ethnocultural Empathy (SEE; Wang et al., 2003); the Intercultural Sensitivity Inventory (Hammer, Bennett, & Wiseman, 2003); the Intercultural Sensitivity Index (ISI; Olson & Kroeger, 2001, as cited in Abbe et al.); the Intercultural Development Inventory (IDI; Hammer et al., 2003); the Cross-Cultural Adaptability Inventory (CCAI; Kelley & Meyers, 1995); and the Cultural Intelligence Scale (CQS; Earley & Ang, 2003).

Although each of these self-report measures is worthwhile and validated for the purpose for which it was designed, most were developed with civilians in mind, and not specifically for military personnel. However, the very real and important differences between the military and other domains cannot be overlooked. These include the fact that early termination of an assignment is not an option; there exists a power differential between military members and the local population; and finally, our military personnel are under continual threat of attack from foreign nationals (Selmeski, 2007). Additionally, the outcome criteria used to validate these instruments is often adaptation and/or adjustment to living in another culture (Matsumoto et al., 2001; van der Zee & van Oudenhoven, 2000). This is not the goal of the military, who are a culture unto themselves and are there to accomplish a specific mission. For these reasons, a measure designed for the military, based on the mission-specific performance criteria found to be critical to mission success, was developed.

CROSS-CULTURAL COMPETENCE IN MILITARY CONTEXTS

To uncover the specific performance dimensions of a particular domain, interviews with subject matter experts (SMEs) is recommended (Borman, 1991). Therefore, we conducted in-depth interviews with nine recently deployed military service members (Ross, 2008). All nine participants relayed important observations as to what they considered to be the dimensions of mission-specific performance and of these, several had sufficient experience to consider themselves competent in terms of cross-cultural interactions. Whereas the findings were not based on extensive coding and inter-rater reliability, their qualitative analysis offered an initial content validation effort linking the performance dimensions found in the literature to the mission-specific performance criteria found in the field. Specifically, Ross (2008) found that relationship-building was mentioned a total of 68 times by the nine interviewees. The behaviors associated with relationship building have also been examined in the literature with regard to 3C (Cui & Van Der Berg, 1991, as cited in Abbe et al., 2007). In a sample of Peace Corps teachers working in Ghana, a performance factor emerged that included both teaching and interpersonal relationship items (Smith, 1966), where it was found that conveying warmth toward students, showing consideration toward the local adults, and displaying tact were the most important elements of effective performance. The interviews also revealed that successfully influencing, persuading, and negotiating with foreign nationals, as well as presenting oneself appropriately during interactions, were most often associated with effective cross-cultural performance (Ross, 2008). Such behaviors are likely to lead to the type of short-term rapport-building necessary to move about safely in a threatening environment as well as to lay the foundation for longer-term relationships.

Therefore, via an integration of the interview data and literature review, the developers of the 3CI proposed the following constructs for a measure of military 3C (Ross, Thornson, McDonald, & Arrastia, 2009; Thornson, Ross, & Cooper, 2008): (1) Cross-Cultural Openness; (2) Cross-Cultural Empathy; (3) Willingness to Engage; (4) Self-Efficacy; (5) Emotional Self-Regulation; (6) Cognitive Flexibility; (7) Self-Monitoring; (8) Low Need for Cognitive Closure; and (9) Tolerance of Ambiguity. The items were adapted or revised from existing validated scales that represented each of the proposed nine dimensions, or were written based upon the interview data. This procedure yielded an initial item pool of 144 items, not including five response distortion items.

INSTRUMENT DEVELOPMENT

DATA COLLECTION 1

The initial 144-item 3CI was uploaded to the Defense Equal Opportunity Management Organizational Climate Survey (DEOCS), an electronic survey routinely administered to all services across ranks and geographic locations. After completing the DEOCS, personnel were given the option of participating in the research. The total number of completed surveys analyzed was 792.

Demographic Data

Of those participants who reported gender, 486 were male (75.8%) and 155 were female (24.2%). The ages ranged from 18 to 40 years of age, with 67 participants between 18 and 20 years of age (10.5%); 220 participants between 21 and 24 years of age (34.3%); 179 participants between ages 25 and 29 years of age (27.9%); 114 participants between ages 30 and 35 years of age (17.8%); and 61 participants between the ages of 36 and 40 (9.5%). Of the 607 participants who reported their pay grade, 154 reported a pay grade between 1 and 3 (24%); 324 reported a pay grade between 4 and 6 (50.5%); 76 reported a pay grade between 7 and 8 (11.9%); 14 reported being at a pay grade between 9 and 10 (2.2%); 22 reported being at a pay grade between 11 and 13 (3.4%); and 17 reported being at a pay grade between 14 and 15 (2.7%). Of the 528 total participants who reported their Branch of

Service, there were 17 participants in the Air Force (2.7%); 181 participants in the Army (28.2%); only 1 participant in the Coast Guard (0.2%); 149 participants in the Marine Corps (23.2%); 179 in the Navy (27.9%); and only 1 reported being in an Other Military Service (0.2%).

Exploratory Factor Analysis

An exploratory factor analysis (EFA) was carried out (N =792), using SPSS Version 12.0 and specifying principal-axis factoring (PAF) as the extraction method (cf. Gorsuch, 1983; Nunnally and Bernstein, 1994). Based on the resulting scree plot (Cattell, 1966) and interpretability, six factors were retained and rotated to simple structure using an oblique rotation (i.e., Oblimin with Kaiser Normalization), which converged in 30 iterations. Items were screened on the basis of their rotated factor patterns. The items with the lowest factor loadings (< .30) and those that cross-loaded onto other factors were discarded. The six factors appeared interpretable and accounted for 28.7% of the total variance.

Examination of Scale Properties

Following classical test theory, reliability was assessed based on the correlations between the individual items that make up the scale and the variances of the items (Nunnally & Bernstein, 1994). Cronbach's coefficient alpha and item-total correlations were examined and those items with low item-total correlations (< .30) were discarded. In addition to this empirical approach, a rational approach was taken so as not to merely seek a high coefficient alpha, which can be achieved simply by having items with maximally similar distributions (Nunnally & Bernstein), but also by examining the content of each item. Five more items were eliminated, resulting in a final 80-item scale, yielding six factors.

DATA COLLECTION 2

In order to re-examine the factor structure of the scales developed in Data Collection 1, as well as to further select the items, the 80-item 3CI was uploaded to the DEOCS. After completing the DEOCS, personnel were again given the option of participating in the research. The total number of usable inventories collected was 4,840.

Demographic Data

Of the 4,840 total participants, 3,872 were male (80%) and 968 were female (20%). The ages ranged from 18 to 40 years of age, with 592 participants between 18 and 20 years of age (12.2%); 2,032 participants between 21 and 24 years of age (42%); 1,130 participants between ages 25 and 29 years of age (23.2%); 672 participants between ages 30 and 35 years of age (13.9%); and 414 participants between the ages of 36 and 40 (8.6%). All but 98 participants reported their pay grade. Of those who reported their pay grades, 1,278 reported a pay grade between 1 and 3 (26.9%);

2,416 reported a pay grade between 4 and 6 (50.9%); 518 reported a pay grade between 7 and 8 (10.9%); 150 reported being at a pay grade between 9 and 10 (3.2%); 208 reported being at a pay grade between 11 and 13 (4.4%); and 172 reported being at the highest pay grades, between 14 and 15 (3.6%). Of the 4,026 participants who reported their Branch of Service, there were 63 participants in the Air Force (1.6%); 1,634 participants in the Army (40.6%); 276 participants in the Coast Guard (6.9%); 705 participants in the Marine Corps (17.5%); and 1,348 participants in the Navy (33.5).

Confirmatory Factor Analysis

Confirmatory factor analysis (CFA) was carried out using LISREL (version 8.30; Jöreskog & Sörbom, 1999) in order to confirm the factors determined by the exploratory analysis. The confirmatory analysis was carried out on approximately two-thirds of the total sample (N = 3,000), henceforth referred to as the confirmatory sample. We specified the measurement model on the basis of the pattern of item–latent factor relationships found in the exploratory step. Specifically, for each item, the path from its respective latent factor (i.e., regression weight for the factor or path coefficient) was allowed to be freely estimated while the paths from other factors were constrained to be zero. We examined the extent to which the model fit the data by using a combination of several fit indexes (i.e., Chi-square, the goodness of fit index [GFI], the root mean square error of approximation [RMSEA], and the standardized root mean square residual [SRMR], and the comparative fit index [CFI]). We eliminated 22 items on the basis of the magnitudes of their loadings on the assigned factors. The model showed reasonable fit (Chi-square = 18,975.94, df = 1,580, p < .01; GFI = .82; RMSEA = .061; SRMR = .058; CFI = .82), confirming the factor structure determined in the exploratory analysis. Thus, 58 items were selected to represent the six factors.

Cross-Validation to Confirm Factors

A confirmatory factor analysis was carried out on the remaining one third of the total sample (N = 1,840), henceforth referred to as the cross-validation sample, using LISREL (version 8.30; Jöreskog & Sörbom, 1999). We specified the measurement model on the basis of the pattern of item–latent factor relationships found in Step 3. Again, for each item, the path from its respective latent factor (i.e., regression weight for the factor or path coefficient) was allowed to be freely estimated while the paths from other factors were constrained to be zero. We examined the extent to which the model fit the data by using a combination of several fit indexes (i.e., Chi-square, the goodness of fit index [GFI], the root mean square error of approximation [RMSEA], and the standardized root mean square residual [SRMR], and the comparative fit index [CFI]). The model showed an acceptable fit (Chi-square = 9,714.23, df = 1,580, p = .00; GFI = .85; RMSEA = .053; SRMR = .057; CFI = .87) and the loadings of all the items are reasonably large (all higher than .40). Thus, the 58-item scale was confirmed to represent the six factors.

CONCLUSIONS

PRELIMINARY INTERPRETATION OF FACTORS

The final 3CI consists of 58 items to assess the six hypothesized dimensions of 3C: (1) Cultural Adaptability; (2) Determination; (3) Tolerance of Uncertainty; (4) Self-Presentation; (5) Mission-Focus; and (6) Engagement. It must be kept in mind that these interpretations are preliminary pending criterion-related data collection to link the dimensions of 3C to important performance criteria. The first factor, the Cultural Adaptability factor, is comprised of items that were originally designed to assess several of the predictors found to be associated with 3C in the literature, such as the willingness to engage with other cultures, self-efficacy, cross-cultural empathy, self-monitoring, openness, and cognitive flexibility. An example item is: "When dealing with people of a different ethnicity or culture, understanding their viewpoint is a top priority for me." Therefore, it is hypothesized that those scoring high on this factor would be adaptable across most types of cross-cultural interactions, especially those requiring diplomacy, an open mind, and an ability to empathize with those from other cultures. This scale is positively and significantly correlated with the Determination, Mission-Focused, and Engagement scales (see Table 1).

The second factor is the Determination factor, which seems to represent those who are determined and focused on reaching their goals as well as able to tune out distractions, whether internal thoughts and feelings, or external events. An example item from this scale is: "After an interruption, I don't have any problem resuming my concentrated style of working." Therefore, a person scoring high on this dimension would probably be someone who is determined and confident in his or her ability to reach goals, solve problems and arrive at solutions quickly. This scale is significantly and positively correlated with the Cultural Adaptability, Mission Focus and Engagement scales (*see* Table 1).

Factor III is the Tolerance of Uncertainty factor. This scale may indicate greater comfort in ambiguous situations. An example item from this scale, which is reverse-scored (greater agreement signifies less tolerance) is: "I like to have a plan for everything and a place for everything." Therefore, this factor might be expected to predict those who would perform better in cross-cultural interactions that involve a high level of ambiguity. Contrary to expectations, this scale was negatively correlated or uncorrelated with the other scales (*see* Table 1).

The Self-Presentation factor is comprised of four items which were originally designed to assess the ability to self-monitor. A sample item from this scale is: "In different situations and with different people, I often act like very different persons." This scale was also negatively correlated or uncorrelated with the other scales (*see* Table 1).

The Mission-Focus factor is comprised of items that indicate someone who is focused, rule-oriented, and a team player. This person is likely to be high in conscientiousness. An example item is, "I think that having clear rules and order at work is essential for success." This scale is significantly correlated with all other scales, except for the Tolerance and Self-Presentation scales (*see* Table 1).

Finally, the Engagement factor is made up of items indicating the willingness to engage with others, openness and the ability to self-regulate one's emotions. An example item from this scale is: "Even after I've made up my mind about something, I am always eager to consider a different opinion." This scale is positively and significantly correlated with all other scales except Tolerance of Uncertainty and Self-Presentation (*see* Table 1).

Table 1 Correlations Among the Scales

Scale Dimension	1	2	3	4	5	6
1. Cultural Adaptability	--					
2. Determination	.46**	--				
3. Tolerance of Uncertainty	-.13**	.07**	--			
4. Self-Presentation	-.05**	-.19**	-.06**	--		
5. Mission Focus	.58**	.48**	-.29**	-.17**	--	
6. Engagement	.55**	.50**	-.10**	-.12*	.73**	--

** Correlation is significant at the 0.01 level (2-tailed).

EXAMINATION OF SCALE PROPERTIES

For these analyses, we used the entire sample (N = 4,840) to estimate the internal consistency reliability (i.e., Cronbach's coefficient α) of scores on the resulting scales for the six factors determined in the previous steps (*see* Table 2).

Table 2 Scale Means, Standard Deviations, and Internal Consistency Reliabilities

Scale Dimension	Scale Mean	Standard Deviation	Cronbach's Alpha
Cultural Adaptability *(18 items)*	4.78	.96	.94
Determination *(7 items)*	4.21	.86	.70
Tolerance of Uncertainty *(11 items)*	3.16	.82	.84
Self-Presentation *(4 items)*	3.01	1.19	.75
Mission Focus *(7 items)*	4.71	.92	.88
Engagement *(11 items)*	4.31	.87	.88

We also estimated the correlations of the scales with the demographic variables (*see* Table 3).

Table 3 Correlations Between Scales and Demographic Variables

Scale	Gender[a]	Pay Grade[b]	Age Range[c]
Cultural Adaptability	-.02	.04**	.10**
Determination	.00	.15**	.20**
Tolerance of Uncertainty	.02	-.01	-.03*
Self-Presentation	-.02	-.19**	-.29**
Mission Focus	-.01	.14**	.24**
Engagement	-.01	.08**	.17**

** Correlation is significant at the 0.01 level (2-tailed).
[a] Male=1, Female=2
[b] (1-3)=1, (4-6)=2, (7-8)=3, (9-10)=4, (11-13)=5, (14-15)=6.
[c] (18–20)=1, (21–24)=2, (25–29)=3, (30-35)=4, (36-40)=5, (40+)=6.

NEXT STEPS

In order to validate the 3CI, criterion data will be collected in the form of supervisory ratings of observed behavior in the field. A well-developed tool will support decisions about training, education, and operations.

REFERENCES

Abbe, A., Gulick, L.M.V, & Herman, J.L. (2007). *Cross-cultural competence in Army leaders: A conceptual and empirical foundation.* U.S. Army Research Institute for the Behavioral and Social Sciences, Study Report 2008-1. Arlington, VA: U.S. Army Research Institute for the Behavioral and Social Sciences.

Borman,W.C. (1991). *Job behavior, performance, and effectiveness.* I/O Handbook.

Catell, R.B., (1966). The scree test for the number of factors. *Multivariate Behavioral Research, 1,* 245-276.

Costa, P.T, Jr., & McCrae, R.R. (1992). Normal personality assessment in clinical practice: The NEO Personality Inventory. *Psychological Assessment, 4,* 5-13.

Earley, P.C., & Ang, S. (2003). *Cultural intelligence: Individual interactions across cultures.* Palo Alto, CA: Stanford University Press.

Gorsuch, R.L. (1983). *Factor analysis.* Hillsdale, NJ: Lawrence Erlbaum Associates.

Hammer, M.R., Bennett, M.J., & Wiseman, R.L. (2003). Measuring intercultural sensitivity: The Intercultural Development Inventory. *International Journal of Intercultural Relations, 27,* 421–443.

Jöreskog, K.G., & Sörbom, D. (1999). *LISREL 8.30 and PRELIS 2.30.* Chicago: Scientific Software International.

Kelley, C., & Meyers, J. (1995). *The Cross-Cultural Adaptability Inventory.* Minneapolis, MN: National Computer Systems.

Krulak, C.C. (1999, Jan.). *The Strategic Corporal: Leadership in the Three Block War.* Retrieved from the Web October 2, 2009.

Available from:
http://www.au.af.mil/au/awc/awcgate/usmc/strategic_corporal.htm

Matsumoto, D., LeRoux, J., Ratzlaff, C., Tatani, H., Uchida, H., Kim, C., & Araki, S. (2001). Development and validation of a measure of intercultural adjustment potential in Japanese sojourners: the Intercultural Adjustment Potential Scale (ICAPS). *International Journal of Intercultural Relations, 25,* 483-510.

Nunnally, J.C., & Bernstein, I.H. (1994). *Psychometric Theory* (3rd ed.). New York: McGraw Hill.

Olson, C.L., & Kroeger, K.R. (2001). Global competency and intercultural sensitivity. *Journal of Studies in International Education, 5,* 116-137.

Ross, K.G. (2008, May). *Toward an operational definition of cross-cultural competence from interview data.* Patrick AFB, FL: Defense Equal Opportunity Management Institute (DEOMI).

Ross, K.G., & Thornson, C.A. (2008, March). *Toward an operational definition of cross-cultural competence from the literature.* Patrick AFB, FL: Defense Equal Opportunity Management Institute (DEOMI).

Ross, K.G., Thornson, C.A., McDonald, D.P., & Arrastia, M.C. (2009, February). *The development of the CCCI: The Cross-Cultural Competence Inventory.* Paper presented at the Conference Proceedings of the 7th Biennial Equal Opportunity, Diversity and Culture Research Symposium, Patrick AFB, FL.

Selmeski, B.R. (2007). *Military cross-cultural competence: Core concepts and individual development.* Kingston: Royal Military College of Canada Centre for Security, Armed Forces, & Society.

Smith, M.B. (1966). Explorations in competence: A study of Peace Corps teachers in Ghana. *American Psychologist, 21,* 555-566.

Stringer, K.D. (2009, February-March). Educating the strategic corporal: A paradigm shift. *Military Review* (pp. 87-95). The Combined Arms Center: Fort Leavenworth, KS.

Thornson, C.A., Ross, K.G., & Cooper, J. (2008, June). *Review of the literature and construction of a measure of cross-cultural competence.* Patrick AFB, FL: Defense Equal Opportunity Management Institute (DEOMI).

van der Zee, K.I., & van Oudenhoven, J.P. (2000). The Multicultural Personality Questionnaire: A multidimensional instrument of multicultural effectiveness. *European Journal of Personality, 14,* 291-309.

Wang, Y.W., Davidson, M.F., Yakushko, O.F., Savoy, H.B., Tun, J.A., & Bleiern, J.K. (2003). The Scale of Ethnocultural Empathy: Development, validation, and reliability. *Journal of Counseling Psychology, 50,* 221–234.

Zbylut, M.R., Metcalf, K.A., McGowan, B., Beemer, M., Brunner, J.M., & Vowels, C.L. (2009, June). *The human dimension of advising: Descriptive statistics for the cross-cultural activities of transition team members.* Fort Leavenworth, KS: U.S. Army Research Institute for the Behavioral and Social Sciences.

Chapter 11

Democracy's Sacred Opinions and the Radicalization of Islam in the Twentieth Century

Jonathan W. Pidluzny

School of Public Affairs
Morehead State University
Morehead, KY 40351, USA

ABSTRACT

This paper argues that the lesson of Iraq is that American policymakers know too little about their own constitutional democracy, and still less about the special impediments to its successful dissemination in the Islamic world. A more accurate theoretical understanding of the social requisites of *liberal* democracy is a prerequisite of a responsible U.S. foreign policy in the Middle East going forward.

Keywords: Liberal democracy, Iraq, Islam, Alexis de Tocqueville, Syed Qutb

INTRODUCTION

Rarely in the annals of history has it been so difficult to determine *casus belli* as it is in the case of America's 2003 invasion of Iraq. The war to overthrow Saddam Hussein's regime will be remembered as one that appeared necessary in the aftermath of the 9/11 terrorist attacks (to most Americans and to lawmakers on both sides of the aisle), but which seemed less and less justifiable as its costs mounted and the public's expectations about the war were, little by little, disappointed. Even today, almost seven years after the initial invasion, no single account of what really led the Bush Administration to invade a country halfway around the world can claim widespread acceptance. Concerns about Saddam Hussein's Weapons of Mass Destruction (WMD), Islam-inspired terrorism, and democracy promotion all figured importantly, but the relationship between these rationales was never perfectly clear. To be sure, the dangers associated with Saddam Hussein's WMD program were publically emphasized as a legitimate, sufficient, and indeed the primary, justification for invasion and regime change in the months leading up to the invasion, and understandably so. Americans' tolerance for the species of risk presented by "the crossroads of radicalism and technology" was lower than ever in the wake of 9/11 (The White House, 2002, p.12). It was widely believed Saddam Hussein could not be trusted (he had made every effort to evade U.N. sanctions and weapons inspections for more than a decade), and every national intelligence service of consequence was under the impression Iraq possessed significant WMD production capacity and stockpiles (largely because Saddam Hussein and his lieutenants actively perpetuated that myth for strategic reasons of their own).

And yet, from the beginning, other justifications for invasion were being discussed privately. Some in the administration saw war with Iraq as an opportunity to reassert American military primacy and to make an example of a relatively powerful leader, an example that would resonate throughout the Middle East and the wider Islamic world. Self-styled realists including Dick Cheney, Donald Rumsfeld, and his Under Secretary for Policy, Douglas Feith, reportedly argued that regime change in Iraq would encourage other states to take a harder line on terrorists and their supporters operating within their borders. Somewhat perversely, they believed the invasion of Iraq would exert a widespread behavior-altering effect *precisely because* Iraq had *not* been involved in the 9/11 terrorist attacks (Feith, 2008, 15). A demonstrated willingness to invade a country perceived as presenting a gathering, though not yet imminent, threat *preemptively*— for harboring extremists, or in Iraq's case, developing dangerous technologies in contravention of United Nations resolutions—would, it was argued, establish a powerful incentive for the leaders of states like Iran and Syria (and even nominally allied states like Egypt, Jordan, Saudi Arabia, and Pakistan) to crack down on anti-American extremism actively and voluntarily, and (in the case of Iran and Syria) give up their WMD ambitions permanently. Planners understood that terrorists willing to martyr themselves cannot be deterred, but they believed the states in which terrorists were organizing could be induced to police the problem more aggressively. "You're either with us or against us" was the message. The invasion of Iraq was, on this rationale, intended to put an overthrown tyrant's face to the threat for the sake of credibility: prevent threats to America from gathering within your borders or your regime too will be changed.

Changed to what? Lastly, but most important for present purposes, an influential cadre persuaded President Bush that regime change in Iraq would provide an historic opportunity to build a moderate constitutional democracy in the heart of the Arab-Islamic world. True, the emphasis the administration placed on Iraq's democratization increased dramatically as Coalition forces combing the country appeared less and less likely to uncover the major WMD stockpiles that had been emphasized pre-invasion (Feith, 2008, 477). It was not, however, as some defenders of the Bush Administration have asserted, merely a rhetorical shift. The underlying democratization rationale was first discussed at the highest levels within days of 9/11 attacks—not as a sufficient justification for the invasion of any particular country, but as a significant opportunity presented by intervention abroad, an end result that would help justify invasion (in Afghanistan and beyond) *post-hoc*, and hopefully attack terrorism at its so-called root cause over the long term (Feith, 2008). Pentagon planners believed preventing another 9/11 required "changing the way *they* live" (p. 71). A crude version of democratic peace theory was operating in the background: liberal democracies do not go to war with one another, and tolerant commercial republics do not produce religious fanatics willing to blow themselves up. Fully six months before the invasion of Iraq, the 2002 National Security Strategy established as one of its pillars the promotion of "modern and moderate government, especially in the Muslim world" (The White House, 2002, p. 1). A number of the President's speeches in 2002 and 2003 expressed the goal in even more audacious terms.

Ultimately, America's ill-fated approach to Iraq's reconstruction and reconstitution, which did so much to open a space for the insurgency, was in large part the result of the overly ambitious and ill-conceived democracy-promotion end game. The replacement of Saddam Hussein's brutal tyranny with a stable, pluralistic, prosperous, and tolerant constitutional democracy was expected to have significant "spillover" effects in the region; changing Iraq's regime would be the harbinger of a long-overdue Arab spring. Or so it was hoped. Instead, the power vacuum created in Iraq by American intervention unleashed forces Saddam Hussein had only managed to suppress at great effort and by brutal means, revealing in tragic fashion precisely how imperfectly policymakers and military planners understand the regime they sought to disseminate in Iraq and beyond. At once, the occupation of Iraq revealed powerful impediments to democratization today present and growing stronger in the Arab-Islamic world, but which policymakers simply failed to perceive in advance.

UNDERSTANDING "REGIME"

For all the discussion of regime change, it is startling the term was so badly misunderstood. A "regime" is much more than an institutional structure and system of laws. No regime can be understood absent an inquiry into its purported purpose. Ostensibly similar institutional and legal frameworks can be employed to achieve entirely different ends: to make possible every individual's self-directed pursuit of happiness, to conquer the known world, to bring about an equality of conditions, to instantiate God's law, etc. More important than any particular constitutional provision is the way of life a regime exists to protect and promote by the totality of those provisions. Success ultimately depends not on the structures and laws themselves, but on the spirit that guides them. Whether a regime achieves its professed purpose therefore depends most importantly on factors external to its rules, institutions, and constitution: above all, the character of the people.

As Samuel Huntington explains, an individual's most sacred opinions, convictions, and habits are in some sense an artifact of, and constitute his membership in, a community that is generally broader than a single village, town, or state: his civilization. Civilization "is the broadest cultural entity," and refers to "the overall way of life of a people"; it denotes "a culture writ large" (Huntington, 1996, pp. 41-43). Communal dedication to notions of right and wrong, good and evil, noble and base, decent and obscene, permitted and impermissible, beautiful and ugly, worthy and worthless—judgments in the context of which an individual defines happiness and the sort of life it is worth living or aspiring to—determines the operation of a regime's laws and institutions in practice. "[T]he world's major civilizations," another scholar explains, "are more or less coterminous with its major religions and, much more roughly, with its major races." (Codevilla, p. 50).

What, then, is the relationship between civilization and regime? If "regime" refers to the overall political organization and purpose of a political community, "civilization" refers to the collection of influences that are extraneous to the regime, influences that have antecedently formed the subject matter, the people, regimes attempt to organize. These guiding opinions and social practices can only be changed against considerable resistance. While regimes can exert a steady and potentially transformative effect on the character of their citizens, rulers, and subjects—even to the point of affecting the tenor of the civilization they overlap over time—civilization-level forces, especially religion and other sacred beliefs, almost always exert the stronger influence. Thus, they affect the kinds of regime that are suited to a given people, and the manner in which this or that set of laws and institutional arrangements will operate in a given time or place. As Angelo Codevilla cogently puts it, "Civilizations set the bounds within which regimes exercise their powers over human habits" (Codevilla, 1997, p. 50).

It follows that distinct regimes and civilizations are more and less flexible. Some civilizations will be amenable to, supportive of, a variety of forms of political organization. The sacred beliefs that define others will mandate specific forms of political arrangement and thwart the establishment of others. Alien structures and institutions (for instance, elections and new freedoms) can be implanted in soil from which they did not naturally spring. They will not, however, achieve their professed ends (for instance, tolerant political life, the noble use of liberty, the protection of minorities, the rule of law, and equality before it) unless the character of the people animates the regime's legal structures in just the right way. More likely, alien institutions and laws will be co-opted in service to ends glorified by the civilization's dominant opinions.

Every student of politics knows that the species of democracy worth aspiring to demands more than elections and majority rule. Where constitutional democracy functions as it does in the West today, it is the fruit of a rare and delicate union. The liberal temperament of the people and a democratic political arrangement are mutually dependent, *vital co-requisites*; freedom and equality are established and secured where they intersect. Put another way, liberal, limited, and stable government is not an inevitable outcome of free elections and participatory institutions. Free elections can just as easily lead to tyranny of the majority and the prosecution of minorities. For government according to the will of the majority to be tolerant, just, and good, the people must first be tolerant, just, and good.

In Iraq, the institutions and privileges of the new political regime—of democracy: free elections, new rights and liberties—have been used to destabilize the country and to empower intolerant factions determined to employ state authority for narrow parochial ends. Tragic illustrations abound. After Saddam Hussein's overthrow, prominent Shiites used their influence with American administrators to push for de-Baathification so thorough that many believed it tantamount to de-Sunnification. In addition to fueling the insurgency, this early abuse of power stands in the way of political reconciliation today, seven long years into Iraq's reconstruction. Even more egregious, the Shiites and the Kurds used the Constitutional Convention (where they were overrepresented as a result of the Sunni boycott of Iraq's first election) to build a radically decentralized state that would advantage the Kurdish and Shiite sections at the cost of using a once-in-a-generation opportunity to build a united Iraq. Prominent religious leaders have also sought political influence through elections. Ayatollah al-Sistani actively employed his religious authority over Iraq's Shiite majority, to the point of issuing *fatwas* or religions decrees, in order to build the powerful Shiite bloc that has dominated Iraqi politics since the country's first election. It has successfully achieved the relaxation of the minority protections Americans fought for on the basis that the will of a majority should never be frustrated in a democracy. One consequence: Iraq's governorates are not subject to the human rights provisions of its new Constitution.

Abuses of process in Iraq were at times much more brutal. Shiites in government uniforms used their control of the interior ministry and state militias to terrorize (even to slaughter) innocent Sunnis in response to insurgent attacks that targeted Shiites and their holy sites. The insurgency was finally suppressed at high cost, but the blatant use of state authority for narrow partisan ends persisted. Nuri Maliki's government has employed state resources to build partisan voter turnout mechanisms and marginalize key rivals; to schedule the execution of Saddam Hussein on a Sunni holy day in direct violation of Iraqi law; to deny Sunnis who helped Americans route Al-Qaeda in Al-Anbar incorporation into Iraq's massive security apparatus on fair terms; and in January of 2010, to support a blatantly partisan decision by the Shiite-dominated Accountability and Justice Commission disqualifying 500 candidates, most of them Sunni, in the looming national election. Sunni parties prepared to withdraw from the 2010 elections altogether in response.

New liberties have caused almost as much harm to Iraq's social state. The end of Saddam Hussein's systematic censorship of the press led to the immediate proliferation of satellite dishes and local media outlets. The result: inflammatory anti-American propaganda flooded into Iraqi households instantaneously doing not a little to increase domestic support for the insurgency; every political interest meanwhile (many of which turned violent in Iraq's darkest days) rushed to create instruments of misinformation in a country with virtually nonexistent libel laws. Most problematic of all, perhaps, the end of Saddam Hussein's brutal system of fear and oppression, and the extension to Iraqis of new freedoms of movement and association, led to an unanticipated explosion of radical Islam in Iraq (it had, thitherto, been driven underground). In sum, coupled with a steady influx of foreign fighters, the new liberties regime change in Iraq afforded Iraqis provided the oxygen that allowed the insurgency to burn out of control.

DEMOCRACY'S "SACRED OPINIONS"

What built a social character suited to uniting participatory institutions, individual freedoms, and moderate, liberal, government in the West? The ideas promulgated by Enlightenment thinkers over centuries, buttressed by New Testament Christianity as interpreted—and liberalized!—by men like John Locke, Benedict Spinoza, and Martin Luther, constitute the modern West's most important character-imparting influences. Though we do not sufficiently appreciate it, the way of life our constitutional arrangement guarantees depends for its endurance on airy nothings—opinions, ideas, habits, and social practices universally imbibed by citizens with the air, simply by living in the regime. We believe that all men are created equal; a just regime is therefore one in which all are equal before the law and entitled to the same privileges of political participation. We have confidence in the human intellect and believe that an individual's right to pursue happiness as he or she personally defines it is sacrosanct; a just regime is therefore tolerant, and religious authority durably separated from temporal authority. We believe certain rights are inalienable; the powers of government are therefore limited, our rights, liberties, and property secure from state encroachment without due process of law.

Alexis de Tocqueville, perhaps the greatest student of democracy, made this point with particular emphasis in *The Old Regime*. He argued that the democratic political revolutions that occurred in Europe and America in the eighteenth century *could not* have occurred in the fifteenth for the simple reason that the ideas and social practices so essential to liberal democracy were not yet capable of taking hold in men's minds: "[f]or doctrines of this kind [the natural rights of man] to lead to revolutions, certain changes must already have taken place in the living conditions, customs, and *mores* of a nation and prepared men's minds for the reception of new ideas" (Tocqueville, 1955, 13). When Tocqueville famously declares in *Democracy in America* that the democratic revolution sweeping Europe is "irresistible," he is referring to the inevitable political impact of Europe's new political consciousness (Tocqueville, 2000, p. 400; c.f. pp. 6-7). Locke taught that human beings had natural rights and that a separation of church and state was indispensable to the integrity of both; Voltaire, that superstitions adhered to on faith and promulgated by the Church had to be jettisoned in favor of the free human intellect; Rousseau, that the sacred truth of modern times is that the legislative authority resides in the people, and legitimate government established only by social contract; and Spinoza, that the Gospels properly understood support, and indeed demand of Christians, a form of government that is limited and liberal.

As important, the ideas resonated, ultimately vanquishing the convictions that had so long sustained the old aristocratic regimes. They resonated and stuck in hearts and minds because the new opinions that support popular government were advanced under social-political circumstances that made them seem true, self-evident, to the population at large. Not unimportantly, the newly depoliticized Christianity of the Gospels a tremendous moral authority— supported, and may even be the first source of, modern liberal government: all men are equal in the eyes of God and capable by their own lights of discovering the road to heaven; charity, forgiveness, and neighborliness are virtues of character; impartiality is a prerequisite of just judgment; a distinction between the obedience properly owed to Caesar and to God is explicitly sanctioned and so, a separation between Church and state; the notion of an immortal soul is a bulwark against the individualism and materialism egalitarian democracy naturally nourishes but which threaten to subordinate the wellbeing of the community to that of selfish individuals.

THE "RESURGENCE" OF ISLAM

In the Middle East, diametrically countervailing ideational trends are today discernable, the result of ominous winds that have been gathering for the better part of a century. Whereas Enlightenment thinkers marshaled the authority of Christianity in support of liberal democracy, the most influential modern Islamic theorists have conspired with the political turbulence of the twentieth century (in which the West has had more than a hand) to radicalize and politicize Islam among a small but vocal, growing, and sometimes violent, segment of the population.

It must be acknowledged that fundamentalist Islam is the product of great and noble minds. Theorists of formidable intellect and learning devoted their lives to reviving and purifying Islam in order to build just and pious communities. Hasan al-Bana is an important early leader of the modern revival. An Egyptian schoolteacher by profession, and active politically during the early twentieth century, his most important legacy is the organization he founded in 1928, the Muslim Brotherhood. He conceived of its project in expressly anti-Western terms, as a direct response to, and repudiation of, the moral-political outlook of the North-Atlantic states. In the name of fidelity to Islam's founding tenets, al-Bana explicitly rejects virtually every distinctive feature of liberal democracy: nationalism, state sovereignty on a secular basis, the confinement of religion to a limited political sphere, unfettered capitalism, a constitutional separation of powers, the notion that social and political equality mandate government according to consent in which participation is widespread, the primacy of the individual and his rights, the notion that individuals are entitled to pursue happiness as they themselves define it within a very generous sphere protected by law.

The Islamic revival he sought to catalyze would, he believed, exceed in scope both the French and Russian Revolutions (Rosen, 2008, p. 117). Al-Bana aimed not simply to lay down new organizing laws; he sought nothing less than to bring Egypt's moral, intellectual, and political life into harmony with a purified interpretation of Islamic law, which is to say, to subordinate all aspects of life to strict religious decrees. In an essay entitled "Our Mission," al-Bana explains that "Islam is an all- embracing concept which regulates every aspect of life, adjudicating on every one its concerns and prescribing for it a solid and rigorous order" (al-Bana, 2006, p. 61). He envisioned building an Islamic state on the most solid foundation possible: millions of hearts and minds dedicated to the literalist species of Islam he sought to rehabilitate. For al-Bana recognized (in direct contradistinction to those who led the U.S. endeavor to build a liberal democracy in Iraq!) that the possibility of successful political reform rests first and foremost on soul-craft: where the opinions and social habits internalized by the citizens (or subjects) of the regime do not support it, no form of government can persist except by repression and force. For this reason he emphasized education and the importance of widespread proselytizing. The Muslim Brotherhood, an organization of global reach today, was explicitly tasked with the "shaping of fully Islamic personalities"—of shaping souls now to prepare society for *bottom-up* political reform in time (Rosen, 2008, p. 118). Thanks to virtually limitless Saudi funding, schools and mosques inspired by the Brotherhood's aims (and often beholden to its leadership) exercise a near-monopoly over the education of young Muslims in vast communities throughout the Islamic world, its reach extending even to neighborhoods in London, Paris, and Toronto. Issam al-Aryan, one of the group's leaders, recently explained that "reforming the Muslim individual, the Muslim home and the Muslim society" leads to "restoring the international entity... and ends with being masters of the world through guidance and preaching (Altman, 29).

Syed Qutb, a prominent member of the Muslim Brotherhood in Egypt mid-century, did more than any other individual to radicalize Al-Bana's teachings. He was imprisoned and later executed in 1966 by Nasser's regime for his political activities and the extremist views his popular works were popularizing. Though Qutb's political influence during his lifetime was impressive, his ideas have exerted a much greater effect in the decades since his death. He influenced Al-Qaeda's top leaders directly: Ayman Zawahiri was one of his students, and through the latter's mentorship, Qutb's thought exerted a deep influence on Osama Bin Laden as well. Like al-Bana, Qutb worked for the establishment of an Islamic state or states governed entirely according to *sharia*. Triumphant assertions that Islam is a totalitarian legal code litter Qutb's works: Muslims should "arrange [their] lives solely according to... the Book of God"; "From [the Koran] we must also derive our concepts of life, our principles of government, politics, economics, and all other aspects of life"; "people should devote their entire lives in submission to God, should not decide any affair on their own" (Qutb, pp. 11-15, 41). As such, there can be no separation of Church and state, no freedom of action for individuals. Laws not derived from sacred books can never be legitimate: "The basis of the message is that one should accept the Shari'ah without any question and reject all other laws in any shape or form. This is Islam" (p. 30). Confidence in the human intellect and the endeavor to free minds from received dogma, everything the Enlightenment achieved in the West, has no place in Qutb's Islam. Speaking of "the way [religion] is to be founded and organized," he emphasizes the importance of "implanting belief and strengthening it so that it seeps into the depths of the human soul"; or as he later puts it, "belief ought to be imprinted on hearts and rule over consciences" (pp., 26, 29).   What can only be called indoctrination is, according to Qutb, "essential for its [religion's] correct development" (p. 26). The aim: complete intellectual submission. Freedom as we understand the term is anathema to Qutb's Islam; in his words, "the spirit of submission is the first requirement of the faith. Through this spirit of submission the believers learn the Islamic regulations and laws with eagerness and pleasure" (p. 27).

What the Reformation accomplished in the West is, similarly, dismissed out of hand: Islam "abhors being reduced" to an individual's private relationship with God; "it cannot come into existence simply as a creed in the hearts of individual Muslims, however numerous they may be" (p. 34). On the contrary, Islam's laws must be established in practice exactly as they were revealed (p. 2). Qutb dismisses the endeavor to modernize and liberalize Islam as "a vulgar joke," the dangerous fruit of contemporary Godlessness (p. 37). Toleration of other religions on equal terms is expressly prohibited: God "made Islam a universal message, [and] ordained it as the religion for the whole of mankind"; non-Muslims must therefore convert, submit to Muslim rule and pay a tax, or find themselves at war (pp., 9, 48). He even explains why the Koranic verse "there is no compulsion in religion" does not prohibit working to "annihilate all those political and material powers which stand between people and Islam" (p. 51).

Everything that stands in the way of Islam's establishment—including the West—must be destroyed as the precondition of widespread submission to Islamic law. If it is not sufficiently clear that this brand of Islam is not an appropriate ideational support for liberal democracy, Qutb explains that what the West values above all, freedom, is actually worthless on a proper understanding of the term. Furthermore, far from preserving the rights and liberties of the citizenry, democratic government is the worst form of tyranny and slavery insofar as it represents universal enslavement to laws made by vain and selfish human beings acting as tyrants on earth. Submission to divine law, on the other hand, "is really a universal declaration of the freedom of man from servitude to other men and from servitude to his own desires, which is also a form of human servitude" (p. 51). Freedom does not require the relaxation of religion's influence in the political sphere or widespread political participation in government to which the citizens willingly consent. On the contrary, true freedom demands the annihilation of the distinction between religion and politics, because, in Qutb's words, "the implementation of the Shari'ah of God," a totalitarian religious code instantiated by force on earth, is the real prerequisite of "freeing people from their servitude to other men" (p. 52).

Qutb agrees with al-Bana that education is of utmost importance. Unlike al-Bana, however, he also perceived a pressing urgency that justified more extreme measures.   He taught that the contemporary Middle East could be declared *jahilliya*—plagued by a ubiquitous ignorance reminiscent of Arabia before the Prophet Muhammad, a condition the Prophet himself had first to alleviate before fully Islamic communities could be built. The West's corruption of everything— the ubiquitous nihilism and materialism spread by economic and military conquest without historical precedent—constituted a powerful impediment to the proselytizing approach al-Bana preferred; in fact, it represents, for Qutb, an imminent threat to Islamic civilization. For this reason, Qutbists fiercely oppose the region's corrupt rulers as well as virtually all modern Koranic interpretation and commentary. That the Islamic world has fallen into a condition of barbarity and ignorance justifies armed *jihad* against Arab states and their rulers (for despoiling Islam by adopting the outer trappings of modernity), foreign powers operating on holy soil (for supporting the corruption), even adherents to apostate strains of Islam and members of other religions (for the crime of spreading disbelief). They believe that contemporary conditions mandate *jihad* insofar as the prevailing barbarism makes a return to wholesome Islamic life impossible by gentler means; it becomes a religious duty to remove impediments to the dissemination of a purified Islam. The traditional seat of Islamic civilization is the first concern. Ultimately, however, Islam "strives... to abolish all those systems and government which are based on the rule of man over men," chief among these, the Western democracies (p. 54).

CONCLUSIONS

The revival has been a tremendous success. It cannot be overemphasized that only a minority of Muslims ascribe to Qutb's Islam. Unfortunately, their fanaticism is sufficient to stand in the way of meaningful democratization in the Middle East. Samuel Huntington's observation that this "Islamic Resurgence" is an event "at least as significant" as the French, Russian, and American Revolutions is perhaps the most important, if little appreciated, observation contained in his seminal *The Clash of Civilizations* (p. 109). Why does fundamentalist Islam resonate today? Contemporary modes of communication have contributed. As Ayatollah Khomeini incited the Iranian Revolution from Paris by cassette tape, so have Sunni Islamists disseminated their views as widely as they have thanks to the indispensable help of modern technologies, most importantly, satellite television and the internet's infinite reach. More important, perhaps, are two centuries of economic stagnation and humiliating military defeats at the hands of Israel, America, and the West. Just as the sociological factors Tocqueville identifies in *The Old Regime* prepared European and American minds for the ideas presented by Reformation and Enlightenment thinkers, so has the ignominious weakness of the contemporary Arab-Islamic world increased the power of fundamentalist Islam over hearts and minds in the region. As Fouad Ajami explains, memories of the Crusades, Ataturk's betrayal of Islam, the Six Days War (and more recently, American intervention in Iraq and support for Israel) have, taken together, "created a deep need for solace and consolation, [for which] Islam provided the needed comfort" (Ajami, 1992, p. 71). Islamic fundamentalists have leveraged persistent Arab angst masterfully to make "an eloquent and moving case" that the Arab world has declined so far from its apogee "because [Arabs have] lost their faith and bearings" (p. 61). The solution: renewed commitment to a purer, militant, Islam.

How should the West respond? Given the demanding social requisites of liberal democracy, and the obstacles to its dissemination in the Middle East, it is time for U.S. policymakers to renew their commitment to foreign policy realism.

REFERENCES

Ajami, Fouad. (1992), *The Arab Predicament*. New York: Cambridge University Press.

al-Bana, Hasan. (2006), Our Mission. In *Six Tracts of Hasan al-Bana*. (Majmu At Rasa Trans.). I.I.F.S.O.

Altman, Israel Elad. (2007), The Crisis of the Arab Brotherhood. [Electronic version]. *Current Trends in Islamist Ideology, volume 6*, 29-47.

Codevilla, Angelo. (1997), *The Character of Nations*. New York: Basic Books.

Feith, Douglas. (2008), *War and Decision*. New York: HarperCollins.

Huntington, Samuel. (1996), *The Clash of Civilizations and the Remaking of the World Order*. New York: Simon & Schuster.

Tocqueville, Alexis de. (2000), *Democracy in America*. (Harvey Mansfield Trans.). Chicago: University of Chicago Press

Tocqueville, Alexis de. (1955), *The Old Regime and the French Revolution*. (Stuart Gilbert Trans.). New York: Doubleday.

Qutb, Syed. *Milestones*. (SIME Journal Trans.). Retrieved from http://majalla.org/

Rosen, Ehud. (2008), The Muslim Brotherhood's Concept of Education [Electronic version]. *Current Trends in Islamist Ideology, volume 7*, 117-133.

The White House. (2002). *The National Security Strategy of the United States*.

CHAPTER 12

Theories of Regime Development Across the Millennia and their Application to Modern Liberal Democracies

Murray S. Y. Bessette

School of Public Affairs
Morehead State University
Morehead, KY 40351-1689, USA

ABSTRACT

The apparent permanence of present relations and arrangements makes understanding the significance of short term changes difficult. Over time the life-development of civilizations is recognized readily – like men, civilizations come into being, mature, and perish from the earth. Underlying this rise and fall of civilizations is the rise and fall of regimes. The development of a regime also inevitably ends in change and death. Philosophers have sought to explain this development with an eye to prolonging the good stages and shortening the bad. Perhaps the two most important attempts are those of Plato and Machiavelli. The Machiavellian analysis in particular is relevant to regime maintenance and change in republican polities. When combined with Nietzsche's analysis of institutions, a proper understanding of the cycle of regimes as it relates to liberal democracies can be acquired and may provide grounds for its continuing maintenance.

Keywords: Plato, Machiavelli, Nietzsche, cycle of regimes, liberal institutions, civilizational development, heredity exhaustion, value relativism, liberal democracy

INTRODUCTION

The apparent permanence of present political relations and arrangements makes understanding the long term significance of short term political changes difficult. Over a sufficiently extended timeline, the life-cycle of civilizations is recognized readily: like men, civilizations too come into being, mature, and perish from the earth. The rise and fall of civilizations is connected to the rise and fall of regimes. The development of every regime inevitably also ends in change or death – every regime eventually comes to an end in its transformation into a regime of a different type.

The question of political impermanence has long held the attention of both political practitioners and political theorists. Over the course of the history of philosophy, philosophers have sought to explain this cycle of regimes with an eye to prolonging the good stages and shortening the bad. Perhaps the two most important such attempts are those of Plato, provided in the *Republic*, and Machiavelli, provided in *Discourses on Livy*. The operation of the underlying mechanism identified by both philosophers – heredity, or rather, hereditary exhaustion – is explained best in contemporary times by Nietzsche in both *Beyond Good and Evil* and *Twilight of the Idols*. The following essay will begin by sketching the cycle of regimes and its implications for founding as treated by Plato and Machiavelli, before preceding to elaborate Nietzsche's conception of hereditary exhaustion. It then culminates in a brief diagnosis of the crisis faced by modern liberal democracies.

PLATO

The discussion of the cycle of regimes in the *Republic* takes as its starting point the city founded in speech by the elderly philosopher Socrates and his young interlocutors Glaucon and Adeimantus, as such it behooves us briefly to examine its origin, genesis, and nature. The intention of the founders here in the *Republic* is to locate justice within the city so as then to identify by analogy justice in the individual (*Republic*, 369ea-b). According to the dialogue, "a city ... comes into being because each of us isn't self-sufficient but is in need of much" (*ibid.*, 369b). The city that fulfills these mutual needs – food, shelter, and clothing – is later named "the true city" (*ibid.*, 372e), perhaps indicating that every city *qua* city must do this, regardless of what else it does. Socrates elides from this list of necessary needs that which is perhaps of the utmost necessity: sleep, or rather, security while sleeping (*ibid.*, 369d). It is not until this healthy city becomes feverish with luxury that the founders find it necessary to add "a whole army" (*ibid.*, 374a). The transition from the city of pigs to the luxurious city is immanent within the city's foundation: mere necessity is barely living (and perhaps procreating). Anything more than that which keeps you above flat-lining is always and already directed toward some understanding of the Good, even if the Good is understood simply as the comfortable. In short, there may be no line of mere need; or, as Rousseau's *First*

Discourse demonstrates, that which is a luxury today inevitably becomes a necessity tomorrow (*cf. Oeuvres*, 1-107). It is because the Good is immanent within the needful that Aristotle says of the city, "while coming into being for the sake of living, it exists for the sake of living well" (*Politics*, 1252b28-30).

That the needful implies the Good; that living well requires more than merely living; that by nature resources are subject to scarcity, together all point forward to the problem of war: if we are to live well, soldiers are necessary both to acquire resources and to protect such as are acquired. Luxury and war are connected not only by the fact that luxury requires war (or at least the threat of war), but also by the fact that war (or at least the preparation for war) requires luxury, that is, the freedom from necessity known as leisure, which requires that others work so that you do not have to work. Thus, the guardians of the city in speech are provided with leisure for military and intellectual exercises, for war and education. While the object of military training is obvious – victory in war and the preservation of the regime – the object of intellectual cultivation is less obvious, at least at first glance: does one seek to cultivate the Virtue of the man or the virtue of the citizen? The educational curriculum of the *Republic* is not that which leads to a philosophic cosmopolitanism grounded in knowledge of truth; it leads, rather, to a dogmatic patriotism grounded in a noble lie (*Republic*, 414bc-415c). The goal is not Virtue simply, but virtue tied to this particular city. Thus, we can see that it too is directed toward the perpetuation of the regime. The goal is to cultivate within the ruling class the following opinions: the laws of the regime are good, the interests of the regime are identical to the interests of the ruling class, and, therefore, the regime is worthy of the love of the guardians. The result of such successful cultivation is the perpetuation of the law, that is, the perpetuation of that which defines the regime and that which the guardians guard most of all. In short, real devotion to the common good in the best regime as detailed in the *Republic* is guarding the form of the regime, that is, preserving, protecting, and defending the fundamental law, because it is best (*ibid.*, 424b).

With the foregoing in mind we are now prepared to move forward to the discussion of the cycle of regimes here in the *Republic*. Plato's Socrates identifies five types of regime, each dominated or defined by the goal or end to which they are devoted: the best regime, aristocracy, is devoted to virtue; timocracy is devoted to honor; oligarchy is devoted to money; democracy is devoted to freedom; and tyranny is devoted to power, or as Hobbes would say, to the "perpetuall and restlesse desire of Power after power, that ceaseth onely in Death" (*Leviathan*, 70). Plato's Socrates refers to the final regime as noble, in the sense of well-bred (*gennaia*) and not fine (*kalos*), because it originates in the degeneration of justice, it grows from the seed of the best regime (*Republic*, 544c). Each regime is defined by that which moves or attracts the soul of the ruling type, meaning the most significant political leaders as well as the most admirable forms of life. Moreover, the transitions are shown to be the result of weakness on the part of the ruling class (*ibid*, 545d). In other words, because a regime is defined by who rules, the decline of a regime is the result of the decline in the rulers. Revolution, therefore, is argued in the *Republic* as a matter of dissention or faction among the ruling class, not as though there are other regimes lying in wait. The rulers are treated as making the laws so as to perpetuate their own way of life; their decline results from both a weakening of the will and a failure of faith in the goodness of their governing principle, the goodness of that to which the regime is devoted. Timocracy, then, transforms into oligarchy when the ruling sons come to equate honor and wealth (549d). Oligarchy, in turn, becomes democracy when the oligarchs begin to permit anything as long as it increases their wealth (*ibid.*, 555c). Democracy, finally, becomes tyranny when the democrats fail to distinguish high and low, when equality becomes equality in everything and inequality in nothing (*ibid.*, 557b-558d; *cf. Politics*, 1301b30-39; *Twilight*, X.48). Notice, however, that aristocracy does not degenerate into timocracy because the rulers think too much; in fact, its degeneration is argued to be a literal result of misbegotten sons arising from the limits of the rulers' knowledge (*Republic*, 546b-e). So as to elide the similarity between the rule and the life philosopher and the tyrant respectively, Plato's Socrates does not depict the full revolution of the cycle of regime, leaving it to the reader to examine the subject of the return to aristocracy from tyranny (*cf. Beyond*, 9 and 211). The key to completion of the cycle is the philosopher-king. This third, largest, and most ridiculous wave of paradox is introduced thusly,

> Unless ... the philosophers rule as kings or those now called kings and chiefs genuinely and adequately philosophize, and political power and philosophy coincide in the same place, while the many natures now making their way to either apart from the other are by necessity excluded, there is no rest from ills for the cities, ... nor I think for human kind, nor will the regime we have now described in speech ever come forth from nature, insofar as possible, and see the light of the sun (*Republic*, 473d-e).

In summation, the founding in deed of the city in speech rests upon the fortunate and unlikely coincidence of philosophy and power (*cf. Laws*, 710c). This final transition would result not from the waning of the vitality of the ruling class, but from the waxing thereof, that is, it would result from the hierarchical ordering of the egalitarian chaos found within the soul of the ruling class.

MACHIAVELLI

Machiavelli is known for exhorting men to imitate the great political men of the past as he himself imitates the great authors of the past; he is known for calling on them to imitate rather than admire "the most virtuous works the histories show us, which have been done by ancient kingdoms and republics, by kings, captains, citizens, legislators, and others who have labored for their fatherland" (*Discourses*, I.Preface.ii). The greatest political act, that which obtains the greatest respect by providing the greatest common good – security – is founding a city. Machiavelli is explicit where Socrates is silent in the *Republic* (369dff.). Machiavelli, therefore, places this highest example worthy of imitation – or, as Socrates characterizes it, "the most perfect of all tests of manly virtue" (*Laws*, 708d) – before his addressees in the first chapter of the *Discourses* (*cf. Prince*, VI). The heading indicates that the chapter will speak of the universal "Beginnings of Any City Whatever" – that Machiavelli adds, "and What Was That of Rome," raises the possibility that Rome is an exception to these universal beginnings. This proves to be the case. While "all cities are built either by men native to the place where they are built or by foreigners," both a foreigner and a native built Rome: Aeneas chose the site; Romulus ordered the laws (*Discourses*, I.1.i; *cf.* I.1.v). In mentioning the example of Numa, another native son, near the end of the first chapter (*ibid.*, I.1.v), Machiavelli introduces the founder as maintainer, for he reordered the laws of Romulus when they were found to be insufficient "for such an empire" (*ibid.*, I.11.i). The example of Rome establishes that for men of quality "opportunity" need not be identical to that which fortune provided to Moses, Cyrus, Romulus, and Theseus, that is, one need not be the builder of a city to be a founder thereof. For both Plato and Machiavelli, "accident" and "opportunity" are the same thing when examined from the standpoint of the founder. Machiavelli is much more explicit than Plato, who only hints at this conclusion by substituting "opportunity" for "accident" in the list of things that pilot human activity in the *Laws*. That the philosophers are thinking of the same situations is also apparent in the examples chosen: war, harsh poverty, disease, and bad weather (*cf. ibid.*, I.39 and II.5; *Prince*, VI; and *Laws*, 709a-d). Perhaps it is more appropriate to speak here of re-founders; every founding is a re-founding, at least insofar as it begins with and reforms people who are preexisting and, thus, formed by a particular regime. The matter limits the form that may be impressed thereupon, the orders that may be arranged therein, or, at the very least, the modes by which this can be completed (*cf. ibid.*, I.17; I.18; I.25; I.26; I.55). Matter that is homogeneous will "accept laws and regimes different from their own" only with great difficulty, even if the laws and regimes are wicked (*Laws*, 708c-d).

The second chapter of the *Discourses* further develops the example of Rome as a republic that was given laws "by chance and at many different times" (*Discourses*, I.2.i). In such a city there is "some degree of unhappiness" for it "is forced by necessity to reorder itself" (*ibid.*). This task is difficult and dangerous, "because enough men never agree to a new law that looks to a new order in a city unless they are shown by a necessity that they need to do it" – this is why a people must be constrained for the founder to be free (*ibid.*; *cf.* I.1.4; and *Laws*, 708b-c). The question of maintenance is connected to that of degeneration (or corruption) and regeneration (or founding) and, thereby, to the question of the cycle of regimes. Here in the second chapter, Machiavelli gives an account of the rise and fall of cities. He reiterates that security is the origin of cities (*Discourses*, I.2.iii; *cf. Leviathan*, 86-90; *Second Treatise*, 269-282; *Oeuvres*, 134-164). The people who have gathered together place one at their head. This elective principality persists until it becomes hereditary – Machiavelli later provides the reason for this change: from "the fact that the fathers of such have been great men and worthy in the city, ... it is believed that their sons ought to be like them" (*Discourses*, III.34.ii). The change from choice to succession represents a diminished freedom on the part of the people: they no longer can choose "the one who would be more prudent and more just" (*ibid.*, I.2.iii). The result is that "at once the heirs began to degenerate from their ancestors" and "tyranny quickly arose" (*ibid.*). The rebellion of the people together with the powerful establishes an aristocracy, which in turn falls prey to heredity, whereby the government of the few arises, resulting in rebellion and the birth of the popular state (*cf. ibid.*, I.16.1 and I.40.v). Heredity proves to be the undoing of this state as well, for "once the generation that had ordered it was eliminated ... it came at once to license, where neither private men nor public were in fear, and each living in his own mode, a thousand injuries were done every day" (*ibid.*, I.2.iii). The corruption of the popular state provides the matter for a new beginning, for the people again finds itself dispersed and, like beasts, without common standards. It is here that Machiavelli for the first time explicitly introduces the good man and equates him with the founder: "So, constrained by necessity, or by the suggestion of some good man, or to escape such license, they returned anew to the principality" (*ibid.*; *cf.* I.58.iv). The example depicts the action the good man takes when faced with a dispersed (*i.e.*, thoroughly corrupt) people: he reintroduces a political or civil way of life by moving them from the licentious form of popular government to principality – he does not reintroduce the popular form (*cf. Republic*, 557d). The good man also provides new common standards. Plato's Socrates would note that the suggestion of some good man is incapable of constraining the people for "if someone proposes anything that smacks in any way of slavery, they are irritated and can't stand it" (*ibid.*, 563d; *cf. Twilight*, X.39; *Measure*). In other words, Socrates would stress that it is the necessity that results from their licentiousness that is necessary for the opportunity to arise. While Machiavelli is silent at first regarding the content of the suggestion that the good man makes, Plato's Socrates is not: the prospective tyrant "hints at cancellations of debts and redistributions of land" (*Republic*, 566a). Or as Machiavelli phrases it later, he offers to act as did King David, "who filled the hungry with good things and sent the rich away empty" (*Discourses*, I.26.i; *cf.* I.55.v).

The existence of the cycle raises the question of whether or not it is possible to prolong the good and to abridge the bad intervals of it, as well as the further question of whether or not the good man must wait until the people are thoroughly corrupt before he can or must act. The next mention of the good man provides an answer to the latter: "The desires of free peoples are rarely pernicious to freedom because they arise either from being oppressed or from the suspicion that they may be oppressed. If these opinions are false, there is for them the remedy of assemblies where some good man gets up who in orating demonstrates to them how they deceive themselves" (*ibid.*, I.4.i). The good man acts to check the desires of a free people that are pernicious to freedom, that is, he acts to preserve their free character. The acceptance of the good man's orations appears to rest not on the recognition of his goodness, but on the fact that he is "a man worthy of faith" (*ibid.*). Obviously, "faith" appears to point toward a religious foundation of authority (*cf. ibid.*, I.54). Moreover, it is the failure of the people's collective faith that leads to the degeneration of the regime; it is their lack of common standards which results in their inability collectively to see the good man as worthy of faith. Thus, when license is mistaken for liberty; when idiosyncrasy is mistaken for independence; when free development is mistaken for authenticity, there is no people with which to work, there are only individuals who require restraint (*cf. Twilight*, X.41).

NIETZSCHE

Unlike either Plato or Machiavelli, Nietzsche does not treat of the cycle of regimes. This is largely the result of his radical re-understanding of life (*i.e.*, nature) as preservation-enhancement, which in a sense is progressive as opposed to cyclical, and implies not revolution and reversal, but overcoming and rebirth (*Beyond*, 10). Life is progressive not in the sense that it has "a necessary direction or character" (Zuckert, 8), but rather insofar as it can cultivate and manifest greater complexity and capacity, insofar as living things can expand their horizons. Note preservation-enhancement expresses the Nietzschean position that the 'what' and 'how' of Being is will to power and eternal recurrence respectively (*ibid.*, 13 and 36; *cf. Nietzsche I*, 7, and 19). If life is preservation-enhancement, where enhancement is primary, and "Every enhancement of the type 'man' has so far been the work of an aristocratic society" (*Beyond*, 257), then aristocracy is the most natural form of regime in the sense of most in harmony with nature. As a result of this conception of nature, Nietzsche's politics and philosophy are both radically aristocratic (*cf.* Detwiler).

Like both Plato and Machiavelli, however, Nietzsche sees that the character of a regime is defined by the perspective of the ruling class. An aristocratic society is one "that believes in the long ladder of an order of rank and differences in value between man and man, and that needs slavery in some sense or other" (*Beyond*, 257). Such a society not only believes in this order and difference, but seeks to bring it into being through action. Out of its actualization develops what Nietzsche terms a "pathos of distance" (*ibid.*). What is crucial in this concept is the willingness, the capability, the strength, or the power to recognize, endure, and maintain distinctions, and to see these distinctions actualized in political and social institutions. What is crucial is the "power to organize – that is, to separate, tear open clefts, subordinate and superordinate" (*Twilight*, X.37) – the exercise of which requires the willingness to evaluate and judge (*cf. ibid.*, X.3). The pathos of distance is similar if not identical to Nietzsche's conception of freedom (*ibid.*, X.38). Freedom is that which frees us from the overwhelming concern with others and ourselves; it is that which allows us to be indifferent to the lives of individuals; it is that which arises from the recognition of something truly higher and more valuable, "something above and beyond" (Zuckert, 8). In short, freedom is the recognition that something is worth killing and dying for.

Nietzsche contrasts his view of freedom with that of we moderns in his "*critique of modernity*," wherein he observes "Democracy has ever been the form of decline in organizing power" (*Twilight*, X.39). Institutions as a form of organization require "a kind of will, instinct, imperative, which is anti-liberal to the point of malice: the will to tradition, to authority, to responsibility for centuries to come, to the *solidarity* of chains of generations, forward and backward in infinitum" (*ibid.*). Our liberal, democratic politics and 'culture' displays that "The whole of the West no longer possesses the instincts out of which institutions grow, out of which a *future* grows [...] One lives for the day, one lives very fast, one lives very irresponsibly: precisely this is called 'freedom'" (*ibid.*). Institutions shape and constrain behavior and, thus, are directly opposed to 'freedom' as understood by we moderns, to *laissez aller*. "That which makes an institution an institution is despised, hated, repudiated: one fears the danger of a new slavery the moment the word 'authority' is even spoken aloud" (*ibid.*; *cf. Republic*, 563d; *Measure*). The authority of an institution is the authority of tradition; it is rooted in nature, but proved in past practice. An institution is itself evaluative (*i.e.*, valuing), it declares that such and such is the best form of organization for the thing in question. Institutions, moreover, are social, political phenomena, they preserve and promote the particular valuation of the community. The liberal character of a liberal institution comes not from the character of the institution itself, but from the effect that it has. For an institution to be liberal, it must promote true freedom; it cannot merely empower choice. Nietzsche clarifies the issue thusly, "These same institutions produce quite different effects while they are still being fought for; then they really promote freedom in a powerful way. On closer inspection it is war that produces these effects, the war *for* liberal institutions, which, as a war, permits *illiberal* instincts to continue" (*Twilight*, X.38). It is only as an idea, that is, as something deemed to be worth killing and dying for, that liberal institutions promote freedom.

Modern constitutions, customs, and 'institutions' are soft – and precisely this softening has been hailed as moral progress. Nietzsche directly raises the question of *"Whether we have become more moral"* in the longest aphorism of *Twilight of the Idols* (*ibid.*, X.37). This lengthy aphorism begins by relating the utter disbelief of Nietzsche's contemporaries that he could fail to recognize "the undeniable superiority" of our age, the self-evident truth that liberal society is more moral than any prior society, "the real *progress* we have made here" (*ibid.*). The question of progress, of what counts as real advancement lies at the heart of the aphorism. Nietzsche's contemporaries, "we moderns," claim precisely this prize on behalf of our age and ourselves – to represent the advancement of light into and over the dark ages. Nietzsche, on the other hand, identifies "that lavish squandering and fatal age of the Renaissance [...] as the last great age; and we moderns, with our anxious self-solicitude and neighbor-love, with our virtues of work, modesty, legality, and scientism – accumulating, economic, machine-like – [are identified] as a weak age" (*Twilight*, X.37). The disagreement between these evaluations of modernity is personified in Cesare Borgia, in whether he is praised or blamed (*cf. Beyond*, 197; *Ecce*, III.1; *Antichrist*, 46, 61; *Will*, 871). The answer to the question of what counts as real advancement for Nietzsche is rooted in nature: it is the enhancement of strength and capacity. The stronger can do all that the weak can do and more, thus, our incapacity to tolerate Renaissance conditions, even in thought, is proof of a later, weaker, more delicate, more vulnerable constitution. Our morality, our customs and institutions, then, are simply the result of our "later constitution, [which is] weaker, more delicate, more vulnerable:" the "softening of customs [...] is a consequence of decline; the hardness and terribleness of customs, conversely, can be a consequence of an excess of life. For in that case much may also be dared, much challenged, and much *squandered*" (*Twilight*, X.37). The hardness and terribleness of customs is most readily apparent in what we moderns would term hard-heartedness. "To be indifferent – that too is a form of strength – for that we are likewise too old, too late." Rather, "Our morality of sympathy," our "*morality of pity*," deems insensitivity and indifference to the trials and tribulations, the suffering and distress of other human beings inhumane (*cf. Amérique II*, 232-233). This humaneness of the West precludes its affirming western values, for such affirmation is coterminous with the assertion that said values are worth killing and dying for, which is to say, worthy of inflicting pain upon others and suffering it oneself; however, insofar as license is mistaken for liberty and all choices are regarded as equal, the only ground of choice-worthiness becomes individual (read idiosyncratic) choice itself, meaning nothing can justify common commitments. And so, like France's corrupt aristocracy, the entirety of the West now seems ready to sacrifice "itself to an extravagance of its own moral feelings" (*Beyond*, 258).

CONCLUSION

While the orders of degeneration proposed by Plato and Machiavelli differ, the underlying mechanism that drives the transformation is the same: heredity. It is the necessary succession of generations that leads to the inevitable decline of the regime, as each subsequent generation suffers from hereditary exhaustion. The Machiavellian analysis in particular is relevant to questions of regime maintenance and change in republican polities. When combined with Nietzsche's observation that the quest for liberal institutions is ennobling, while their establishment undermines the strength required for their continuance – a fact most readily visible in the degeneration of liberty into license as individuals in society 'let themselves go' – a proper understanding of the cycle of regimes as it relates to liberal democracy can be acquired and may provide grounds for its continuing maintenance.

Plato, Machiavelli and Nietzsche all identify the last gasp of democratic regimes with radical individualism, meaning a lack of common standards and values. The western commitment to the idea of cultural relativism is the natural outgrowth of the western commitment to freedom understood are empowered choice. Insofar as there are no standards within society to differentiate between choices, for every choice is choice-worthy because chosen, there are no standards to differentiate between societies, thus, all societies or cultures are deemed to be of equal worth. While the politically pernicious implications of value relativism are subdued within western societies by its very pervasiveness, they are evident in the relations of western and non-western societies. The West perceives a saintly virtue in its unwillingness assertively to defend western values as good, as choice-worthy, as better than the alternatives. Non-western societies, on the other hand, not only see their own culture as good, choice-worthy, and better than the alternatives, but also perceive western reluctance to assert the worth of western culture as vindicating their judgment. They do not perceive virtue in our unwillingness to defend, protect, preserve, and perpetuate our way of life; rather, they rightly perceive our weakness, they perceive the fact that our unwillingness merely masks our incapacity. We should not be surprised when they say to each other, "Those men are ours. For they are nothing" (*Republic*, 556c-e; *cf. Beyond*, 257). The question, however, remains: are they correct in their assessment?

REFERENCES

Aristotle, and Lord, C. trans. (1984), *The Politics*. The University of Chicago Press, Chicago, Illinois.

Detwiler, B. (1990), *Nietzsche and the Politics of Aristocratic Radicalism*. The University of Chicago Press, Chicago, Illinois.

Heidegger, M. and Krell, D.F. trans. (1979), *Nietzsche Volume I: The Will to Power as Art*. Harper Collins, San Francisco, California.

Hobbes, T. and Tuck, R. ed. (1991), *Leviathan*. Cambridge University Press, New York, New York.

Locke, J., and Laslett, P. ed. (1960), *Two Treatises of Government*. Cambridge University Press, New York, New York.

Machiavelli, N., and Mansfield, H.C. trans. (1985), *The Prince*. The University of Chicago Press, Chicago, Illinois.

Machiavelli, N., and Mansfield, H.C. trans. (1996), *Discourses on Livy*. The University of Chicago Press, Chicago, Illinois.

Nietzsche, F., and Kaufmann, W. trans. (1954), "The Antichrist" in *The Portable Nietzsche*. Penguin Books, New York, New York.

Nietzsche, F., and Kaufmann, W. trans. (1954), "Twilight of the Idols" in *The Portable Nietzsche*. Penguin Books, New York, New York.

Nietzsche, F., and Kaufmann, W. trans. (1967), Will to Power. Vintage Books, New York, New York.

Nietzsche, F., and Kaufmann, W. trans. (1968), "Beyond Good and Evil" in *Basic Writings of Nietzsche*. Modern Library, New York, New York.

Nietzsche, F., and Kaufmann, W. trans. (1968), "Ecce Homo" in *Basic Writings of Nietzsche*. Modern Library, New York, New York.

Plato, and Bloom, A. trans. (1968), *The Republic of Plato*. Basic Books, USA.

Plato, and Pangle, T. trans. (1980), *The Laws of Plato*. The University of Chicago Press, Chicago, Illinois.

Rousseau, J.J. (1964), *Oeuvres complètes, tome III*. Gallimard, Paris, France.

Shakespeare, W. and Lever, J.W. ed. (1965), *Measure for Measure*. The Arden Shakespeare, London, England.

Tocqueville, A. de (1961), *De la démocratie en Amérique II*. Gallimard, Paris, France.

Zuckert, C.H. (1996), *Postmodern Platos: Nietzsche, Heidegger, Gadamer, Strauss, Derrida*. The University of Chicago Press, Chicago, Illinois.

Chapter 13

Relativism and Its Consequences for the West

Stephen J. Lange

Institute for Regional Analysis and Public Policy
Morehead State University
Morehead, KY 40351, USA

ABSTRACT

The West has neglected to pay attention to its intellectual roots and has failed to keep current its intellectual defense. In particular, value relativism undermines the West's ability to defend itself to citizens, friends, detractors, and enemies. This paper outlines the elements of Western civilization, the development of relativism, and the consequences for the West. A thorough examination of relativism is necessary to elucidate the character and depth of the intellectual crisis we face and its relationship to the motivations of our enemies, the means by which to combat them, and the hearts and minds of our own citizenry as we seek to sustain our fight against terrorism.

Keywords: Relativism, values, reason, Western civilization, the West, terrorism

INTRODUCTION

International terrorism and the foreign regimes that support it have threatened the safety and security of citizens and their property, the stability and integrity of the government, the protection of the economy, critical infrastructure, key resources, and civil society as a whole. Viewing this range of targets comprehensively, we can say that what is being threatened is our entire way of life in the West. These objectives are clearly stated by al Qaeda (U.S. Department of Justice 2002). Is the West able to combat this threat, not militarily and politically, but in the hearts and minds of those involved?

For some time now, the West has neglected to pay attention to its intellectual roots and, as a result, it has failed to keep current its intellectual defense. Over the course of the last 200 years, the principled foundations of Western civilization have been profoundly weakened by the rise of relativism in virtually every area of Western intellectual activity. The origins of relativism include convergent streams from the natural sciences, mathematics, social sciences, and humanities. These streams have culminated in a widespread doubt of the ability of reason to guide human action; that is, reason is now held to be incapable of determining the correct ends or goals of our actions. Domestically, this manifests itself in waning patriotism and deep skepticism toward elected officials and all branches and levels of government, thereby undermining confidence in the goodness of our regime. It has further resulted in a self-deprecating posture among our intellectuals, such that many believe we are to blame for most of the international conflict we face and even deserve the attacks of terrorists. Our intellectual failure to grasp the roots of our civilization, our history, and our political philosophy—and thereby to understand what is good and choice-worthy about our way of life in comparison to the other fundamental alternatives—has undermined Western civilization from within. Many Western intellectuals are no longer willing (or perhaps able) to defend Western civilization to our detractors and enemies. At the same time, the hearts and minds of our citizenry are no longer united in their support for their own civilization.

This paper outlines the elements of Western civilization, the development of relativism, and the consequences for the West. Value relativism is an important component in the decline of Western civilization in that it undermines the West's ability to defend itself to citizens, friends, detractors, and enemies. Relativism enervates the West's ability to maintain an effective fight against terrorism over a prolonged period and our ability to sustain our way of life internally.

WHAT IS WESTERN CIVILIZATION?

A civilization is "the highest cultural grouping of people and the broadest level of cultural identity people have short of that which distinguishes humans from other species. It is defined both by common objective elements, such as language, history, religion, customs, institutions, and by the subjective self-identification of people" (Huntington 1996, 43). The West refers "to what used to be called Western Christendom. The West is thus the only civilization identified by a compass direction and not by the name of a particular people, religion, or geographical area" (Huntington 1996, 46). More specifically, the West is founded upon "three distinct traditions: (1) the classical culture of Greece and Rome; (2) the Christian religion, particularly Western Christianity; and (3) the Enlightenment of the modern era" (Kurth 2003/2004, 5). The West is the embodiment of intellectual tensions between these traditions, between classical philosophy and Christianity, between classical political philosophy and modern political philosophy, and between modern philosophy and Christianity. From ancient Greece, in particular, the West inherits philosophy, the love of wisdom understood as the attempt to understand the truth about nature, ourselves, and the universe as a whole through the application of unaided reason. "As children of the West, the philosophic spirit, if not philosophy itself, is something deeply rooted in our civic beings—we are rational, questioning, searching, reasoning people" (Watson 2005, xviii). This philosophic spirit has been met with a strong Christian tradition, which has engaged philosophy in dialogue and caused philosophy to engage with it. This dialogue, tension, and struggle among these three traditions has been the source of the intellectual energy and spiritual strength of the West, for:

> Man cannot live without light, guidance, knowledge; only through knowledge of the good can he find the good that he needs. The fundamental question, therefore, is whether men can acquire that knowledge of the good without which they cannot guide their lives individually or collectively by the unaided efforts of their natural powers, or whether they are dependent for that knowledge on divine revelation. (Strauss 1953, 74)

The great conversation between these traditions over what constitutes the authoritative source of guidance has been essential to the West, animating its intellectual life and providing the architecture for our civilization.

More recently, "the principle identity of people in the West is now defined by an ensemble of liberal ideas: (1) liberal democracy; (2) the free market; (3) the open society; and (4) and individualist culture. These ideas...are seen as the essence of Western civilization at the present time..." (Kurth 2006, 58). At the heart of the West is "a vision of society and political order" (Scruton 2002, x) and the cornerstone of the Western political order is the social contract, the individual rational choice to join society and consent to be ruled by its laws. This public rationality is at the core of the West and it relegates religious belief to the private sphere, which effectively separates society from the state. As a result, "Western civilization is composed of communities held together by a political process, and by the rights and duties of the citizen as defined by that process" (Scruton 2002, 16). The social contract derives from modern political philosophy, which is the foundation for modern liberalism. As can be seen from this overview of what constitutes the West, our civilization does not so clearly refer to a race, a religion, or a geographical territory as do the other civilizations that Huntington enumerates (Sinic, Japanese, Hindu, Islamic, Orthodox, Latin American, and African). Western civilization is more an idea, more conceptual.

A war of ideas, therefore, aims at the heart of Western civilization. Secretary Chertoff's remarks reminded us of this when he warned that the terrorists we now face are not the same as the "political terrorists in the past. ... They seek the ultimate domination in many countries. Their goal is a totalitarian, theocratic empire, a religious empire...." What we face is "an ideological threat" and we are engaged in "a battle of ideas" (Chertoff 2007). Is the West, this constellation of traditions and ideas, of sufficient intellectual health to meet this challenge?

As Huntington reminds us, "While civilizations endure, they also evolve. They are dynamic; they rise and fall; they merge and divide; and as any student of history knows, they also disappear and are buried in the sands of time" (Huntington 1996,

44). Observations that Western civilization is in decline have been persistent for more than 100 years, from Nietzsche (1886) and Spengler (1918) to Huntington (1996) and Joseph Ratzinger (2006 [now Pope Benedict XVI]). One part of this decline has been the distrust of reason and the embrace of relativism, which enervates the West's ability to explain, justify, and defend itself both to itself and to others.

WHAT IS VALUE RELATIVISM?

Value relativism is "a doctrine that holds that all judgments of value are 'subjective' in the sense that they are relative to the time, culture, or personality of the subject who makes them" (Malcolmson, Myers, and O'Connell 1996, 20). Such values include judgments of what is good, bad, right, wrong, just, unjust, moral, immoral, noble, base, etc. Value relativism claims that judgments about all such things as these are relative; that is, they are related to either the opinions or feelings of an individual, the prevailing legal system, the culture, or the historical epoch in which one lives. The fundamental relativity of such judgments to either individuals or to society entails that there is no knowledge that can be acquired to determine the objective truth about these things. As a result, such judgments are subjective; value judgments cannot be established as either true or false. "Value relativism holds that there are no universally and permanently true answers to the great questions of human existence" (Malcolmson, Myers, and O'Connell 1996, 20). The most important things that we need to know in order to live our lives well cannot be known; there is no knowledge to be had about them.

This view is not unknown in classical political philosophy, but it was not the prevailing view until comparatively recently. It is acknowledged in Plato's *Republic*, for example, but rejected (340a-341a). In contrast to classical philosophy, the relativity of values was articulated in early modern political philosophy more directly and assertively by Thomas Hobbes among others:

> But whatsoever is the object of any mans Appetite or Desire; that is it, which he for his part calleth *Good*: And the object of his Hate, and Aversion, *Evill*; And of his Contempt, *Vile* and *Inconsiderable*. For those words of Good, Evill, Contemptible, are ever used with relation to the person that useth them: There being nothing simply and absolutely so. (1651, 39)

This is a clear statement of the relativity of so-called value judgments, in this case making our judgments about good and evil relative to individuals' desires, feelings, or passions. Hobbes' *Leviathan* was published in 1651, and while the context of this passage must be interpreted in light of the work as a whole in order to discern Hobbes' true intention, this is nonetheless a clear articulation of what today has become commonplace. It is so unquestionably accepted that it is now the orthodox view.

That this view is not self-evident and that there are plausible alternatives to it is reflected in the fact that its ascendency is comparatively recent, most people and most philosophers throughout history accepting that judgments of good and bad, just and unjust, noble and base are based upon either fact and reason or divine revelation. Those alternatives cannot be described and examined here, but let it suffice it to say that judgments in these areas can be fully rational, that is based on facts and synthetic reasoning (Craig 1994, 326-336). For present purposes, let us examine the impact of this view for the West today.

THE CONSEQUENCES OF RELATIVISM FOR THE WEST

In so far as value relativism asserts our judgments as to what is good, just, and noble are subjective and relative to individuals or societies, then what constitutes a good way of life, a just system of government, and a noble cause for which one might reasonably make personal sacrifice, are not rooted in any rationally defensible or otherwise authoritative foundation. "Various names have been given to this school today: post-enlightenment thinking, post-modernism, 'weak thought,' deconstructionism. The labels have changed, but the target is always the same: to proclaim that there are no grounds for our values and no solid proof or argument establishing that any one thing is better or more valid than another" (Pera 2006, 11). As a result, it is impossible to defend the American republic, liberal democracy, or Western civilization upon any objective, universal grounds. Our belief in freedom, equality, human dignity, and all other individual rights are only that, our beliefs, which are in principle not amenable to any rational defense. And since our citizenry, public intellectuals, and many academics embrace value relativism (albeit not consistently), they are ineffective defenders of our way of life.

Contemporary political theory has largely embraced value relativism and so may serve as an example of the difficulty at hand. Regarding the modern notion of freedom as living as one likes, the theorists of liberalism make clear that there is a limit to such freedom—as much as is compatible with that of others (Berlin 1958, 8-11). It cannot be the case, for example, that even if we embrace the supposed relativity of all values, we can be completely indifferent to whether foreigners or even our fellow citizens employ their freedom to subvert our government and way of life to establish communism, fascism, or a theocracy. It is for this reason that public education in a liberal democracy must ensure that people have some understanding of—and commitment to—genuine freedom.

Yet many are uncomfortable with such efforts, and Isaiah Berlin's arguments in this case are both representative and instructive. He concedes there is merit to a conception of liberty as genuine self-rule and acknowledges that individual freedom is not,

> even in the most liberal societies, the sole, or even the dominant, criterion of social action. We compel children to be educated, and we forbid public executions. These are certainly curbs to freedom. We justify them on the ground that ignorance, or a

barbarian upbringing, or cruel pleasures and excitements are worse for us than the amount of restraint needed to repress them. This judgment in turn depends on how we determine good and evil, that is to say, on our moral, religious, intellectual, economic, and aesthetic values; which are, in their turn, bound up with our conception of man, and of the basic demands of his nature. (1958, 54-55)

Berlin also holds that the "ends of man are many, and not all of them are in principle compatible with each other.... The necessity of choosing between absolute claims is then an inescapable characteristic of the human condition" (1958, 54). But herein lies the problem for Berlin: we must choose but we cannot do so in a wise or principled way. We must "choose ends without claiming eternal validity for them," which results in a "pluralism of values" (1969, 172). It is for this reason that Berlin cannot accept anything other than the "negative" concept of liberty as non-interference. So while we may "compel children to be educated" in a particular way, it is not in a way that will be ultimately defensible. Even in a liberal education that seeks to provide its beneficiaries with the full range of alternatives, the ability to choose among them, and the wherewithal to live by the choice made, the result will never be anything less than a bewildering array of choices and lives in Berlin's view. Pluralism is the necessary outcome. Berlin tacitly acknowledges the unsatisfactory character of this outcome when he suggests that we must bravely accept the inadequate ground for our choices. In elaborating upon this, he reveals more clearly the root of his position. For he seeks to assure us that our

principles are not less sacred because their duration cannot be guaranteed. Indeed, the very desire for guarantees that our values are eternal and secure in some objective heaven is perhaps only a craving for the certainties of childhood or the absolute values of our primitive past. 'To realise the relative validity of one's convictions,' said an admirable writer of our time, 'and yet stand for them unflinchingly, is what distinguishes a civilized man from a barbarian.' To demand more than this is perhaps a deep and incurable metaphysical need; but to allow it to determine one's practice is a symptom of an equally deep, and more dangerous, moral and political immaturity. (1958, 57)

What are we to do then, when confronted with challenges to our way of life, challenges to liberalism itself? We are to stand unflinchingly in support of our values. Upon what ground are we to stand in support of them? Upon the ground that our values are our choice, they are values we have given to ourselves.

This is essentially the position of well known political theorists such as John Rawls and Richard Rorty, when they argue that there is no need to provide philosophic or psychological foundations for liberal democracy, that we may simply take liberal democracy as a given not requiring justification. Rorty approvingly cites Rawls:

Rawls thinks that "philosophy as the search for truth about an independent metaphysical and moral order cannot provide...a workable and shared basis for a political conception of justice in a democratic society." So he suggests that we confine ourselves to collecting, "such settled convictions as the belief in religious toleration and the rejection of slavery" and then "try to organize the basic intuitive ideas and principles implicit in these convictions into a coherent conception of justice."

As Rorty notes, "This attitude is thoroughly historicist" (1991, 180). To the question of "whether there is any sense in which liberal democracy 'needs' philosophical justification at all" Rorty answers "no" (1991, 178-179). He elaborates:

> To refuse to argue about what human beings should be like seems to show a contempt for the spirit of accommodation and tolerance, which is essential to democracy. But it is not clear how to argue for the claim that human beings ought to be liberals rather than fanatics without being driven back upon a theory of human nature, on philosophy. I think we must grasp the first horn [of the dilemma]...[and] grant that historical developments may lead us to simply *drop* questions and the vocabulary in which those questions are posed. (1991, 190; italics in original)

This is why Rorty affirms Rawls' contention that we attempt to stand firm behind our settled convictions. However, to do so in the self-conscious absence of any rational, principled attempt to defend them is, as in Berlin's case, to recognize them as mere values "without claiming eternal validity for them." This situation is part of the larger crisis of the West as identified by Nietzsche and further discussed by Heidegger (1977).

Heidegger agrees with Nietzsche that our crisis is nihilism, that "The world appears value-less" (Heidegger 1977, 69). Heidegger explains:

> [T]he terms "God" and "Christian god" in Nietzsche's thinking are used to designate the suprasensory world in general. God is the name for the realm of Ideas and ideals. This realm of the suprasensory has been considered since Plato, or more strictly speaking, since the late Greek and Christian interpretation of Platonic philosophy, to be the true and genuinely real world. In contrast to it the sensory world is only the world down here, the changeable, and therefore the merely apparent, unreal world. ... [Nietzsche's] pronouncement "God is dead" means: The suprasensory world is without effective power. It bestows no life. ... If God as the suprasensory ground and goal of all reality is dead, if the suprasensory world of the Ideas has suffered the loss of its obligatory and above all its vitalizing and upbuilding power, then nothing more remains to which man can cling and by which he can orient himself. (1977, 61)

Nietzsche's solution to this crisis, according to Heidegger, is to have man face up to the truth that all values have always been posited by man himself (rather than given to man by something other than man and higher than man). Having faced up to this, man must now create new values for himself consciously and deliberately. As Heidegger observes, though, "through this making conscious, the inquiring gaze is directed toward the source of the new positing of values, but without the world's regaining its value at all in the process" (1977, 69). That is, as soon as we recognize that the source of the new values is nothing other than ourselves, or as soon as we recognize something as a mere value, its obligating power is diminished, it is no longer authoritative for us, it no longer commands our allegiance (Heidegger 1977, 103-108). The position of Berlin, Rawls, and Rorty is inadequate then. In the terms used by Berlin, the very recognition that our values are merely values, having no "eternal validity", means that they are not "sacred" at all. Such recognition, combined with the impossibility of discerning true values, leaves us full of self-doubt, unable to distinguish civilized man from barbarian. For to have no grounds for one's convictions but to "yet stand for them unflinchingly" is not "what distinguishes a civilized man from a barbarian" as Berlin suggests. At best it is mindless dogmatism; at worst it is self-conscious nihilism. Either of these is capable of being the basis for deeds that everyone, including Berlin, would deem to be unquestioningly barbaric.

This short foray into contemporary political theory is sufficient to show how the embrace of the supposed relativity of all values leaves us with no principled, rational grounds upon which to resist those who seek to subvert our government and way of life. We are left merely to assert our way of life as good simply because it happens to be ours; but that is utterly unpersuasive to those who understand themselves to have firm principles for their beliefs, whether rooted in reason or divine revelation. The conflict we face from terrorism is driven by "fundamental issues, including rival conceptions of truth and justice" (McAdams 2007, 23), but value relativism makes it impossible even to engage seriously in discourse and debate over these rival conceptions—"dialogue will be a waste of time if one of the two partners to the dialogue states beforehand that one idea is as good as the other" (Pera 2006, 45). In the face of unequivocal attacks on America and the West, for example that we are unjust, corrupt, and evil, it is utterly ineffective to respond that all such claims are merely subjective value judgments. Such relativism leads to the view that "one man's terrorist is another man's freedom fighter...[which] is to suggest the impossibility of any kind of moral judgment" (Watson 2006, 1). This is part of what many Muslims find repugnant about the West (Scruton 2002, 15).

In addition to enervating our ability to defend ourselves intellectually abroad, relativism has a deleterious effect at home. Both in Europe and in America there is a significant sector of society that claims "that America was asking for it, articulating self-flagellating theories for the West and a sympathetic understanding of the terrorists' motivations" (Pera 2006, 42). Joseph Ratzinger describes "a peculiar Western self-hatred that is nothing short of pathological. ... All that it sees in its own history is the despicable and the destructive; it is no longer able to perceive what is great and pure" (2006, 78-29). This "ideological animus against the West" includes public intellectuals as well as the press (Alt 2006).

Relativism has also infused itself into every major division in the academy, the humanities, social sciences, and natural sciences. There is a prevailing view among theorists across disciplines that all knowledge is ineradicably dependent upon the biases of individuals and society: "knowledge is a human construct, grounded in the facts of our social life" (Trigg 1985, 28). That is, all knowledge is relative to the social setting in which it is acquired and, as a result, there is no means by which one may ascertain the objective truth about reality. By 1962, Thomas Kuhn had articulated this view in *The Structure of Scientific Revolutions*, arguing that in the choice between scientific paradigms "there is no standard higher than the assent of the relevant community" of scientists (1970, 94). Furthermore, the choice of paradigms is "constitutive" of reality itself, determining what questions are asked, what problems are identified, what facts are observed, what instruments are used, and what knowledge results. Consequently, "when paradigms change, the world changes with them" (1970, 111). There is no objective truth because we have no objective access to reality. All knowledge is relative therefore. The one universal and permanent truth of the human condition is that human beings cannot know any universal, permanent truths. Despite the incoherence of this view when applied reflexively—that it results in an obvious self-contradiction—the view remains dominant. It is perhaps for this reason that Ratzinger claimed that "Relativism...in certain respects has become the real religion of modern man" and that it "is the most profound difficulty of our day" (2004, 84, 72).

In fact, relativism is bound up with certain views about modern science, which have in turn contributed to the weakness and decadence of the West.

> The truly fundamental problem here, however, is the dogmatic acceptance of one of the most disastrous intellectual mistakes in the history of the universe: the "fact-value" distinction that has accompanied the saturation of everyday life by "scientism." This utterly misbegotten way of construing everyday experience—which now so thoroughly pervades contemporary thought and discourse that people today cannot discuss anything important without repeated references to these murky somethings called "values"—is traceable to the popular misunderstanding that modern science provides the paradigm of a genuine, *objective* knowledge precisely because it deals exclusively with "facts," that science is "value free." And that, by contrast, principles of right and wrong, good and bad, noble and base, decent and obscene are regarded as incorrigibly *subjective* "value judgments," and as such, inadequate grounds for ruling any contrary "values" as simply wrong, much less intolerable. (Craig 2006, 89)

A thorough examination of relativism, then, necessitates an examination of certain aspects of modern science and the consequences for the West in its intellectual defense of itself. Such an examination of relativism in all of its guises is necessary to elucidate the character and depth of the intellectual crisis we face and its relationship to the motivations of our enemies, the means by which to combat them, and the hearts and minds of our own citizenry as we seek to sustain our fight against terrorism.

At America's founding we were able to turn to the world, out of "a decent Respect to the Opinions of Mankind," and boldly enunciate fundamental truths about humanity together with the principles upon which all legitimate government should be based. Two centuries later, are we still able to do so? "Does this nation in its maturity still cherish the faith in which it was conceived and raised? Does it still hold those 'truths to be self-evident'?" (Strauss 1953, 1). And now, more generally, does this nation still believe in truth, in the very possibility of being right or wrong on the most fundamental questions of human existence?

REFERENCES

Alt, Robert. 2006. Media bias in Iraq: The battle for the hearts and minds of Americans. In *The West at war*, ed. Bradley C.S. Watson, 199-210. Lanham: Lexington Books.

Berlin, Isaiah. 1958. *Two concepts of liberty: An inaugural lecture delivered before the University of Oxford on 31 October 1958*. Oxford: Clarendon Press.

_____. 1969. *Four essays on liberty*. New York: Oxford University Press.

Chertoff, Michael. 2007. Remarks to the European Parliament. Department of Homeland Security. Released May 15, 2007.

Craig, Leon H. 1994. *The war lover: A study of Plato's* Republic. Toronto: University of Toronto Press.

_____. 2006. The liberal regime under attack. In *The West at war*, ed. Bradley C.S. Watson, 73-98. Lanham: Lexington Books.

Heidegger, Martin. 1977. *The question concerning technology,* trans. William Lovitt. New York: Harper Torchbooks.

Hobbes, Thomas. 1651. *Leviathan*. Edited by Richard Tuck. New York: Cambridge University Press, 1996.

Huntington, Samuel P. 1996. *The clash of civilizations and the remaking of world order*. New York: Simon & Schuster.

Kuhn, Thomas S. 1970. *The structure of scientific revolutions*. 2nd ed. Chicago: University of Chicago Press.

Kurth, James. Western civilization, our tradition. *The intercollegiate review*. Fall 2003/Spring 2004, 5-13.

_____. 2006. Western identities versus Islamist terrorism: Liberals and Christians in the new war. In *The West at war*, ed. Bradley C.S. Watson, 53-72. Lanham: Lexington Books.

Malcolmson, Patrick, Richard Myers, Colin O'Connell. 1996. *Liberal education and value relativism.* Lanham, MD: University Press of America.

McAdams, A. James. 2007. *The crisis of modern times*. Notre Dame: University Press of Notre Dame.

Nietzsche, Friedrich. 1886. *Beyond good and evil.* In *Basic writings of Nietzsche,* trans. Walter Kaufmann, 191-435. New York: Modern Library, 1968.

Pera, Marcello. 2006. Relativism, Christianity, and the west. In *Without roots: The West, relativism, Christianity, Islam,* eds. Joseph Ratzinger and Marcello Pera, 1-49. New York: Basic Books.

Plato. *Republic,* trans. Allan Bloom. New York: Basic Books, 1968.

Ratzinger, Joseph. 2004. *Truth and tolerance: Christian belief and world religions,* trans. Henry Taylor. Fort Collins, CO: Ignatius Press.

Ratzinger, Joseph. 2006. The spiritual roots of Europe: Yesterday, today, and tomorrow. In *Without roots: The West, relativism, Christianity, Islam,* eds. Joseph Ratzinger and Marcello Pera, 51-80. New York: Basic Books.

Rorty, Richard. 1991. *Objectivity, relativism, and truth,* Vol. 1. New York: Cambridge University Press.

128

Scruton, Roger. 2002. *The West and the rest: Globalization and the terrorist threat*. Wilmington, Delaware: Intercollegiate Studies Institute.

Spengler, Oswald. 1918. *The decline of the West*. New York: Vintage Books, 1990.

Strauss, Leo. 1953. *Natural right and history*. Chicago: Chicago University Press.

Trigg, Roger. 1985. *Understanding social science*. Oxford: Basil Blackwell.

U.S. Department of Justice. 2002. *Reference for terrorist acts*. http://www.usdoj.gov/ag/trningmanual.htm/ (accessed January 17, 2002).

Watson, Bradley, ed. 2005. *Civic education and culture*. Wilmington: ISI Books.

Watson, Bradley, ed. 2006. *The West at war*. Lanham: Lexington Books.

Chapter 14

Factors of Destabilization and Collapse: A Comparative Study of the Roman and British Empires and the Consequences for Western Civilization

Michael W. Hail

School of Public Affairs
Morehead State University
Morehead, KY 40351, USA

ABSTRACT

This research examines the major factors that contributed toward the decline of the British and Roman Empires and the implications for the U.S.; including economic, security, cultural, and political factors. The comparative historical approach yields substantive findings for modeling the relational forces shaping regime decline in the West today. The cyclical nature of interdependent factors reflects that the most significant causal factor is related to theoretical challenge and conflict from ideologies and philosophy.

Keywords: Comparative Systems Analysis, Cultural Analysis, Empire, Human Systems Analysis, Ideologies, Societal-level Analysis, Western Civilization

INTRODUCTION

There are many beginnings and endings in the cycles of world power. Our understanding of these cycles of power is limited and usually follows a theoretical premise that hegemony is derivative of military and economic power, yet these are results rather than causes. There is great need for expanding our horizon of understanding. The macro-political forces that created the circumstances for establishing military and economic dominance have nationalism and ideology as precursors of power. And the relationship of these societal forces to underlying political culture, as transmitted by communities, families, and the institutions that govern them, is the essential network of values that requires careful analysis. The theoretical foundations of the political system, the structure of education, the capacity for innovation and the cohesion of the public philosophy of the regime are all significant variables in the larger equation of the national interest. But of those, none is more important than the political philosophy.

For the present research, Western Civilization is the societal-level question of concern. Western Civilization begins in Athens, and from Greek and Roman civilization emerges the greater Western Civilization. And the epochs of hegemony and conquest, of vanquishment and decline, follow cycles of politics that situationally determine the prospects for leadership in any given era. The fate of the West is considered by some, like Spengler and Hutchens, to be the fate of civilization itself. But there is more than one civilization, though how many is open to debate. Eric Voegelin and Leo Strauss argue there are but two great civilizations, East and West, whereas Spengler and Toynbee are for several civilizations. It is assumed that understanding the fate of the West in the present circumstance requires comparative analysis with past epochs, and that this comparison be based upon relational factors that affect leadership and the use of power.

WESTERN CIVILIZATION: CRISIS AND THREATS TO THE EXISTING WORLD ORDER

Western Civilization is the highest culture of the greatest of all civilizations in human history. Western Civilization, or the West, has alternately been termed Christendom, *Pax Britannia*, the Allies, and the "free" world. The United States succeeded Great Britain at the end of the World Wars as the leader of this civilization. In terms of global dominance, as Leo Strauss noted, "In 1913, the West – in fact [the U.S.] together with Great Britain and Germany – could have laid down the law for the rest of the earth without firing a shot."(Strauss, 2) Politically, economically, and militarily, the West was at a zenith as the U.S. ascended into leadership. Yet, the West was in crisis, resulting in a decline of culture and of faith, and at its core, the civilization was experiencing the decay of self-critique. The

West was self-doubting, unsure of its future and with collective regret and angst over its past, and immersed in a present of materialism and Epicurean over-indulgence. It was in terms of internal forces that the West faced the most serious threat. At the heart of this phenomenon, the West was experiencing a profound philosophical crisis, and it was one it had faced for some period of time.

When did the crisis of the West begin? What are its causes? Oswald Spengler diagnosed "the Decline of the West" as comprising "nothing less than the problem of Civilization [itself]," and this, he said, is "one of the fundamental questions of all higher history."(Spengler 24)   Spengler understood Western Civilization as the expression of culture, of the "soul-life."   "THE standpoint from which to comprehend the economic history of the great Cultures is not to be looked for on economic ground.   That which we call national economy today is built up on premises that are openly and specifically English."(Spengler, 398)   English economics from Adam Smith and English philosophy from Hobbes lay the foundation for the modern framework of regime identity that circumscribes leadership for the West today.   Yet as the British Empire declined and the American century ascended, the enduring features of the regime remained English political philosophy and Anglican political culture.   These were definitive at the time of the founding of the American republic, and as Gordon Wood noted, the Americans came "to believe that they alone among the peoples of the Western world understood the true principle of representation."(Wood, 1)   It was their Lockean natural right, their reclamation of their rights as Englishmen that was the basis of the American colonists' fight for independence, not a foundation built on some utopian theory.

The American Revolution was soon followed by the French Revolution and continental wars.   The growing perils of extreme violence and mass destruction were again the consequences and results, but it was the sprit of radicalism that had been unleashed.   This extreme radicalism was grounded in Enlightenment philosophy and various social utopian theories, but the results were mass political action and violence that transformed into ideology and political programs that demanded rights of all types that could be exploited for strategic advantage and ultimately to be utilized for internal self-destruction in the West.   These new mass cultural ideologies were the basis of what would transform into modern terrorist organization and extremist networks of the present.   Edmund Burke, and other scholars of the West, recognized that civilization was threatened by the masses, the forces of mob rule, of what we later would call mass culture or popular culture. Western Civilization experienced a crisis of political philosophy, the crisis of modernity, just as it achieved global hegemony over other world civilizations.

There can be no escape from the cycle of decline the West has faced from at least the French Revolution to the multi-cultural wars of the "post-modern" present, except through a return to the foundations of the West. As noted scholar Dr. Leo Strauss remarked, "However much the West may have declined, however great the dangers to the West may be, that decline, that danger, nay, the defeat, even the destruction of the West would not necessarily prove the West is in a crisis: the West could go down in honor, certain of its purpose. The crisis of the West consists in the West's having become uncertain of its purpose."(Strauss, 3) James Kurth concurs: "Post-modern ideologues have engaged in a compulsive anti-Western project both in Europe and America. They have been joined by their post-colonial counterparts in the non-Western world. Together, they have formed a grand alliance against Western civilization, and they seek to obliterate it everywhere around the world, and especially within the West itself."(Kurth, 12) As these scholars have noted, the crisis of the West is as much an internal crisis as an external one. These forces are also not mutually exclusive; external forces and internal forces arrayed against the West are interdependent and mutually reinforcing. Understanding the foundations of the West, particularly those of the U.S. founding and constitutional system, and their relationship to the values of the West is essential to contextualize the factors of decline that have parallels from the Roman and British experiences.

FACTORS OF DESTABILIZATION AND COLLAPSE: LESSONS FROM THE EMPIRES OF ROME AND GREAT BRITAIN

The Romans were the first to comprehensively conquer the known world. They defeated the other major world powers at the time; Corinthian, Carthaginian, Macedonian, Persian, Gaul, and Greek. The Roman Empire's period of world conquest is traditionally dated from the victory in 168 BC over Perseus at Pydna. The Romans built their empire through struggles against their neighbors over a long period of nation-building, dating from the sack of Rome (*Vae victis*) by the Gauls in 390 BC. By 387 BC Rome was rebuilt and in 377 walls were erected around the city. Prior to this nation building, political changes brought an end to the Roman monarchy in 510 BC with the deposing of Tarquin the Proud and the establishment of a Roman Republic. The plebeians chose their own tribune as early as 471 BC, by 450 BC the plebeians are playing a formal legislative role, and in 445 BC inter-marriage between plebeians and patricians is allowed. Trade agreements like that established with Carthage in 348 BC created foreign intercourse that combined with social equalitarian movements to create an emerging societal tension between elites and masses. Some important considerations for the political development of the Roman Empire are in the admission of plebeians to the priesthood in 300 BC and then the full political equality between patricians and plebeians in 287 BC. This period of expanding political rights for the people and eroding rights by the elites parallels the expansion of trade, growth of the population of Rome by migration, technological advances, and ultimately, expansion of political and military power resulting in world conquest only 219 years after rebuilding from the *Vae victis*.

The first Roman coins were introduced in 338 BC and the first Appian aqueduct completed in 312 BC. The technological advantages were significant, and exceeded all known in human history by any civilization before and for at least a millennium since. The Romans, and historians thereafter, correctly view the expanding population, exponential advances in technology, and the military capacity that resulted from these factors as significant factors of Roman success. But an important precursor is the development of philosophy and the political theory to provide governance capacity at home and abroad. The role of Greek philosophy in the developments is most profound. "The Romans captured Greece, but culturally Greece took captive its fierce [Roman] conqueror."(Langer, 219) And as Donald Kagan has noted, "The Romans had even fewer hesitations about the desirability of power and the naturalness of war than the Greeks. Theirs was a culture that venerated the military virtues, a world of farmers, accustomed to hard work, deprivation, and subordination to authority."(Kagan, 571) And the establishment of the public philosophy of Rome by theorists like Cicero places the model with statesmanship, the just and responsible rule of the wise, in accordance with the *leges regiae*.

Cicero (106 BC -43 BC) is a Platonist who builds upon earlier work by Polybius (204 BC -122 BC) on mixed constitutionalism and statesmanship and the latter influence he provides Marcus Aurelius (122 AD). As noted by Roman historians such as Edward Gibbon, the assassination of rulers and the murder of political enemies, were reflections of the break down of Roman law in the century after Marcus Aurelius. The civil disorder followed a period of disaggregation of law and political philosophy, a crisis in the legitimacy of the theoretical foundations of the governing order. As this expanded into political crises of violent change at home, defeat for a mercenary army abroad, and a devaluation of the monetary system, Rome would fall in stages from 200 AD until 395 BC when Theodosius dies as the last Roman ruler of a united Empire. The Byzantines will variously fight and establish Eastern Roman control for another 500 years as chaos and defeat befall the remains of the former majestic Roman Empire and in 950 AD the "Dark Ages" begin. As Rome falls, one of its northern provinces in Britain begins a long march to unification and future empire. The English under Alfred the Great, King of England in 871 AD, are unified, recapturing London from the Danes, and establishing shires and feudal governing structures by 900 AD. The English moved slowly to consolidate political unity in the British Isles over the next 400 years, through Crusades, invasions, and epidemics.

By the Age of Discovery, the relationship of England to the Dutch was transforming with the advent of international financial systems for global markets. "Industrialization began in a group of islands off the north-west coast of Europe some few centuries ago, with what is usually described as the British Industrial Revolution. ....What seems clear is that it was the leading edge of a change of European dimensions and that nothing else like it was taking place at the time."(Kemp, 1) "A technological attitude to knowledge, an extreme readiness to apply science in immediately practical ways, eventually became one of the principal characteristics which distinguished western civilization, the civilization originally of Europe, from other great civilized societies. The unprecedented power which it produced eventually led Europe from Reconnaissance to worldwide conquest."(Perry, 15) "Britain by 1914 was a net importer of the products of the second industrial revolution. The pressures compromising British independence and freedom of action as a great power in the world were clearly in evidence before the turn of the [20th] century. ...The problems of 'culture' and 'civilization' assumed their places in the modern intellectual debate of the 1870s and 80s."(Shannon, 13)

The British had a system of colonies, markets, and a financial infrastructure that required imperialism just as the public philosophy of government and the values of British society delegitimated it. "The corollary of natural aristocracy was natural servitude, since the more perfect should hold sway over the less. The Aristotelian theory, was made to constitute a general mandate for civilized peoples 'who require, by their own nature and in their own interests, to be placed under the authority of civilized and virtuous princes and nations, so that they may learn from the might, wisdom, and law of their conquerors to practice better morals, worthier customs, and a more civilized way of life." The transformation in England was dramatic.

Within a century, England would no longer embrace imperialism, no longer have an Empire, and no longer lead Western Civilization, but the West would hold tremendous influence even in decline. "The expenditure patters of society throughout the world are becoming westernized, breaking down the indigenous social patterns and so leading to modern habits" which "tend to encourage inflationary monetary systems. Thus, the worldwide expansion of money has been partly caused by, but has far exceeded, the vast expansion of population."(Davies, 8) The monetarist theoretical basis of the emergent economic system, based upon British political philosophy from, among others, Adam Smith and David Hume, established a complimentary foundation for the economic system that would support the political system of global imperialism.

As table one reflects, changes in population and hard monetary factors reveal an inelasticity that British banking overcame as it revolutionized world finance. "The emergence of a genuine money economy from the eleventh century onward served gradually to transform medieval society."(Langer, 451) "The flow of gold and especially silver from Spanish colonies greatly increased Europe's supplies of the monetary metals, at least tripling them in the course of the sixteenth century."(Cameron, 107) Yet it was the innovation in credit markets that allowed the English banking system to transcend the Dutch just as they transcended the Spanish before them. This innovation in monetary policy combined with the innovations in technology from the industrial revolution to establish British dominance for almost 300 years. Yet it the crisis for British imperialism was not the imbalance in either wealth or military forces, but it was the imbalance in public spirit, in the promulgation of the values under which the political authority was established. These theoretical foundations were shaken by the forces of the French Revolution, the forces of the crisis of modernity.

Table 1: Population and Monetary Factors

(Davies 2002 and Butler 2000)

	Population in millions	Gold Per Capita oz per person	Silver Per Capita oz per person	Silver to Gold ratio
2000 BC	80	0.1	1	10:01
1200 BC	90	0.22	2.2	10:01
0600 BC	100	0.4	4	10:01
0300 BC	150	0.4	4.5	11:01
0500AD	300	0.59	6.9	12:01
1500 AD	500	0.6	8.2	14:01
1800 AD	980	0.46	7.6	16:01
1900 AD	1,650	0.49	8	16:01
1950 AD	2,520	0.79	9.21	12:01
1975 AD	4,000	0.74	7.48	10:01
2000 AD	6,000	0.72	6.82	9:01
2010 AD	6,200			16:01

IMPLICATIONS FOR MODELING SOCIETAL-LEVEL CHANGE AND CIVILIZATION

There are a number of important findings from this research for informing better meta-analytic models and more accurate and empirically sound formal models. First, there must be a more robust theoretical framework developed to operationalize the significant movements of theory, ideology, philosophy, and religion against the more quantitative variables from economics and military science. Variable specifications for theory, ideology, philosophy, and religion are typically unsophisticated and vectors to measure change and construct forecasting models are in need of very significant improvement. The role of faith, reflected in such factors as voluntary military service, religiosity, monetary value instruments, and political institutions must be more formally articulated. These are just some of the important implications for the work on societal level factors and the causal theory that underlies the measures for systems analysis.

This comparative study suggests that over 200 years prior to the Roman Empire and British Empire being ascendant international powers, there was a period of political theory that established or re-established legitimacy for the political system that was a crucial precursor to nation building and expansion of power and influence, hard and soft power capabilities. The situational factors often diagnosed as the most relevant at the end or collapse of this regime order are actually misdiagnosed according to prevalent change factors at the end of each cycle.

Table 2: Civilization Change Factors and Situational Decline

Economic	Political
Expansionist Trade	Establishment of Foundational Political Philosophy and Public Theory
Monetary Policy Moves from Traditional Stores of Value to Innovative Instruments Often followed by Hyperinflation	Maintenance Becomes Uneven and Forces of Regime Repudiation foster Social Revolutions

CONCLUSIONS: THE FATE OF THE WEST

"Nationalism, however inescapable, is simply no longer the historical force it was in the era between the French Revolution and the end of imperialist colonialism after World War II."(Hobsbawm, 169) Without nationalism, without a public philosophy that honors the past and legitimates the future of the West, there can be only limited progress, and all that remains is the continuation of a rear guard action for defense of the philosophical foundations of Western Civilization.

Leo Strauss remarked of the present crisis, "as regards modern political philosophy, it has been replaced by ideology: what originally was a political philosophy has turned into an ideology. This fact may be said to form the core of the contemporary crisis of the West."(Strauss, 2) Strauss goes on to note the East threatens the West unlike any epoch since earliest time.

Today, so far from ruling the globe, the West's very survival is endangered by the East as it has not been since its beginning. From the Communist Manifesto it would appear that the victory of Communism would be the complete victory of the West – of the synthesis, transcending the national boundaries, of the British industry, the French Revolution, and German philosophy – over the East. We see that the victory of Communism would mean indeed the victory of originally Western natural science, but surely at the same time the victory of the most extreme form of Eastern despotism. (Strauss, 3)

The Eastern despotism Strauss refers to here is one of political theory reduced to political ideology. There remains an ever stronger threat to the West from Eastern ideologies and totalitarian regimes that foster them, from Chinese Communism to Islamic Jihad. "The war against extremist Islam is as much an ideological war as the cold war ever was. And despite all our successes on the battleground, the ideological struggle against extremist Islam is one we are losing – that is, when we bother to wage it at all."(Frum and Perle, 147) If we are to win this battle, The West will need not only a united home-front but also to be persuasive abroad. Carnes Lord observes this point by stating, "At the end of the day, however, the terror war will not be won without creating –and sustaining— new friends and allies of the U.S."(Lord, 2)

Allies require maintenance, just as the values of Western Civilization require the fiduciary conservation of the highest things, one that higher education has performed in the West at least since Plato founded the Academy. "We consider education as a process in which we discover and begin to cultivate ourselves, we may regard it as learning to recognize ourselves in the mirror of this civilization. I do not claim universality for this image of education; it is merely the image (or part of it) which belongs to our civilization."(Oakeshott, 188) The crisis of modernity is a fundamentally philosophical one that has parallel crises in the foundations of the West, future and present – the education of future generations and the statesmanship capacity of current public leaders. The need for understanding the present forces of decline and their implications are of urgent importance.

Additionally, there remains much important work ahead for even the prospect of a next cycle of Western ascendancy. "It may be that we are awaiting a great change, that the sins of the fathers are going to be visited upon the generations until the reality of evil is again brought home and there comes some passionate reaction, like that which flowed in the chivalry and spirituality of the Middle Ages. If such is the most we can hope for, something toward that revival maybe prepared by acts of thought and violation in this waning day of the West."(Weaver, 187) Careful study of Western epochs of decline and the comparative analysis of regime forces hold

138

hope for policy alternatives, yet the greatest challenge remains the rebirth of classical political rationalism and a renaissance of Greek and Roman philosophy like that which regenerated the West in prior times of peril.

REFERENCES

Bloom, A. (1968). *The Republic of Plato*. New York: Basic Books, 1968.

Butler, Marion. (2000)   editorial from March 19, 2000:
    http://www.gold-eagle.com/editorials_00/mbutler031900.html

Cameron, R.   (1993).   *A Concise Economic History of the World, 2nd ed*. New York: Oxford University Press.

Davies, Glyn. (2002). *History of Money from Ancient Times to the Present Day*. Cardiff: University of Wales Pres.

Frum, D. and Perle, R. (2003). *An End to Evil*. New York: Random House.

Hobsbawm, E.J. (1992). *Nations and Nationalism Since 1870*. Cambridge: Cambridge University Press.

Huntington, S.P. (1993) "The Clash of Civilizations?" *Foreign Affairs*. 72(3) Summer 1993.

Hutchins, R.M.   (1952)   *The Great Conversation*.   Chicago: Encyclopedia Britannica.

Kagan, D. (1995) *On The Origins of War*. New York: Doubleday.

Kemp, T.   (1983)   *Industrialization in the Non-western World*.   New York: Longman.

Kurth, J.   (2003)   "Western Civilization, Our Tradition."   *The Intercollegiate Review*. Fall 2003/Spring 2004.

Langer, W. L. (Ed). (1968). *Western Civilization*. New York: Harper & Row Publishers.

Lord, C.   (2006). *Losing Hearts and Minds?*   Westport, Connecticut: Praeger Security International.

Oakeshott, M. (1962). *Rationalism in Politics and Other Essays*. Indianapolis: Liberty Fund.

Pangle, T.L. (1992) *The Ennobling of Democracy*. Baltimore: Johns Hopkins University Press.

Perry, H.C. (1963). *The Age of Reconnaissance*. Cleveland: the World Publishing Company.

Tuck, R., Ed. (1996) Hobbes, Thomas. *Leviathan*. New York: Cambridge University Press.

Shannon, R. (1974). *The Crisis of Imperialism*. London: Paladin.

Schmitt, Gary J. and Abram N. Shulsky. (1999) "Leo Strauss and the World of Intelligence (By Which We Do Not Mean *Nous*)." In Deutsch and Murley, eds. *Leo Strauss, the Straussians, and the American Regime*. Lanham: Rowman & Littlefield Publishers.

Spengler, O. (2006) *The Decline of the West*. C.F. Atkinson, Trans. New York: Vintage Books.

Strauss, Leo. (1964). *The City and Man*. Chicago: Rand McNally & Co.

Shulsky, A.   (2002)   *Silent Warfare: Understanding the World of Intelligence*. Washington, DC: Potomac Books Inc.

Watson, B.C.S. (2006) *The West at War*. Lanham: Lexington Books.
Weaver, R. (1948). *Ideas Have Consequences*. Chicago: University of Chicago.
Wood, G. (1969) *Representation in the American Revolution*. Charlottesville:
    University Press of Virginia.

CHAPTER 15

Authoring By Cultural Demonstration

Webb Stacy[1], Joseph V. Cohn[2], Kevin Sullivan[1], Diane Miller[1]

[1]Aptima, Inc.
12 Gill Street
Woburn, MA 01801, USA

[2]Defense Advanced Research Projects Agency
3701 North Fairfax Avenue
Arlington, VA 22203, USA

ABSTRACT

Modern military missions require soldiers to communicate effectively and at a personal level with people whose cultures, languages, lifestyles, and beliefs are very different from their own. Constructing suitable, game-based, culturally relevant scenarios to prepare soldiers for effective communication is difficult, time-consuming, and expensive. This chapter describes a method to simplify the creation of cross-cultural scenarios, Authoring By Cultural Demonstration (ABCD). ABCD allows authors to construct discrete cultural vignettes to represent key points in the story. It then guides the author to generalize those vignettes using an approach similar to Lakoff's (1987) radial categories applied to gesture, facial expression, language, and other cultural abstractions. ABCD connects the generalized vignettes into a cultural envelope that constitutes the scenario. Trainees' actions during scenario execution are then monitored to ensure that they stay within the envelope, so that they will encounter the cultural training opportunity that the author intended.

Keywords: Cultural training, cross-cultural scenarios, game-based training, scenario authoring, scenario engineering, radial categories

Per 5 C.F.R. 3601.108 and DoD JER 5500.7-R, 2-207 "The views, opinions, and/or findings contained in this article/presentation are those of the author/presenter and should not be interpreted as representing the official views or policies, either expressed or implied, of the Defense Advanced Research Projects Agency or the Department of Defense."

INTRODUCTION

This is a game of wits and will. You've got to be learning and adapting constantly to survive.
--General Peter J. Schoomaker, USA, 2004, quoted in Army FM 3-24, *Counterinsurgency*

Increasingly, military personnel require a sound and current understanding of the culture in which they are embedded. Such understanding eases local transactions, enables the US military to work with, rather than against, civilian communities, and ultimately multiplies operational effectiveness dramatically. Because both the local culture and our understanding of it are constantly evolving, learning about and adapting to culture are crucial.

Warfighters need to be able to interact with host-nation persons who are likely to recognize them as outsiders, but will appreciate their efforts to understand a culture that is different from their own; they need to avoid giving offense and demonstrate acceptance and respect for the culture's beliefs, customs, and way of life. Computer-based training using games has the potential to help military personnel and others prepare to work with persons from different cultures, and can provide cost-effective solutions to meet many needs (Keeney et al., 2009).

Unfortunately, constructing suitable game-based scenarios is difficult, time-consuming, and expensive.

For one thing, specialized personnel are required to create synthetic assets including terrain, buildings, roads, vehicles, and characters; cultural subject matter makes the creation of human avatars especially difficult if they are to provide realistic nuances such as gesture, facial expression and speech. Different specialized personnel are then required to create the scripts that will execute the scenario. Their work gets especially onerous as the number of potential logic branches—each representing a different scenario choice—explodes. Further, it is rare that personnel with special skills in digital asset creation, or in script programming, will deeply understand the content of the cultural lessons, and this leads to an inevitable disconnect between the intended and the actual content in the scenario. As if all this weren't enough, going through the production cycle for a game can take many months; by the time the cultural training is available, it may be stale.

There is a need to improve and accelerate the creation of culturally aware models of human entities for socio-cultural training simulations and games. More specifically, what is needed is a means by which deployed personnel with no special training in game development can easily create scenarios that express what they have learned in interactions with persons from the host nation – to generate their scenarios in a form that can be used in a gaming environment to teach others. Marines returning from patrol, for example, might take half an hour to express what they have learned from the day's interactions with the local sheik. Doing this, though, requires a major advance in scenario authoring.

We are addressing these challenges by developing Authoring By Cultural Demonstration (ABCD). ABCD extends existing scenario generation technology to accommodate cultural concepts, runs scenarios in a COTS environment called Game Distributed Interactive Simulation (GDIS), available from Research Networks, Inc., and is intended for use by deployed personnel.

ABCD STRATEGY

When warfighters come back to base having had a meaningful cross-cultural experience, the memory trace of that specific experience is fresh in their minds. By recounting that memory trace to others, whether by telling a story or by creating a replay of the incident in a virtual environment, they can effectively convey what happened, stopping and commenting at critical points. This can be a valuable learning experience for the listeners and viewers.

The trouble is that, by its nature, this is passive learning: the audience can only watch and listen. Modern learning theory posits that that both active and passive learning is required for maximum effect (Schwartz & Bransford, 1998; Schwartz, 2008), but this requires both the passive approaches and a means to provide the active experience. Leveraging the memory trace to create a scenario that has some degree of freedom for the trainee—that allows the trainee to take actions and make decisions in the scenario—gives the trainee the opportunity to actively participate and is, therefore, an effective multiplier of training effectiveness.

To accomplish this, ABCD adapts technology from an ongoing project at Aptima called CROSSTAFF. CROSSTAFF is a tool suite that uses Flight Data Recorder (FDR) data and events defined by Military Flight Operations Quality Assurance (MFOQA) to create flyable simulator scenarios. A particular focus of CROSSTAFF is to use the flight logs of safety mishap data to create scenarios that will train pilots to avoid the mishap in future flights.

CROSSTAFF FUNDAMENTALS

Figure 1 shows the steps involved in the creation of a mishap-related scenario with CROSSTAFF. The *original flight path* (a) is provided from FDR data and MFOQA events, either from the actual aircraft or from a re-creation in a simulator. The FDR data and MFOQA events describe the exact state of the aircraft, its instruments, and its controls during the ill-fated flight.

On the flight path, the author identifies key events of two kinds (1) *events* that were out of the pilot's control that led up to the incident (in blue in Figure 1b), and (2) pilot *decision points* that were key to causing or avoiding the incident (in red.)

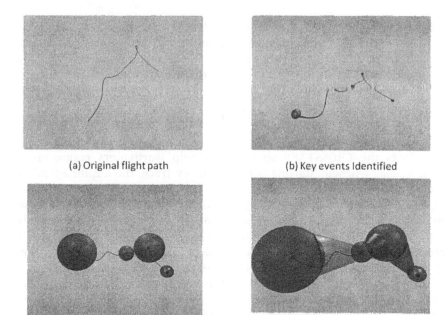

(a) Original flight path   (b) Key events Identified

FIGURE 1. Scenario Creation in CROSSTAFF.

The next step is an important one: to *generalize* the key events and decision points, shown in Figure 1(c) by metaphorically turning discrete 3D points into larger 3D spheres. The generalization step is an acknowledgement that a safety incident will never happen exactly the same way twice—the location, altitude, speed, or other aspects of the situation may be different from the original mishap, but it will still be the same kind of mishap. For each event and decision point, CROSSTAFF lets the scenario author specify what is important for recreating the mishap and what is not.

Suppose, for example, that the aircraft was at an altitude of 22,937 feet for the second (blue) key event. To recreate the mishap, it may only have been necessary that the altitude be greater than 15,000 feet in order for the incident to occur. This is a generalization. Identifying factors that had no effect on this particular mishap—say, weather or terrain—is another form of generalization. If some particular aspect of the flight must be exactly the same to recreate the mishap—if it can't really be generalized—the author notes that as well. Other space-and-time-related specifics can also be generalized, as might specific states of instruments, controls, and other aircraft in the scenario. In effect, generalization turns specific events and decision points in the MFOQA log into 4D regions in space-time.

The final step, Figure 1(d), connects those 4D regions into a *continuous envelope*. This constitutes the scenario. As long as pilots stay within this envelope during the training mission, they will encounter the key set of circumstances involved in the mishap, and they will arrive at the mishap's critical decision points. They will be actively flying in a scenario based on, but not exactly the same as, the original mishap. They will have an opportunity to experience the mishap's circumstances and to take action accordingly. A beneficial side effect is that every use of the scenario will potentially be different. As a result, with repeated runs through the scenario, pilots will encounter a variety of conditions under which the mishap might occur, giving them a broader experience base.

GENERALIZING CROSS CULTURAL VIGNETTES

ABCD adapts the CROSSTAFF approach to help warfighters create cross-cultural lessons learned. Starting with a memory trace, the author identifies (and creates) discrete vignettes representing key cultural events and decision points. This is accomplished by using the GDIS level editor and digital asset library to choose terrain and avatars, position and animate them appropriately, and by adding dialog when necessary. We call this a vignette, and it corresponds to CROSSTAFF's events and decision points. The author then generalizes the vignettes and connects them to create a scenario envelope.

A major challenge in doing this is the generalization step. Generalizing culturally salient information is a more complex task than generalizing CROSSTAFF's spatiotemporal information. While the latter is easily expressed on numerical scales, there is no obvious corresponding set of numbers for the former. ABCD needs to generalize over many dimensions such as gesture, facial expression, language, and other cultural abstractions. What is needed is a mechanism that can organize cultural information, whether it is linguistic or non-linguistic.

To address this problem, we use radial categories (Lakoff, 1987) as a framework for generalization, as shown in Figure 2.

Cognitive scientists have known for some time that a key aspect of human perception is the categorical knowledge people bring to the task of filtering and interpreting the physical stimuli that impinge on their senses (Biederman, 1981; Biederman, Rabinowitz, Glass, & Stacy, 1974; Barsalou, 2004). Hence, it is natural to approach the problem of organizing cultural phenomena by leveraging recent findings in human categorization.

Lakoff's (1987) theory of radial categories has successfully accounted for many categorization phenomena, both in the laboratory and in existing natural languages. The term was inspired by empirical work on the human conceptual system (Rosch & Mervis, 1975, and many others; see Murphy, 2004 for an excellent, thorough review.) This body of empirical research has established rather conclusively that the human conceptual system is not definitional; that is, that human categories exhibit properties that cannot be explained by a taxonomic system, where each concept is a node represented by a simple necessary-and-sufficient definition.

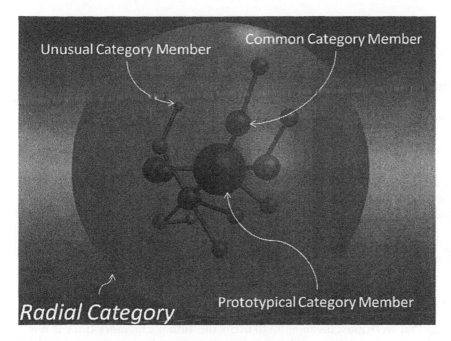

FIGURE 2. The Structure of Lakoff's Radial Categories.

One of the most important phenomena that cannot be explained with a definitional system is that category members show strong typicality effects: some instances are better members of a category than others. A common example is that a robin is considered by most people to be more typical of birds than is an ostrich. This typicality shows up in many ways in the psychology laboratory, as well as in linguistic research across cultures (e.g., Jurafsky, 1996; Archard & Niemeier, 2004; Narayan, 2008.) For example, the problem of categorizing hot and cold entities across cultures has proved particularly difficult, but Narayan (2008) was able to automatically construct culture-specific radial categories for hot and cold foods with impressive accuracy across multiple cultures (including English, Tamil, Javanese, Chinese, Hindi, and Persian.)

The categories are deemed radial because a category starts off with a central concept, and as a person encounters more ambiguous and difficult-to-classify instances, variants of the central concept are added, sometimes to the central concept, sometimes to other variants. For example, the category mother includes not only the prototype—a married woman who conceived, gave birth to, and is raising a child—but also many variants, such as stepmother, birth mother, adoptive mother, surrogate mother, and others too complicated to have a name, such as female legal guardians who don't personally care for the child. In this way, the variants of the category radiate outward from the central concept.

The same logic can be applied to other cultural, but non-linguistic, phenomena. There are prototypical bows and there are less common bows; there are prototypical smiles and there are less common smiles; and there are prototypical tones of voice that are assertive, and others that, while still assertive, are less so.

It is interesting to note that categories, structured this way, are considerably more robust to new knowledge and experience than definition-based categories. As new meanings of mother are encountered, it is a simple matter to create a new variant within the category. It is quite another to modify the definition to encompass all previously encountered instances, or to create new categories and to try to decide where existing instances belong. In ABCD, the ideal is to have native members of the culture provide the radial structure that they use to understand to categorize gestures, facial expressions, and other semantic information. But even if it is the warfighters returning from patrol that provide the categorizations, the categories will need restructuring far less often in the face of new experience, when organized by radial categories.

THE ABCD APPROACH

Extending the CROSSTAFF approach with radial categories results in the ABCD approach shown in Figure 3. Because humans tend to encode and remember the things that were most meaningful to them, the author's memory trace is not a high-fidelity sequential log of the events they encountered; it is not directly analogous to CROSSTAFF's FDR data or MFOQA events. For this reason, in ABCD we start by capturing *key events and decision points*, shown in Figure 3(b). Key events are more complex than in CROSSTAFF, so we have labeled

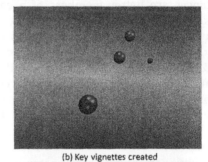

(a) No need to for initial scenario walkthru   (b) Key vignettes created

(c) Vignettes generalized using radial categories   (d) Cultural envelope for scenario generated

FIGURE 3. ABCD Approach to Scenario Generation.

them vignettes. The author uses a simple game level editor to select terrain, buildings, vehicles, avatars, and other assets out of a prebuilt library, and then positions them and indicates any key gestures or facial expressions (also using the library). If language and speech play a role, users also indicate that.

The next step is for the author to *generalize* the key vignettes using radial categories, as in Figure 3(c). Radial categories structure the libraries. If the author chooses a prototypical smile for a certain vignette, the library will present other facial expressions that have some level of membership in the category "smile," and the author will select those that work in the scenario. This is true for gestures and for semantic textual dimensions, and perhaps for other cultural pragmatics such as tone of voice.

To assist with the generalization, we are developing a formal representation for radial categories that will be used to structure culturally relevant gestures, facial expressions, and other culturally relevant phenomena. We are also designing a mechanism that leverages the radial categories to assist users in generalizing the vignettes. Authors can choose a central concept when constructing the original "plot point" vignette; then, during generalization, ABCD presents other members of the category radially and asks which members fit the vignette and which don't. Authors are able to do this for all aspects of the vignette that they feel are generalizable. The union of these judgments provides a "cultural enlargement" of the vignette that, as described above, will be used to create the cultural envelope that is the scenario.

Finally, ABCD connects the generalized vignettes into a *cultural envelope*, as shown in Figure 3(d), that constitutes the scenario, and the trainee's actions are monitored as the scenario runs to ensure that the trainee stays within the envelope and encounters the cross-cultural training opportunity.

IMPLEMENTING ABCD

Figure 4 shows the architecture for ABCD. There are two major sets of components. The first set is used at authoring time, to define cultural behaviors and interactions with native populations. The second set of components is used during the cultural training session to ensure that trainees stay within the cultural envelope and therefore actively encounter the situation the author had in mind.

The ABCD Authoring component combines the vignette editing capabilities of the GDIS game editor with the library of cultural artifacts, which include gestures, tone of voice, facial expressions, and other pre-built cultural phenomena. Authors create discrete vignettes describing their experience, and then generalize their experience using radial categories as well as more conventional spatio-temporal means (such as those used in CROSSTAFF). ABCD then connects the discrete generalized vignettes to create a cultural envelope, forming the basis for the cultural lessons learned scenario.

148

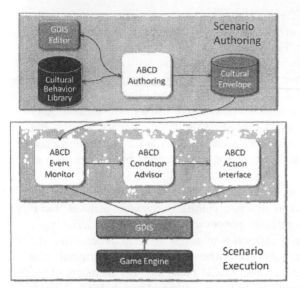

FIGURE 4. ABCD Components.

Because of the generalization, every trainee's path through the scenario is potentially different, but the methodology guarantees that scenario conditions will steer the trainee into the circumstances the author had in mind. When the game is played—that is, when the scenario is executed—ABCD monitors conditions to determine if the trainee is within the scenario envelope. If not, ABCD advises the instructor of the situation, recommends actions to be taken, and provides the instructor with a means to put those actions into effect in the game. In future versions of ABCD, we intend to provide an optional mechanism to take those actions automatically, either because the instructor prefers that the system operate this way or because no instructor is available.

CONCLUSION

ABCD meets a real need—training cross-cultural lessons in game-based environments in a way that is guaranteed to be fresh and current, because those lessons will be authored by warfighters as they themselves experienced the situations and learned the lessons. ABCD enables authoring by people who are not game engine or computer programming specialists. Because it enables lesson authors to generalize their experience, it allows them to directly and meaningfully expand the circumstances under which that lesson applies, while at the same time providing experiential variation if trainees choose to experience the lesson more than once. We believe that ABCD has the potential to be an important component of the learning and adapting that Gen. Schoomaker highlighted as so important for the warfighter.

REFERENCES

Archard, M. & Niemeier, S. (2004). *Cognitive Linguistics, Language Acquisition, and Pedagogy*. Berlin: Mouton de Gruyer.

Barsalou, L.A. (2004) Situated conceptualization. In H. Cohen & Lefebvre, C. (Eds.) *Handbook of Categorization in Cognitive Science*. St. Louis: Elsevier.

Biederman, I. (1981). On the semantics of a glance at a scene. In M. Kubovy and J.R. Pomeranz (Eds.), *Perceptual Organization*. Hillsdale, NJ: Lawrence Erlbaum Associates.

Biederman, I., Rabinowitz, J. C., Glass, A. L., & Stacy, E. W. (1974). On the information extracted from a glance at a scene. *Journal of Experimental Psychology*, **103**, 597-600.

Jurafsky, D. (1996). Universal tendencies in the semantics of the diminutive. *Language*, 72(3), 533-578.

Keeney, M. J., Neal Reilly, S. Beltz, B. C., Reiter-Palmon, R., & Weyhrauch, P. (2009, April). CAATE: A system for developing culturally aware agents for training environments. Chapter under review for E. G. Blanchard & D. Allard (Eds.) *Handbook of Research on Culturally-Aware Information Technology: Perspectives and Models*.

Lakoff, G. (1987) *Women, Fire, and Dangerous Things: What Categories Reveal about the Mind*. Chicago: Chicago Press.

Murphy, G.L. (2004). *The Big Book of Concepts*. MIT Press.

Narayan, S. (2008). The thermal qualities of substance: A cross-cultural account. *Proceedings of the 2008 International Conference on Cognitive Science*.

Rosch, E. and C. Mervis (1975). Family Resemblances: Studies in the internal structure of categories. Cognitive Psychology 7, 573-605.

Schwartz, D. L. (2008). Why direct instruction earns a C- in transfer. Presentation to research panel on adolescent learning. National Science Foundation, Washington DC.

Schwartz, D.L. & Bransford, J.D. (1998). A time for telling. *Cognition and Instruction* **16**(4), 475-522.

CHAPTER 16

Using Behavioral Science Principles to Train Small Unit Decision Making

V. Alan Spiker[1], Joan H. Johnston[2]

[1]Anacapa Sciences, Inc.
Santa Barbara, CA 93101

[2]Naval Air Warfare Center
Training Systems Division
Orlando, FL 32826

ABSTRACT

Combat tactical profiling provides U.S. Marines with techniques for observing and categorizing human behavior that can be used in cross-cultural decision making for Irregular Warfare. This paper describes how behavioral science principles are at the root of the effective instruction provided by Profiling and Cue Recognition training at Camp Pendleton.

Keywords: behavioral science principles, culture, decision making, learning, memory, motivation

INTRODUCTION

The Profiling and Cue Recognition (PCR) instruction at Camp Pendleton trains U.S. Marine infantry units in tactical profiling skills using advanced techniques in human behavioral pattern recognition analysis. PCR is a set of skills and techniques that a profiler uses (e.g., cues and indicators of behaviors) to spot people and events *before* the situation becomes lethal. Staying "left of bang" by constructing these behavior profiles in a proactive fashion is now considered to be a protective element for small units, and is every bit as important as body armor and weaponry; as a result, behavior profiling techniques have become an invaluable addition to the small unit's tactics, techniques, and procedures (TTPs). The 5-day PCR instruction is combined with a 5-day course on combat tracking and optics-based observation and has been reported as effective for cross-cultural decision making based on self-reports and Lessons Learned from units returning from Iraq and Afghanistan.

PCR instruction is conducted as a classroom and field exercise course on how to assess hostile intent based on piecing together patterns of information collected on human physiological and social behaviors, and social indicators in the physical terrain (e.g., human heat signatures, familiarity in social interactions, types of body language, and signs or symbology). The concept of a behavioral profile – that would characterize persons of interest (POIs), high value individuals (HVIs), high value targets (HVTs), leaders, body bombers, or whoever – is based on years of experience from undercover police work. Combat Profiling has its scientific roots in the social sciences – principally psychology and anthropology – where the behavior profiles generated are essentially prototypes of what to look for. These prototypes are less rigid than a feature-listing of templates and thus can be updated readily as the insurgents' own TTPs change (e.g., how Improvised Explosive Devices (IEDs) are placed).

That PCR training is effective is now widely accepted, but *why* the course is effective is less well-established. Identifying reasons for PCR training effectiveness is important as multiple efforts are underway to significantly increase trainee throughput. Expansion options include converting to digital media, employing a larger cadre of instructors, and capturing live instruction through video. But whatever options are pursued, it is essential the key ingredients in the PCR "recipe" for creating effective behavioral profiles are not lost.

In early 2009, the second author observed a portion of the course and noted course elements were clearly responsible for the high levels of student buy-in and engagement—experienced instructors well-versed in "Marine-speak," fast-paced delivery, humor, and extensive use of graphics and photos. She noted one important element was the course's strong grounding in science, as its instructional delivery, pedagogy, and content are a veritable tour de force in anthropology, sociology, biology, logic, and psychology. Discussions with the lead instructor led to an understanding that he had designed the course based on training strategies he had leveraged from research, including the tactical decision making under stress program (Cannon-Bowers & Salas, 1998). Therefore, to facilitate PCR extensibility without sacrificing its enduring value, the authors conducted a formative study to test our assumption that a number of learning and behavioral science principles had been adopted, even inadvertently, to fashion the PCR instruction. By behavioral science, we refer to the various sub-disciplines that comprise the study of individual and group behavior, beginning with basic principles of perception, conditioning, and learning; and branching out to more complex paradigms such as concept formation, memory, attention management, decision making, and social learning. Since behavioral science principles apply to all people, they are acultural, making them useful for cross-cultural applications, a point repeatedly emphasized in PCR training.

This paper presents the study findings in three sections. First, we describe the methods used to observe course delivery and "capture" its scientific underpinnings. We then illustrate some of the key behavioral science principles supporting PCR instructional delivery, content, and pedagogy. We conclude with suggestions for extending PCR instruction while maintaining its behavioral science roots.

METHOD

The PCR five-day course, which is taught to 38-40 students, consists of three days of academic instruction and two days of field exercises. The first author employed a naturalistic observation methodology during two successive offerings of the PCR course at Camp Pendleton in October 2009. These two evolutions provided a broad sampling of trainee characteristics, as the first unit was composed of senior weapon school instructors, while the second unit was from an intact infantry battalion. The researcher supplemented his observations with probe questions of instructors and students. This was done on a not-to-interfere basis, when breaks between the academic topics afforded an opportunity to solicit amplifying information from the instructors and gauge student reaction to the instruction.

During the field exercises, the senior author accompanied the students as they occupied one of three Observation Posts (OPs) established to observe a mockup village where role-players acted out scripted scenarios. Students were divided into three squads that rotated OPs across scenarios. The researcher divided his time to observe each group in one of the OPs, where he sat behind the group as they made their observations. From this vantage point, he could observe each student, listen to their radio communications, and note their verbal inter-communications. Since several hours were spent at each OP, the researcher had time to ask the Subject Matter Expert (SME) instructor amplifying questions about the scenario and obtain process information on how the group was doing as the scenario unfolded. He also attended the debrief after each scenario, which was led by one of the SME instructors who gave the group feedback on their profiling performance.

The final exercise had the Marines deploy as two ground maneuver squads with the remaining squad divided into two groups to occupy two of the OPs. The senior author accompanied one of the maneuver units into the village, taking notes along the way. Once inside, the researcher divided his time among several of the SMEs who were acting as field judges during the exercise. From this vantage point, the researcher could monitor multiple communications and get a balanced perspective on how each of the subgroups of students was doing.

Throughout academics and exercises, the senior author recorded his observations on a notepad and a structured behavioral observation checklist (BOC). Designed by the second author, the BOC is a 33-item list of instructional objectives organized around seven categories of critical thinking and decision making skills. Based on lessons learned from the Navy's Tactical Decision making Under Stress program (Johnston, Poirier, & Smith-Jentsch, 1998), BOC indicates, on a 3-point scale, the extent to which critical thinking and decision making behaviors and strategies are addressed in the course. While the BOC could be used to measure student performance, in this application it was used to structure observations of course value and content validity.

The researcher's observation notes and BOC data were reviewed after course end to identify behavioral science principles that were in evidence during the academics and field exercises. Some 28 principles were prominent, with 15 noted in the academics and 13 in the exercises. A technical report (Spiker & Johnston, 2010) presents these findings in detail. Below, we provide highlights from the report by describing how 12 of these principles figured prominently in the instructional delivery, content, and pedagogy of PCR.

APPLYING BEHAVIORAL SCIENCE PRINCIPLES

INSTRUCTIONAL DELIVERY

The unique delivery of PCR instruction can be characterized as fast-paced, highly visual, emotion- and humor-laden, and resulting in extremely high levels of student buy-in and high student ratings (Spiker & Johnston, 2010). Four principles from behavioral science, illustrated below, stand out as underlying factors: social learning theory, emotional memory, dual-coding theory, and storytelling.

Social learning theory holds that learning can occur from observing a high-valence role model, where behavior is most likely to change when the observer can closely identify with a model and when the observer has high self-efficacy, i.e., believes he/she has the capability to behave like the model. High identification and self-efficacy produce high levels of motivation and learning (Bandura, 2001). Social learning theory provides a solid account of the high levels of motivation instilled by the Combat Profiling instructors, who clearly establish themselves as high-valence role models in several ways. First, prior to class, they engage in informal discussions with the students on the latest events in Operation Enduring Freedom or Operation Iraqi Freedom, where it is quickly evident that the instructor thoroughly understands the problems Marines are facing, where their sole mission is to increase the chances that they will return home safely after their next deployment. Second, the instructor solidifies the students' respect by providing an unusually dense and rapidly-presented discourse, without notes, on some technical topic of interest to Marines, such as weapons, ammunition, optics, or TTPs. Third, the instructor gets the students laughing through a combination of self-deprecating humor and jokes, often profane, that are just plain funny. At this point, the students are fully engaged, not only because they know the material they are about to receive is important, but because they are drawn to the instructor.

Research on *emotional memory* supports the view that memories for emotional events, whether personal or public, are more vivid, longer lasting, and more detailed, than memory for neutral events. Sometimes referred to as "flashbulb" memories, events surrounding a very emotional event, such as the World Trade Center bombing or the Challenger Disaster, have been found to evoke profound memories years after they were originally experienced (Reisberg & Heuer, 2004). While the physiological basis for such superior memories lies in the amygdala, the psychological processes underlying this effect (e.g., greater rehearsal immediately after the event) are still being studied. Nevertheless, couching a to-be-remembered event in an emotional context adds to the weight of that event and increases the likelihood that it will be retrieved later. PCR instructors make extensive use of this principle in their training. The ten-minute, highly emotional description of what it is like to be beheaded is a case in point, in which students will undoubtedly never forget the story nor the reasons why the perpetrators of these atrocities were allowed to go free due to poor profiling practices at the time.

Dual-coding theory posits that verbal and visual information are processed on distinct channels, where unique representations are formed to create stored knowledge (Paivio, 1969). The results of these processes then organize future incoming information to facilitate retrieval. This theory provides the basis for use of multimedia presentations, where learning is aided when the visual and verbal information complement, rather than conflict or compete with one another (Mayer, 2009). PCR makes extensive use of dual-coding by coordinating the presentation of graphics and verbal narration; they avoid overwhelming the student with information by minimizing the use of text-laden PowerPoint. The pace of delivery is rapid fire, with the instructor talking quickly while he walks about the room. He puts his hands on the shoulders of various students, sometimes to kid them or ask them a direct question. Like an entertainer "working a room," it is very engaging to students and certainly not like typical instruction. Many humorous references are made and students frequently laugh out loud. The language is often profane, which is absolutely appropriate for a Marine audience. Sometimes, the instructor will call a student up to the front of the class for a demonstration, adding a personal feel to the class, making it difficult for any one student to hide or fade out.

Storytelling is an instructional method used in a variety of disciplines with obvious roots throughout millennia that telling stories has been a primary way that key principles are illustrated and values passed down from generation to generation. As an information medium, storytelling is now used widely in such fields as the military, dentistry, law, general medicine, and business to build the analytic prowess of trainees and students (Andrews, Hall, & Donahue, 2009). Within PCR, instructors make extensive use of storytelling to introduce each new topic and even subtopics. The stories are often highly vivid and detailed, and are typically told in a way that students can easily relate to their own personal experiences. Sometimes the story is about a police case, or it could be drawn from everyday life, such as discussing a photo taken at a food court. The stories are important because they personalize the information, draw the student in, and lay a foundation for them to integrate their own experiences with the technical information they are about to receive.

CONTENT

A considerable amount of technical content concerning behavioral pattern recognition is presented during the three days of academics. These are reinforced and amplified during the PCR field exercise scenarios. Spiker and Johnston (2010) describe more than a dozen principles of psychology that are tied directly to course content. Below, we illustrate the use of: prototype learning, cue consistency theory, stimulus redintegration, superstitious conditioning, learning to learn, and habituation.

Prototype learning is a theory of concept learning in which people acquire concepts or categories based on identifying representative or typical instances of the concept rather than having to identify all defining features in a template. A key notion is that some instances are more central to the concept than others (Rosch, 1973, 1975). Prototypes predict faster learning and more adaptive performance when the subject's experience with instances of the concept changes. Since insurgents are constantly changing their TTPs, students are taught in the PCR course to identify such core concepts as Vehicle Born Improvised Explosive Devices (VBIEDs), IEDs, body bombers, and HVTs as prototypes. This is heightened by the use of heuristics, where baselines can be established by considering "it looks like X" (e.g., a man looking at map because he is lost). This would then represent the best-guess prototype of some event or behavior. Later, they would look for anomalies or deviations from that baseline. Teaching a template-based identification (exhaustive feature listing) strategy would be useless because it would rapidly become outdated.

Cue consistency theory posits that constellations of cues will reveal internal consistency across modality and time to reflect the inherent stability in behavior that can be attributed to actors in a given domain (e.g., bargaining behavior). According to Kelley (1973), one should see both consistency in the direction and information-value of a given cue over time, as well as a comparable consistency within the pattern of cues itself. PCR makes extensive use of the cue consistency assumption as the instructors tell students to wait, and look for multiple cues in a given situation, as they attempt to create a behavior profile of insurgent activity, such as vehicle-born sniper platforms or IEDs. By waiting for a cluster of cues, the profiler will be less likely to "jump to conclusions" and have more confidence in his judgment once he decides to act.

Hullian theory was an early attempt at a psychological theory that resembled the physical sciences in its reductive analysis of complex stimuli into simpler elements (Hull, 1943). While this approach has been largely abandoned in favor of human cognition theories, aspects of the theory are still highly useful. One such byproduct is the principle of *redintegration* which holds that *all* the features of a stimulus acting on the individual at or near the time that a response is evoked tend to each acquire the capacity to evoke the same response (Hull, 1930). This little-cited, yet powerful principle, explains the ability to recognize parts of a stimulus when the entire stimulus is not in view. Within PCR, the redintegration principle is ever present as students are taught how to identify a terrorist planning cycle by finding clues on one stage and then inferring the rest of the activity. In our opinion, it is this approach toward part-stimulus recognition that makes profiling such an effective tool for Marines, as it simplifies the observational process and provides manageable subgoals for establishing baselines and identifying anomalies.

Superstitious conditioning is a variant on Skinner's operant conditioning procedure in which organisms respond consistently with accidental vs intentional delivery of reinforcement (Skinner, 1948). Hypothesized to underlie all of superstitious behavior, the accidental pairing of a response with reinforcement will increase the chances of that response occurring in the future even when the reinforcement is absent in later occasions. In PCR, this mechanism underlies the interpretation of phenomena such as criminals wearing certain "lucky" clothes when they engage in criminal acts or when insurgents use the same routes or vehicles that have been effective in the past. Profilers capitalize on these repeat behaviors, both to establish baselines and identify cues in a pattern that can be profiled.

Learning to learn is the phenomenon in which subjects who are given a set of related yet different tasks (e.g., visual discrimination) exhibit faster rates of learning as they progress through the series. Originally demonstrated by Harlow (1949), the interpretation is that subjects are learning more than what the relevant cues are; they are actually learning a strategy (or a learning set) that allows them to master each new task faster than the previous one. A typical component of this learning set is a win/stay-lose/shift strategy, in which subjects will continue with a given response strategy until it no longer leads to reinforcement. The win/stay-lose/shift strategy figures prominently in the PCR explanation of insurgent behavior, in which terrorists continue to use a certain strategy, such as type of wiring in a bomb or sniper tactic, as long as it has worked in the past. The learning to learn strategy provides a rational basis for understanding how behaviors are repeated across similar – though not identical – situations.

Habituation is the phenomenon where organisms' responses to novel stimuli dissipate with time and repeated exposure if nothing notable happens following that stimulus. The habituation process is biologically-based, does not require conscious awareness, and occurs in all species (Groves & Thompson, 1970). As discussed by PCR instructors, overcoming this process is one of the most important challenges in-theater as troops that go on "routine" patrols, where nothing happens, habituate to their environment and fail to notice cues that would have been helpful. The field exercises contain discrete periods where "nothing" happens, to provide students training in techniques (e.g., putting down the binoculars and using the naked eye for awhile) to avoid habituating to (or "painting out") potentially significant cues.

PEDAGOGY

The pedagogical approach taken throughout PCR academics and field exercises is best described as crawl-walk-run. The instruction on each topic begins with fairly basic examples to illustrate, for example, one of the profiling domains. Subsequent examples are more complex, bringing in additional variables and domains to consider. For example, in covering the topic of making legal-ethical-moral decisions, the instructor begins with the fairly black-and-white issue of making legally correct decisions. However, as he transcends into the grayer areas of, first, ethical and then, moral decisions, the complexity underlying decision making in this context builds, and the students begin to realize that it is not easy to make such decisions. Within this basic foundation, a number of behavioral science principles provide scaffolding support. Below, we illustrate two: imperfect performance during training and mental rehearsal.

In a highly influential paper, Schmidt and Bjork (1992) summarized research showing that traditional methods of training, in which students are given large amounts of feedback to keep performance near 100%, are ineffective for ensuring high levels of transfer of these skills to the operational environment. Instead, *imperfect performance during training* is preferred since traditional training, with continual feedback and various "crutches" to avoid mistakes, is actually counter-productive for preparing trainees to perform outside the training environment. This fact is recognized in PCR as students are actually expected to make mistakes during all the exercises, including the final exercise. Rather than have artificial scenarios, where perfect performance is guaranteed, all scenarios are stocked with many complex events to ensure that students are exposed to the various cues needed for profiling. The debriefs are designed to provide feedback on these errors, where students are guided to arrive at their own insights about their mistakes and how to correct them.

As the name suggests, *mental rehearsal* is a systematic method of imagining the performance of a particular task between occasions where the task is physically performed. It has been used effectively in a number of skill areas as a way to increase the amount of practice a trainee receives without the fatigue that can accompany massed physical practice (Schmidt & Bork, 1992). Mental practice aids performance in such tasks as chess, theater, track and field, typing, knot tying, and pursuit-tracking, among others (Annett, 1989). In PCR, instructors encourage students during the OP scenarios to mentally practice by visualizing the profiling cues, either individually or together. While this mental rehearsal is intended to provide the students with additional practice, a secondary purpose is to instill a readiness to profile once the next scenario begins.

FUTURE DIRECTIONS

We close by recommending how to "grow" PCR training so it reaches more students and enables skill transfer to the operational environment, while retaining the behavioral science principles that make it effective. For example, supplemental or bridge training should be made available on select skills once the PCR course is over. A promising method is *game-based training* (GBT) to promote learning for students using their inherent interest and prior use of games to instill motivation and engage in repetitive practice during leisure time (Hays, 2005). Higher order skills could be facilitated through GBT which would deliver basic principles of feedback,

158

practice, and associative learning needed for skill development. One such skill is learning to switch fields of view, where profilers must develop an internal timing sense to know when to switch between observation with optics and with the naked eye. Switching lets observers balance the risks of channelized attention and loss of situation awareness. Another skill aided by GBT is making a positive identification (PID) of key events, places, and people using decision criteria to ensure sufficient balance of speed and accuracy. A video- or machinima-based (Conkey, 2009) game for progressive development of combat reporting skills would be valuable post-PCR. GBT would allow practice and reinforce the skills that underlie effective teamwork, such as a common vocabulary, appropriately phrased and timed messages, monitoring probes to ensure that information is both shared and understood, and accurately reporting the results of one's observation in a clear, organized, and effective manner.

Embedding capability within deployed units via expert trainers, mentors, and peers is a crucial strategy to ensure transfer of KSAs to the operational environment. *Special, highly-concentrated train-the-trainer sessions should be conducted* with a core cadre of trainers who would then inculcate profiling knowledge back into their units. This plan would create a new generation of instructors who would ensure that the essential profiling KSAs are retained and spread throughout the tactical forces. Such a course would leverage and even amplify the behavioral science principles discussed here and in Spiker and Johnston (2010). *Combat profiler SMEs should accompany units on range exercises.* This supplemental training would occur outside the regular PCR course, where profiling KSAs would be incorporated into other courses. For example, at Camp Pendleton, several of the profiling SMEs are embedded with Marines on field exercises in weapons and tactics courses. The SMEs are present solely to provide profiling expertise, offering suggestions on how profiling cues can be used to improve their Marine field craft. In essence, they act like coaches or mentors just as they do during the OP scenarios. This appears to be a very efficient and cost-effective way to expand the reach of Combat Profiling while not siphoning resources from the main course.

Finally, a cost effective option is to *cultivate the profiling "naturals" that emerge from each class.* We noted in our observations that, in each class, one or two Marines were especially adept at profiling, volunteered information, and often took the lead during scenarios for their unit. In short, they seemed to be "naturals" at the science and art of profiling. Follow-up inquiries revealed that these naturals emerge in every class, suggesting that such individuals should be singled out for Advanced Profiling training and, when sent back to their units, would have a force-multiplying effect on profiling KSAs throughout the service.

REFERENCES

Andrews, D.H., Hull, T.D., & Donahue, J.A. (2009). Storytelling as an instructional method: Descriptions and research questions. *The Interdisciplinary Journal of Problem-based Learning. 3*(2), 6-23.

Annett, J. (1989). Training skilled performance. In A.M. Colley and J.R. Beech (Eds.), *Acquisition and performance of cognitive skills* (p. 61-84). New York: John Wiley & Sons.

Bandura, A. (2001). Social cognitive theory: An agentic perspective. *Annual Review of Psychology, 52,* 1-26.

Cannon-Bowers, J.A., & Salas E. (Eds.) (1998) *Making decisions under stress: Implications for individual and team training.* Washington, DC: APA.

Conkey, C. (2009, December). Machinima technology: Serious games lite. *Proceedings of the 29th Interservice/Industry Training Systems and Education Conference [CD-ROM].* Orlando, FL

Groves, P.M. & Thompson, R.F. (1970). Habituation: a dual-process theory. *Psychological Review, 77*(5), 419-450.

Harlow, H.F. (1949). The formation of learning sets. *Psychological Review, 56,* 51-65.

Hays, R. T. (2005, November). *The effectiveness of instructional games: A literature review and discussion.* Naval Air Warfare Center Training Systems Division (Technical Report Number 2005-004): Orlando, FL.

Hull, C.L. (1930). Knowledge and purpose as habit mechanisms. *Psychological review, 37,* 511-525.

Hull, C.L. (1943). *Principles of behavior.* NY: Appleton-Century-Crofts, Inc.

Johnston, J.H., Poirier, J., & Smith-Jentsch, K.A. (1998). Decision making under stress: Creating a research methodology. In J.A. Cannon-Bowers and E. Salas (Eds), *Making decisions under stress: Implications for individual and team training* (pp. 39-59). Washington, DC: APA.

Kelley, H.E. (1973). Attribution theory in social psychology. In D. Levine (Ed.), *Nebraska symposium on motivation, 15,* 192-238. Lincoln, NE: University of Nebraska Press.

Mayer, R.E. (2009). *Multi-media learning* (2nd ed.). New York: Cambridge University Press.

Paivio, A. (1969). Mental imagery in associative learning and memory. *Psychological Review, 76*(3), 241-263.

Reisberg, D. & Heurer, F. (2004). Memory for emotional events. In D. Reisberg and P. Hertel (Eds.), *Memory and emotion* (pp.3-41). NY: Oxford Univ. Press.

Rosch, E.H. (1973). Natural categories. *Cognitive Psychology, 4,* 328-350.

Rosch, E.H. (1975). Cognitive reference points. *Cognitive Psychology, 7,* 532-547.

Schmidt, R.A. & Bjork, R.A. (1992). New conceptualizations of practice: Common principles in three paradigms suggest new concepts for training. *Psychological Science, 3* (4), 207-217.

Skinner, B.F. (1948). 'Superstition' in the pigeon. *Journal of experimental psychology, 38,* 168-172.

Spiker, V.A. & Johnston, J.H. (2010, January). *Limited objective evaluation of combat profiling training for small units.* Naval Air Warfare Center Training Systems Division Draft Special Report: Orlando, FL.

Chapter 17

Training Decision Making for Small Units in Complex Cultural Contexts

William Ross, Jennifer Phillips, Clark Lethin

Cognitive Performance Group
3662 Avalon Park East Blvd. Suite 205
Orlando, FL 32828, USA

ABSTRACT

The purpose of this paper is to report on applied research used for marking and embedding cultural cues within a promising prototype training solution and to describe the impact of cross-cultural training on small unit performance of mission essential tasks based on this training. Evidence is mounting for an operational cross-cultural competence capability that is demonstrated by highly effective small units. The requirement is expressed by operational commanders who report that it is not possible to emerge victorious in counterinsurgency struggles if the forces in contact do not engage with non-combatant civil populations and win their support. One approach that has been adopted to produce high quality small unit training is under development as part of a USJFCOM funded initiative, a Joint Concept Technology Demonstration called the Future Immersive Training Environment (FITE). Most recently, the requirements for cross-cultural training were expressed by the Commander, International Security Assistance Force (COMISAF) in his counter insurgency (COIN) Guidance and as part of his Tactical Directive. In these directives, COMISAF recognizes that strategic goals will be accomplished not by killing the enemy, but by influencing the population—his overriding operational imperative. The goal of immersive training systems was to provide the learner with a realistic, cognitively authentic experience that can be transferred to an operational setting. This required a shift in perspective as well as adaptations to individual and group mental models of tactical problem contexts like COIN or Stability Operations. An independent assessor determined that decision skills and cultural training took place. These conclusions were based on survey data as well as post-training critical incident interviews for verifying changes in individual and group mental models as a result of training.

Keywords: Cross Cultural Cues, Decision Skills, Future Immersive Training Environment, Small Unit Performance, Training Effectiveness

INTRODUCTION

As our adversaries embrace the tenets of asymmetric warfare to achieve their goals, we must continually evaluate how we think about and engage our adversaries. U.S. General Purpose Force (GPF) and Special Operations Forces will be challenged to adapt their doctrine, training, and leader development to prepare for Irregular Warfare challenges. The GPF on the land, predominately U.S. Army and Marines as well as Special Operations Forces, most often confront these adversaries in the form of small unit actions. These U.S. small units operate on the "edge of chaos" against an adaptable and unorthodox enemy. To be as adaptable and to succeed, U.S. small units must learn to be efficiently adaptable in dynamically complex settings.

The emerging operational environments will require highly skilled and adaptive forces that must think about how to negate the effectiveness of adversaries' efforts across the entire threat spectrum while winning the support of civilian populations across the entire range of military operations. This focus requires small units to go beyond training immediate action drills and battle drills to acquiring and practicing the skills used to make decisions, recognizing patterns. sensemaking within problem contexts, and learning and using perspective taking. These learning outcomes affect small unit performance in novel situations like those where cross-cultural interactions are not only necessary but also critical to success

The needs for cross-cultural competence have come under study in recent years especially by researchers working with the ground forces (see for example, Abbe, Gulick, & Herman, 2007; Thornson, & Ross, 2009). Cross-cultural competence has been defined as

> The ability to quickly and accurately comprehend, then appropriately and effectively act in a culturally complex environment in order to achieve the desired effect, without necessarily having prior exposure to a particular group, region, or language.
>
> Selmeski, 2007, p. 12

The development and practice of these abilities have become a part of a Joint Concept Technology Demonstration (JCTD), which has been sponsored by the Joint Forces Command (JFCOM) and supported by a team of developers and researchers. At the core of the JCTD is a mixed-reality training system that leverages advances in simulation-generated objects, animatronics, and cognitive systems engineering. The goal is to practice decision making and sensemaking skills in a cognitively authentic and instrumented setting, where small units experience problem solving in operational contexts like those that make up Stability Operations and COIN environments. In this manner, small units developed the abilities to adapt efficiently through guided practice in tactical as well as metacognitive skills, often as a result of receiving feedback on their performance from highly skilled coaches who possessed mature, highly developed mental models of complex problem settings.

The purpose of our effort was to examine the current roles and challenges faced by small units and their leaders who operate at the edge of chaos. Members of small units must rely on their mental models of operational contexts in order to make sense of cues and factors, recognize patterns or gaps in information, all the while acting in a moral, legal and ethical manner. Their goals might shift rapidly from direct fire tactical engagements with an insurgent to hosting a meeting with village leaders, often doing both concurrently. While neutralizing the insurgent is important, favorably influencing the local population is more so. To address the need to rapidly transition from context to context, a new training solution was required and we proposed a model for individual cultural development that included cognitive, affective, and behavioral skills. We believed that these skills could be practiced, developed and assessed within an immersive training environment.

This initial study examined how to develop training cases where cultural capabilities could be developed through authentic representations of critical cues and factors. To determine whether these capabilities were a result of FITE training, we proposed a series of competency analyses, which were based on critical incident interviews that were part of the post-training assessment process. In this manner we sought to find evidence that individual mental models of the operational contexts had been shaped by the training experience. This evidence was part of the narratives that individuals expressed following the training and represented in their decision demand tables. Further study is needed to determine how and whether these mental models transferred to operational contexts and to better understand how they can be updated in lieu of operational experience.

METHOD

PARTICIPANTS

The proposed sample consisted of 13 Marines and 10 Army enlisted participants who were members of infantry squads grades E3 to E6. The participants were drawn from a small population of individuals who were trained during the Future Immersive Training Environment (FITE) JCTD, which was conducted in August-September 2010. A demographic survey was constructed and administered to select individuals who expressed a willingness to participate in the interviews.   Ten interviews were conducted and all interviews were deemed usable and transcribed. We achieved a sample size consistent with our plan, which was designed to include one-third of those participating in the training.

PROCEDURE

Two data collection trips were made in conjunction with the mixed reality and augmented reality demonstrations which were conducted within the Mixed Reality Infantry Immersive Trainer (IIT) at Camp Pendleton, CA. All interviews were conducted at the IIT at the completion of the small unit training event, which consisted of three battle runs. The interview procedure included demographic information, task diagrams, team ranking, and critical incident elicitation.

ANALYSIS

The analysis of the interviews consisted of two parts. First the interviews were transcribed to avoid biases in recall or notes taken. Secondly, several sweeps of the data were conducted in order to answer several research questions. The first pass through the data was designed to identifying the mission requirements that made up the battle run.   A second sweep was used to determine which experiences stood out as most critical to the learning experience. The third sweep was to conduct a thematic analysis of the cultural factors of the narratives to isolate the learning outcomes which participants reported. The analysts coded the number of instances the skills (i.e., cognitive, affective, and behavioral) were reported by the participants while training. These reports were organized and presented as a decision demand table for later comparison with the source documents. Once the data had been analyzed the analysts identified themes that characterized the training experience. Before the beginning of the thematic analysis, the analysts agreed what constituted themes by 1) reviewing the definitions of the learning outcomes and 2) calibrating their work by cooperatively reviewing two transcripts used for analysis.

FINDINGS

DEMOGRAPHICS

Marine infantry squads are made up of 13 members, an E6 squad leader, three E4 or E5 team leaders, and nine riflemen, E3 and below. Three Marine infantry squads participated in the JCTD. Army infantry squads are made up of 10 members, and E6 squad leader, two E5 team leaders and eight riflemen, E4 and below. Three Army infantry squads participated in the JCTD. All participants were drawn from these squads, five from each Service.

TRAINING CASE DESIGN

At the core of the training experience, we developed a series of training cases that were intended to provide the learners with a cognitively authentic experience which was mission relevant. Each case was designed to develop a cultural-cognitive ability through decision making, perspective taking, and sensemaking or pattern recognition in a dynamically complex environment. By dynamically complex, we meant that the conditions for training were the result of individual actions and decisions. There was a recognizable cause and effect relationship between the decision and outcomes. The tactical situations were shaped by multi-tasking, time pressure, information quality and quantity and new missions.

    The training cases were painstakingly crafted and were based on interviews and narratives of experienced Marines or Soldiers, who had served in Afghanistan. Case development was systematic and centered on training and mission task requirements, both kinetic and non-kinetic. Each case was comprised of several components: a decision demand table, a use case model, a narrative, and a master sequence event list (MSEL).

    A decision demand table was a synthesis of information derived from the narratives of highly experienced individuals who reported critical incidents during a series of interviews. Central to the decision demand table were the dilemmas or decisions which the individual reported, and the interviewer recognized as a valued training experience. A valued training experience was one that was reported across several subjects and appeared to challenge novices, thus requiring practice and strategies for achieving success. For the cultural factors, we sought to learn the key elements of declarative and conceptual knowledge which affected the performance. These factors, combined with individual tacit knowledge, were the bases for critical thinking that took place in a situation. The training cases included this knowledge base as part of the planning process or as embedded cues or factors in the learning environment.

LEVELS OF LEARNING

For the FITE JCTD, we proposed that several levels of learning would operate concurrently to produce a learning outcome. While the principal aim was to provide practice in decision making, planning, and problem solving, the training design would have been incomplete without a cultural component typical of COIN and Stability Operations missions.

We proposed three levels of learning would occur within the FITE: 1) comprehension and recall, 2) analysis and application, and 3) synthesis and evaluation. These levels occurred concurrently and required that each squad member contribute to the learning activity as form of team learning. Within each of these levels, we looked for learning outcomes that would suggest that learning had taken place. For the cultural knowledge and capabilities learning requirements, we portrayed the relationships among the levels of learning and learning outcomes as illustrated in Figure 1.

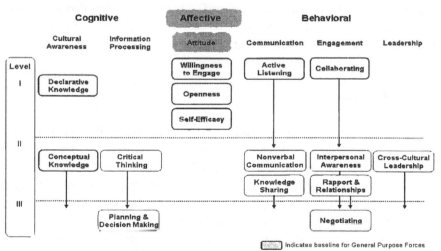

Figure 1. Developmental Model for Cross-Cultural Capability within Small Tactical Units. (Source: Ross, K., Bencaz, N., Thornson, C., and MacNulty, C., Framing Cross-Cultural Competence Learning Outcomes in a Developmental Model, Feb 2010, in press)

The Competency Model provides a framework for understanding the key competencies which interact to support cross cultural tasks. Preparation of the General Purpose Force like the Marine and Army infantry squads who participated in the JCTD was accomplished by practicing a sub-set of skills (shown in grey above). In addition, the squads receive additional practice in Critical Thinking and Planning & Decision Making, which are higher order abilities.

The Competency Model for Cross Cultural Competence also suggests the type of instruction that is typically associated with each level. Level 1, Comprehension and Recall outcomes are best achieved through reading, lecture and coaching; Level 2, Analysis and Application, through facilitated discussion, demonstration, and practical exercise, and Level 3, Synthesis and Evaluation, through problem solving, case studies and interactive simulation

CONTENT ANALYSIS OF COMPETENCE FACTORS

All the transcribed interviews were analyzed for the factors of competence that had previously been identified during our work with the Army. The analysis grouped cognitive, affective, and behavioral skills. Key competencies and learning outcomes were identified, and the analysts aligned them with levels of learning. Some additional factors were added by the analysts. This effort was a preliminary, exploratory analysis to identify trends and not an effort to establish an unequivocal coding scheme and obtain inter-rater reliability. One session was conducted to calibrate the three analysts across the categories. Factors not identified in the Air Force data are still included in the table to facilitate comparison with the Army findings. Shaded cells indicate refinement or addition of new categories.

Cognitive- Themes

The first category of themes is Cognitive. Perspective-taking (the ability to see events as another person sees them) was a large factor as it was in the findings. Simple awareness of cultural differences was the other largest factor expressed by those we interviewed to prepare the training cases. The sample, generally in high contact roles and awareness, was quickly surpassed as other more complex skills and knowledge were attained and valued in their missions.

Other themes we found reflected a nuanced cognitive perspective that might be labeled "systems thinking" and related to a Big Picture Mentality. These factors were not just indicative of cultural awareness but served as an active process to identify and understand "drivers" in the situation. These respondents talked about understanding the social dynamics and organizational issues with the situation.

They described actions such as observing a community to understand family and community dynamics in terms of who is afraid of whom, who is looked at as a leader, informal rank structure, local power structure and family functioning.

Understanding and using a one's knowledge of cultural values when 1) making an assessment of civil considerations to learn what factions, tribes or other important groupings of people are physically located where; 2) understanding the capacity of people to locally undertake or maintain reconstruction projects ("skill and will assessment"); and 3) understanding how local history affects individual perceptions. This theme was identified in the data; however, it was not possible to activate it in the simulation setting which was available.

Much of our research in support of the JCTD was based on people who have deep and rich experiences in Afghanistan and/or Iraq. The ability to adapt to a variety of non-kinetic challenges during a combat patrol was critical. However, practice to achieve a level of cross-cultural competence necessary for multiple non-kinetic contingencies at the squad level was rare. This type of cultural training was not just to obtain competence in one region but where current conflicts demand it. It required a cognitive ability to identify and synthesize a lot of information quickly in preparation for patrols or to conduct a successful engagement with the civil population. The ability to rapidly prepare for a new location is a skill we have not yet thoroughly investigated or mastered.

Affective- Themes

Cross-cultural competence was a "contact sport" and required constant interaction to grow and to succeed in many current roles and missions. Willingness to engage was the primary attitude identified in the sample. Simply being willing to go out and spend time with people and talk to them as long as needed to get the job done was the primary factor in success. Obviously, this attitude must be backed up with the interpersonal skills and cultural knowledge or the ability to observe and gain the knowledge.

The second most important attitude components were an openness and appreciation of new cultural experiences and general openness and respect for people. Cultural openness was also the second most important attitude in the findings and general open-mindedness was also highly discussed by the participants. Cultural openness and open-mindedness were described as being non-judgmental to people doing things differently than we would do them which includes respect for those people—for their experience and position in life, regardless of their illiteracy, poverty or other current circumstances.

Behavioral-Themes

Relationship building and communication skills were the primary elements discussed in the sample. These two other themes were discussed as taking actions to make local development or reconstruction belong to the local people. It was not enough to simply understand, at a cognitive level, what the local people are willing and able to do. One must work to encourage and set the conditions for local participation and ownership and use local expertise to ensure success. The most effective small units demonstrated the ability to go out and visit widely and learn who was in charge of what and how they function. These types of presence patrols were part of the squad's mission to build relationships and understanding the network of agencies, government officials and others who were part of the social fabric of the village setting. This theme was identified in the data; however, it was not possible to activate it in the simulation setting which was available.

The final skill is communication. Many people in the sample seemed to equate the highest level of cultural competence with language proficiency. The civil population and Security Forces were highly impressed by language skills. Language skills can create mission success even when an interpreter was present. This theme was identified in the data; however, it was not possible to activate it in the simulation setting which was available.

CONCLUSION

While the FITE JCTD was intended to provide small units with a facility where they could train decision making, problem solving and other cognitive skills, it also proved an effective setting for developing cross-cultural competence. In the study, we concluded that neither military specialty nor rank can predict levels of cultural competence levels or competence requirements. The current environment requires broad cross-cultural training because nearly anyone can end up in a situation where cultural interaction must occur. The importance of cross-cultural competence and the role in which our military forces interact with the civil population changes as the mission dictates and the phase our forces find themselves. In this study, there was also a lot of variability in competence demonstrated by the participants, a relatively homogeneous group of infantrymen. We sensed through their actions and reports that each participant recognized the importance of cross-cultural competence. The simulation setting was limited in meeting learner expectations.

Division of cross-cultural skills into categories such as interpersonal/communication; awareness of cultural values social and organizational factors; and, more low-level general cultural awareness and knowledge might help to organize training more effectively. The cultural cues were represented adequately. The physical environment was augmented by a small cadre of highly trained role players, who were fluent, native speakers.

The generation of intelligent avatars, natural language interfaces, and fluent native role players in combination with cognitively and culturally authentic training cases help to fill the gap in developing a cultural capability at the small unit level. The requirement for guided practice and coaching in advance of the lived experience would benefit small unit members who must operate at the edge of chaos and efficiently adapt to operational challenges.

REFERENCES

Abbe, A., Gulick, L. M. V, & Herman, J. L. (2007). *Cross-cultural competence in Army leaders: A conceptual and empirical foundation.* U.S. Army Research Institute for the Behavioral and Social Sciences, Study Report 2008-1. Arlington, VA: ARI.

Hardison, C. M., Sims, C. S., Ali, F., Villamizar, A., Mundell, B., & Howe P. (2009). *Cross-Cultural Skills for Deployed Air Force Personnel—Defining Cross-Cultural Performance.* Rand Monograph, http://www.rand.org.

Selmeski, B. R. (2007). *Military cross-cultural competence: Core concepts and individual development.* Kingston: Royal Military College of Canada Centre for Security, Armed Forces, & Society.

HQ International Security Assistance Force Letter (6 Jul 2009), COMISAF's Tactical Directive. Counterinsurgency guidance and imperatives for influencing non-combatant populations. (NATO/US UNCLASSIFIED)

HQ International Security Assistance Force Letter (30 Aug 2009), COMISAF's Initial Security Assessment for the Secretary of Defense. (Redacted- US UNCLASSIFIED)

Ross, K.G., Thornson, C.A., & Arrastia, M. C. (2009). *Final Report: Development of the Cross-Cultural Competence Inventory (3CI).* Directed by Dr. Daniel P. McDonald, DEOMI Technical Report. Available through Cognitive Performance Group.

Ross, K.G., Thornson, C.A., & Bencaz, N. (2010). *Final Report: Framing Cross-Cultural Competence (3C) Learning Outcomes in a Competency Model.* Directed by Dr. Daniel P. McDonald, DEOMI Technical Report. Available through Cognitive Performance Group.

170

Phillips, J. and Ross, W., (2010). *Summary Report*: Process for the Development and Validation of Cognitively Authentic Small Unit Training Cases. (Manuscript).

Chapter 18

FITE - Team Training for Cross-Cultural Decision Making

Jeffery Wilkinson[1], David Holness[2], William Giesey[1]

[1]MYMIC LLC
1040 University Blvd, Suite 100
Portsmouth, VA 23703, USA

[2]Naval Air Warfare Center
Training Systems Division
12350 Research Parkway
Orlando, FL 32826, USA

ABSTRACT

The Future Immersive Training Environment (FITE) provides military instructors with an increased set of enablers to train close combat tasks in a realistic fully immersive cross cultural training environment that creates and reinforces complex decision making skills. Effective decision making under these conditions comes from cognitive dominance which enables the agile leader and high performing unit to maintain situational understanding and work inside the enemy's decision cycle to defeat that enemy.

FITE is incorporating two instructional methods: Team Dimensional Training (TDT) and Think Like a Leader (TLAL) to bring focus to the importance of team behavior and adaptive thinking skills to enhance decision making within the cross-cultural context. FITE is revising the Adaptive Thinking Training Methodology from the Army Research Institute's Think Like a Commander research into the Think Like a Leader (TLAL) system with a revised focus on the dismounted warrior

172

and small unit team.

Both TDT and TLAL are introduced to the trainees early in the FITE training sequence. More importantly instructors are trained on both methods and taught to link the critical themes of each method to key events in the cross cultural scenarios as they facilitate After Action Reviews (AARs). This reinforces the importance of both adaptive thinking and team performance in the conduct of complex cross cultural decision making. The following is an overview of the instructional approaches and the strategies used to integrate them into the FITE program.

Keywords: Cross-cultural Decision Making, FITE, Adaptive Thinking, Team Dimensional Training, Think Like a Leader

INTRODUCTION

Recent Joint Forces Command (JFCOM) Commander's Guidance includes the requirement to "...improve our capability in the irregular fight." This straight forward statement is expanding the focus of the traditional training transformation role of JFCOM down to not only the Special Operations Force/Infantry unit, but to any military command or unit involved in counterinsurgency and stability operations as well. The Future Immersive Training Environment (FITE) Joint Capabilities Technology Demonstration (JCTD) mission is to support the development of immersive training technologies that enhance the maturation of agile, high performing leaders and units. This mission comes with the intent that the first firefight should be no more difficult than the last training simulation. Cognitive dominance is the centerpiece to rationalize investments in these capabilities.

Cognitive dominance in this context is demonstrated by "...full mastery of complex situations confronting enemy commanders who know little of U.S. operations and are losing control of their own forces (Pirnie & Gardiner, 1996).". Information technology does not provide a commander with cognitive dominance but is just one factor in an equation that is human performance centric and focused on exploiting information dominance (Hamid, White, & Gibson). Cognitive dominance sets the conditions for the agile leader and high performing unit to maintain situational understanding and work inside the decision cycle of the enemy commander to effectively apply lethal non-lethal power to defeat that enemy.

The FITE provides military trainers with an increased set of enablers to train squad level close combat tasks in a realistic cross-cultural fully immersive training environment. This training creates the conditions to reinforce complex decision making skills. FITE is focused on the technologies and methods that create cross-cultural conditions in support of human-centric decision activities at the individual and team levels. Effective decision making under these conditions comes from cognitive dominance supported by effective team performance.

Training for cognitive dominance requires more than information awareness augmented by information technology. It includes deliberate practice intended to internalize processes, methods, and proactive information flow behavior by members of the team. Cognitive dominance also requires adaptive thinking. Adaptive thinking is the behavior of an individual who is confronted by unanticipated circumstances during a planned military operation. The presentation of cross-cultural events that include voice interaction among the trainees and between the trainees and role playing avatars enables FITE to drive complex scenarios that challenge team skills, adaptive thinking, sense making and decision making at all levels.

A training system that enables one to simply experience complex cross-cultural scenarios and the challenge of decision making under those conditions is not necessarily efficient and may not be effective. To maximize the potential for learning of both team performance and adaptive thinking skills within the cross-cultural context, FITE is incorporating two instructional methods: Team Dimensional Training (TDT) and Think Like a Leader (TLAL). FITE is adapting, for use with dismounted squads, the Army Research Institute's (ARI) TLAC method into the Think Like a Leader system.

TEAM DIMENSIONAL TRAINING

Team Dimensional Training (TDT) is an empirically-based team strategy that facilitates team self-correction and effective team processes to improve team performance. The requirement to explore team training came as a result of the tragic USS Vincennes incident in which the U.S. warship mistook a civilian jetliner for an attacking Iranian fighter plane. After the USS Vincennes accident, the Office of Naval Research sponsored a 15-year research program named Tactical Decision Making Under Stress (TADMUS), which investigated the question 'what makes a team of experts into an expert team?' One primary TADMUS result was that four teamwork processes and 11 subcategories were related to effective decision making and expert team performance (Smith-Jentsch, Zeisig, Action & McPherson, 1998). These four teamwork processes include:

- Information Exchange
- Communication Delivery
- Supporting Behavior
- Initiative/Leadership

Information Exchange involves knowing what information to pass to whom and when. The subcategories of Information Exchange are: using available sources, passing information before being asked, and providing situation updates.

Communication Delivery focuses on how information is shared. The subcategories of Communication Delivery are: using correct terms, providing complete reports, using brief communications, and using clear communications. Supporting Behavior entails compensating for other teammates in order to achieve team objectives. The subcategories of Supporting Behavior are: correcting team errors, and providing and requesting backup. Initiative/Leadership encourages anyone on the team to demonstrate behaviors of leadership, not just the formal leader. The subcategories of Initiative/Leadership are: providing guidance and stating priorities (Smith-Jentsch, Jentsch, Payne, & Salas, 1996).

TDT provides a standardized method of guiding teams to self-correct and set goals during After Action Reviews (AAR) to improve performance. TDT also reduces the workload of instructors and empowers trainees to build more autonomous teams. Most importantly, TDT does not substitute for, but complements, the tactical training already in place.

TDT is traditionally delivered through a TDT Workshop conducted over a period of 2-3 days, depending on the number of participants. Participants are usually instructors, facilitators, or team leaders. The workshop has been developed to instruct participants on the dimensions, TDT cycle, and the facilitation skills needed to conduct an effective AAR. The TDT Workshop consists of two sections—the classroom presentation and the role-play exercises. During the classroom section, the participants are presented with information about the history behind TDT and the research tied directly to its development. Then, they are introduced to the four dimensions through various interactive activities. The classroom section concludes with an overview of the TDT training cycle and a discussion on facilitation skills that reinforce team self correction. The skills learned and practiced during the classroom are applied and evaluated in high-fidelity role-play scenarios with actual teams-in-training. Each participant typically receives two occasions to use the facilitation skills during the training cycle. The TDT instructors guide this process by providing feedback on the briefing performance and participants to set goals for the subsequent training cycle.

THINK LIKE A LEADER

The new Think Like a Leader approach to training Adaptive Thinking skills to dismounted warriors is based on the effective Think Like a Commander program developed by the Army Research Institute and employed with great success in the Armor School (Shadrick, Crabb, Lussier, Burke, 2007). The term Adaptive Thinking has been used to describe the cognitive behavior of one who is confronted by unanticipated circumstances during the execution of a planned military operation (Lussier, Ross, & Mayes, 2000). It refers to the thinking the leader, Marine, or Soldier must do to adapt operations to the requirements that are unfolding. Adaptive Thinking is a behavior. It involves the skilled application of knowledge under challenging performance conditions; it is not the knowledge itself. The ability to apply this knowledge expertly under battle conditions, however, demands training and practice, until the framework of thinking becomes automatic. To address this requirement a training method called The Adaptive Thinking Training Methodology (ATTM) was developed. The ATTM served as the foundation for ARI's TLAC program. The TLAL training program, based on the ATTM, was created to take full advantage of the capabilities presented by the FITE system to enable trainers to focus on this critical skill as part of the employment of the FITE system.

Because success on the battlefield will depend on the ability to think creatively, decide promptly, exploit technology, adapt easily, and act as a team, there is a need to create effective methods and techniques to ensure that future leaders, Soldiers, Marines and teams possess these qualities. Adaptive Thinking can be thought of as thinking on one's feet framed in terms of the battlefield. In this viewpoint adaptive thinking is not so much a type of thinking (i.e., creative, lateral, or out-of-the-box) but rather it is defined by the conditions under which it occurs. (Lussier, Shadrick, Prevou, 2003). In the FITE system these conditions include cross-cultural engagements that can have significant secondary and tertiary impacts on the tactical mission.

As in all operations, leaders begin with a detailed plan but as it executes, one must constantly make adjustments, alter timing, take advantage of unforeseen opportunities, and overcome unexpected difficulties. In short, one must adjust or adapt the plan. The thinking that underlies these decisions is not made in isolation or in a calm reflective environment. It must be done while performing as a Soldier or Marine: assessing the situation, scanning for new information, anticipating the next critical event, making decisions, and monitoring the progress of multiple activities of a complex plan. On the battlefield, multiple events compete for attention. No easy guidelines can be applied (Lussier, Shadrick, Prevou, 2003).

In many fields where expertise has been systematically studied, including chess, music and sports, development beyond advanced novice level requires large amounts of deliberate practice and good coaching. How does deliberate practice differ from performance or from casual exercise? Here are some characteristics that distinguish deliberate practice (Lussier, Shadrick, Prevou, 2003):

- Repetition. Task performance occurs repetitively rather than at its naturally occurring frequency.
- Focused feedback. Task performance is evaluated by the coach or learner during performance.
- Immediacy of performance. After corrective feedback on task performance there is an immediate repetition so that the task can be performed more in accordance with expert behavior.
- Stop and start because of the repetition and feedback, deliberate practice is typically seen as a series of short performances rather than a continuous flow.
- Emphasis on difficult aspects. Deliberate practice will focus on more difficult aspects.

- Focus on areas of weakness. Deliberate practice can be tailored to the individual or team and focused on areas of weakness.
- Conscious focus. Expert behavior is characterized by many aspects being performed with little conscious effort. Such automatic elements have been built from past performances and constitute skilled behavior.
- Work vs. play. Characteristically, deliberate practice feels more like work and is more effortful than casual performance. The motivation to engage in deliberate practice generally comes from a sense that one is improving in skill.
- Active coaching. Typically, a coach must be very active during deliberate practice, monitoring performance, assessing adequacy, and controlling the structure of training.

A key element of the Adaptive Thinking Training Methodology (ATTM) interaction between the coaches and the learners are what are called themes. These are thinking behaviors (the "how to think" element) that are characteristic of high-level tactical experts. They are the elements of expert tactical thinking that the coaches are observing and the students are modeling. While well known to most military professionals and understood at a conceptual level, these behaviors are often not exhibited during actual exercises. They have not become automatic and thus, when attention is focused, as it should be, on the specific situation confronting them, the behaviors are omitted. This program adapted the TLAC themes and developed the TLAL themes to support small unit tactical decision making:

- Focus on the Mission and Higher's Intent
- Model a Thinking Enemy/Consider the Terrain
- Use All Available Assets
- See the Big Picture/Visualize the Battlefield
- Consider Contingencies/Remain Flexible

This is a good set of themes for individual warriors and small units to focus on for the following reasons. First, these behaviors are characteristic of expert tactical decision-makers. Second, the themes are familiar to most Soldiers and Marines. Third, the set of themes describes thinking actions that can be characterized as "how to think" or "what to think about" rather than "what to think." Fourth, and very importantly, these themes represent thinking behaviors that are relatively consistent over a wide range of situations - both tactical and cross-cultural.

The TLAL themes represent patterns of thought, characteristic of good decision-makers, which are typically lacking at the novice level, particularly when they are performing in a stressful environment. Experienced warfighters must have extensive knowledge and have learned to apply that knowledge skillfully through practice. However, study and practice alone do not ensure expertise. Repetitious performance alone only ingrains habits; it does not necessarily lead to the improvement of performance. Deliberate practice involves the performance of exercises specifically designed to improve performance.

INTEGRATION

Incorporation of the TDT and TLAL constructs into the FITE training system required adjustments in their respective methods of instruction, the content of the FITE structured scenarios, and the conduct of the AAR. Both TDT and TLAL are focused on empowering the individual warrior to be proactive in the face of complex decision events. This directly supports a critical intended purpose of FITE which is to provide a more effective training environment for complex cross-cultural decision making. Taking full advantage of the TDT/TLAL instruction within FITE required identification of key decision points in each scenario and development of specific probes, as applicable, to be used by the trainer in the AAR to facilitate self-assessment and reinforce the behavior themes TDT and TLAL. This a priori development of structured queries that link specific scenario decision events to the training can enable more rapid post exercise development of the AAR and provide the trainer with clear focused probes to help ensure the discussion stays on track.

Modifications were made to the TDT Workshop to prepare the FITE trainees and instructors to utilize the TDT methodology in a virtual training environment. The trainees participated in a condensed version of the classroom presentation that focused on defining the dimensions and the role that TDT would play in the FITE training cycle. After the classroom presentation, the trainees were able to apply their knowledge of TDT by completing the TDT dimensions tabletop exercise.

The FITE instructor training mirrored the content and the scope of the traditional TDT classroom presentation. The instructors were introduced to the background and history of TDT, the dimensions, and the three stages of the TDT cycle: prebriefing, assessing performance, and debriefing teams. The instructor presentation concluded with training on effective facilitation skills. Specifically, they learned how to document and categorize positive and negative behavior examples, orchestrate briefing sessions by recapping key scenario events, asking for concrete examples, and stressing the importance of potential impact. Finally, the instructors were asked to review scripted templates containing TDT probes based on critical events in the FITE scenarios. The probes were designed to be used as a reference document to assist in ensuring that trainees discussed both tactics and teamwork during debriefs. The TDT classroom presentation materials and probes were developed and validated with extensive assistance from Army and Marine Corps personnel.

TLAL was also introduced in a classroom session early in the FITE training sequence. This instruction included an initial focus on the underlying concept of Adaptive Thinking and its relationship to effective decision making under complex conditions. The TLAL themes were introduced and explained in terms of cross-cultural tactical situations where these themes set the conditions for decision events. The TLAL concept instruction was followed by 2-3 practical exercises using Virtual Battle Simulation 2 (military simulator training software) based vignettes that include a tactical situation that is made more complex by cross-cultural cues, events, and environmental conditions. Trainees, as a team, assess each vignette with respect to the TLAL themes with the intent of developing greater situational understanding that will inform their next decision.

The instructors were trained on TLAL in the same fashion as the trainees. These instructors, along with the scenario developers, are critical to the effort to link the TLAL themes to the key events in the scenarios and develop probing questions to be used in the AAR as they facilitate after action reviews. The structured scenarios in FITE are designed to maximize trainee freedom to make decisions that, in turn, will have secondary and tertiary impacts as the scenario unfolds. The use of the term structure in this sense is more about the structure of the events and cues presented to the trainee in a manner that drives cross-cultural decision making and not about a rigid series of events. Structure also refers to the nature of the relationship between the training objectives, the complex events that set the conditions for reaching these objectives, and the a priori development of the AAR probes associated with each event that facilitates trainee and team reflection on the decisions made with respect to these events. These are all critical to a regime of deliberate practice.

EXECUTION

During the rehearsal for the FITE Operational Demonstration, both TDT and TLAL instruction was provided to the Marine squad. The instructor was briefed and the scenario specific AAR probes were reviewed. The instructor was a highly experienced trainer who had a deeply ingrained mental model of the structure and conduct of an AAR and had difficulty adapting to a model that focused on cross-cultural decision making and incorporated the TDT/TLAL themes. The instructor had difficulty collecting examples of TDT behaviors during the FITE scenarios and did not seek out and use the TDT/TLAL probes during the FITE AARs. This rehearsal demonstrated that it is very difficult to rapidly apply new concepts and techniques when conventional ways of doing business are deeply ingrained. The experience provided a number of lessons learned that will be incorporated into future FITE demonstrations.

As an excursion, during this FITE Operational Demonstration rehearsal, in the final exercise the squad leader for the training unit was moved to the trainer role and one of the team leaders was given the squad leader position. As the squad leader was not highly experienced in the conduct of a formal AAR, the FITE support team coached him on the process and provided the TDT/TLAL probes for his use during the AAR. The squad leader identified those areas of focus for the AAR, selected the TDT/TLAL probes to use, and rehearsed with the materials provided by the support team. The squad leader then conducted the AAR and employed the TDT/TLAL probes to facilitate discussion among the trainees and to bring focus to specific events that are well addressed by TDT/TLAL. This excursion did not validate the efficacy of this approach to improve the training of cross cultural decision making but it did demonstrate that TDT and TLAL can serve as a basis for the AARs conducted after training scenarios focused on cross cultural decision making.

FUTURE CONSIDERATIONS

As the experienced instructor had difficulty collecting examples of TDT behaviors during the FITE scenarios and did not seek out and use the TDT/TLAL probes during the FITE AARs the instructor training will be restructured to account for this deficiency by providing more opportunities to practice collecting TDT/TLAL related observations. The structure of the TDT/TLAL probes will be reevaluated to ensure that the instructors are able to facilitate a discussion that focuses on both tactics and teamwork skills. It is likely that the probes are not an issue but instead the ease of identifying which probes to use and not having them readily available in a rehearsed sequence during the AAR may have driven the instructor to fall back on old habits. At the upcoming International Conference on Cross-Cultural Decision Making, this team will report on the results from implementing an updated TDT/TLAL protocol in the FITE Operational Demonstrations.

REFERENCES

Hamid, S., White, I., and Gibson, C., (2001). *"Challenges for Joint Battlespace Digitization (JBD)"*, Defence Evaluation and Research Agency (DERA), Portsdown, Fareham, PO17 5EU, ENGLAND

Lussier, J. W., Ross, K. G., & Mayes, B., (2000). *"Coaching Techniques for Adaptive Thinking"*, proceedings of the 2000 Interservice/Industry Training, Simulation and Education Conference, Orlando, FL.

Lussier, J. W., Shadrick, S. B., Prevou, M. I., (2003). *"Think Like a Commander Prototype: Instructor's Guide to Adaptive Thinking"*, U.S. Army Research Institute (ARI) Research Product 2003-02.

Shadrick, S.B, Crabb, B.T., Lussier, J.W., Burke, T.J., (2007). *"Positive Transfer of Adaptive Battlefield Thinking Skills"*, U.S. Army Research Institute (ARI) Research Report 1873.

Smith-Jentsch, K.A., Jentsch, F.G., Payne, S., & Salas, E. (1996). *"Can Pretraining Experiences Explain Individual Differences in Learning?"*, in Journal of Applied Psychology, 81, 110-116.

Smith-Jentsch, K.A., Zeisig, R.L, Acton, B., & McPherson, J.A. (1998), *"Team dimensional training: A strategy for guided team self-correction"*, In J.A. Cannon-Bowers & E. Salas (Eds.), Making decisions under stress: Implications for individual and team training. (pp. 271-297). Washington, DC: APA.

Pirnie, Bruce and Gardiner, Sam B. (1996), *"An Objectives-Based Approach to Military Campaign Analysis"*, Santa Monica, CA, National Defense Research Institute, RAND Corp., 111 p. (RAND report, MR-656-JS).

CHAPTER 19

Translating Science into Practice: Developing a Decision Making Training Tool

Elizabeth H. Lazzara, Eduardo Salas, David Metcalf, Clarissa Graffeo, Sallie Weaver, Kyle Heyne, William Kramer

Institute for Simulation and Training
University of Central Florida

ABSTRACT

Irregular warfare is ill-defined, ambiguous, and complex; therefore, effective and efficient decision making is a critical component for successful performance. To ensure that Marine's are well equipped with such skills, we have developed a simulation-based training tool for Combat Hunter to supplement existing classroom training. Therefore, the purpose of this chapter is to illustrate the development of this mobile training tool as well as offer a set of best practices for developing decision making training.

Keywords: Decision making, Simulation-based training, Mobile Device

INTRODUCTION

Irregular warfare (IW) is non-kinetic warfare, focused on maintaining influence over a population as opposed to adversarial armed forces or territories. Due to the inherent ambiguity and complexity, situational awareness, cue recognition, and decision making are integral elements of the cognitive competencies required for IW. To thrive under these circumstances, it is essential that today's Marines are competent at recognizing relevant cues and patterns as well as executing decisions efficiently.

To cultivate the skills necessary for IW, the Marine Corps implemented the Combat Hunter Program to assist Marines fighting in Iraq and Afghanistan. Combat Hunter teaches cue recognition and decision making skills inherent in naturalistic settings to approximately 40 students every two weeks in a traditional classroom setting. However, the training is short and learners are bombarded with an excessive amount of knowledge, making retention and transfer of training difficult.

Research suggests simulation as an effective instructional tool to facilitate learning and accelerate expertise. Expertise in cue recognition and decision making can be developed by providing learners with opportunities for practice as well as offering timely, constructive, and diagnostic feedback. Consequently, a simulation component would be an advantageous supplement to the existing Marine Corps training program. Therefore, the purpose of this chapter is twofold: 1) to illustrate a simulation tool designed to improve the current Combat Hunter Training Program and 2) to offer a set of best practices for developing advanced decision making skills.

THE SCIENCE

To begin designing a low-cost training tool to supplement current Combat Hunter Training, we reviewed existing evidence regarding how to best develop and train advanced decision making skills. This review produced a comprehensive list of best practices (see Table 1) organized according to delivery method (i.e., information-based, demonstration-based, practice-based, etc.).

METHODOLOGY

Our method for conducting the review involved searching in psychology literature as well as other relevant disciplines. Key search terms included *critical thinking training, decision making training, situational awareness training, pattern recognition training, cue recognition training, and metacognition training.* We reviewed and coded 80 articles to extract key themes. These key themes that derived from qualitatively reviewing the literature resulted in a list of best practices. The primary areas in which we found best practices involve training critical thinking, cue recognition, decision making, meta-cognition, pattern recognition, situational awareness, and sensemaking. In addition, we also incorporated some general training practices that should be valuable for Combat Hunter Training. We will begin by describing the characteristics of efficient decision makers.

DECISION MAKING DEFINED

Combat hunter involves assessing the situation, making decisions, and generating problem solutions in a naturalistic environment; therefore, it is essential that trainees become efficient decision makers. According to Cannon-Bowers and Bell (1997), effective decision makers are characterized as the following: flexible, quick, resilient, adaptive, risk taking, and accurate. In other words, decision makers must be able to adapt in uncertain, evolving environments. Thus, expert decision makers should have multiple strategies to utilize when responding to different cues and problems. Additionally, decision makers should be capable of making decisions quickly even when the stakes are high. This is particularly important since time pressure is a common characteristic when operating in the natural environment (Orasanu & Connolly, 1993). Furthermore, expert decision makers should be able to complete tasks under stress without experiencing performance decrements (Means et al., 1993). Moreover, decision makers must involve a process of continually assessing and modifying strategies by identifying the appropriate decision strategy and adjusting the strategy accordingly. Lastly, expert decision makers should be capable of determining the risks associated with course of action as well as the weight of errors against any possible gains. Ideally, expert decision makers are capable of performing all of the necessary skills accurately. In order to become an expert decision maker, one must have skills in the following areas: metacognition, critical thinking, cue recognition, pattern recognition, sensemaking, and situation awareness.

Table 1 Decision Making Best Practices

Best Practice		Comments	Source
Training Design			
Best Practice 1: Design training scenarios based on the experiences of experts	o	Identify the cues experts attend to through structured interviews, observations and video analysis, think-aloud verbal reports, and recording of eye movements	Means, Salas, Crandall, & Jacobs, 1993; Salas, Cannon-Bowers, Fiore, & Stout, 2001
	o	Gather "war stories" from experts to develop training scenarios	
	o	Create a blog or wiki in which trainees exchange war stories to reinforce training	
Best Practice 2: Match training	o	Align training in contexts that mirror or closely approximate those of the actual environment to	Cannon-Bowers & Bell,

scenarios to appropriate environmental context	improve critical thinking, cue recognition, and decision making skills.	1997; Cohen et al., 2006, Klein, 1997; Rosen, Salas, Lyons, & Fiore, in press; Oalas et al., 2001
Best Practice 3: Provide trainees with prompts	o Prompt individuals to attend to relevant cues rather than allowing them to discover them for themselves to ensure that trainees learn the correct cues to attend to o Incorporate automated prompts to allow for reflection to facilitate developing metacognitive skills	Gama, 2004; Salas et al., 2001

Information-Based Strategies

Best Practice 4: Provide multiple methods of presenting information	o Present material in multiple formats (i.e. text, diagrams, video) to increase the likelihood of retention o Supplement the text with diagrams (i.e. graphical representations of abstract concepts) when training is text heavy. o Outline training content to highlight the key learning principles to help develop shared mental models and situational awareness	Cannon-Bowers & Bell, 1997; Cohen et al., 2006; Cuevas et al., 2002, 2004; Stout, Cannon-Bowers, & Salas, 1996, 1997

Demonstration-based Strategies

Best Practice 5: Augment written learning objectives with videotaped behavioral demonstration	o Include video demonstrations to assist trainees in identifying key cues to attend to	Mann & Decker, 1984
Best Practice 6: Provide demonstrations of expert behaviors	o Have instructors demonstrate their metacognitive processes (i.e. how they think about a problem), decision strategies, and cues attended to to facilitate the development of trainees' metacognitive skills, decision making skills, and situational awareness	Klein, 1997; Schaw, 1998; Stout et al., 1996, 1997
Best Practice 7: Highlight subtle cues	o Highlight subtle cues through several approaches such as displaying behavior out of context, exaggerating the behavior, repeating the behavior frequently, or including written learning points	Mann & Decker, 1984; Salas et al., 2001

Practice-based Strategies

Dest Praclice 8: Provide trainees with guided practice opportunities	o Have trainer point out the cues, processes, and critical steps that trainees should attend to as the trainee engages in practice. o Guided practice has been identified as key to developing critical thinking skills, cue recognition skills, decision making skills, metacognitive skills, and situational awareness.	Cohen et al., 2006; Means et al., 1993 Osman & Hannafin, 1992; Rosen et al. in press; Salas et al., 1995, 2001; Schaw, 1998; Stout et al.,

184

Best Practice	Description	References
Best Practice 9: Provide multiple practice opportunities	o Match between the cues and patterns trainees are exposed to in practice to the cues and patterns trainees will be exposed to outside of training, o Expose trainees to a variety of contrasting stimuli in order for them to be able to adapt to changing situations	1996/1997). Means et al., 1993; Norman et al., 2007; Salas et al., 2001; 2005; Schmidt, Norman, & Boshuizen, 1990; Stout, 1996, 1997; Zsambok & Klein, 1997
Best Practice 10: Present pictures and have trainees identify the relevant cues	o Determine which cues are significant and should be attended to o Develop skills by having trainees begin with easy pictures and progress to difficult pictures.	Norman, Young, & Brooks, 2007; Shapiro, Rucker, & Beck, 2006

Feedback

Best Practice 11: Provide feedback as the scenario progresses as well as after each training module	o Provide feedback to foster critical thinking training, cue recognition training, metacognitive training, decision making training, pattern recognition training, sense making, and situational awareness. o Feedback should highlight the cues and strategies that are most important.	Behrmann & Ewell, 2003; Cohen et al., 2006; Louis, 1980; Means et al., 1993; Osman & Hannafin, 1992; Rosen et al., in press; Salas et al., 1995, 2001; Stout et al., 1996, 1997

Measurement

Best Practice 12: Incorporate open-ended questions	o Engage trainees in alternative hypothesis generation (i.e. coming up with multiple possible explanations for behaviors instead of relying on only one explanation) to improve decision making	Ark, Brooks, & Eva, 2006; Eva, Hatala, LeBlanc, & Brooks, 2007
Best Practice 13: Incorporate fill-in-the-blank questions	o Include fill-in-the-blank question stems such as "how are ___ related to ___" to increase metacognitive activity	King, 1989, 1992; King & Rosenshine, 1993

Combined, these skills create an expert decision maker, which is fundamental for IW and specifically for the Combat Hunter Program. Consequently, the science behind training these skills should be embedded within any training tool. The following section will detail the Combat Hunter simulation tool, which is rooted in the scientific evidence previously mentioned.

COMBAT HUNTER PROFILING PART TASK TRAINER MINIGAME (CHPPTTM): FUNCTIONALITY AND INTERFACE

CHPPTTM is a lightweight training tool designed to foster cue recognition and decision making. The tool is intended as a supplement to the existing Combat Hunter profiling training, to address its short time frame and large information base by providing a resource for additional learning and reinforcement. Instructors can use CHPPTTM as a supervised classroom exercise or assign students to review additional scenarios outside the classroom. Students can also revisit completed scenarios, or access new ones, for reinforcement and self-evaluation following completion of the training program. Additional opportunities for use as a precursor to lane training, or launched from a virtual situation, are also being explored.

As noted, the design of CHPPTTM was grounded in our scientific review. Trainees are presented with a number of photographs or video clips of potentially dangerous locations and individuals. Use of photographs and videos of real-life situations helps to position the training materials firmly in the real world context as recommended by best practice 2 and also fulfills best practice 10, which advises presenting trainees with pictures and having them identify relevant cues. The variety of scenarios ties back to the research finding that the most effective training provides multiple practice opportunities (best practice 9). Additionally, including static photographs as well as video clips connects back to providing multiple methods of presenting information (best practice 4). After each content item is presented, trainees must identify items of interest, classify them into one or more of six domains, (e.g. atmospheric, proxemic, etc.) and make an action decision (e.g., let go, contact, capture, or kill). The tool then allows students to compare their responses to expert responses by displaying them alongside each other, with cues marked to highlight subtle cues as advised in best practice 7 (Figure 1). A score is assigned by the system for each scene, as well as an aggregate score for the overall scenario based on the percentage of cues that trainees are able to correctly identify.

Figure 1. Evaluation screen

SCENARIO AUTHORING

CHPPTTM scenarios consist of one or more photos or videos chained together in sequence, along with their associated correct response keys. Instructors can update the database with new scenarios, images, videos, and answer keys as desired. The instructor creates a scenario and adds any number of content items. Image or video content is uploaded along with a marked key image—currently created manually, though future versions will allow dynamic tagging of cue areas on the image. The instructor enters correct cue response information for each item, along with a scoring value. Once the content is added, the instructor can restrict access to their scenarios to specific user groups. This allows use by multiple instructors or across branches while retaining control over how widely protected IP or sensitive content is shared. See Figure 2 for screenshots of the CHPPTTM authoring interface.

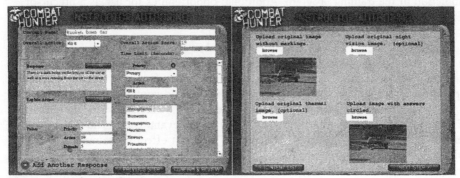

Figure 2. Authoring Screen

Allowing instructors to create their own scenarios presents a fast and efficient method of enacting the first best practice recommendation, to design scenarios based on expert experiences. It also facilitates the addition of new profiling information or scenarios targeting current issues, which will help ensure that the tool will remain up to date.

In addition to flexible scenario development and deployment, the program design offers customization for the type of cue recognition or profiling training. Classification options for scenario responses, such as the six domains or four available action decisions, are stored in the database, thus, enabling the shell to dynamically populate menu items. This allows administrators to swap out cue classification systems or decision frameworks for alternate methods used in different training programs or requirements. This enables expansion to other agencies, either carrying the Combat Hunter profiling concepts outside of the Marines or providing a tool for other situational awareness and profiling frameworks, such as police or FBI crime scene analysis and profiling.

CONCLUSION

CHPPTTM offers a scientifically rooted, simulation-based approach to developing the core competencies essential for IW (i.e., cue recognition, pattern recognition, decision making). Simulation-based approaches to training can supplement existing curriculum by providing vital opportunities to practice and refine new skills. We highlight not only the CHPPTTM itself, but also an approach to developing simulation-based training rooted in the science of training and simulation in order to provide guidance to future training designers and developers.

ACKNOWLEDGEMENTS

This work was supported by funding from US Joint Forces Command via contract with RDECOM STTC (W91CRB08D0015). All opinions expressed in this paper are those of authors and do not necessarily reflect the official opinions of The University of Central Florida, US Army Research Development and Engineering Command, US Joint Forces Command, and the Joint Training Integration and Evaluation Center.

REFERENCES

Ark, T. K., Brooks, L. R., & Eva, K. W. (2006). Giving learners the best of both worlds: Do clinical teachers need to guard against teaching pattern recognition to novices? *American Medicine, 81*, 405-409.

Behrmann, M., & Ewell, C. (2003). Expertise in tactile pattern recognition. *Psychological Science, 14*, 480-486.

Cannon-Bowers, J. A. & Bell, H. H. (1997). Train decision makers for complex environments: Implications of the naturalistic decision making perspective. In C. E. Zsambok & G. Klein (Eds.), *Naturalistic decision making* (pp.99-109). Mahwah, NJ: Lawrence Erlbaum Associates.

Cohen, M. S., Freeman, J. T., & Thompson (2006). Critical thinking skills in tactical decision making: A mdel and a training strategy. In J. A. Cannon-Bowers & E. Salas (Eds.), *Making decision under stress: Implications for individual and team training* (pp.155-189). Washington, DC: American Psychological Association.

Cohen, M. S., & Thompson, B. B. (2001). Training teams to take initiative: Critical thinking in novel situations. In E. Salas (Ed.), *Advances in human performance and cognitive engineering research* (Vol. 1, pp. 251-291). Greenwich, CT: JAI.

Cuevas, H. M., Fiore, S. M., Bowers, C. A., & Salas, E. (2004). Fostering constructive cognitive and metacognitive activity in computer-based complex task training environments. Computers *in Human Behavior, 20*, 225-241.

Cuevas, H. M., Fiore, S. M., & Oser, R. L. (2002). Scaffolding cognitive and metacognitive processes in low verbal ability learners: Use of diagrams in computer-based training environments. *Instructional Science, 30,* 433-464.

Druckman, D. & Bjork, R. A. (1994). *Learning, remembering, believing: Enhancing Human Performance.* Washington, DC: National Academy Press.

Endsley, M. R. (1995). Toward a theory of situation awareness in dynamic systems. *Human Factors, 37*(1), 32-64.

Eva, K. W., Hatala, R. M., LeBlanc, V. R., & Brooks, L. R. (2007). Teaching from the clinical reasoning literature: combined reasoning strategies help novice diagnosticians overcome misleading information. *Medical Education, 41,* 1152-1158.

Flavell, J. H. (1979). Metacognition and cognitive monitoring: A new area of cognitive-development inquiry. *American Psychologist, 34,* 906-911.

Ford, J. K., Smith, E. M., Weissben, D. A., Gully, S. M., & Salas, E. (1998). Relationships of goal orientation, metacognitive activity, and practice strategies with learning outcomes and transfer. *Journal of Applied Psychology, 83*(2), 218-233.

Gama, C. (2004) Metacognition in interactive learning environments: The reflection assistant model. In *Proceedings of 7th Conference on Intelligent Tutoring Systems,* 668-77. Berlin: Springer-Verlag.

Halpern, D. F. (1996). *Thought and knowledge: An introduction to critical thinking.* Hillsdale, NJ, England: Lawrence Erlbaum Associates, Inc.

King, A. (1989). Effects of self-questioning training on college students' comprehension of lectures. *Contemporary Educational Psychology, 14,* 366-381.

King, A. (1992). Facilitating elaborative learning through guided student-generated questioning. *Educational Psychologist, 27,* 111-126.

King, A & Rosenshine, B. (1993). Effects of guided cooperative questioning on children's knowledge construction. *Journal of Experimental Education, 61,* 127-148.

Klayman, J. (1988). Cue discovery in probabilistic environments: Uncertainty and experimentation. *Journal of Experimental Psychology: Learning, Memory, and Cognition, 14,* 317-330.

Klein, G. (1989). Recognition primed decisions. In W. Rouse (Eds.), *Advances in man-machine systems research* (Vol. 5, pp. 47-92). Greenwich, CT: JAI Press.

Klein, G. A. (1993). A recognition-primed decision (RPD) model of rapid decision making. In G. A. Klein, J. Orasanu, R. Calderwood, C. E. Zsambok (Eds.), *Decision Making In Action: Models and Methods* (pp. 138-147). Westport, CT, US: Ablex Publishing

Klein, G. (1997). An overview of naturalistic decision making applications . In C. E. Zsambok & G. Klein (Eds.), *Naturalistic decision making* (pp. 49-59). Mahwah, NJ: Lawrence Erlbaum Associates.

Klein, G., Moon, B., & Hoffman, R. (2006, Jan 1). Making sense of sensemaking 1: alternative perspectives. *IEEE Intelligent Systems* .

Klein, G., Phillips, J. K., Rall, E. L., Peluso, D. A. (2007). A data-frame theory of sensemaking. In R. Hoffman (Ed.), *Expertise out of context* (pp. 113-155.). New York: Lawrence Erlbaum Associates.

Louis, M. R. (1980). Surprise and Sensemaking: What newcomers experience in entering unfamiliar organizational settings. *Administrative Science Quarterly, 25,* 226-251.

Mann, R. B., & Decker, P. J. (1984). The effect of key behavior distinctiveness on generalization and recall in behavioral modeling training. *Academy of Management Journal, 27,* 900-910.

Means, B., Salas, E., Crandall, B. & Jacobs, T.O. (1993). Training decision makers for the real world. In G. Klein, J. Orasanu, R. Calderwood & C.E. Zsambok (Eds.), *Decision-making in action: Models and methods* (pp. 306-326). Norwood, NJ: Ablex.

Norman, G., Young, M., & Brooks, L. (2007). Non-analytical models of clinical reasoning: The role of experience. *Medical Education, 41,* 1140-1145.

Orasanu, J. M. (1990). *Shared mental models and crew decision making* (Tech. Rep. No. 46). Princeton, NJ: Princeton University, Cognitive Sciences Lab.

Orasanu, J., & Connolly, T. (1993). The reinvention of decision making. In G. A. Klein, J. Orasanu, R. Calderwood, & C. E. Zsambok (Eds.), *Decision making in action: Models and methods* (pp. 3-20). Norwood, NJ: Ablex Publishing.

Osman, M. E. & Hannafin, M. J. (1992). Metacognition research and theory: Analysis and implications for instructional design. *Educational Technology Research and Development, 40*(2), 83-99.

Rosen, M.A., Salas, E., Lyons, R., & Fiore, S.M. (in press). Expertise and naturalistic decision making in organizations: Mechanisms of effective decision making in complex environments. In G.P. Hodgkinson, & Starbuck, W.H. (Eds.), *The Oxford Handbook of organizational decision making: Psychological and management perspectives.* Oxford, UK: Oxford University Press.

Salas, E., Cannon-Bowers, J. A., Fiore, S. M., & Stout, R. J. (2001). Cue-recognition training to enhance team situation awareness. In M. McNeese, E. Salas, & M. Endsley (Eds.), *New trends in cooperative activities: Understanding system dynamics in complex environments* (pp. 169-190). Santa Monica, CA: Human Factors and Ergonomics Society.

Salas, E., & Priest, H.A. (2005). Team Training. In N. Stanton, A. Hedge, K. Brookvis, E. Salas, & H. Hendrick (Eds). *Handbook of Human Factors and Ergonomics Methods.* (pp.44-1 through 44-7). London: Taylor & Francis.

Salas, E. Prince, C. Baker, D. P., & Shrestha, L. (1995). Situation awareness in team performance: Implications for measurement and training. *Human Factors, 37,* 123-136.

Shapiro, J., Rucker, L., Beck, J. (2006). Training the clinical eye and mind: using the arts to develop medical students' observational and pattern recognition skills. *Medical Education, 40,* 263-268.

Schaw, G. (1998). Promoting general metacognitive awareness. *Instructional Science, 26,* 113-125.

190

Schmidt, H. G., Norman, G. R., & Boshuizen, H. P. A. (1990). A cognitive perspective on medical expertise: Theory and implications. *Academic Medicine, 65*, 611-621.

Stout, R. J., Cannon-Bowers, J. A., & Salas, E. (1996/1997). The role of shared mental models in developing team situational awareness: Implications for training. *Training Research Journal, 2*, 85-116.

Zsambok & G. Klein (Eds.) (1997). *Naturalistic decision-making.* Hillsdale, NJ:LEA.

Chapter 20

Implications of Physiological Measures of Stress for Training Cross-Cultural Decision Making Skills

Erica D. Palmer, David A. Kobus

Pacific Science & Engineering Group
San Diego, CA 92121, USA

ABSTRACT

The current climate of irregular warfare and counterinsurgency operations brings the importance of cross-cultural decision making to the forefront. Decisions made in this context have significant implications for both safety and mission success, thus effective training in decision making (DM) is more crucial than ever. Because such decisions are very often made under stress, a key question concerns the effects of introducing stressors during training. Training under stress could be detrimental if it inhibits the learning process. Yet, experiencing stressors during training may serve to inoculate trainees, helping them habituate to stressors and perform better under stressful conditions. Determining effects of stress on DM training has important implications for training design. For example, what (if any) kinds of stressors will provide benefits that transfer to operational performance? What "level" of stress is optimal? How much exposure to stressors is needed, and when in training should exposure be provided to maximize benefits?

Crucial to answering these questions will be the ability to accurately measure trainees' stress levels. While questionnaires and subjective reports have some value, they are susceptible to demand characteristics, and other biases. Thus, more objective measures are also needed. A variety of physiological measures of stress are available, each with its own strengths and limitations. Stress hormone levels, for example, may be obtained fairly unobtrusively through saliva samples. However, timing of samples impacts results and it may not be feasible to obtain samples at optimal times relative to key events. Electrophysiological monitoring is another potential approach. While sensor systems have traditionally been restricted to laboratory settings due to cumbersome and sensitive equipment, recent technological advances support viable and potentially valuable tools for use in a wider range of environments. Emerging technology supports systems that are robust and easy to wear for ambulatory operation in physically demanding (including immersive) settings, and that can collect/clean/process data in real-time. Currently, such systems are being fielded to collect data in several sensor modalities (eg, EEG, EOG, EMG, ECG), providing gauges for stress-related aspects of performance such as workload.

Also critical to determining the role of stress in training DM will be the type and availability of training environments in which various stressors can be produced, controlled, and repeated in the context of DM scenarios. Camp Pendleton's Infantry Immersion Trainer (IIT) is one example that could meet these requirements. Stressors ranging from live role players using local customs and languages, to simulated IEDs, RPGs, and small arms fire can be built into scenarios designed to train and provide DM practice. The stressors and scenarios can be tuned to the specific training objectives and experience/skill level of trainees. The IIT lends itself well to outfitting trainees with sensor systems for monitoring individuals and small units in relation to scenario events. In addition to identifying environments for this kind of training and monitoring, it will be essential to address multiple challenges in developing metrics to assess training outcomes and the impact of stress on training.

Besides investigation of the impact of stress on DM training, physiological measures can be leveraged to help increase trainees' awareness of stress and their ability to manage it. A system that provides feedback to the trainee will help him better understand the effect of stress on mission performance, and could provide the basis for an adaptive training system tailored to individual users or small teams. For example, physiological monitoring would enable the system to determine whether sub-optimal performance is due to stress or to other factors (eg, lack of specific knowledge/skills/abilities), and to automatically customize training and mitigations accordingly. In addition, physiological measures of stress could be related to operational constructs familiar to trainees, such as Cooper's Color Code (CCC), which is used by the USMC to reflect an individual's level of awareness or engagement. Potential physiological correlates of levels of CCC have been identified, and validation of the relationships among CCC, performance, stress, and physiological measures could provide a valuable source of feedback and adaptive training. This training would help individuals to be more aware of their status (and that of others), and to maintain an optimal level of engagement for DM.

Physiological measures of stress have great potential for enhancing efficiency and effectiveness of training. This paper will expand upon considerations described above, providing a discussion of the possible roles and application of physiological measures of stress in training cross-cultural DM.

Keywords: Stress, Decision Making, Training, Physiology

INTRODUCTION

Warfighters operate under some of the most physically and psychologically stressful conditions possible. They sustain operations for long periods of time, often in harsh climates, with insufficient nutrition/hydration, while sleep deprived and physically fatigued. They operate in surroundings that may be unfamiliar in terms of geography, culture, and language. They are required to multi-task while remaining vigilant and situationally aware, and to make multitudes of decisions with far-reaching consequences, under pressure of time and threat to their safety. In irregular warfare and current counterinsurgency (COIN) operations, many of these decisions require understanding and consideration of cultural issues. Particularly in current COIN operations, winning the trust and support of the local populace is crucial. These decisions, made under stress, therefore impact mission success and public perception, as well as safety of personnel.

The question of how best to prepare warfighters to make decisions under stress is a focus of the training and research communities. One obvious possibility is to provide opportunities for warfighters to practice making decisions under stressful conditions. This approach however, raises a number of key questions with important implications for the design of training. For example, what (if any) kinds of stressors will provide benefits that transfer to operational performance? How can these stressors be introduced into training, and controlled by trainers? What "level" of stress is optimal? How much exposure to stressors is needed for training (or for inoculation), and when in the training continuum should exposure be provided to maximize benefits?

Any attempt to answer these questions will require accurate measurement of trainees' stress levels. While questionnaires and subjective reports have some value, they are susceptible to demand characteristics, and other biases. Physiological measures provide some promising possibilities for obtaining more objective, quantitative measures of stress. Further, some physiological measures may be leveraged for augmenting and enhancing training in decision making. This paper first provides a brief overview of physiological measures and recent technological advances that can be used for measurement of stress. Following this overview is a discussion of how these measures may be applied to help answer questions about training decision making under stress, and ultimately to enhance such training.

PHYSIOLOGICAL MEASUREMENT OF STRESS

SENSOR SYSTEMS

A wide range of tools exists for measuring stress in the laboratory, however there are additional requirements and considerations for measurement of stress in training situations. Ideally, the same tools and measures would be available for use across a variety of training environments, from seated computer-based applications, to standing but stationary virtual environments, to full-mobility immersive mixed reality environments. In order to record physiological data in each of these environments, the sensor systems must be worn during task performance and able to record during all the physical activities required by the training task. Therefore, the systems must be rugged, quick and easy to don and remove by the novice user, comfortable, and as lightweight and transparent to the user as possible. The systems must also include a means for reduction of noise and artifacts that are introduced into the data either by the training environment or by movement and activity of the user. In addition, sensor and data collection systems must not interfere with the training environment, either mechanically or electronically.

Increasingly, such sensor systems are becoming available. For example, QUASAR has developed two revolutionary technologies for fielded bioelectric measurements. The first is based on purely capacitive contact that operates on the skin or through fabric (Matthews, McDonald, Fridman, Hervieux, and Nielsen, 2005), and the second is a hybrid electrode that uses both capacitive and weak resistive contact and operates through hair (Matthews, McDonald, Anumula, and Trejo, 2006). Both types of sensors have adequate sensitivity and bandwidth for all types of electrophysiological measurement (e.g. electrooculogram (EOG), electrocardiogram (ECG), electroencephalogram (EEG)). Furthermore, neither system requires skin preparation, gels, or skin abrasion. Both systems can be worn comfortably for extended periods of time with no adverse skin reaction and can be easily donned by a novice (Matthews, McDonald, Hervieux, Turner, and Steindorf, 2007).

In order to maximize transparency of the sensor system to the user, it is desirable to integrate the sensors into the user's own gear. To this end, QUASAR has integrated a wireless EEG system with through-hair high impedance sensors into a military helmet. The sensors are mounted in a harness designed to fit under the Kevlar helmet of a soldier. In addition, they have designed an ECG shirt for unobtrusive, real-time ECG data collection (Hervieux, Matthews, and Woodward, 2007). The system has been successfully evaluated in the field. QUASAR has also integrated its capacitive ECG sensors into military body armor. A number of solutions have also been developed for noise suppression and artifact removal.

This type of cutting edge technology, while offering the greatest range and ease of use, is not necessarily required for all training applications. For example, as part of Spiral 1 of the Future Immersive Training Environment (FITE) Joint Capability

Technology Demonstration (JCTD), Pacific Science & Engineering (PSE) has been successfully collecting heart rate/variability data using Suunto brand heart rate monitoring belts during training scenarios in which users wear individual virtual reality systems. Data can be collected across the scenario, possibly serving as an objective measure of immersion, and can also be related to specific scenario events that have been time-marked in real-time in the data collection software. Users donned the belts prior to donning their virtual reality gear (which includes a vest), and reported no discomfort in using the belts for the duration of the demonstration scenarios (up to one hour). The feasibility of collecting data in a similar fashion in an immersive, fully mobile mixed-reality environment will be assessed during Spiral 2 of the FITE JCTD.

STRESS MEASURES

Stress affects physiological functions (as well as behavior) through a balance of sympathetic and parasympathetic regulation. Previously demonstrated indicators of stress responses to stimulated situations include heart rate (HR), heart rate variability (HRV), respiration, and the galvanic skin response (GSR). Mental stress is also known to reduce parasympathetic cardiac control, and is typically indicated by a decrease in low frequency power of the HRV signal (Salahuddin, Cho, Jeong, and Kim, 2007; Langewitz, Ruddel, and Schachinger, 1994). New physiological measures of stress are being developed. For example, as part of a project investigating the feasibility of developing a closed-loop neuroergonomic system for adaptive training (PSE and QUASAR), a Continuous Cognitive State Monitor (CCSM) based on continuous EEG features was explored for use in seated and fully mobile training environments. The CCSM is designed to classify electrophysiological data collected during defined cognitive tasks and predict user cognitive states such as mental stress, workload, fatigue, and engagement.

Pacific Science & Engineering is currently developing an additional indicator of stress by relating both cognitive task requirements and physiological measures (such as HR and respiration) to Cooper's Color Code (Cooper, 1989). Cooper's Color Code (CCC) is used to reflect an individual's level of awareness or engagement. It is often taught to warfighters as a way to identify and maintain the appropriate status during periods when vigilance is critical (observation posts and certain types of patrols), and during combat. Thus, CCC represents an operational construct familiar to some users, and to which physiological and neurophysiological data can be related. The relationships between CCC levels and HR have been postulated previously (Grossman and Christensen, 2004). Work is currently underway to collect data in a range of training environments and to begin validation of the relationship between CCC levels and physiological measures as an indication of stress and various aspects of human performance. Early results have suggested that the CCSM is one measure that correlates well with the levels of CCC.

STRESS AND TRAINING DESIGN

EFFECTIVE USE OF STRESSORS

The objective, quantifiable measures of stress that can be provided by physiological data are a key source of information that can inform the design of decision making training. For example, it is possible to measure physiological responses to a variety of different types of stressors, ranging from physical stressors such as high temperatures to psychological stressors such as making difficult decisions under pressure, trying to interact with locals speaking an unfamiliar language, or hearing the sound of weapons fire. The measured responses can help to identify which stressors are effective within a given training environment. It may be the case that some stressors elicit strong responses in one environment (e.g., in a mixed reality facility) but not in another (e.g., in a computer-based virtual simulation). Not only does this information help to design training that is optimized for in a given environment, but it also provides a basis for assessment of the effects of training on performance and whether they transfer across training environments – and ultimately to operational performance.

Another training consideration that can be informed by physiological measures of stress is what "level" of stress is optimal for a given training application. It has often been demonstrated that an inverted U relationship exists between stress and performance; too little or too much can negatively impact performance, whereas the best performance is elicited under moderate amounts of stress (cf., Yerkes and Dodson, 1908). The optimal level of stress during training may vary considerably based on experience or skill level, training tasks and objectives, individual differences, and a host of other factors. Fortunately, many training environments allow the level of stress to be titrated by altering the number and/or intensity of stressors. Physiological measures can quantify the stress experienced by an individual during training, and these data can be correlated with performance data to help determine the range of stress levels that produce the greatest effect on performance in a given training application.

Related to the consideration of stress level is the question of how much exposure to stressors is required, and when in the course of training the stressors should be introduced so as to maximize their benefits. Again, the answers to these questions may depend on a number of factors. For example, if the goal is to provide training in making a variety of decisions under stress, numerous opportunities to experience stressors may be desirable, and those experiences may be most beneficial if the trainee is far enough along in training that he has mastered many of the basic skills required by decision making scenarios. Conversely, if stress inoculation is the goal, a single exposure early in training may be all that is required. As discussed above, quantitative physiological measures can be correlated with performance data relevant to the training application in order to determine when and how much stress is optimal. It would be possible to document, for example, the learning curve across multiple exposures to a particular set of stressors and identify the point at which performance effects plateau.

CHARACTERIZATION OF TRAINING ENVIRONMENTS

Also key to the design of training for decision making under stress is the identification of suitable training environments, the specific types of stressors that can be introduced within those environments, and the ability to control, titrate, and repeat the intensity and number of stressors experienced by trainees. One candidate environment is the individual worn virtual reality environment currently being demonstrated as part of the FITE JCTD. Through the VBS2 (Virtual Battlespace 2 by Bohemia Interactive Australia) based scenarios, artificial intelligence, and operator controlled avatars, a wide range of stressors can be produced and there is a high level of control over the stressors. Another example of a potentially suitable environment is Camp Pendleton's Infantry Immersion Trainer (IIT). Stressors ranging from live role players using local customs and languages, to simulated IEDs, RPGs, and small arms fire can be built into scenarios designed to train and provide practice in making decisions under stress. While the degree of control is not as great as in a virtual system, the stressors and scenarios can still be tuned to the specific training objectives and experience/skill level of trainees.

Both the FITE system and the IIT lend themselves well to outfitting trainees with sensor systems for monitoring individuals and small units in relation to scenario events. Very similar scenarios can be used across these training environments, as well as in seated simulations, providing a good basis for comparison of physiological and performance measures across a range of training platforms/environments. In addition, physiological measures of stress will help to determine the degree to which experiences in immersive environments can produce stress that may be comparable to real world situations, and will help capture individual differences in responses to stress. As future training environments are identified and characterized, it will be possible to better address challenges in developing objective metrics to assess training outcomes and the impact of stress on training. Physiological measures show great promise for helping to meet these challenges by providing quantitative, objective data in domains where subjective measures continue to be a primary method of evaluation.

AUGMENTED TRAINING

There is a clear role for physiological data in informing training design and providing measures of the impact of stress on decision making training. Feedback based in physiological measures of stress can also be provided to trainees to help them better understand the effects of stress on mission performance. The ability to collect physiological data in real-time offers even greater application by allowing the data to be leveraged to provide augmented, adaptive training tailored to individual users. In the most basic application, stress responses could be monitored so that stressors during a training evolution could be titrated to keep an individual in his optimal stress level range. More advanced applications are also possible. For example, real-time physiological monitoring would enable the system to determine whether sub-optimal performance is due to stress or to other factors (eg, lack of specific knowledge/skills/abilities), and to automatically customize training and mitigations accordingly. Building on the correlation of physiological measures with levels of CCC described earlier, ongoing feedback on CCC level could be provided to individuals during training. This feedback may help individuals to be more aware of their status (and that of others), and to maintain an optimal level of engagement and awareness for decision making.

CONCLUSIONS

Physiological measures of stress hold great potential for providing objective data to improve the efficiency and effectiveness of training in a number of domains. Their application to training in cross-cultural decision making is especially pertinent given the current emphasis on COIN operations in Afghanistan. Various widely available basic technologies, such as wireless heart rate monitors, are ready and suitable for many applications. Cutting edge technology with improved portability, wearability, and transparency to the user will continue to expand the range of training environments in which such data can be collected, as well as the potential applications of the data. From designing training, to assessing its impact, to providing adaptive augmented training, physiological measures can reveal a wealth of information to help better prepare warfighters for making decisions under the kinds of stressful conditions that they will face in an operational setting. Providing the most effective training possible in this domain is critical not only to mission success, but also to enhancing the safety of both warfighters and the civilian populace.

REFERENCES

Cooper, J. (1989). *The principles of personal defense*. Boulder, CO: Paladin Press.

Grossman, D.A., and Christensen, L.W. (2004). *On combat: The psychology of deadly conflict in war and peace*. Belville, IL: PPCT Research Publications.

Hervieux, P., Matthews, R., and Woodward, J.S. (2007). *Garment incorporating embedded physiological sensors*.

Langewitz, W., Ruddel, H., and Schachinger, H. (1994). "Reduced parasympathetic cardiac control in patients with hypertension at rest and under mental stress." *American Heart Journal*, 127, 122-128.

Matthews, R., McDonald, N.J., Fridman, I., Hervieux, P., and Nielsen, T. (2005). "The invisible electrode – zero prep time, ultra low capacitive sensing." Presented at *11th International Conference on Human Computer Interaction*, Las Vegas, NV.

Matthews, R., McDonald, N.J., Anumula, H., and Trejo, L.J. (2006). "Novel hybrid sensors for unobtrusive recording of human biopotentials." Presented at *Proceedings of 2nd Annual Augmented Cognition International Conference*, San Francisco, CA.

Matthews, R., McDonald, N. J., Hervieux, P., Turner, P.J., and Steindorf, M.A. (2007). "A wearable physiological sensor suite for unobtrusive monitoring of physiological and cognitive state," *Conf Proc IEEE Eng Med Biol Soc*, 5276-81.

Salahuddin, L., Cho, J., Jeong, M.G., and Kim, D., (2007). "Ultra short term analysis of heart rate variability for monitoring mental stress in mobile settings." *Conf Proc IEEE Eng Med Biol. Soc*, 4656-4659.

Yerkes, R. M. and Johnson, J.D. (1908). "The relation of strength of stimulus to rapidity of habit formation." Journal *of Comparative Neurology and Psychology*, 18, 459-482.

Chapter 21

Training Tactical Decision Making Under Stress in Cross-Cultural Environments

David A. Kobus[1], Gregory Williams[2]

[1]Pacific Science & Engineering Group
9180 Brown Deer Rd
San Diego, CA 92121

[2]Cubic Corporation
9333 Balboa Avenue
San Diego, CA 92123-1589

ABSTRACT

This paper discusses an approach to better prepare Marines and Soldiers for Counterinsurgency (COIN) operations. Methods in enhanced observation (EO), combat profiling (CP) and operational cultural (OC) need to be systematically combined to better understand foreign human terrains. Advanced EO and CP methods are currently taught to the Marines as part of the Combat Hunter program. This training begins with an understanding of human perception and identifies the advantages and disadvantages of human sensory systems. Training is also conducted on advanced optics and devices that extend human capabilities. The CP portion of the course relies on human behavior pattern recognition and can be applied to people, events, or vehicles, anywhere in the world. Combat Profiling trains Marines to identify potential cues that fall into one of six categories, or domains for further analysis and how to develop a baseline level of activity for each of these domains. Many of these cues are culture free. However, a critical component of this training is to understand the operational culture to better recognize the context in which the behaviors occur. The premise is to provide Marines with a method to quickly make legal, moral, and ethical decisions. One goal of this training is to make Marines and Soldiers more aware of their surroundings, better understand the context, and be able to identify oddities as indicators of potential threat.

Keywords: Combat Profiling, Perception, Affordances, Anomaly Detection

INTRODUCTION

Training for tactical decision making requires the formation of a collective thought process using critical thinking skills and military doctrine to develop solutions and evaluate outcomes. In the current operational environments small unit leaders must not only possess a firm grasp of infantry tactics, but also understand the principles of counterinsurgency (COIN) versus conventional military operations as shown in Table 1 (Zeytoonian, et. al. 2006). Many of the procedures that are generally followed under conventional operations have different information resource requirements for COIN decision making. Leaders at the squad and platoon level often find themselves in unique and seemingly abnormal situations that require a clear understanding of the oppositions Tactics, Techniques and Procedures (TTPs), culture, and environment in order to develop more accurate tactical solutions. Behavior patterns of the civilian populace, culture, terrain, economy, religion and politics all make up the peripheral conditions that need to be considered by small unit leaders to achieve objectivity during planning and decision making. In addition, each of these considerations must be taken into account while making decisions under a variety of stressors (e.g., fatigue, overload, anxiety, etc.) experienced by Marines in combat.

Table 1. Differences between conventional operations and COIN.

Characteristic	Conventional Ops	COIN
Battlespace	Physical Terrain	Human factors – Demographics, culture, tribes, clans, etc
Effects	Politics not considered	Politics are central
	Linear	Asymmetric
	Effects of physical terrain and weather	Effects of Infrastructure, services, jobs, and media
Threat	Order of Battle	Networks (cells)
	Doctrinal Templates	Enemy TTPs
	Military in focus	Insurgents / Supporters / Gen Pop.
COA	Event Templates	Pattern, link analysis, networks
Targeting	Order of Battle	Pattern, incident analysis
Collection	At a stand-off range	Personal contact
	Military communications	Personal Communications
	Ops executed with intel	Ops conducted to create intel

In 2006, casualty reports for Marines showed a marked increase in direct fire casualties. This dramatic increase in precision fire engagements of personnel indicated that the military was being hunted. Further analysis of the casualty reports indicated that the attacker often blended into the crowd or was *hiding in plain sight*. At the bequest of General Mattis, the Marine Corps Warfighting Laboratory (MCWL) developed the Combat Hunter program. This training was developed to train Marines to become better hunters, proactively seeking out and identifying the enemy while remaining undetected (MCWL report, 2008). In addition, these skills proved to be especially important to have when conducting COIN operations. More recently the Center for Advanced Operational Culture Learning (CAOCL) has identified "Operational Cultural as those aspects of culture that influence the outcome of military operations: conversely the military actions that influence the culture of an area of operations" (Dunne, 2009, page 10).

Training for such operations has been evolving over the past several years and has required a much different mindset from conventional operations training. The terrain of concern is no longer only physical, but now requires a better understanding of the human terrain. One aspect of this training is to better understand not only the tactics and mindset of the insurgent, but to also understand the human landscape in which they operate. This landscape establishes the context in which they fight which is heavily influenced by religion and culture. A vital characteristic of this training is to learn the what, when, where, why and how of advanced observation skills.

TRAINING COMPONENTS

This paper describes not only the purposes behind specific training requirements for COIN operations, but to also outline why certain components of training, that are currently isolated, need to be combined to best prepare Marines for using these skills appropriately.

ENHANCED OBSERVATION

Every day we are exposed to a plethora of sensory information in a complex world. The amount of information we are exposed to, would overload any one of our senses, if we were required to process each piece of information. To protect ourselves from such overloads our sensory systems allow us to use selective filters to attend to information we think is most important, or information that captures our attention. In essence we learn to unconsciously ignore most of what goes on in our environment. So, this ability to limit what we "see" to is a blessing, but it is also a curse. It allows us to obtain detailed information of what we attend to, but at the cost of missing something else in the environment that could be important. Knowing these naturally occurring limitations is important component for observers to learn.

During this training Marines learn to effectively use day, night, and thermal optics to observe and survey their operating areas. Binoculars, spotting scopes, rifle combat optics, night vision devices, and infrared and thermal imagers all extend the range of human eyesight. During the course of training both long range (observation post) and short range (while patrolling) observation skills are taught. The shorter range skills are especially important during COIN operations.

Marines are also taught the proper observation techniques, theory, and methods for improving memory. It is critical to teach infantrymen how and what to look for when making observations because, *"what we see depends mainly on what we look for"* (quoted from John Lubbock). These skills are the foundation upon which combat tracking and profiling skills are built. They are taught that most people "look" but it takes some skill to "see" what may be important in a scene. For example, they are taught that color, movement, and texture are three of the critical components to focus on when observing. They are trained to explore for differences in these aspects to identify things that do not blend in with surrounding terrain or vegetation. The detection of such anomalies will help them identify where they need to focus their search and attention more closely. This skill is critical to accurately direct the observer to focus their attention to specific aspects of what they are viewing to extract details to identify the cause of the anomaly.

BEHAVIOR PROFILING

Once Marines know "how to see" they need to learn what to look for. This training method defines and classifies behavioral cues that serve as anomalies for identifying potential threats in the environment. The method works with all cultures but is further enhanced by understanding the cultural context. The Combat Profiling skill set is the art of reading the human terrain using primal skills we all have that are related to our ability for making approach/avoidance decisions.

This method, called Combat Profiling, relies on human behavior pattern recognition and can be applied to people, events, or vehicles, anywhere in the world. This method is one of the main tenets of the Combat Hunter training program. Combat Profiling trains infantrymen to identify potential cues that fall into one of six categories, or domains; Heuristics, Proxemics, Geographic's, Atmospherics, Biometrics, and Kinesics and to develop a baseline level of activity for each of these domains. When deviations in the baseline are detected and they are in conflict with the current environmental context, the probability of potential threat is greater. A rule of thumb used by Marines is that when more than three anomalies deviate from baseline (across all of the domains), they should decide quickly how to respond, whether to question, capture, or kill the cause of the anomalies while considering the context and relevance of anomalies. If the anomalies can be explained, then they can let it go. The premise is to provide Marines with a method to quickly make legal, moral, and ethical decisions. One goal of this training is to make Marines more aware of their surroundings and consider oddities as indicators of potential threats. This training provides a real-time, eyes-on method for detecting potential threats that are "hiding in plain sight".

The science supporting Combat Profiling is in concert with the ecological / evolutionary social psychology premise of a perceptual affordance management system for identifying affordances (anomaly in baseline) that are directly related to fitness-relevant categorical inferences (a threat to physical safety). The affordance management system like Combat Profiling is a component of reading the human terrain. Neither relies on ethnic, cultural, or religious components, but rather is concerned with identifying specific behavior traits that may trigger the "fight or flight" mechanism within most humans.

However, the key to truly understanding behavior is to know the context in which it is produced. Contexts vary as a function of culture. The terms "high context" and "low context" have been used to describe broad-brush cultural differences between societies (Hall, 1959). High context cultures (or groups) refer to where people have close connections over a long period of time. In these cultures, many aspects of cultural behavior are not made explicit because most members know what to do and what to think from years of interaction with each other. Low context refers to cultures (or groups) where people tend to have many connections with others, but of shorter duration. In low context cultures, behavior and beliefs are spelled out explicitly so that when interacting in the cultural environment people know how to behave.

This distinction has only recently being considered by the military as an important training component (Dunne, 2009). Yet, this type of cultural training has not been fully incorporated in the observation or profiling training. A basic understanding of the culture is paramount for fully understanding behavior. This point is even of greater importance when conducting operations in a culture with high context. Some of the characteristics of a high context culture are:

- Less verbally explicit communication, less written/formal information
- More internalized understandings of what is communicated
- Multiple cross-cutting ties and intersections with others
- Long term relationships
- Strong boundaries - who is accepted as belonging and who is considered an "outsider"
- Knowledge is situational, relational.
- Decisions and activities focus around personal face-to-face relationships, often around a central person who has authority.

Conversely a low context culture is more direct and openly provides information to understand the expected behavior. In this type of culture the rules are clearly laid out. Some of the characteristics of a low context culture are:

- Rule oriented, people live by external rules
- More knowledge is codified, public, external, and accessible.
- Sequencing, separation--of time, of space, of activities, of relationships
- More interpersonal connections of shorter duration
- Knowledge is more often transferable
- Task-centered. Decisions and activities focus around what needs to be done, division of responsibilities.

Cultures differ greatly in their contexts. While these terms are sometimes useful in describing some aspects of a culture, one can never say a culture is "high" or "low" because societies all contain both modes. "High" and "low" are therefore less relevant as a description of a whole people, and more useful to describe and understand particular situations and environments. However, in the current operational environment most culture contexts that are encountered are tribal, or high context in nature. High contexts are more difficult to enter, or relate to, if you are an outsider. This is because outsiders don't carry the context information internally, and can't instantly create close relationships.

CONCLUSIONS

The premise of this training follows the philosophy "every Marine a collector". The method is to "See", "Assess" and then "Communicate" information to everyone on the team and to higher command to provide a dynamically evolving situation awareness of the environment. This paper describes how two key components of training are used to train Marines to become more effective in combat decision making. In addition, it provides a rational for further encapsulating the understanding of cultural contexts in future training.

Through this training Marines are trained to detect changes from sensory baseline as a warning, but it is equally important for them to understand the nature of the culture where they are being deployed and know the similarities and differences of the local cultures to their own. Behaviors may easily be misinterpreted without fully understanding the cultural contexts. Such an understanding would lead to quicker decisions regarding the action required of detected anomalies. Such training ensures that Marines are more aware of surroundings and which oddities to consider as indicators of potential threats. This training provides a real-time, eyes-on method for detecting potential threats that are "hiding in plain sight". This ability enhances the dynamic awareness of Marines in theater, a key component to success in any operational area.

REFERENCES

Dunne, J.P. (2009). Cultures are different. *Marine Corps Gazette*, 93 (*2*), 8-15.
Hall, E. T. (1959). *The Silent Language*. New York: Doubleday and Co., Inc., 240 pp.
Marine Corps Warfighting Laboratory Report 2008.
Zeytoonian, D. et. al., (2006). COIN operations and intelligence collection and analysis. *Military Review,* Sept-Oct, 30-37.

CHAPTER 22

Intertemporal Reasoning and Cross-Cultural Decision Making

Donald G. MacGregor[1], Joseph Godfrey[2]

[1]MacGregor Bates, Inc.
1010 Villard Avenue
Cottage Grove, Oregon 97424

[2]WinSet Group, LLC
4031 University Drive, Suite 200
Fairfax, Virginia 22030

ABSTRACT

Elements of economic theory are examined to understand the Euro-American cultural assumptions encoded in their formulation. The research develops a framework by which to understand how Euro-American beliefs relevant to economic theory might be understood or misunderstood in other cultures, with the goal of facilitating inter-cultural dialog. The framework is constructed using an approach based on decomposing key economic concept into their component elements and relating those elements to research that has identified important cultural views that influence how these components are matched or mismatched by cultural beliefs. We take as our starting point the notion of Net Present Value and the Time Value of Money demonstrating how this single, but central, concept from investment analysis encodes cultural assumptions regarding the linearity of time, equity, individual property, wealth creation. The results are applied to non-western cultural beliefs, particularly in Arabic cultures.

Keywords: Cross-cultural decision making, net present value, risk analysis

INTRODUCTION

As western nations engage other cultures with the intent of establishing productive relationships they bring concepts, theories and methods to aid or support those cultures in improving their economic infrastructure. These methods are rooted in western economic theory and its derivatives, including risk assessment, decision analysis, multi-attribute utility, and game theory to name but a few. They extend as well to modern finance and banking, including credit and insurance.

In this paper we introduce an approach to understanding cross-cultural decision making through the lens of western economic theory. Economic science has developed powerful techniques for modeling and aiding important social decisions. For example, techniques derived from economic theory, such as the capital asset pricing model, have proven useful for evaluating (and making decisions about) capital improvement projects at both a national and a local level. Our approach is to identify key economic concepts and principles. These are decomposed into their social, cultural and psychological components and are then evaluated in light of research from a number of disciplines, including cross-cultural psychology, cultural anthropology and comparative religion.

The guiding framework for the evaluation is based on two concepts. One is *cultural congruence*, defined here as the compatibility between the underlying concepts or rationality associated with a theory or method and the cultural context within which the method is applied (e.g., Resnicow, et. al., 2000). A lack of congruence is revealed when social and cultural precepts are contradictory to the underlying rationality of western economics.

A second concept is that of *intertemporal* reasoning. We use the rubric of intertemporality to refer to psychological, social and cultural processes that are engaged when people are called upon to either integrate past experiences or project forward to future (and unrealized or imagined) experiences as a basis for indicating a preference, desire or likelihood in the present (Trope and Liberman, 2003; Forster, Friedman and Liberman, 2004). For example, the concept of risk in economics and finance theory is based on intertemporal reasoning processes that summarize and project past experiences into potential consequences for the future. Likewise, the process of reconstructing a cultural history for the purposes of evaluating the appropriateness, desirability, and/or risk of a present action calls upon intertemporal reasoning processes that involve the accumulation and integration of historical events or decisions to arrive at a reference point for current actions. Our working hypothesis is that research in cultural anthropology, psychology and economics can be used to inform and guide the appropriate use and development of methods of formal analysis in the context of non-western cultures. This hypothesis leads us to a methodology that involves examining the concepts underlying western economic methods in light of their cultural congruence and how intertemporal reasoning is influenced by cultural factors.

METHOLOGICAL APPROACH

The methodological approach is based on the fundamental precept that methods of analysis based on western economics require for their inputs a set of *thought experiments*. These can take the form of direct assessments or questions (e.g., probability or desirability of a future event) or they involve implicit judgments (e.g., social or cultural acceptability of a financial prospect such as insurance). Acceptance of the output of the analysis is predicated upon acceptance of the validity of the thought experiments, and the validity of the logic applied to the results of the thought experiments (e.g., MacGregor and Slovic, 1986). Key to these thought experiments are notions about time and the relationship between time and preferences, and the effects of intertemporal reasoning on cognitive processes (Trope and Liberman, 2003; Forster, Friedman and Liberman, 2004). Thus, to apply these methods requires a pre-existing set of cultural factors (e.g., beliefs, tenets) that are consistent with what is required from those on whose behalf the method is applied.

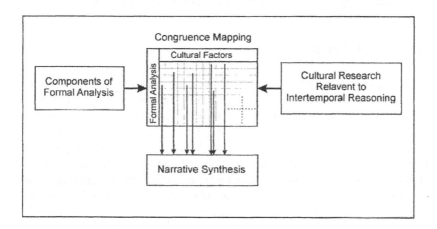

FIGURE 1.1 Congruence mapping approach.

To improve our understanding of western economic principles applied in non-western cultures, we develop a framework for mapping the relationship between cultural perspectives and the elements of economic theory. To explore the connections between practically linked but disciplinarily distinct areas (such as economic theory and cultural anthropology) we require a methodology based on the techniques of *grounded theory* to induce relationships from theories and data obtained from a range of domains (Strauss and Corbin, 1994). Essentially, grounded theory advocates the development of formal knowledge and theories from the substantive analysis of interconnections, essentially developing theory from theory (Glaser and Strauss, 1967). Grounded theory is operationalized here by developing a mapping between economic principles in terms of a collection of theories and applied techniques, and cultural knowledge. The mapping can be conceptualized as a two-dimensional matrix, the rows of which represent component elements of economic theory, and the columns of which represent culture-based logic (Figure 1).

The cells of the matrix represent points of connectivity that reflect either cultural congruence, contradiction, or some amalgam of the two. The mapping provides a platform for the development of a narrative interpretation. We illustrate the basics of our approach with a brief analysis of the concept of Net Present Value (NPV) that is central to western economics and finance. We first decompose NPV into its component elements and then discuss these elements in terms of concepts drawn from the general literature on Arabic culture, including the principles of Islam.

THE CONCEPT OF NET PRESENT VALUE (NPV) IN WESTERN AND NON-WESTERN CULTURES

HISTORY OF NET PRESENT VALUE

The modern notion of NPV is based on the time value of money (TVM) and dates from Fibonacci's *Liber Abaci* in 1202 (Goetzmann, 2003). In this work Fibonacci introduced discounting the future value of cash flows to arrive at their *present value*. Hindu mathematicians anticipated Fibonacci by nearly 700 years. As early as 500 CE the Hindu mathematician Āryabhata solved interest rate problems, but did not develop the concept of present value. Fibonacci built directly on the work of the Arabic mathematician Al Khwārizmī, but with a different focus. Whereas Al Khwārizmī, focused on dividing inheritance among family members, Fibonacci focused on dividing capital among merchants. The concern for mercantile applications and public finance appear to be distinguishing characteristics of uses in the West of TVM in contrast to the East, notably China and India. (Goetzmann and Rouwenhorst, 2005)

DECOMPOSITION OF NET PRESENT VALUE

Assuming a fixed rate of interest, the net present value of a stream of cash flows is represented by the general formula

$$P = \sum_{k=0}^{N} \frac{F_k}{(1+i)^k},$$

where the index, k, represents an interval of time. During these intervals the interest rate, i, is constant, but the cash flow, F_k, is allowed to vary per interval. The present value of all these cash flows is given by the sum and denoted as P. Each component in the formula embodies assumptions with cultural significance. We enumerate each component and then subsequently discuss their potential cultural implications.

Time Interval

The most fundamental requirement for TVM is linear time. It must be possible to identify regular intervals of time and that these proceed in a successive manner, i.e., a linear, monotonic sequence. That this is a non-trivial requirement can be seen in cultures that redistribute property on a cyclic basis, e.g., the Year of Jubilee. The progressive view of time, especially one that envisions a future of ever increasing wealth, is a strong but common assumption of western economics. Critiques of this assumption from within the western economic framework (e.g., Malthus) illustrate this point. Even those sympathetic to Malthus, may regard this concern as s "horizon problem," and continue to rely on linear, progressive time for routine economic/financial calculation.

Cash Flow

The concept of cash flow involves several social constructs. First and most obvious is a *medium of exchange*, an intermediary instrument that enables exchanges of goods without direct barter. In order for exchange to be meaningful and possible, the goods must have a recognized value mediated by the medium of exchange. For an instrument to serve as a medium of exchange certain properties are necessary: it must carry and retain value, it must be recognizable, transportable, divisible, and difficult to counterfeit (Gwartney and Stroup, 1997). Money, however, is rarely of value in and of itself. Paper money, for example, carries and retains value, but only in its capacity to mediate exchanges and has no intrinsic value. Gold, by contrast, can serve as money but also as store of value, i.e., as something valued in itself. Our principal concern is with money that is not a store of value.

For a medium of exchange to function, social constructs are required, notably property rights (Grossman, 2002). It must be possible for two parties to separately own goods that are exchanged via intermediate transactions involving money. Ownership consists in the ability and the power, to determine the disposition of a good – its creation, destruction, modification, or exchange for other goods. The enforcement of this power, in turn, requires sufficient social organization to recognize and respect property rights on a routine basis. And finally, the social organization must be sufficiently stable that goods can be exchanged over and extended time period, so that one may consider the possible variation of value between goods in the period between which the exchange of these goods is realized, in order for the time value of money to be relevant. Thus, we require a stable social system with a mechanism to enforce basic property rights, in the sense that agents can and do benefit from commerce in goods they own.

Interest

A fundamental problem in a mediated exchange of goods is the possibility that value of the goods changes between the time when the exchange is agreed to and completed. Fibonacci addressed this problem by introducing the notion of *present value* by which sums of money at different points in time are incommensurable. To compare the value of a good in the present with another good in the future, one must transform the future value into a present value. The transformation is accomplished by removing (or adding) to the future value it's change in value from the present, ΔP, to wit

$$F = P + \Delta P.$$

The change in P can be expressed by some multiplicative factor, r, the "effective interest rate" for the period in question, so that

$$F = (1 + r)\, P.$$

If there is a defined interval, the effective interest rate can be expressed using an interest rate per interval, i, such that

$$F = (1 + i)^N P.$$

For r much smaller than 1, $r \ll 1$, $i \approx r/N$.

The important question for this paper is how interest arises. In western cultures, interest arises from a number of sources. For example, a sufficient condition for commerce is that two parties to an exchange value the goods received in the exchange. This can be achieved when each party is "better" at producing his or her product than the other party – so that each benefits more from the exchange than from producing the goods on their own. In short, there is a division of labor. This division of labor gives rise to interest rates, as a form of rent. Interest rates in western economies represent the general demand for capital. According to this picture the interest rate in effect measures the degree of and demand for innovation (division of labor) in a market.

Interest rates can also be based on uncertainties associated with the future value of a good due to a variety of factors, including incomplete knowledge about the future and future conditions that would affect the good. In this sense, interest can be seen a reflection of risk in transactions involving future exchanges. This presents a challenge to non-western cultures when interest is derived from either speculative activities *(maysir)* or when the future value of a good is subject to factors beyond human control.

COMPARATIVE SYNTHESIS OF ARABIC CULTURAL FACTORS AND NET PRESENT VALUE

For the sake of this paper, our comparative synthesis of Arabic cultural factors associated with NPV is based on Islamic principles that relate to economic and financial transactions. A strict prohibition in Islamic cultures is placed on *ribā*, the "predetermined interest collected by a lender that the lender receives over and above the principles amount it has lent out" (Schoon, 2007 pg. 27).[1] Thus, Islamic principles present a challenge to the notion that money can have a time-variant value.

The underlying concern in Islam derives from a strong cultural adherence to social justice as prescribed in the Qur'án (e.g., Schuon, 1998; Sells, 1999). This expresses itself as a cultural concern that no party to a transaction be disadvantaged by the transaction. In a fair barter exchange, for example, both parties receive goods of comparable value. Each party benefits from the exchange. Likewise, in the loaning of money, the creditor is expected to share in the risks faced by the debtor. For the creditor to profit in the form of interest, without exposure to any of the debtor's risk, is both unjust and constitutes *gharar*. Essentially, *gharar* is the uncertainty associated with transactions and may come in several forms, including uncertainty about the receipt of payment, social worth of the transaction and ignorance on the part of the debtor with respect to the properties of the transactions (e.g., El-Gamal, M A., 2001). Within the concept of *gharar* complex prohibitions exist that require careful consideration before risk-bearing transactions can be undertaken in a manner consistent with the principles of Islam. Violation of these principles constitutes "trading in risk."

Islam has no difficulty charging rent for the use of commodities, provided these rents do not exploit the renter. As rents are permitted under Islam, a non-Islamic economist may find the prohibition on *ribā* puzzling. Isn't money just a commodity and so available for rent? Ahmad and Hassan (2006) argue that Islam rejects the view of money as a commodity. Money is regarded strictly as a medium of exchange, lacking any inherent value. Money can be neither hoarded nor wasted. It must be used to realize its value and is not a culturally-acceptable store of value in itself.[2]

---

[1] It should be noted that Christianity prior to the 16th century had similar prohibitions.

[2] Specifically Ahmand and Hassan (2006) argue that money has no intrinsic utility, being unable to fulfill any human need directly. It lacks any other qualities except its role as a medium of exchange. Other commodities are posited as multifaceted. And finally money is

As a short hand, *ribā* forbids the exchange of money for money – where the only difference is time. In this respect the Islamic theory matches views such as those of William of Auxerre (ca. 1220) who asserted that the "the usurer acts contrary to natural law, for he 'sells time, which is common to all creatures'" (Goetzmann and Rouwenhorst, 2005, p.5). Within the time value of money we can view the interest rate as a clock: within each period (i.e., clock tick) interest compounds. Indeed, under continuous compounding the expression for compounding $(1+i)^N$ becomes e^{it}. In this form the interest rate is manifestly conjugate to time, having units of inverse time.[3]

Within Islamic cultures the difficulties associated with time come in part from the relatively imprecise manner in which intertemporal reckoning occurs, some of which may be reflected in and attributable to the Arabic language itself (e.g., Patai, 2007). The tendency within Arab cultures to conceptualize history in terms of broad stages or periods rather than discrete events (as is in the case in western cultures) is antithetical to the precise and highly-resolved sense of time characteristic of western society. In this sense, we can regard western economics and finance as having commodified time by identifying its properties and value, thereby rendering it suitable for trade. Indeed, it would not be an excessive extrapolation to regard trading in time as trading in risk.

Despite its reservation concerning trading in time, Islam is not inimical to the notion of present value. Indeed Islamic finance recognizes that an asset's value can change over time and so the comparison of the value of assets over time requires a comparison in terms of present value. Present value, however, is computed using a discount rate, and this does present significant conceptual and theoretical challenges for Islamic finance. As a practical matter, these do not appear to be insurmountable as evidenced by the evolution of Islamic banking along the lines of culturally-acceptable Qur'ánic principles. However, Muslims, concerned to avoid *ribā*, have not deposited their wealth in banks for investment, preferring to maintain their wealth in assets that are true and culturally-acceptable stores of value such as precious metals, stones, and the like (Kula 2008). From a western economic perspective this suggests that Islamic cultures have significantly underutilized their capital, thereby foregoing significantly higher standards of living. A similar thesis has been propounded by Hernando de Soto (2000), where he argues that in failing to secure property rights (a key assumption of TVM) for their peoples, non-western countries have similarly forgone opportunities for creating wealth. This brings into sharp focus the challenges faced by non-western economies to realize their potential for commerce and wealth creation.

---

not to be localized in a transaction. Money should facilitate a transaction, but should not be essential to a transaction.

[3] That time is not a commodity one can buy and sell was a common understanding between Islam and medieval Christianity. Western Christianity, however, began to abandon its prohibitions on usury in the 16 century, when Henry VIII of England relaxed the prohibitions on usury with his "Acte Agaynst Usurie" (Makower, 1972; pg. 454).

CONCLUSIONS

This paper has presented a framework for examining cross-cultural rationalities. The framework, based on cultural congruence and intertemporal reasoning, was applied to a concept central to western economics and finance, that of Net Present Value (NPV). The decomposition of NPV revealed that embedded in its logic are assumptions about intertemporal reasoning and culture-bound conceptualizations of the time value of money. Using this decomposition as lens through which to view Arab culture, we find that non-western views about time interact with views about the culturally-appropriate ways to define and accumulate wealth. These views differ sharply from those in the west in that they place strong emphasis on social justice, set in place cultural prohibitions against certain types of financial transactions, and reflect culturally-defined views about risk that may carry over into a variety of decisions that include, but are not necessarily limited to, those involving economics and finance.

ACKNOWLEDGEMENT

This work was funded by the Office of Naval Research, Human Social Cultural Behavioral Program, under Contract Number N00014-09-C-0570 to MacGregor Bates, Inc.

REFERENCES

Ahmad, A. U. F. and Hassan, M. K. (2006). The time value concept of money in Islamic finance. *American Journal Of Islamic Social Sciences,* 23(1), 66-89.

Ahmad, A. U. F., and Hassan, M. K. (2007). Riba and Islamic banking. *Journal of Islamic Economics, Banking and Finance*, 3(1), 1-33.

De Soto, H. (2000). *The Mystery of Capital (2000)*. New York: Basic Books.

El-Gamal, M. A. (2001). *An economic explication of the prohibition of gharar in classical Islamic jurisprudence.* Paper presented at the 4th International Conference on Islamic Economics, Leicester, UK, 13-15 August 2000. (Downloaded from http://www.ruf.rice.edu/~elgamal/files/gharar.pdf).

El-Gamal, M. A. (2003). Interest and the paradox of contemporary Islamic law and finance, *Fordham International Law Journal*, 27(1), 108-149.

Forster, J., Friedman, R. S., and Liberman, N. (2004). Temporal construal effects on abstract and concrete thinking: Consequences for insight and creative cognition. *Journal of Personality and Social Psychology*, 87 (2), 177-189.

Glaser, B.G., and Strauss, A.L. (1967). T*he discovery of grounded theory: Strategies for qualitative research*. New York: Aldine.

Goetzmann, W. N., (2003). *Fibonacci and the Financial Revolution.* Yale ICF Working Paper No. 03-28. Available at SSRN: http//ssrn.com/abstract=461740.

Goetzmann, W. N. and Rouwenhorst, G. (2005). *The origins of value*. New York: Oxford University Press.

Grossman, H. I. (2002), Make us a king: Anarchy, predation, and the state. *European Journal of Political Economy*, 18(1), 31-46.

Gwartney, J. D. and Stroup, R. D. (1997). *Economics*. Orlando, FL: The Dryden Press.

Kula, E. (2008). Is contemporary interest rate in conflict with Islamic ethics? *KYKLOS*, 61(1), 45–64.

Lichtenstein, S., and Slovic, P. (2006). *The construction of preference*. New York: Cambridge University Press.

MacGregor, D., and Slovic, P. (1986). Perceived acceptability of risk analysis as a decision-making approach. *Risk Analysis*, 6, 245-256.

Makower, F. (1972). *The Constitutional History and Constitution of the Church of England*. New York: Burk Franklin.

Resnicow, K., Soler R., Braithwaite R., Ahluwalia J., and Butler J. (2000). Cultural sensitivity in substance use prevention. *Journal of Community Psychology*, 28, 271-290.

Strauss, A., and Corbin, J. (1994). Grounded theory methodology: An overview. In N. K. Denzin & Y. S. Lincoln (eds.), *Handbook of qualitative research (pgs. 273-285)*. Thousand Oaks, CA: Sage.

Trope, Y., and Liberman, N. (2003). Temporal construal. *Psychological Review*, 110 (3), 403-421.

Schoon, N. (2007). Islamic finance: risk management challenges and the impact of Basel II. *Global Association of Risk Professionals*, **37**, 27-37.

Schuon, F. (1998). *Understanding Islam*. Bloomington, IN: World Wisdom

Sells, M. A. (1999). *Approaching the Qur'án*. Ashland, OR: White Cloud Press.

CHAPTER **23**

Cultural Influences Associated with Adversarial Recruitment

Lora Weiss, Elizabeth Whitaker, Erica Briscoe,
Ethan Trewhitt, and Margarita Gonzalez

Georgia Tech Research Institute
Georgia Institute of Technology
Atlanta, GA 30332, USA

ABSTRACT

The recruitment of individuals to participate in adverse behaviors often involves tactics that exploit cultural underpinnings to help persuade the individual to participate in the activity. The exploitation of aspects that culturally resonate with the individual may be what makes the final determination for the individual to 'tip to the other side'. Cultural aspects include beliefs, attitudes, and values that an individual acquires when being raised in a particular culture, yet limited research has been conducted in methods to analytically assess how cross-cultural influences may affect the recruitment process. Cross-cultural influences affect interactions between members of disparate cultural groups. As an example, consider members of Afghanistan's Taliban, who are predominantly Pashtun, trying to recruit from a subgroup of Pakistanis who are Punjabi. This chapter presents a modeling approach for capturing some of these cultural aspects and enabling analysis to support a cross-cultural understanding of the influences that may affect an individual's decision to support adversaries.

Keywords: Cultural Influences, Influence Diagrams, Recruitment Modeling, System Dynamics Modeling

218

INTRODUCTION

Recruiting individuals to support adversarial behavior has been on-going throughout the ages, whether it was to support the Crusades in Medieval times, the street gangs of today, or the more recent events of terrorism. Established adversaries target individuals, usually young men, to convert them into supporters of behaviors such as terrorism. Factors such as finances, family life, and religious beliefs are often considered (Bellerby 2009). Cross-cultural influences may have a key role in this recruitment process as national, tribal, and ethnic borders become more porous. Recruiters may exploit aspects of the targeted individual's culture to help persuade him into participation. Recruitment can occur in a number of ways, but usually an individual moves through a set of states, which at a high-level can be labeled as being a member of the General Population, the Grey Population, and an Actively Engaged Population. Cultural factors that may affect an individual's vulnerability to recruitment include collectivism (e.g., group norms, value of family) (Hofstede 2009), shaming to influence an individual to perform an act of revenge to restore honor (Wilson 2008), the expectation of submissive and obedient subordinates (Wunderle 2006), and the tendency toward a fatalistic belief (Grunow 2006). An example of how culture differences lead to violent crime at the beginning of the war in Bosnia, is that individuals who had 'mixed parentage' (a Serbian and Croatian parent) felt they had to demonstrate loyalty to one side or the other (pick a culture), and one way they could do this was by committing a violent crime against the other side. Ethnic, religious, and political conflicts are also aspects of the tilt. If a recruiter from one cultural is aware of aspects of other cultures, then that recruiter has a potentially larger pool of individuals from which to recruit. These cross-cultural influences are therefore of high interest to those seeking to curtail the recruitment.

Quantitative models allow exploration of different aspects of cultural influences and may provide indicators that increase one's propensity to being radicalized, recruited, deradicalized, or disengaged from the adverse behavior. Such tools support the discovery of trends, where cultural influences could be deciding factors that lead people to become involved in or abstain from the behavior. These models are valuable in that they provide a means to analyze and experiment with the impact of potential influences on population behavior (Zacharias et al. 2008). Subject matter experts (SMEs) often provide an interpretation of social behaviors and applicable psychological theories (e.g., Crenshaw 2000) and their insights support modelers in selecting approaches to represent a given social, cultural, or behavioral interpretation.

This chapter presents a modeling construct to explore the space of potential influences on those who may become involved in adversarial behavior before they have successfully acted. It presents a core model that incorporates submodels of recruitment, including radicalization, deradicalization, engagement, and disengagement. As such, cultural influences can be explored through variations in these submodels. The modeling methodologies include mind maps for preliminary knowledge engineering, system dynamics models (Sterman 2000) to represent overall system behavior, and influence diagrams to show causal relationships between different aspects of culture and society. The causal relationships are incorporated as submodels and are where a majority of the cross cultural interactions are captured. Figure 1 presents a summary of the approach, where the material in this chapter emphasizes the content in the blue boxes.

Specifically, information is acquired from multiple sources, including open literature, SMEs, doctrine, and reported scenarios. This information is captured via knowledge engineering methods and incorporated into various model types. The model components create a modeling environment with specific model instantiations, which are then subjected to evaluation by developers and SMEs. These models can be adapted and updated as new uses and new information are obtained. A report by Weiss (Weiss et al. 2009) details this modeling cycle for the specific application of understanding perpetration of improvised explosive devices (IEDs), complete with analyses that can be performed using such a construct. This chapter describes the front-end information associated with instantiating the modeling aspects for recruitment of terrorists in support of IED perpetration.

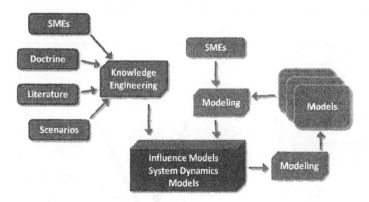

Figure 1. Model construct. Blue boxes are the emphasis of this chapter.

KNOWLEDGE ENGINEERING USING MIND MAPS

For SME information capture, a knowledge elicitation document was developed (see Weiss, et al. 2009 for details of the survey instrument). The document was a structured questionnaire used to interview SMEs and gather specific information including motivations, goals, beliefs, purposes, the environment, etc. This information was translated into mind maps for knowledge representation. Mind mapping is a semi-structured technique for initial representation and organization of knowledge. Figure 2 depicts a portion of one mind map showing how related concepts are interconnected via common elements. Mind maps provide a visualization of concept relationships by showing hierarchical connections between textual concepts. For this research, in addition to literature sources, seven SMEs from the US and UK were interviewed to create a collection of mind maps. Once the domain knowledge was formally structured, it was constructed into various representations to support multiple aspects of the modeling.

Figure 2 shows how Iraqi Nationalists may relate to Foreign Jihadis to collectively support an Iraqi Insurgency. Tribal feuds, which have cross-cultural influences internal to their own conflicts, may have reason to support (or not) the larger Iraqi Insurgency. Traditions and values may influence practices in recruitment for adversarial actions. The Iraqi Insurgency may maintain its customary practices, yet to effectively recruit, it may alter its methods to attract individuals from other cultures or foreign entities, e.g., from other jihadi elements or across tribes, but where the cross-cultural recruitment methods are tailored to the context of the targeted group. It is these influences between different aspects of culture and society that modeling seeks to represent so that tools can be developed to understand aspects of the cultural that affect the overall recruitment process. By capturing information in a structured manner, computational methods to represent relationships between entities can be more readily devised and implemented. This chapter describes incorporating information regarding IED perpetration into system dynamics models and influence diagrams to allow for analysis within this domain.

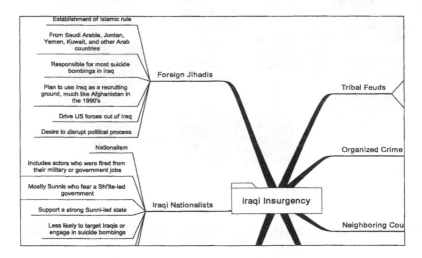

Figure 2. Mind map for preliminary knowledge structuring (portion)

SYSTEM DYNAMICS MODELS

System dynamics models are executable models that represent the dynamic behavior of complex systems over time (Sterman 2000). They use stocks and flows to represent system elements and relationships. Stocks represent inventories of accumulated entities (e.g., IEDs, people), while flows represent how entities move between stocks. Figure 3 presents a simplified schematic of a core IED perpetration model, with three focus areas: Materials and Supplies, IED Inventory, and Active Insurgents. Stocks are indicated with boxes. Flows are indicated with double-lined arrows. Clouds take the place of stocks and represent the world outside the scope of the model. The third line is of utmost interest in this chapter and is subsequently details; however, it is presented within the larger system context to understand that people's actions are coupled with materials and events. The core model is then used as a foundation from which submodels or model expansions are incorporated, and it encompasses many aspects of the larger terrorism and recruitment domain.

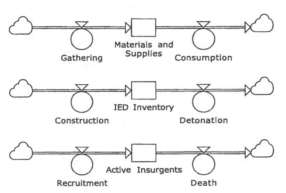

Figure 3. Simplified core model with three focus areas.

Materials and Supplies Focus Area. Figure 3 shows a single stock, Materials and Supplies, which represents the inventory of materials and supplies available to insurgent groups. The input flow, Gathering, represents actions that cause materials to accumulate. The output flow, Consumption, represents material used during IED construction. More details are behind this model (Weiss 2009), but for a recruitment exposition, it is kept simple. Other behavioral aspects may include gathering illicit items that cannot be readily purchased or financial backers of base materials (National Research Council 2008).

IED Process Focus Area. The IED process is presented in the center of Figure 3 and expanded in Figure 4. It contains five stocks representing how IEDs that move through phases from being constructed to emplaced to detonated. In practice, the process is varied, but this model is a generalized representation that SMEs felt was reflective of the process. Stepping through Figure 4, a typical IED is constructed either for a particular attack or stored for future use. Once it is constructed it moves into inventory, such as a warehouse, or a less conventional storage approach (e.g., distributed throughout the community). An IED may also be held by an individual with little knowledge of the item's actual purpose. It then moves from inventory to being emplaced and eventually triggered. At any point, counter-IED methods may destroy. This disruption detours the IED flow and deposits it in the Disrupted IEDs stock. Insurgent motivations for participating in this process are represented in the IED Motivation submodel. These motivations are included as one of many submodels that allow influences to be assessed.

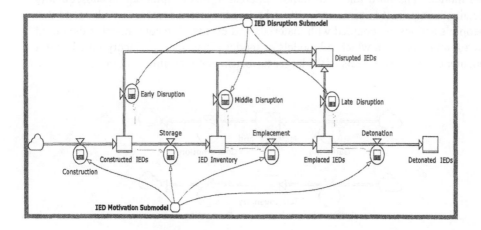

Figure 4. IED Process model components.

Active Insurgents & Personnel Focus Area. This section of the model is where the main influences associated with recruitment are incorporated and assessed. Understanding the behavior of people involved in IED activities includes understanding when and where they may be susceptible to being recruited or radicalized. The recruitment process results in several levels of categorization: the General Population, the Grey Population, and Active Insurgents. Each of these groups is represented as a stock within the system dynamics model. See Figure 5, which shows the transition of an individual from the General Population into a sympathizer (a member of the Grey Population susceptible to further radicalization) and then into an Active Insurgent. While the indoctrination and recruitment of insurgents is a nuanced and multi-faceted process (Gerwehr & Daly 2006), the model initially simplifies this so that the critical aspects can be identified. Figure 5 shows how the flows (radicalization, recruitment, deradicalization, and disengagement) are controlled by submodels. The core model sees the output of each submodel as a single value that influences subsequent stocks and flows. There are many points within the process where an individual may be influenced in deciding whether or not to engage in adversarial behavior. Elements that affect such decisions are captured in these submodels. By allowing submodels to plug-in or to be swapped with other submodels, the system allows for differing views, influences, cultural factors, or perspectives to be incorporated and assessed.

INFLUENCE DIAGRAMS FOR SUBMODELS

Swappable submodels allows for the development, modification, and reuse of model components as modules within the model. A submodel based on a particular set of assumptions about cultures or behaviors can be replaced by a different submodel for analysis or refinement or to incorporate differing views experts may have. This research leverages influence diagrams to create submodel influences that represent behavior. An influence diagram is a graphical representation of a group of causal relationships and offers a method to couple the essential elements of a situation, including decisions, uncertainties, and objectives, by describing how they influence each other. This section describes two of the submodels that support the core model and where cultural and cross-cultural influences can be incorporated. The two submodels are the Population Radicalization-Deradicalization Submodel and the Insurgent Recruitment-Disengagement Submodel.

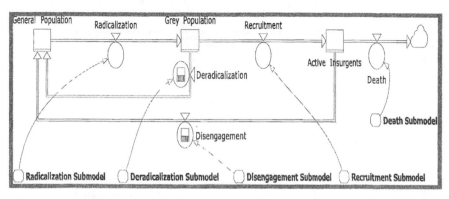

Figure 5. Insurgent personnel model components.

Radicalization and Deradicalization Submodel. Radicalization is defined in many ways and means different levels of involvement to different people. For this exposition, it represents the transition of a person from within the General Population to the Grey Population. This occurs when a previously neutral person has taken a position of sympathy for insurgent beliefs. Insurgent groups achieve this through various means, such as spreading broad propaganda supporting their goals, by using community roles as an influence, etc. When a person holds a positive view of the insurgents' goals and tactics, that person is vulnerable for recruitment. This stage has less outward manifestations and may exist to varying degrees. It may occur as an overt decision by the participant or it may be a gradual process in which a person becomes more sympathetic to an insurgent group.

Some of the known factors that contribute to this conversion include: poverty, ignorance, injustice, and bad governance (Shemella 2009). The range of factors can be roughly categorized as external and internal. Internal factors include variables such as strength of personal ideology, the increase of self-esteem that results from joining a cause, and the resentment that is felt toward such matters as military presence and the loss of loved ones in military engagements (Gerwehr & Daly 2007). External factors include the influence felt by both pro- and anti-insurgency religious leaders, family and peer pressure. More general factors that affect individuals' behaviors can be grouped into four main categories: (i) Camus variables include those related to the morality of a population (e.g., is the population particularly violent or does common religious doctrine condemn violence), (ii) Dewey variables consist of social aspects, such as freedom of speech, (iii) Smith variables represent economic variables within the society (e.g., GDP), and (iv) Maslow variables cover quality of life factors, such as the availability of water and medical care. The collection of these factors can be combined using an influence diagram (see Figure 6), which then supports assessing the broad influence on radicalization. This grouping of factors is based on Bartolomei, et al. (Bartolomei et al. 2004). Deradicalization occurs when the attitudes of an individual are moderated from the radical views to the more mainstream views of the general population.

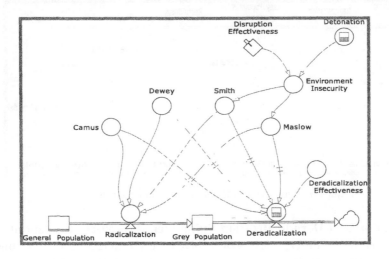

Figure 6. Population radicalization and deradicalization submodel

Recruitment and Disengagement Submodel. Recruitment and disengagement represent the voluntary or coerced actions of persons joining or leaving the insurgency. As a person becomes an active participant, this person is considered recruited. This may be an overt decision by the participant, or it may be a gradual process in which an insurgent group eases a sympathizer into increasingly more severe tasks. Foreign insurgents are more likely to use force to coerce a potential recruit into joining the insurgency, since they do not have as deep an understanding of the local culture, while local insurgents are more likely to use social and cultural ties to influence a potential recruit. These discrete avenues for recruitment can by modeled using separate submodels and integrated into the larger model. The model then considers a person to be recruited when he or she is actively involved.

Disengagement occurs when someone has left the group of active insurgents and reduces the number of active insurgents. The Recruitment and Disengagement submodel is presented in Figure 7, where the variables surrounding the Recruitment and Disengagement flows represent influences that drive those decisions.

Some amount of disengagement occurs naturally, as many terrorists become disillusioned with their life as a terrorists, either because the lifestyle is limiting, because they realize the futility of the effort, because it is unprofitable (Bergen), or because they realize that they have exceeded their moral limits (Kershaw, 2010). This natural 'flow' from active insurgent could be augmented, perhaps by promoting ideas that undermine the stature of terrorist leadership. The promotion of ideas, akin to the western concept of marketing, has several potential applications in terms of increasing disengagement. For example, many Iraqis, while angry at the U.S. coalition for the state of insecurity of their communities, are also disenchanted with insurgents over the destruction and bloodshed that has resulted from their violent tactics (Clark 2005). Psychological studies have shown that hardened terrorist would likely reject such a technique and cognitive response theory suggests that their attitude might become even more steadfast in the face of these types of persuasive materials (Cragin & Gerwehr 2005; Tess & Conlee 1975).

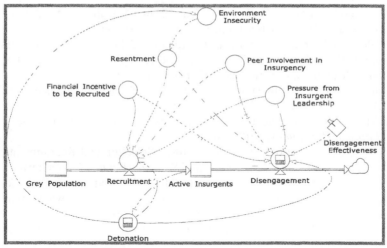

Figure 7. Insurgent recruitment and disengagement submodel

CONCLUSIONS

This chapter focuses on component modeling of socio-cultural and cross-cultural behaviors related to terrorist recruitment and the motivation to engage in adversarial behaviors. The approach combines computational and social science research to understand activities and influences within a population. The benefit is that, although integrated models will not precisely predict who will become recruited, they can provide insight into the relative importance of factors and influences, e.g., it may be perceived that the best intervention point is to influence the General Population before they are radicalized, but if a large part of the population is inherently radicalized, there may not be much benefit in working with the general population. A more effective approach may be to address the flow from the Grey Population to the further radicalized stage of Active Insurgent. This modeling construct may also expose previously unconsidered aspects of the problem so that insight is gained on issues that may not otherwise be considered. The integrated components then provide a construct for analysis of recruitment deterrents and potential intervention points associated with adversarial behavior.

REFERENCES

Bartolomei, J., Casebeer, W., & Thomas, T. (2004). *Modeling Violent Non-State Actors: A Summary of Concepts and Methods.* United States Air Force Academy, Institute for Information Technology Applications, Colorado.

Bellerby, R. (2009) *How Crusaders Were Recruited in Medieval Times*, available at http://highmiddleages.suite101.com/article.cfm/how_crusaders_were_recruited_in_medieval_times

Bergen, P. (2007). *War of Error.* The New Republic. Available at http://www.newamerica.net/publications/articles/2007/war_error_6154

Clark, M. (2005). "Observations on Local Insurgents and Foreign Fighters in Iraq: an Interview with Mark Edmond Clark." Interviewed by Columbia International Affairs Online. Columbia International Affairs Online.

Cragin, K. and Gerwehr, S. (2005). *Dissuading Terror. Strategic Influence and the Struggle Against Terrorism.* RAND Monographs.

Crenshaw, M. (2000). The Psychology of Terrorism: An Agenda for the 21st Century. *Political Psychology, 21*(2), 405-420.

Gerwehr, S. & Daly, S. (2006). Al-Qaida: Terrorist Selection and Recruitment. In D. Kamien (Ed.) *The McGraw-Hill Homeland Security Handbook.* McGraw-Hill.

Grunow, Carl D. 2006. "Advising Iraqis: Building the Iraqi Army," *Military Review.* Available at: www.army.mil/professionalwriting/volumes/volume4/october_2006/10_06_1_pf.html

Hofstede, Geert. 2009. *Cultural Dimensions.* Available at: http://www.geert-hofstede.com/

Kershaw, S. (2010). *The Terrorist Mind: An Update.* The New York Times.

National Research Council. (2008). *Disrupting Improvised Explosive Device Terror Campaigns.* Washington, D.C.: National Academies Press.

Shemella, P. (2009) *Reducing Ideological Support for Terrorism.* Strategic Insights. Vol III, Issue 2. Naval Postgraduate School. Monterey, California.

Sterman, J. (2000). *Business Dynamics: Systems Thinking and Modeling for a Complex World* (1st ed.). Boston: McGraw-Hill/Irwin.

Tess & Conleee. (1975). *Some effects of time and thought on attitude polarization.* Journal of Personality and Social Psychology, vol 31.

Weiss, L., Whitaker, E., Briscoe, E., Trewhitt, E., (2009). *Modeling Behavioral Activities Related to Deploying IEDs in Iraq*. Georgia Tech Research Institute, Technical Report #ATAS-D5757-2009-01, Atlanta, GA.

Wilson, James S. 2008. "Iraqi Sunnis' Cultural Receptivity of Islamic Extremists' Beliefs." Maxwell Air Force Base. Available at: https://www.afresearch.org/skins/rims/.../display.aspx

Wunderle W. D. (2006). *Through the Lens of Cultural Awareness: A Primer of US Armed Forces Deploying to Arab and Middle Eastern Countries*. Combat Studies Institute Press.

Zacharias, G., MacMillan, J., & Van Hemel, S. (Eds.). (2008). *Behavioral Modeling and Simulation: from Individuals to Societies*. National Research Council.

Chapter 24

An Evidence-Based Framework for Decision Making in Culturally Complex Environments

Lisa Costa

The MITRE Corporation
7515 Colshire Drive
McLean, VA 22102
lahc@mitre.org

ABSTRACT

How do we assess the short- and long-term ramifications of our actions in cultures different from our own? We require input from multiple and disparate sources for understanding the dynamics of such complex environments. These multiple sources can yield large amounts of data and drive the need for frameworks that support defensible assessments that help reduce data bias while increasing decision making transparency. The framework described in this paper employed multiple data sources feeding qualitative, quantitative, and computational models, methods, and tools supporting a nation-state-level strategic analysis. The result was a set of prioritized actions to meet operational objectives. Perhaps more important, however, were the observations made along the way about the capacity of United States and/or coalition staff to perform such assessments in current operational environments.

Keywords: Evidence-Based Analysis, Bias Reduction, Subject Matter Expert Judgment, Socio-Cultural Decision Science, Strategic Analysis

INTRODUCTION

U.S. Army Major General Michael Flynn, the top military intelligence officer in Afghanistan, recently observed, "Eight years into the war in Afghanistan, the U.S. intelligence community is only marginally relevant to the overall strategy. Having focused the overwhelming majority of its collection efforts and analytical brainpower on insurgent groups, the vast intelligence apparatus is unable to answer fundamental questions about the environment in which U.S. and allied forces operate and the people they seek to persuade" (Flynn et al., 2010). But how do we assess the short- and long-term ramifications of our actions in cultures different from our own? Whether those actions fall within one, or span the spectrum of, the instruments of U.S. national power and that of our coalition partners, there is often a significant difference between how we see ourselves, how we see others, and how others see us. It is often difficult to take the pulse of any nation, especially one that is highly factionalized. We often rely on articles, news sources, and subject matter expert (SME) judgment as our barometer, but have poor means by which to fuse and weigh various opinions, findings, probabilities, and their biases in context to events (Tversky and Kahneman, 1982). In addition, our own biases incentivize us toward assumptions and conclusions that skew any general ability we may have to forecast an outcome (Fukuyama, 2007).

FRAMEWORK

Much as a legal team must piece together eyewitness accounts, deposition remarks, phone records, financial transactions, and other key data to identify inconsistencies in stories, timelines, and geographic locations, military analysts and planners must be able to do the same – but on a much larger, more complex scale, understanding both current and historical context as they piece together critical understanding of a dynamic, culturally complex environment. As in the case of the legal team, military analysts must strive to understand "ground truth" in environments where evidence may be anecdotal, inferred, and emotionally charged. As social scientists can attest, triangulation (the practice of cross-checking findings with multiple data points and sources) on broad scales is difficult (Wilkinson, 2008), and achieving some level of understanding of the complexities of a situation may be the best possible outcome that can be realized.

The framework depicted in Figure 1 shows a generic repeatable framework for achieving greater understanding of complex environments and developing and prioritizing courses of action based on that understanding. The purpose of the framework is to help move the decision maker away from unaided intuition and provide transparency, as well as historical accountability, into the data used to drive the decision-making process.

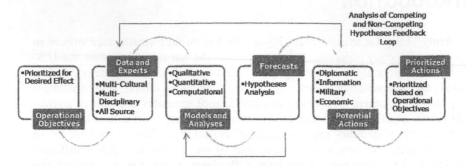

Figure 1.1 Strategic Assessment Framework

The framework was used to conduct a detailed strategic analysis of the nation of Sudan in 2007, a geographic region considered to be a "complex" state. That is, Sudan possesses porous borders, has many competing internal and external interests, and is representative of many countries in Africa where there is a lack of vetted intelligence analysis products available to military operational planners. Africa is also where seven of the top ten failed states are located (FSI, 2009).

Operational Objectives – The process was driven by the operational objectives, or outcomes, for the effort. The operational objectives for this study were to develop actionable recommendations, placing less emphasis on the military aspects of the Diplomatic Information Military and Economic (DIME) process to: (1) reduce human suffering, (2) promote regional peace and stability, and (3) deny safe haven for international terrorists and criminals. No assumptions were made about the causation of observed conditions; rather, the operational objectives sought to raise the study to the level of developing actions that could address the root causes of the observables and not simply to address observed behaviors that were based on the root causes (Renfro and Newcomb, 2004).

Data and Experts – Once the operational objectives were established, data and experts were brought together. Approximately 80 people from 26 organizations across academia, industry, and the U.S. government came together to form a geographically distant and multidisciplinary team. Representatives from the Department of State, USAID, academia, and the U.S. intelligence community provided background information to all participants to set the stage for an initial baseline understanding of the challenges in Sudan's diverse physical environment, demography, economy, ethnography, and geo-political landscape. The set of SMEs assembled on the Sudan included anthropologists, historians, aid workers, members of the Sudanese diaspora, statesmen, economists, computer scientists, military analysts and planners, modelers, cognitive psychologists, mathematicians, and policy makers. Members of the team were representative of the factionalized nature of Sudan, and that diversity often caused passions to run high. As a result, data points were as diverse as the individuals. SME judgment was augmented with a massive data-mining effort, a formalized analytic effort to support requests for information (RFIs), and an elicitation process for interviewing senior leaders intimately familiar with the Sudan to provide a strategic perspective of the issues in a consistent manner and to find some levels of commonality among the various factions that might be exploited (Numrich, 2007). All of these sources were used to triangulate root causes and resulting behaviors.

Models and Analyses – Models, methods, and tools (MMTs) encompassed qualitative, quantitative, and computational approaches that supported analyses at the regional, national, sub-national, group, and individual levels. Some of the models included the following:

- A quantitative econometric-based model measuring the probability of civil war focusing on greed and grievance hypotheses at the national level (Collier and Hoeffler, 2005) (Bankes, 2002)

- A quantitative regressive dynamic system of equations for measuring relative political capacity identifying macro political and econometric model at regional and sub-national levels (Feng et al., 2000)

- A quantitative structural system of equations for measuring the dynamics of relative political capacity identifying macroeconomic structure at the regional, national, and sub-national levels (Kugler et al., 2005) (Arbetman-Rabinowitz and Johnson, 2007)

- A computational dynamic system of equations agent-based model for measuring the likelihood of conflict and conflict intensity by simulating structural preconditions of interstate conflict residing in notions of power parity and policy dissatisfaction at the national and sub-national levels (Tammen et al., 2000)

- A computational state stability dynamics model measuring violent dissident level and anti-regime acts using concepts of non-linear dynamics, feedback loops, state capacities and loads, and tipping points at the sub-national and group levels (Choucri et al., 2006)

- A qualitative and computational socio-cultural agent simulation model for measuring individual and group social identities and attitudes simulating the effect that events have on group and individual opinions, attitudes, and identities (Grier et al., 2008)

- A qualitative and computational partially observable Markov decision process agent simulation for measuring individual and group actions, beliefs, and behaviors (Marsella et al., 2004) (Pynadath and Marsella, 2005)

- A qualitative and computational game and decision theory model for measuring individual and group position and support for issues by simulating stakeholder and decision-maker influence, relationships, and negotiations (Abdollahian et al., 2006)

Additional analyses included multiple social network analyses, a center of gravity analysis, a water use study, and an insurgency modeling activity. Some of the tools used for these analyses included Carnegie Melon University's Organization Risk Analyzer (ORA), MetaCarta's geosearch and geotagging applications, Teleport Very Large Exploration (VLX) webspider, SAP Business Objects entity extraction and part of speech tagger, Analytic Technologies' UCINET social network analysis software, Convera's Retrievalware for search and discovery, and a MITRE Corporation-developed pipeline manager, SQL database, virtual collection builder and other analytic prototypes. Massive amounts of all source unstructured and structured data were processed (part-of-speech tagged, entity extracted, statistically analyzed, geocoded, temporally coded, and normalized) for this effort.

Forecasts – Seven specific possible root causes that could affect the three operational objectives were defined through the previous steps. These were not only seven specific causes in which the end user was interested, but they were representative of the level of causation that was of interest (e.g., the Government of Sudan's perceived legitimacy). From an analytic perspective, these seven root causes constituted seven independent variables, while the three operational objectives noted above constituted three dependent variables. When combined this resulted in twenty-one hypotheses that were tested.

Potential Actions – From the original twenty-one tested hypotheses, a set of DIME actions was constructed and tested. These DIME actions were also treated as hypotheses, were tested, and fed back iteratively through the framework process as seen in the feedback loop on Figure 1. It became apparent at this stage of the framework that the three operational objectives driving the assessment should have been weighted (i.e., it is entirely feasible that to ensure stability, one may have to be willing to turn a blind eye to the humanitarian crises or vice versa). This weighting of objectives would have added an additional dimension to the richness of the prioritization of actionable recommendations and should be considered in any subsequent use of the framework. Likewise, timing and resources needed to achieve effect should be included. To generate potential actions and test them across a set of potential futures, at minimum the components of time and resources are required. From a temporal perspective, severely restrictive economic sanctions may be successful in the short term to meet an objective, but may end up to be the wrong strategy for the long term. Likewise, the amount of resources required to achieve a specific percentage of an objective needs to be calculated to determine its overall cost and return on investment. Without taking these factors into consideration, potential actions lack real-world operational relevance unless the point of the effort is to determine what it would take to achieve an objective, no matter what the time or cost.

Prioritized Actions – The result of this assessment was a set of prioritized DIME actions, driven by the operational objectives. The DIME actions were prioritized by performing sensitivity analysis of the findings based on differentiating data input values from multiple, disparate sources (e.g., models, SME judgments, data mining, etc.). An analytical approach developed by The MITRE Corporation, called topological hypotheses analysis toolkit (TopHAT™), provided analysts the ability to measure the "truthfulness" of each hypothesis being tested as a function of the strength of disconfirming or confirming evidence and evidence veracity (i.e., source). This approach used algorithms supporting Analysis of Non-Competing Hypotheses (ANCH) rather than Analysis of Competing Hypotheses (ACH) (Heuer, 1999) as DIME actions are rarely executed individually but in concert with one another in a synchronized fashion. Other approaches which may be useful at this point are those that after a set of DIME actions are prioritized, allow the planner to perform sensitivity analysis by revealing combinations of actions that produce extreme outputs and identifying the actions that are most robust and which minimize overall regret; that is, the actions that perform well over the widest range of possibilities (Lempert et al., 2003).

EVIDENCE-BASED ANALYSIS: A CRITICAL APPROACH FOR SOCIO-CULTURAL UNDERSTANDING

The framework described above is one that uses evidence-based analysis, treating potential courses of action as hypotheses, examining confirming or disconfirming evidence to rate the validity or truthfulness of a hypothesis. It is comparable to practices in evidence-based medicine, evidence-based policy, and evidence-based management where we use science, and not expert judgment, to determine what practices are most effective. In the case of evidence-based analysis, the notion is that certain analytic practices and methods help zero in on truth while others do not. As a quick example, it is very robustly established that when making a forecast or prediction, the accuracy of the forecast of a group of experts that results from an unstructured meeting of those experts is the same as randomly choosing one of the expert's pre-meeting forecasts and going with that forecast. Simply taking the average and not letting the experts meet, does robustly better than letting them meet. Most things that actually work in terms of evidence-based analysis are counterintuitive, which makes it difficult to convince organizations to adopt them (Lehner and Cheikes, 2009). Yet it is the very nature of their counterintuitiveness that reveals potential, critical, unintended consequences in courses of action.

In his paper about intelligence in Afghanistan, MG Flynn wrote, "Select teams of analysts will be empowered to move between field elements, much like journalists, to visit collectors of information at the grassroots level and carry that information back with them to the regional command level. These items will integrate information collected by civil affairs officers, PRTs [Provincial Reconstruction Teams], atmospherics teams, Afghan liaison officers, female engagement teams, willing non-governmental organizations and development organizations, United Nations officials, psychological operations teams, human terrain teams, and infantry battalions, to name a few" (Flynn et al., 2010).

Socio-cultural understanding is both an excellent and necessary objective. As seen in Figure 2, it is a critical cross-cutting component to many different mission types.

Figure 1.2 Socio-cultural Knowledge Base (Costa, 2008)

In developing a proposed socio-cultural knowledge taxonomy for AFRICOM, fourteen different taxonomies were used to develop a working concept for capturing socio-cultural knowledge. The final high-level categories of the proposed taxonomy included Communications, Economy, Geography, Government, People, Security, Transportation, Transnational Issues, and U.S. Relations. Each of the high-level categories had several primary subcategories, and for each primary subcategory, there were many secondary subcategories. Taking the high-level category of Security as an example, the primary subcategories included Armed Insurgents, Criminal Networks, Critical Incidents, Indigenous Police Force, Military, Sponsors/Supporters of Non-secure Environments, and Unarmed Radical Groups. Taking just the primary subcategory of Indigenous Police Force, the secondary subcategories of that entry were broken down further still into affinities; capabilities; corruption; leadership; legitimacy; locations; order of battle; power; projection; recruiting; structure; and training, tactics, techniques, and procedures (Costa, 2008). Capturing this level of detail for understanding and the changing values of data over time will require not only sustainable collection, processing, and dissemination infrastructures, but significant human resources.

As learned in these efforts, the level of engagement discussed in the Flynn paper will result in numerous and varied data points, both from the perspective of the data providers and the data collectors. From these myriad data points, military staff will have to deconflict and seek a common understanding. Reaching a common understanding includes an interpretation process which needs to be captured and communicated to provide transparency and, due to multiple redeployments, historical accountability into the data and thought processes used to drive decisions. Part of capturing this interpretation process is to attempt to reduce individual and group bias including those introduced by the varying backgrounds and experiences of the individuals involved in the process. Types of cognitive bias include hindsight bias, confirmation bias, and selection bias. These types of cognitive bias can be exacerbated by additional personal bias such as an individual's culture, language, experience, geography, gender, religion, politics, and other backgrounds.

Additionally, and perhaps most critically, the cadre of experts needed to achieve an in-depth socio-cultural understanding of an environment simply does not reside within most organizations. For example, a key lesson learned from the Sudan Strategic Assessment was that the socio-cultural models that were most useful were the least likely to be able to be run and interpreted by anyone other than their considerably expert developers. Because of this constraint, socio-cultural modeling is best done in unclassified environments and then moved to other environments if necessary. Finally, building the trust and understanding that ad hoc, multidisciplinary, multicultural teams need to operate effectively does not occur overnight.

Think of the differences between a field-trained anthropologist, a mathematician, and a military planner. Each discipline has its own unique lexicon and style of operating or attacking a problem. Sustaining a capability that would allow deep looks into culturally complex environments may require establishing a more enduring and robust capability within the U.S. Government, building and maintaining significant relationships with academia and industry to provide similar assessments and advice to operational planning. Such a capability would address critical quick-reaction contracting needs, SME vetting, human social-cultural-behavioral education and training for staff, transition hub for proven social science methods, models, and tools (MMTs), and being the focal point, or center of excellence, for socio-cultural understanding for multiple mission sets.

The establishment of such a capability is achievable, but it represents a technical, cultural, systems engineering, security, policy, and integration challenge. What we have learned through this effort is that it must be approached with the full application of the scientific method. A recent report by JASON (an independent group of scientists that advise the U.S. Government on matters of science and technology) which looked at the nation's ability to anticipate and assess the risk of "rare events" (e.g., terrorist attainment and/or use of weapons of mass destruction) specifically noted that "the complexity of social sciences problems has led some to advocate the suspension of normal standards of scientific hypothesis testing, in order to press models quickly into operational service. While appreciating the urgency, JASON believes such advice to be misguided" (JSR-09-108, 2009). Evidence-based analysis is the application of the scientific method to analysis, and the introduction of analyzing socio-cultural phenomena requires that much more rigor in dealing with the biases discussed earlier. In fact, the socio-cultural drive to reduce bias from the Sudan Strategic Assessment decision process was driven largely by what were viewed as deficiencies in Operation Iraqi Freedom in using what was seen as an extremely biased source, Iraqi exiles, (oftentimes referred to as the "Chalabi effect") for much of the situational assessment in Iraq. It was also driven by the desire to avoid the 7,000 mile screwdriver – a pejorative term used to describe the predilection of some military commanders in the past to micromanage operations from the security of Washington, D.C., with little understanding of the native environment (Sprouli and Kesler, 1992).

CONCLUSIONS AND WAY AHEAD

The framework described in this paper can help MG Flynn's vision become a reality. There is much work to do to, but it is achievable. This framework has been proven to be effective to capture and provide transparency into the vast amount of various data points and sources needed for understanding a socio-culturally complex environment. The framework must be pared with advanced approaches in information science (e.g., data extraction, part-of-speech tagging, geographic coding, temporal coding, source provenance, etc.) to automatically process and visualize the data, allowing both producers and consumers of the data to better understand nuances and opinions driving planning and "what if" scenarios. While models, methods, and tools are important, they cannot replace analytic rigor. Combined, they can help decision makers make sense of a complex problem space and shape actionable recommendations. It is critical that expectation management prevent decision makers from expecting pinpoint predictions from this type of process. Socio-cultural knowledge is critical, but it is yet one resource in our model of interacting maturely, knowledgably, and assuredly in a multicultural environment. Decision makers instead should expect from socio-cultural insight an understanding of the dynamics of a problem and view socio-cultural knowledge, along with social science modeling, as tools in the overall toolkit that moves the decision maker further away from unaided intuition (Armstrong and Greene, 2005).

REFERENCES

Abdollahian, M., M. Baranick, B. Efird, and J. Kugler. "Senturion: A Predictive Political Simulation Model." Defense & Technology Paper #32. Center for Technology and National Security Policy, National Defense University. 2006.

Arbetman-Rabinowitz, M. and K. Johnson. "Relative Political Capacity: Empirical and Theoretical Underpinnings." Presented at Claremont Graduate University School of Politics and Economics 2007 Conference on Political Economic Indicators. 2007.

Armstrong, J.S. and K.C. Green., "Demand Forecasting: Evidence-Based Methods." *Strategic Marketing Management: A Process-Based Approach.* Eds. Luiz Moutinho and Geoff Southern. Cengage Learning Business Press. Hampshire, UK. 2009.

Bankes, S.C. "Tools and Techniques for Developing Policies for Complex and Uncertain Systems." Proceedings of the National Academy of Sciences, 99. May 2002. pp. 7263-7266.

Choucri, N., C. Electris, D. Goldsmith, S. Madnick, D. Mistree, J.B. Morrison, M. Siegel, and M. Sweitzer-Hamilton. "Understanding & Modeling State

Stability: Exploiting System Dynamics." Proceedings of the 2006 IEEE Aerospace Conference, Big Sky, Mont. 2006.

Collier, P. and A. Hoeffler. *Understanding Civil War: Evidence and Analysis, Vol 1.—Africa.* World Bank. Washington, D.C.

Costa, L. "Taxonomy of Analytic and Geospatial Layers to Support Africom Intelligence and Operations Planning." The MITRE Corporation. 2008.

Failed State Index (FSI). Foreign Policy and The Fund for Peace <http://www.foreignpolicy.com/>. 2009.

Feng, Y., Kugler, J., and P. Za. "The Politics of Fertility and Economic Development." *International Studies Quarterly.* Vol 44 Issue 4. 2000. pp 667-693.

Flynn, M.T., M. Pottinger, and P. Batchelor. "Fixing Intel: A Blueprint for Making Intelligence Relevant in Afghanistan." Center for a New American Security (CNAS) <http://www.cnas.org/node/3924>. 2010.

Fukuyama, F. *Blindside: How to Anticipate Forcing Events And Wild Cards in Global Politics.* Brookings Institution Press. Washington, D.C. 2007. p 2.

Goodman, P.S. "China Invests Heavily in Sudan's Oil Industry: Beijing Supplies Arms Used on Villagers." *The Washington Post.* 23 December 2004. p A01.

Grier, R., B. Skarin, A. Lubyansky, and L. Wolpert. "SCIPR: A Computational Model to Simulate Cultural Identities for Predicting Reactions to Events." Aptima. <http://www.aptima.com/publications/2008_Grier_Skarin_Lubyansky_Wolpert-2.pdf>. 2008.

Heuer, R. *Psychology of Intelligence Analysis.* Center for the Study of Intelligence, Central Intelligence Agency. <https://www.cia.gov/library/center-for-the-study-of-intelligence/csi-publications/books-and-monographs/psychology-of-intelligence-analysis/PsychofIntelNew.pdf>. 1999.

Kugler, J., M. Abdollahian, M. Arbetman-Rabinowitz, Y. Feng, P. Zak, K. Johnson, and R.L. Tammen. "Identifying Fragile States." Final Report for the Defense Advanced Research Projects Agency (DARPA). 2005.

JASON Study JSR-09-108. "Rare events." FAS. <http://www.fas.org/irp/agency/dod/jason/rare.pdf>. 2009.

Lehner, P., and B. Cheikes. "Cognitive Support to Analysis: Better Methods Aided by Software." Briefing to The MITRE Corporation Board of Trustees, Tampa, FL. 2009.

Lempert, R.J., S.W. Popper, and S.C. Bankes. "Shaping the Next One Hundred Years: New Methods for Quantitative, Long-Term Policy Analysis." The Rand Corporation. RAND MR-1626-RPC. 2003.

Marsella, S.C., D.V. Pynadath, and S.J. Read. "PsychSim: Agent-Based Modeling of Social Interactions and Influence." *Proceedings of the International Conference on Cognitive Modeling.* Pittsburgh, PA. 2004. pp. 243 248.

Nickerson, R. "Confirmation Bias: A Ubiquitous Phenomenon in Many Guises", *Review of General Psychology,* Vol 2 No 2. 1998. pp 175-220.

Numrich, S. "Conversations on Sudan." Institute for Defense Analyses. 2007.

Pynadath, D.V. and S.C. Marsella "PsychSim: Modeling Theory of Mind with Decision-Theoretic Agents." *Proceedings of the International Joint Conference on Artificial Intelligence.* 2005. pp. 1181-1186.

Renfro, R. and D. Newcomb. "Air Force Capability Review and Risk Assessment (CRRA) Analytic Methodology." Air Force Studies and Analysis Agency AFSAA/SAPT. Washington, D.C. 2004.

Sproull, L. and S. Keisler. *Connections: New Ways of Working in the Networked Organization*. MIT Press .Cambridge, MA. 1992. p 117.

Tammen, R.L., J. Kugler, D. Lemke, C. Alsharabati, B. Efird, and A.F.K.Organski. *Power Transitions: Strategies for the 21st Century*. Seven Bridges Press, LLC. New York. 2000.

Tversky, A. and D. Kahneman. *Judgment under Uncertainty: Heuristics and Biases*. Cambridge University Press USA. New York. 1982.

Wilkinon, Emily. "Monitoring and Evaluation (M&E) Sourcebook: Methods and Process: Processing and Analyzing Data." Provention Consortium. <http://www.proventionconsortium.org/?pageid=69>. 2008.

Williamson, R.S. "Sudan: From the Diplomatic Front Line in Search of Peace." Heritage Lecture #1105. The Heritage Foundation. 2009.

ACKNOWLEDGEMENTS

The author wishes to thank the many individuals and organizations that supported the Sudan Strategic Assessment for the US Government. Particular appreciation goes to Dr. Paul Garvey, Dr. Mark Maybury, Mr. Jerry Cogle, and Dr. Gary Klein of The MITRE Corporation for valuable discussion and feedback on earlier drafts of this paper.

Chapter 25

A Multi-Scale Model of Cultural Distinctions in Technology Adoption

Erica Briscoe, C.J. Hutto, Carl Blunt, Ethan Trewhitt, Lora Weiss,
Elizabeth Whitaker, Dennis Folds

Georgia Tech Research Institute
Atlanta, GA 30332-0822, USA

ABSTRACT

This chapter presents research on the benefits of multi-scale modeling to capture cultural and behavioral influences, using the adoption of new technology as a domain of demonstration. The spread of new technology depends on the decisions of individuals with diverse beliefs, traditions, and dispositions, which can be modeled using various approaches. The work here uses an agent-based model to represent factors at an individual level, and a system dynamics model to represent societal influences on technology acceptance. These models exchange information by having the society-level model generate outputs that are trends and which are used as inputs to the individual-level agent-based model, while the aggregation of individual behaviors supply reciprocal input to the higher-level society model. This type of federated modeling construct provides an analysis tool for examining the potential implications of economic and technology-related policies that may affect human behavior rising from one's cognition, social interactions, and culture. The approach is also extensible to modeling decision-making by individuals in other activities beyond technology acceptance.

Keywords: multi-scale modeling, agent-based modeling, system dynamics, culture, e-commerce

INTRODUCTION

When modeling individuals as part of a group or society, a single modeling paradigm may be inadequate to capture the nuances and interactions within the domain. Individual behaviors are often modeled using agent-based models, which are able to represent causal relationships among individuals using various methods, such as collections of rules (Bonabeau, 2002). This allows personality and other internal attributes of an individual to be incorporated within the model. Society-level interactions, on the other hand, are often modeled using paradigms that focus on trends among large numbers of people and can be incorporated using methods such as system dynamics modeling (Sterman, 2000). This allows high-level behaviors to be represented that are not apparent among individuals or even groups of individuals. Each modeling paradigm is suited for a particular level of interaction, so that an ideal system integrates multiple approaches to model at multiple scales.

This chapter describes a system that uses an agent-based model to encompass the detailed behavior of individuals and their immediate network of relationships, as well as an influence-based system dynamics model to show how society-level influences function interactively and with respect to the individual agent. The system combines these two model types with specific interaction points to create a single integrated model of human behavior and culture. The integrated model is then applied to the adoption of a technology to determine how the characteristics of individuals affect the use and prevalence of a new technology in a society.

Modern society is distinguished by the constant and rampant introduction of new technology, including cell phones, laptops, and the Internet. Many factors contribute to the choice to observe, slowly accept, or quickly adopt a new technology, and these factors often derive from longstanding values, beliefs, and traditions, (or 'culture'). As such, technology adoption and usage often exhibit varying patterns (e.g., Straub, Keil, & Brenner, 1997). The spread of a new technology typically depends on the decisions of varied individuals, each with his own characteristic set of beliefs, traditions, and dispositions. Although the decision process has often been described at the intra-individual level of cognition, it is subject to social and cultural influences at both the interpersonal level as well as the societal level. Thus, it is a confluence of several factors at each of these levels of human socio-cognitive and cultural interaction which contribute to the overall decision process associated with an individual (and aggregately, society) adopting the new technology. Relevant literature in cognitive psychology, social psychology, economics, and sociology is abundant, independently describing and modeling, at various levels of detail, the processes and mechanisms for intra-individual, interpersonal, and societal levels of human socio-cognitive and cultural behavior. In comparison, there is a relatively small body of literature that attempts to develop a multi-scaled view of the phenomena.

A critical factor in technology adoption is whether the product provides a substantial benefit, but culture also plays an important role. Behavioral norms may resist change and eventually override any technological effects (Watson, Ho, & Raman, 1994). Also significant are societal factors and individual characteristics such as user appeal (Carr, 1999). Numerous theories have been developed to explain and model the adoption of technological innovations. Macro-level theories focus on broad, population-wide influences where innovation encompasses a wide range of technologies and economic practices. Micro-level theories focus on individual adopters of a specific innovation. Although prominent theories exist to explain the spread of innovations (e.g., the technology adoption lifecycle, [Beal, Rogers & Bohlen, 1957; Rogers, 1962] and the Technology Acceptance Model [Davis, 1986]), it has been noted that many theories lack an adequate account of the various types of influence (e.g., Davis, et. al., 1989). Newer, composite theories may prove to be more predictive. The Unified Theory of Acceptance and Use of Technology (UTAUT) combines eight of the most prominent technology acceptance models (Venkatesh, et al., 2003).

Starting with the UTAUT as the theoretical basis for integration, the research in this chapter presents a multi-scale model to represent influences on the adoption of technology at three interrelated levels of human socio-cognitive and cultural behavior. At the intra-individual level, a person's decision making process is captured using an innovative approach called the Cognitive Network Model (CNM), in which cognitive belief structures are represented as a network in an agent-based model. The Socio-Cognitive Network Model (SCNM) extends the CNM to the interpersonal level, capturing influences resulting from interactions between individuals in a multi-agent model. At the societal level, a system dynamics model is used to represent macro-level influences of technology acceptance. These models are linked so as to exchange information by having macro-level influences feed into the micro-scale agent-based models, while the aggregation of individual behaviors supply input to the macro-scale system dynamics model. This interaction between component models at multiple scales is significant as it allows for the manipulation of variables specific to particular cultures at both an individual and societal level.

THE COGNITIVE NETWORK MODEL

The Cognitive Network Model (CNM) approach is an innovative method of quantitatively characterizing specific cognitive mechanisms associated with human social-cultural decision making at the individual agent level. The fundamental tenet of CNM is the application of network analysis techniques as a basis for characterizing and understanding human decision making resulting from the diffusion of information. In essence, CNM represents decision making in terms of the emergent interactions of a set of parameters associated with beliefs. Thus, beliefs are the foundational element of cognitive network modeling.

Beliefs are often represented as propositions and can be thought of as subjective probability estimates of an object having a particular attribute (Fishbein and Ajzen 1975). In processing information (e.g., "raisins are a healthy snack"), a belief exists when the human agent assesses the probability that the information is true. From a hierarchical category perspective, "raisins" might be defined as inclusive of the category "healthy" (raisins are healthy) or exclusive of the category (raisins are not healthy). From a network perspective, beliefs might be represented as network nodes, with links representing the relationships between nodes. In the CNM approach, a belief proposition is described as a pairing of cognitive concepts to which a degree of perceived truth is assigned. For example, the proposition "raisins are healthy" pairs the concept of "raisin" with "healthy". Although this is consistent with (and derived from) existing theory, the model presented here differs in some important respects. To represent beliefs and provide insight into how to predict the influence of new information, this chapter represents beliefs through three main quantitative parameters:

Veracity: This represents the general strength of the belief held by an agent. Quantitatively, veracity is represented as a value between 0 and 1 indicating the degree to which an agent accepts a given belief proposition as being true, where 0 indicates the agent believes the proposition is not true at all (proposition is completely rejected), and 1 indicates complete belief in the truth of the proposition.

Epsilon (ε): Here, each belief has an associated epsilon representing the malleability of the veracity of the belief; epsilon is the degree of acceptable variance/differences in the veracity value of a belief that an agent will accept without arousing the defense. Thus, epsilon captures the interval limits on either side of current veracity for which new values of veracity are outright accepted/rejected (new values of veracity for given belief propositions might be based on information received from an external source such as another agent or contextual evidence)[1]. Epsilon may not be uniform in each direction; changes in veracity may be more readily accepted for one direction versus the other.

Defense: Defense measures the degree to which the agent tolerates/considers changes in veracity that fall outside epsilon; it includes how strongly one might argue for the veracity of a belief (or against its veracity if an alternative belief is supported). Quantitatively, defense is a value between 0 and 1, where 0 indicates a weak strength of defense value (high degree of tolerance for beliefs outside of ε) and 1 indicates a strong defense (intolerance for beliefs outside of epsilon). In other words, the strength of the defense of the belief is a value that indicates the degree to which a person rejects beliefs that are different from (conflict with) their current belief and fall outside its associated epsilon range. If the agent's defense of the belief is low, differing/conflicting information is less likely to be rejected outright without first modifying one or more belief parameters. The defense value affects the degree to which an existing belief is amenable to change via (a) adjusting epsilon, (b) adjusting the strength of veracity, (c) adjusting the strength of the defense, or (d) a combination of belief parameter adjustments[2]. The defense value is

---

[1] When considered in conjunction with veracity, ε is a partial indication of confidence: small values correlate with higher confidences; larger values correlate with lower confidences.

[2] The strength of defense is influenced not only by the belief to which it is attached, but by relationships within the belief network (i.e., the degree to which a change in the relevant

the mechanism by which the CNM is able to model both rational and irrational cognitive behavior.

If beliefs are conceived as estimates of the accuracy of information, then the model needs to support changes to targeted beliefs within an individual agent's belief system (termed a Cognitive Belief Network), either through shifts in strength of veracity of the belief ("maybe raisins are not as healthy as I thought"), shifts in epsilon surrounding the veracity of the belief, or changes in the strength of the defense of the belief. Thus, the belief parameters of veracity, epsilon, and defense are the mechanisms by which cognitive network modeling quantitatively characterizes many of the cognitive processes associated with human decision making. Among the means for making adjustments to these belief parameters is the introduction of interventions (alternative beliefs) which effect a shift in proposition referents (e.g., "raisins are poisonous").

THE COGNITIVE BELIEFS NETWORK

The lowest level of analysis in CNM comes from studying the emergent properties associated with the Cognitive Beliefs Network (CBN). The CBN (FIGURE 1) is conceptualized according to precepts familiar to areas of graph theory and network science: vertices (nodes/points) represent individual beliefs held by an agent, and edges (links/lines/arcs) represent connections/relationships between beliefs; they are usually weighted and may be directed (see Diestel (2005) for an introduction to graph theory).

The CBN is the collection of all beliefs and associated parameters held by an individual agent. The CBN structure evolves over time, as new beliefs are added and/or parameters associated with prior beliefs change. Following the work of Bourdieu (1977) and Kameda, Ohtsubo, and Takezawa (1997), we introduce the concept of *cognitive capital* as the emergent features of interest at this level of network modeling. Within CNM, cognitive capital refers to the relative influence power of a belief node in the CBN; it is a composite score as

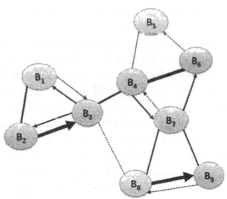

FIGURE 1. Conceptual rendering of the Cognitive Beliefs Network (CBN)

belief influences change in other beliefs) as well as individual difference variables that operate more generally (e.g., we might expect a person high in the personality attribute of "openness" to be more accepting of changes in beliefs and therefore have an overall tendency of lower defense values across the belief network).

measured by network metrics for cognitive prestige (a compound score based on various measures of network centrality), and cognitive importance as measured by the eigenvector centrality (weighted according to relative scores for cognitive prestige of related vertices; connections to nodes with higher cognitive prestige scores contribute more the importance score of the node in question)[3].

In this manner, the CNM becomes a framework for studying decision making resulting from agents' processing of information. Cognitive network modeling is then a matter of (a) selecting the beliefs of interest for the chosen domain, (b) defining the relevant belief parameter values, and (c) describing the governing functions for dependencies between beliefs as well as other intra-personal influences[4].

From this framework, it follows that CNM is able to capture the concept of culture by adjusting one or more of the following in the CBN: (1) the relative strengths of individual belief parameters for belief nodes determined to be culturally germane, (2) the weighted magnitude of the CBN edges, and (3) the relationships defining the intra-network and external influence forcing functions.

THE SOCIO-COGNITIVE NETWORK MODEL

The Socio-Cognitive Network Model (SCNM) is the next level of the multi-scale modeling spectrum described in this chapter. Socio-cognitive network modeling is extremely relevant to the study of human social-cultural decision making: just as the CBN does not function in isolation within an individual agent, neither do agents act in isolation – a number of external environmental and social influences interact with the agent to affect the agent's CBN. At the SCNM level, network vertices are individual human agents, and network edges are the cognitive ties, communication links, and social relationships between them. Within the current context, cognitive ties generally refer to the extent of agreement between the CBNs of multiple agents. A central premise of SCNM is that the strength of the cognitive ties (degree of agreement between agents' CBNs) affect the degree to which agents influence–and are influenced by–one another's beliefs during social interactions.

The SCNM relies heavily on the principles of social cognitive theory (Bandura, 1986), which subscribes to a model of emergent interactive agency (i.e., triadic reciprocal causation whereby personal factors, behavioral factors, and environmental events all operate as interacting determinants of human decision making). Additionally, Bandura's (2001) characterization of the role of social

---

[3] Other emergent properties of the CBN that have greater relevance in other contexts include: network size, network density, and network cohesion.

[4] An individual human agent's CBN does not function in isolation, but with other, non-cognitive personal determinants such as personality, transient affective states (moods), biological factors (e.g., age, gender), and personal history (e.g., education, work). Although not detailed herein, they are considered an integral part of the CNM, and have been notionally represented in the agent based model described in this chapter.

cognitive theory in mass communications is particularly relevant to modeling the governing functions of information exchange within the SCNM.

The SCNM and CNM are represented physically within an agent based model (see Bonabeau (2002) for a description of agent models). By representing the relevant beliefs for a given domain, as well as the forcing functions associated with the modification of those beliefs, the model can monitor the changes in beliefs (both at the individual and at the aggregate levels) to determine the effects of information as it propagates through the network. The domain of demonstration selected for this chapter is cultural determinants of human decision making within the context of technology adoption (e-commerce), that is, the decision to use e-commerce technology to make an on-line purchase. The set of interrelated beliefs and governing functions selected for modeling purposes are those familiar to researchers in the area of technology acceptance. The specific beliefs that are represented are based on the work by Venkatesh, Morris, Davis, & Davis (2003). Their UTAUT model combines eight of the most prominent technology acceptance models observed in the literature and provides a definitive list of variables that are critically relevant to an individual's decision to adopt a new technology. Additionally, cultural distinctions in technology adoption at this level of modeling are manifested in varying algorithmic functions associated with culturally relevant determinants of interpersonal influence, as well as six cultural value-dimensions as described by Zakour (2004).

SOCIETAL MODEL

A system dynamics approach is used to capture high-level factors that contribute to the adoption of technology (specifically, e-commerce). A system dynamics model is a type of executable model used to represent and understand the dynamic behavior of a complex system over time (Sterman, 2000). Complex systems, whether social, technical, or some combination, often exhibit highly non-linear behavior where the relationship between cause and effect is not intuitively evident. System dynamics models use stocks and flows to represent system elements and their relative influences upon each other. Stocks represent an inventory of accumulated entities (e.g., inventory of cell phones) and are indicated in the e-commerce adoption model using rectangular boxes (see Figure 2). Flows are indicated by double-lined arrows, and show how entities move between stocks or between a stock and a cloud. Clouds, which may take the place of stocks, indicate the world outside the scope of the model and act as either a source or a sink.

Figure 2 depicts a system dynamics model that captures some of the critical variables involved with the adoption of e-commerce. This particular model concentrates on two areas: economic variables related to selling a product on the internet, and the effect of marketing by e-commerce retailers on purchasing trends.

This society-wide model is valuable in that it allows for the exploration of feedback relationships within the system. For example, in this model the demand that exists for a product drives the desire for retailers to offer that product through a particular medium (in this example, through the online e-commerce medium). Because retailers wish to sell goods online, they increase their marketing efforts to make that service known to the buying public, which in turn increases the familiarity that the public feels for the retailer's online presence. As potential customers become more familiar with a retailer's online presence, they become more likely to desire to purchase a product from that retailer, thereby increasing online demand, which completes the loop.

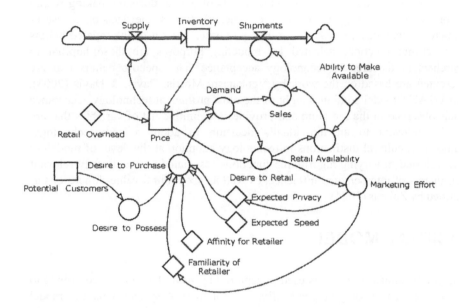

FIGURE 2: A system dynamics model of selected variables related to e-commerce

Significant cultural variables are captured within the system dynamics model. The actual entities that exist within the model, in terms of stock, flows and influences, are representative of relevant aspects of the particular culture that is being modeled. For example, in some cultures, privacy is more highly regarded than others, and differences in the perceived privacy of different retail options may be amplified in society.

MODEL INTEGRATION AND E-COMMERCE EVALUATION

The most obvious benefit to developing and executing an integrated model is that it provides the potential for multiple levels of analysis. For example, one may wish to test a scenario of technology adoption to evaluate the effect that a specific change in individual behaviors might have on societal variables and vice versa. Individuals live in, react to, and affect their environment (society). A scenario could be constructed to investigate how an advertising campaign targeting a particular demographic will influence the e-commerce retailing of a given product. Characteristics about the campaign could be represented in the model in terms of the type of psychological persuasion tactics (marketing) used. The effect of this campaign could then be evaluated at both the individual level (e.g., how the campaign altered the individuals' beliefs), at the inter-personal level (e.g., how the influence of the campaign translated into influence between individuals), and at the societal level (e.g., did the demand for the product cause an increase in the number of retailers competing for e-commerce customers). Of special interest are the relationships between each of these levels. Namely, the influence of individual beliefs on society and culture, and how culture and society work to encourage, reinforce or suppress beliefs at the level of the individual agent.

Perhaps the greatest benefit afforded by this type of model is the flexibility to tailor it to particular populations through the manipulation of cultural variables. Each level of the model provides a representation of cultural variables. The integrated model allows for a demonstration of not only how these variables influence individuals and society independently, but how they interact. For example, an ad campaign targeted at the societal level may concentrate on the increased level of privacy afforded by e-commerce (see FIGURE 2). In this case, privacy is represented in the societal model because it is a relevant characteristic of business interactions in the modeled culture. The expectation of privacy as a society becomes a relevant variable in the individual model and is an influence on an individual's belief about his own acceptable levels of privacy. This in turn affects the opinions concerning the adoption of e-commerce that are communicated toward others. The amount and type of communication with other individuals is also highly dependent on culture. For example, in some cultures it is impolite to speak of personal financial decisions. The type and manner in which relevant information circulates within the social network influences an agent's decision to use an internet retailer. The decision to use e-commerce, as well as the decisions of the network of peers, is aggregated to feed into the high-level model as demand, which, ultimately, affects how marketing campaigns are designed. This interaction among cultural variables would be difficult to capture using a single-scale model.

CONCLUSION

This chapter has presented an integrated, multi-scale modeling approach that captures human cognitive, social, and cultural components of decision-making behavior using the adoption of e-commerce as a domain of demonstration. The exploration of technology adoption grounds the model in a realistic environment and incorporates well-researched psychological models. The federated approach provides an analysis tool for examining the potential implications of economic and Internet-related policies that may affect human perceptions, as informed by one's culture, and is extensible to modeling decision-making by individuals related to their involvement in other activities beyond e-commerce.

REFERENCES

Bandura, A. (1986). Social foundations of thought and action: A social cognitive theory. Englewood Cliffs, NJ: Prentice-Hall.

Bandura, A. (2001). Social cognitive theory of mass communications. In J. Bryant, & D. Zillman (Eds.). *Media effects: Advances in theory and research* (2nd ed., 121-153). Hillsdale, NJ: Lawrence Erlbaum.

Beal, G., Rogers, E. & Bohlen, J. (1957). Validity of the concept of stages in the adoption process. *Rural Sociology, 22(2),* 166-168.

Bonabeau, E. (2002). Agent-based modeling: methods and techniques for simulating human systems. *Proceedings of the National Academy of Sciences of the United States, 99(3),* 7280-7287.

Bourdieu, P. (1977). *Outline of a Theory of Practice.* Cambridge and New York: Cambridge University Press.

Carr, V. (1999). *Technology Adoption and Diffusion.* Retrieved October 12, 2009 from United States Air Force Air War College Gateway to Internet Resources website: www.au.af.mil/au/awc/awcgate/innovation/adoptiondiffusion.htm.

Davis, F. (1986). A Technology Acceptance Model for Empirically Testing New End-User Information Systems: Theory and Results. In MIT Sloan School of Management. Cambridge, MA: MIT Sloan School of Management, 1986.

Davis, F. Bagozzi, R. & Warshaw, P. (1989). User Acceptance of Computer Technology: A Comparison of Two Theoretical Models. *Management Science, 35,* 982-1003.

Diestel, R. (2005). *Graph theory (3rd ed.).* New York: Springer-Verlag.

Fishbein, M. and I. Ajzen (1975). Belief, attitude, intention, and behavior: an introduction to theory and research. Reading, MA: Addison-Wesley.

Kameda, T., Ohtsubo, Y., & Takezawa, M. (1997). Centrality in sociocognitive networks and social influence: an illustration in a group decision-making context. *Journal of Personality and Social Psychology, 73*(2), 296-309.

Rogers, E. (1962). *Diffusion of Innovations.* Glencoe: Free Press.

Straub, D., Keil, D., & Brenner, W. (1997). Testing the Technology Acceptance Model across Cultures: A Three Country Study. *Information & Management,*

31(1), 1-11.

Venkatesh, V., Morris, M.G., Davis, G.B & Davis, F.D. (2003). User acceptance of information technology: Toward a unified view. *MIS Quarterly, 27,* 425-478.

Watson, R., Ho, T. & Raman, K. (1994). Culture: A Fourth Dimension of Group Support Systems. *Communications of the ACM, 37*(10), 45-55.

Zakour, B.A. (2004). Cultural difference and information technology acceptance. Proceedings of the 7th Annual Conference of the Southern Association for Information Systems, 156-161.

CHAPTER 26

An Architecture
for Socio-Cultural Modeling

Guy A. Boy

Florida Institute for Human and Machine Cognition
40 South Alcaniz Street
Pensacola, Florida 32502, U.S.A.

ABSTRACT

The proposed architecture for socio-cultural modeling is based on the concept of social object that has been instantiated in the form of scenarios. Players of these scenarios are called agents who/that have cognitive functions. Anytime a scenario is developed with domain experts, emergent cognitive functions are generated. It is the incremental inductive generation of cognitive function networks that progressively shape purposeful social objects.

Keywords: Socio-cultural modeling, social object, scenarios, cognitive functions.

INTRODUCTION

This paper presents an architecture for socio-cultural modeling. This socio-cognitive and ethnographical approach and method to impact analysis of psychological factors in crisis management is based on the central concept of social actors. Social actors are people immersed in a social context, with various kinds of factors such as individual, cultural, influence and socio-cognitive factors that determine their behavior. This architecture is focused on decision-making and its impact on people in the context of crisis management. An analysis of impact is proposed with respect to contextual factors of the crisis, factors within the decision team and external factors (such as media and people reactions). An ontology of these decision-related factors was derived and will be presented in the paper. It enables the analysis of a crisis and the support of its management. Terrain analysis is an important military task, also in operations other than war. Numerous socio-cultural details have to be observed under time pressure, and patterns have to be distilled from these details. There frequently is insufficient time to store these patterns in either human memory or digital systems. This severely jeopardizes the adequate execution of these tasks. The proposed architecture can be seen as an electronic implementation of cognitive functions that are used by people in the execution of a task in a socio-cultural context. We propose the development of a system that helps observers and analysts with an appropriate external memory support. This is expected to improve socio-cultural learning in various crisis contexts, valorize logistical and structural adaptation, optimize and value changes of limiting habits and attitudes, help in the organization of crisis management, support crisis simulation seminars, take into account experience feedback, cross-fertilize experiences and invent new intervention means.

FROM AGENTS TO SOCIAL OBJECTS

Many authors introduced and discussed the "agent" concept for a long time. We choose Minsky's approach to agent definition (Minsky, 1985, i.e., an agent is a society of agents. This means that agents can be described recursively. An agent can be a human being as well as a group of human beings. Such a group can be a team (i.e., a small highly-connected group of people who know each other very well), an organization (i.e., a larger group of people interconnected through rules, often hierarchical relationships) or a community (i.e., a group of people who share same interests).

I introduced the cognitive function representation to analyze human-automation systems (Boy, 1998). An agent, i.e., a human being or an appropriately-automated machine, has at least one cognitive function. A cognitive function is represented by its role, its context of validity and a set of resources that support it operationally. Cognitive function analyses enable to incrementally improve function allocation among humans and machines. Cognitive functions can be distributed among various groups of agents.

Interaction among agents, and consequently cognitive functions, facilitates the emergence of social objects. Claude Rivière introduced the concept of social object in his essay of sociological epistemology (Rivière, 1969). Three main issues are at stake to better define social objects:

- current research techniques must be improved by adding two kinds of adaptation, i.e., context-sensitive and motivated adaptation;
- validity of the methods must be improved also to insure objectivity;
- research rationale must be clearly defined, i.e., must elaborate social object epistemology.

Social objects are defined by their objectivity, i.e., they must be described impartiality and universality as much as possible. They are usually the result of a consensus among a group of specialists instead of coming from a single expert for example. The best we can do is usually relative universality because we cannot get all opinions from all centuries, countries, experts and so on. Objectivity can be partially obtained by:

- precision, e.g., through measures, axiomatization, predictions;
- explanation, e.g., from understood facts, by seeking causes, thanks to a methodological pluralism, by inducing laws;
- the "total" approach, e.g., through historical anthropology (Macfarlane. 1977).

Whenever we describe a social object, we have a part of subjectivity coming from our own culture and society considered as a norm. It is very difficult to make sure that we reach universal objectivity. This is why several experts and consensus reaching methods are required. However, we are always in a catch-22 situation where the object depends on the method and the method depends on the object. Any social object will be described as we perceive it. Obviously, this description should be useful for further prediction and evolution of interconnected social objects.

A social object emerges from purposeful interactions between agents. There is no social object without action. The social object does not exist as an isolated entity, but it is understood as a network of relations. Taking into account the cognitive function analysis approach, a social object is a persistent network of cognitive functions distributed among a set of agents, e.g., the network of mail cognitive functions distributed among a set of postmen. This why it is crucial to identify the social context in which this cognitive function network emerges. In addition, this network is characterized by a set of individual, influential and socio-cognitive factors that need to be elicited in order to explain the resulting collective behavior (Figure 1).

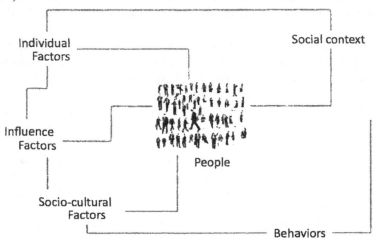

Figure 1. Factors that help describing social objects.

SOCIAL CONTEXT

People tend to "economize" their lives by finding out patterns that enable them to simplify their interrelations among each other. Simplicity is always a matter of frequency and regularity. Regularity tends to establish stable relations in groups. Paradoxically, the identity of a social object is only found when there is a sufficient diversity of subjects who share a common understanding of it, e.g., the identity of a country. Yugoslavia's identity for example was only maintained for many years by a common deliberate social object called communism, which broke into parts when the supervision model became obsolete, and constituent identities came back, e.g., Serbia, Croatia, Bosnia and Kosovo. A social object may emerge from several attributes such as religions, traditions, languages, political beliefs and so on.

More generally, social context is typically defined by cultural factors, beliefs, social evolutions, ability to cope with social changes, social norms, traditions, rituals, myth, customs, social values, moral and obviously history.

INDIVIDUAL, INFLUENTIAL AND SOCIO-CULTURAL FACTORS

Agents in a social context have their own individual differences with respect to the following factors: personality, mental features, physical constitution and characteristics, affectivity, individual behavior, cognitive capacities and social involvement. Some of these factors could be common to several social objects, other are able to discriminate social objects among each other. It is important to elicit them to figure out useful distinctions that enable the identification of emergent social objects.

Influential factors can be developed in the same way. People can be influenced either by themselves or by others. Therefore, influential factors may be internal such as expectations (either people expect too much and their expectations are not fulfilled; or have wrong expectations, for example), accomplishment needs and motivation. Influential factors may be external such as extreme determination due to social pressure, manipulation and history of other people (stronger people may influence weaker people for example), cultural or religious influence, and public opinion via media and propaganda communication.

Socio-cultural factors include cognitive dissonance (e.g., when people do not reason or process information in the same way), egocentrism (i.e., some people think about themselves and do not care about the others), social identity (e.g., depending on their social status, people are likely to produce different behaviors), orthodoxy (i.e., people obey strictly to a doctrine), social perception (i.e., perception of the social context may not be objective but rather subjective, the elicitation of such subjectivity is always useful), bias (i.e., we see with our biased eyes depending on our country, political beliefs and so on), social regulation (i.e., other people reflect on us, experience feedback is constantly re-injected), social representation (i.e., the way we represent our society or another society depends on the representation that we use), and stereotype (e.g., French people shake hands or kiss women any time they meet).

RESULTING BEHAVIOR

Resulting behavior is precisely a social object. Accumulation of factors described above may end up in various kinds of behavior such as discrimination (i.e., special treatment taken toward or against a person of a certain group; it could be based on class or category; discriminatory behaviors usually induce exclusion, racism or rejection), specific social relations (e.g., crime watching, over insurance), and specific status (e.g., specialization of groups of people, delegation of responsibility, authority sharing, accountability).

Of course, each social context, human factor and behavior should be elicited from groups of experts. Let's further define this anthropological approach that supports the incremental definition of social objects.

ARCHITECTURE AND METHODOLOGY

The proposed method is based on the search for types of interaction among agents using knowledge and knowhow elicited from domain experts.

INTERACTION MODELS

The interaction model, which agents typically use, depends on the quality and quantity of knowledge that they have of the organizational environment where they interact among each other.

There are various levels of interaction in a multi-agent system. At the local level, each agent should be able to interact with his/her/its environment in the most natural way. Actions means should be as affordable as possible at the right time, and more generally in the right context. Affordable action means are incrementally constructed, often unconsciously; when they stabilize, they become social objects. As Saadi Lahlou already said "People experience difficulties in interpreting new things for which there is no social representation." (Lahlou, 2010).

The use of social representations in multi-agent communications is the most sophisticated model of interaction that I previously denoted **communication by mutual understanding** (Boy, 2002). However, we should not forget that there are two other possible models of interaction that are **supervision** and **mediation**. In the former, a knowledgeable agent supervises the interaction between the other agents. This happens when interacting agents do not know each other, i.e., when they do not have appropriate social representations. In human-machine interaction, the supervisor could take the form of an instruction manual, a context sensitive help or an expert person. In the latter model, agents may not have social representations of the other agents, but they can communicate through a meditating space. Mediation could be done via other agents usually called facilitators, diplomats, or lobbyist. User-friendly interfaces, and the desktop metaphor in particular, that were developed during the eighties, are typical examples of such meditating spaces. A continuum between these three models of interaction was presented in a recent paper (Boy, 2009).

These three models are typically used to identify the type(s) of interaction that a group of people may have.

THE GROUP ELICITATION METHOD (GEM)

The group elicitation method (GEM) was developed over the years for the last three decades to help people make collective decisions, foster creativity and rationalization for design, as well as other processes involving experts (Boy, 1996). We can see GEM as a discount anthropological method where objectivity is partially but effectively reached because it involves a set of experts who provide viewpoints that are consequently categorized into concepts, these concepts are ordered with respect to an a priori metric. Everybody assess each concepts of the social object; scores are collected and processed (averaged) toward a consensus. Sometimes differences among GEM participants emerge and are further taken into account. GEM produces an ordered set of concepts that can be further used for the construction of a socio-cultural model. Social objects are characterized as cognitive functions distributed among agents. This distribution of cognitive functions is commonly represented by scenarios. The more similar scenarios are found, the more generic scenarios can be induced, and the more these scenarios become stable social objects. Consequently, conducting a cognitive functions analysis (Boy, 1998) consists in developing appropriate scenarios with domain experts.

COGNITIVE FUNCTION ANALYSIS = SCENARIO DEVELOPMENT

Cognitive function networks (Boy, 2010) are incrementally developed through the generation of cognitive functions in both the resource space and the context space (Figure 2). The development of such cognitive function networks is guided by several properties of cognitive function themselves. In the resource space, configuration scenarios are described to improve the rationalization of the allocation of physical and cognitive functions to appropriate and available agents. In the context space, event scenarios (e.g., chronologies in temporal contexts) are described to improve awareness of procedural connections between cognitive functions.

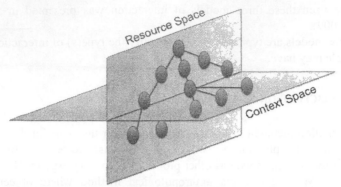

Figure 2. A cognitive function network view in both context and resource spaces.

AN ORGANIZATIONAL MEMORY SYSTEM

Since knowledge is incrementally acquired, emergent social objects in the form of generic scenarios are elicited through an organizational learning mechanism. For that matter, a well-designed experience feedback process should support organizational learning. Experience is captured in the form of episodes, viewpoints and concepts, using the GEM approach as previously described.

A decision model supports the categorization process that transforms and articulates scenarios into more generic socio-cultural models. However, these social objects are not identified unless they are themselves experienced, tested and validated, i.e., it is by using this organizational memory system that social objects are incrementally stabilized.

Such a system will be effectively used if the socio-cultural patterns that it includes are easily traceable, i.e., they need to be easily retrieved, understood and usable.

SCENARIO-BASED LEARNING PROCEDURE

Developing a socio-cultural model turns out to be eliciting and categorizing series of scenarios with domain experts. Each scenario is incrementally added into an organizational memory system (OMS) as follows. Let's assume that we have generated a scenario, which will be called current scenario. There are four cases that can be investigated.

Case 1. The current scenario is not related to any generic scenario in the OMS. Therefore it becomes a new entry in the OMS, and is called a generic scenario.

Case 2. The current scenario is related to one generic scenario in the OMS. Therefore it either is added as an additional description or contributes to the modification of this generic scenario.

Case 3. The current scenario enables to merge several generic scenarios of the OMS.

Case 4. The current scenario enables to split a generic scenario of the OMS into several generic scenarios.

DISCUSSION ON GENERIC EXAMPLES

If the problem is to improve the effectiveness of security measures and procedures in airport passenger terminals, then we need to better understand the various roles, contexts of operations and resources that are relevant in such environments. In other words, what are the main cognitive functions that make sense in such environments. We will then first work with security experts to elicit scenarios that happened in airport terminals and are related to potential terrorism. We will analyze elicited cognitive function both in the resource space and the context space, and incrementally find generic scenarios that we will accumulate and articulate in an organizational memory system.

Along the way, we will use the three interaction models to figure out how agents are organized and actually interact among each other. Identification of these models in each scenario and more importantly in generic scenarios is crucial to better understand current social objects, i.e., effectiveness of the chosen security solutions. These social objects need to be tested either by asking experts to evaluate them or by generating other scenarios that will either confirm or disconfirm them.

Even if we are in the early stages of the development of this architecture, we already applied the approach to crisis management in defense cases in France (DeBrito & Boy, 2004). We deduced a series of psychological factors that directly influence decision-making in crisis situations. This impact analysis involved four dimensions: the crisis itself and its context; the decision-maker him/herself; the decision-making group; and external groups. We analyzed a large-scale accident such as the explosion of the Bhopal chemical factory in India. Regarding the **social context**, several cognitive functions emerged such as processing large quantities of information and data, management of the magnitude and surprise of the situation, real-time information processing and the reaction time issue, management of an off-norm environment, reasoning on uncertain data, time management, management of a huge quantity of people, communication issues, management of decision impact, management of information credibility, management of inter-individual differences and self image. Regarding the decision-maker, **individual factors** emerged such as management of the initial chock, feeling of impotence, urgency, uncertainty, stakes, loss of frame of reference, and guilt. Other **influential factors** were found such as stress, workload, lack of experience and physical condition. The decision-making group was analyzed in terms of experts, their choice and organization, specific knowledge, lack of time, contradictory relationships, familiarity with a specific crisis, management of implicit expert knowledge, power management and so on. Sometimes, it may happen that a decision-making group, even made of the best experts, may be buried into a close thinking (the groupthink syndrome coined by Janis in 1972), the group is over-valorized and the overall thinking is uniform. It

258

may happen that several obvious tests are not performed and the group becomes blind. For example, there is no other hypothesis that the groupthink syndrome to explain the flawed decision of President Kennedy and his advisors to authorize the Bay of Pigs invasion of Cuba on May 17, 1961. Indeed, Kennedy's advisors made excessive efforts to reach agreement without taking into account external advice. They believed that they were more invulnerable than they were. They rationalized their decisions and believed stereotypes about US enemies. They felt increasing pressure to agree with others in the group. Regarding external groups, several cognitive functions were elicited such as management of exogenous differences, management of new work modes, and management of forced cohesion. Media influences were also analyzed, and the following cognitive functions were elicited: management of media reactivity, intrusiveness and incompetence; management of public image; management of journalists' comments and rumors.

CONCLUSION

This paper presents an architecture for socio-cultural modeling. This architecture together with the methodology need to be further tested and validated on several applications. It is based on a preliminary ontology categorized by social context, individual factors, influence factors, socio-cultural factors, and resulting behaviors. This ontology needs to be augmented and further articulated.

It is anticipated that it will be implemented using the conceptual map (CMaps) diagrams (Novak & Cañas, 2008). Cognitive function networks will be graphically visualized in order to enable the analyst to implement the scenario-based learning procedure.

Content-wise, it would be very interesting and useful to continue developing research and knowledge capture on inter-cultural problems, and establish cognitive function networks, a kind of concept map of the society that we plan on studying. When this concept map will be mature enough, it could be used a decision aid.

Finally, since this approach is agent-based, it could be supported by a socio-cognitive simulation that encapsulates the cognitive function representation as first principles, i.e., role, context and resources, and produces the emergence of social objects through the observation of resulting behaviors.

REFERENCES

Boy, G.A. (1998). *Cognitive function analysis*. Ablex/Greenwood, Westport, CT, USA.
Boy, G.A. (1996). The Group Elicitation Method for Participatory Design and Usability Testing. *Proceedings of CHI'96, the ACM Conference on Human Factors in Computing Systems*, Held in Vancouver, Canada. Also in the *Interactions Magazine*, March 1997 issue, Published by ACM Press, New York.

Boy, G.A. (2002). *Theories of Human Cognition: To Better Understand the Co-Adaptation of People and Technology. In Knowledge Management, Organizational Intelligence and Learning, and Complexity, edited by L. Douglas Kiel, in Encyclopedia of Life Support Systems (EOLSS), Developed under the Auspices of the UNESCO, Eolss Publishers, Oxford, UK.*

Boy, G.A. (2009). The Orchestra: A Conceptual Model for Function Allocation and Scenario-based Engineering in Multi-Agent Safety-Critical Systems. *Proceedings of the European Conference on Cognitive Ergonomics*, Otaniemi, Helsinki area, Finland; 30 September-2 October.

Boy, G.A. (2010). *Cognitive Function Analysis in the Design of Human and Machine Multi-Agent Systems.* Handbook of Human-Machine Interaction, Ashgate, U.K.

De Brito, G. & Boy, G.A. (2004). *Study of the impact of psychological factors on decision-making in crisis situations* (in French). EURISCO-DGA Technical Report no. T-2004-014.

Janis, I.L. (1972). *Victims of Groupthink.* New York: Houghton Mifflin.

Lahlou, S. (2010). Socio-cognitive issues in human-centered design for the real world. In G.A. Boy (Ed.) Handbook of Human-Machine Interaction, Ashgate, U.K.

Macfarlane. A. (1977). Historical Anthropology (Frazer Lecture). Cambridge Anthropology, Vol.3, no.3:
http://www.alanmacfarlane.com/TEXTS/frazerlecture.pdf

Minsky, M. (1985). *The Society of Mind.* Simon and Schuster, New York.

Novak, J. D. & A. J. Cañas (2008). The Theory Underlying Concept Maps and How to Construct Them, Technical Report IHMC CmapTools 2006-01 Rev 01-2008, Florida Institute for Human and Machine Cognition, 2008", available at: http://cmap.ihmc.us/Publications/ResearchPapers/TheoryUnderlyingConcept Maps.pdf.

Rivière, C. (1969). *L'Objet Social.* Marcel Rivière et Companie, Paris, France.

CHAPTER 27

Cultural Network Analysis: Method and Application

Winston R. Sieck

Applied Research Associates
Fairborn, OH 45324, USA

ABSTRACT

A method is described for studying in detail the common perspective that members of a culture bring to a situation. The method results in models of the culture that provide a basis for outsiders to begin to frame events from the cultural-insider point of view. The cultural models can then be used to identify priority cultural aspects to emphasize in training, as an aid to anticipating how messages will be interpreted and evaluated by members of the culture, or as a means of diagnosing cultural frictions that impede effective multicultural team functioning. Example applications are presented to illustrate the value of the method.

Keywords: Cultural models, cultural epidemiology, decision making, cultural sensemaking training, cross-cultural communications, multicultural collaboration

INTRODUCTION

The purpose of this chapter is to describe an approach to cultural modeling, cultural network analysis (CNA), and to illustrate some of its applications. Cultural network analysis builds on a foundation of research practices drawn from the fields of cognitive anthropology, cultural and cognitive psychology, and decision analysis. It enhances current cultural research practices by providing a rigorous, systematic method for constructing *cultural models* for groups, organizations, or wider societies. Cultural models derived by CNA are represented graphically as a network of the culturally-shared concepts, causal beliefs, and values that influence key decisions in a particular context (Sieck, Rasmussen, & Smart, 2010). In their most fully developed form, cultural models also convey detailed quantitative information about the prevalence of their specific components. It is useful to scrutinize the need for cultural modeling techniques in general, prior to describing the method in detail.

THE IMPORTANCE OF CULTURAL MODELING

Considerable effort is being devoted to increase understanding across cultures, and to improve the quality of intercultural interactions. These efforts include: a) the design of training to enable individuals to quickly achieve social competence within a host culture; b) the development of communications strategies for selecting and tuning messages that increase persuasive impact within a cultural group; and c) the design and implementation of systems and processes that enable multicultural teams to leverage the advantages of their diversity to solve challenging problems. One approach for improving performance in intercultural encounters is to capture how culturally-competent individuals think within relevant situations. By studying in detail the common perspective that members of a culture bring to a situation, a model of the culture can be constructed that provides a basis for an outsider to begin to frame events from their point of view. The model can then be used to identify priority cultural aspects to emphasize in training, as an aid to anticipating how messages will be interpreted and evaluated by members of the culture, or as a means of diagnosing cultural frictions that impede effective team functioning.

Why have the social sciences not already provided cultural modeling tools? Many cultural psychologists have been working under the assumption that a wide variety of cultural phenomena can be understood in terms of a few key dimensions, such as individualism-collectivism (Hofstede, 2001). The cultural dimensions paradigm focused on collecting data on the proposed "essential" dimensions all over the world to describe national differences. As cognitive approaches to culture have been maturing within anthropology and cognitive science, there has been an increased interest in describing and explaining the origins of complex cultural knowledge (Atran, Medin, & Ross, 2005). The promise of these approaches is to enable scientists to examine cultural concepts at a much finer level of granularity that has been achieved in the past. The fine-grained detail permitted removes the speculation required to apply cultural dimensions to particular situations, and allows the cultural perspective on the situation to become obvious. To bring the promise to fruition, a set of techniques is needed to systematically produce cultural models in common formats for ready inspection and objective interpretation. CNA provides a collection of techniques to do exactly this. To understand the CNA method, however, it is necessary to unpack what is meant by culture in the first place.

THE NATURE OF CULTURE

There is a broad, perhaps natural tendency to talk about culture as if it were a concrete, material thing. It is sometimes described as a thing that people belong to, or like a kind of external substance or force that surrounds its members and guides their behavior. Although it is sometimes difficult to avoid these intuitions, they do not hold up to careful scrutiny. Instead, researchers often informally take nationality as an operational definition of culture. Practically, this allows researchers to study cultural differences by comparing national averages on dimensions. A problem is that numerous studies have consistently found a wide range of variability around the averages. This implies that the nation might provide the wrong level of analysis. The results beg the question - how should we define the boundaries of cultures?

An alternative approach begins by defining culture in terms of the widely shared ideas within a population. Here, "idea" refers generically to concepts, values, beliefs, or other mental representations. To take an example, "cultural" values are simply values that most people within a cultural group agree on as important. Cultural knowledge is knowledge that is shared by most everyone in the population. Taking this conception a step further, it is currently popular within cognitive science to draw on a disease metaphor for understanding cultural ideas, describing the ideas that spread widely through a population and persist for substantial periods of time as especially "contagious" (Sperber, 1996). This suggests that ideas can be studied using some of the same techniques that epidemiologists use to study diseases, and the metaphor is useful in this regard.

OVERVIEW OF CULTURAL NETWORK ANALYSIS

Cultural network analysis is a method for describing ideas that are: a) shared by members of cultural groups, and b) relevant to decisions within a defined situation. CNA discriminates between three kinds of ideas: concepts, values, and beliefs about causal relations. The cultural models resulting from CNA use network diagrams to show how all of the ideas relate to one another. The CNA approach also includes the full set of techniques needed to build cultural model diagrams. This consists of specific methods to: 1) elicit the three kinds of ideas from people in interviews or survey instruments, 2) extract the ideas from interview transcripts or other texts, 3) analyze how common the ideas are between and within cultural groups, 4) align and assemble the common ideas into complete maps. CNA shares many aspects with other approaches to cultural analysis, especially cognitive approaches developed by anthropologists (Garro, 2000). However, it offers some specific aspects as a complete method that distinguishes it from other ways of examining cultures.

EMPHASIS ON DECISIONS

CNA begins by identifying the judgments or decisions of primary interest for study. The decisions chosen arise in specific contexts as defined by critical incidents or scenarios. They are made by members of the cultural group being investigated, typically in a way that is surprising or confusing to members outside the group. Once the key decisions are identified, investigators build models of the cultural ideas that directly influence those decisions. This feature of CNA ensures that the aspects of culture investigated are relevant to practical application, such as training or communications. Some other approaches begin by considering cultural differences that the researchers hypothesize to be important. Then, studies may be conducted to determine their actual relevance, or products developed based on the assumed relevance. This is a riskier, less efficient process than the CNA approach.

EMIC PERSPECTIVE

CNA aims to directly portray cultural ideas in diagrams just as they are expressed by members of the cultural group. This is sometimes referred to as an "emic" or cultural insider perspective, and is contrasted with "etic" or outsider-scientific explanations of beliefs or behavior. For example, etic approaches involving cultural dimensions tend to rely on the scientists' theories concerning the implications of some cultural values on other cultural values. The approach generally results in explanations that mix scientific concepts with accounts that come from people within the culture. CNA instead attempts to describe emic cultural ideas using scientific formalisms. With respect to explanation, CNA operates under a broader theory, known as "cultural epidemiology," that seeks to explain how factors that are external to content knowledge, such as the structural properties of stories or the flow of information in the environment, contribute to the successful maintenance and spread of ideas or models. In the present state of cultural science, a well constructed "emic" cultural model is the most important ingredient for many applications.

IDEA NETWORKS

Cultural ideas are often studied as independent elements, especially in survey questionnaires. For example, researchers might study the extent to which members of different cultures aim to have satisfying relationships with family and friends. Scientists examine correlations between independently measured ideas and interpret the connections between them. CNA takes a different approach. It maps out the cultural ideas in a way that directly shows how they relate to one another from the perspective of members of the cultural group. A key premise of this approach is that cultural knowledge consists of shared networks of ideas. Commonly held causal beliefs provide the connections that bind other ideas together. Although CNA seeks to preserve the content of cultural knowledge as expressed by members of the cultural group, it does represent the idea networks in a common, scientific format. Specifically, CNA employs influence diagrams that have been long used to map out knowledge in decision analysis (Howard, 1989). An example from the domain of romance will help explain how this works. Figure 1 depicts an Arab-American cultural model of romantic relationships. The set of illustrated ideas were extracted from a newspaper article that reported on interviews with Arab-Americans about dating (MacFarquhar, 2006). It is provisional for illustrative purposes.

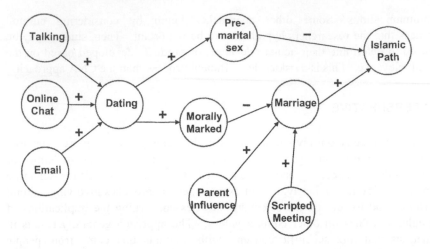

Figure 1. Arab-American cultural model of romantic relationships

Figure 1 depicts a number of ideas using circles, lines, and color. These ideas include simple concepts such as dating and marriage, represented as circles. It also includes causal beliefs, i.e. the antecedents and consequences of romantic activities, such as that dating decreases ones chances of marriage, and of staying on an Islamic path. These are represented as arrows in the figure, with +/- indicating the direction of the causal relation. Finally, desired states or values are portrayed using color. Staying on the Islamic path is a good thing, something one should do. Finding a marital partner is likewise valued. Since dating increases the risk that one will be toppled off of the Islamic path, as well as hampering ones chances of getting married, it should be avoided. Hence, this cultural model is likely to have fairly strong consequences for how members of a cultural group will decide and act.

PREVALENCE ESTIMATION

When epidemiologists study a disease, they tend to be more specific than just asking whether the disease is widespread or not. They ask "how widespread is that disease?" and then estimate its *prevalence* - the proportion of people who actually have it. Such a step is essential for achieving scientifically rigorous descriptions of the current state of a culture. It is also imperative for understanding changes in prevalence of ideas with any precision. CNA differs from earlier related approaches by taking seriously the need to get to that same level of description with ideas. It asks not just whether the idea is "cultural" (that is, widely shared), but takes the next step to determine precisely how prevalent is the idea in numerical terms.

DESCRIPTION OF CNA

Cultural Network Analysis encompasses both qualitative, exploratory analysis, and quantitative, confirmatory analysis. The specific techniques used to achieve each step in the analysis depend on whether the cultural researcher is employing exploratory CNA or confirmatory CNA.

EXPLORATORY CNA

A primary goal of exploratory CNA is to develop an initial understanding of the concepts and characteristics that are culturally relevant within the domain. In exploratory CNA, concepts, causal beliefs, and values are extracted from interviews and other qualitative sources. Semi-structured interviews employ questions intended to elicit antecedents and consequences of concept states, as in the "explanatory models framework" sometimes used in cognitive anthropology (Garro, 2000). Questioning along these lines draws out a more comprehensive set of ideas than would typically be verbalized in standard think aloud procedures, and places particular emphasis on drawing out perceived causal relations. We have also combined this interview approach with "value focused thinking" from decision analysis to elicit values and objectives directly, along with the causal beliefs that link more fundamental values with the means intended to achieve them (Keeney, 1994). Qualitative analysis and representation at this stage yield insights that can be captured in initial cultural models. Influence diagrams are an important representation format for depicting these models, as described and illustrated above.

CONFIRMATORY CNA

Confirmatory CNA serves to test the structure of previously developed qualitative cultural models, as well as to elaborate the models with quantitative data on the prevalence of ideas in the population(s) of interest. In confirmatory CNA, specially-designed structured questionnaires are used to obtain systematic data that can be subjected to statistical analysis. Most questionnaires treat ideas as independent entities, and so do not provide any means for revealing their interrelated, network form. A few studies have attempted to capture first-order causal beliefs. We have begun developing questionnaires that permit the analysis of longer causal belief chains based on exploratory CNA results. Statistical models are employed in confirmatory CNA to assess the patterns of agreement from the "causal-belief" surveys, and derive statistics describing the distribution of concepts, causal beliefs, and values. Mixture modeling is an approach that permits direct segmentation of cultural groups based on clusters of consensus (Mueller & Veinott, 2008; Sieck & Mueller, 2009). Mixture models have been applied in many scientific fields. In cultural modeling applications, the distinct segments resulting from the analysis represent *cultural groups*, i.e., groups defined by the similarity of their ideas. Finally, influence diagrams of the cultural models are constructed in confirmatory CNA that illustrate the prevalence of ideas, as well as the qualitative structure elucidated in exploratory CNA.

APPLICATIONS OF CNA

Cultural network analysis has been used in the development of three kinds of applications so far. It has been used as a method for analyzing cultural knowledge to develop cultural training requirements and content, as a method for designing cross-cultural communications strategies, and as an approach for designing processes and tools that support multicultural collaboration.

CULTURAL TRAINING

Cultural sensemaking training is an approach for building training that provides learners with cultural knowledge relevant to situations in which they will likely be performing (Sieck, Smith, & Rasmussen, 2008). Cultural sensemaking training compares the cultural models for a specific culture and scenario with novice models of the culture. Learning objectives are derived from an analysis of the gaps and inconsistencies between the cultural model and novice expectations. Training products are then developed on how members of that culture think and decide.

The full CNA process was used to develop several cultural models of Afghan decisions for use in a cultural sensemaking training application. Exploratory CNA was conducted first with Afghan expatriates, using scenario-based interviews to trace the decisions from immediate intentions to fundamental cultural values. The resulting qualitative cultural models were used as a reference to develop a questionnaire for use in confirmatory CNA. To provide a concrete example, we focus on a scenario involving a Mullah who was helping to distribute humanitarian assistance supplies. The Mullah was extremely helpful to the team, yet after finishing with the distribution he kept a truckload of the supplies. The Mullah section of the survey consisted of a series of brief vignettes describing possible intended actions, followed by closed-form questions about the objectives and anticipated consequences of those actions. The sequences ultimately led to seven fundamental values that were derived in the exploratory phase of the study: *status, respect, wealth, power, honor, safety, and family approval*. The CNA survey was translated into Dari and Pashto languages, and administered to 405 participants in Afghanistan through structured face-to-face interviews. The data were analyzed using finite mixture models. For brevity, a fragment of a cultural model is presented for only one of the cultural groups (see Figure 2). As illustrated, participants in this group tended to believe the Mullah would use the supplies in his own household, though reasonable proportions felt he would either sell them, or distribute them among the needy in his village. Interestingly, the majority of possible motivations for Mullah actions link to fundamental values of status and respect. The possibility that the Mullah is simply seeking to increase his wealth appears to constitute a

267

minority view among Afghans.

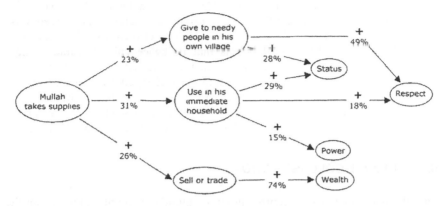

Figure 2. Afghan cultural model of a Mullah's decision making

This and other cultural models were compared with American novice interpretations of the same scenarios to identify training requirements, and ultimately build a set of cultural training materials.

CROSS-CULTURAL COMMUNICATION

Cultural models can serve as a basis for composing culturally meaningful messages in communications campaigns. This application of CNA was demonstrated in a study about terrorists' operational concepts and perceptions of risk in a nuclear smuggling scenario. Terrorists' ability to plan a viable mission that employs nuclear weapons is determined, in part, by their understanding of the regional actors, systems, and processes that afford desired actions or create risk for the intended operation. This understanding was elicited from experts in terrorist culture and military operations. Initial characterizations of terrorist shared knowledge about the acquisition and transport of nuclear weapons were modeled with influence diagrams, emphasizing concepts and causal beliefs that determine the terrorist groups' perceived risk of failure. For example, a common terrorist belief was that "waterway chokepoints" influence the value of "maneuverability," and decreased maneuverability was believed to increase the chances of being interdicted.

In a cultural models diagram of this kind, each concept and causal belief represents an opportunity to affect perceptions, such as increasing the perceived risk of mission failure. Hence, the diagrams provide a systematic basis for determining the content of communications. Messages are created so as to activate concept nodes, which then propagate across perceived consequences to stimulate other concepts. These effects spread through the idea network, ultimately changing the overall perception of risk shared by the group. With this CNA approach, information efforts focus on transmitting the most relevant information to affect perceptions in a way that makes sense within the cultural group's understanding. Candidate themes were recommended for influencing terrorist planners' perceptions of the inherent mission risks based on the cultural models. The results demonstrated the potential value of the CNA methodological framework for characterizing, anticipating, and influencing terrorist understanding and perceptions.

Cultural models also highlight the interrelation between causal beliefs and values, and so can identify causal beliefs that influence cultural value change. Consider the cultural model of Arab-American romantic relationships merely as a convenient example. The cultural model might be changed by focusing on the specific causal chain of beliefs that dating will decrease the chances of marriage. That is, changing the causal belief chain so that dating is seen as increasing the chances of marriage can also affect the relevant value (or attitude) towards dating.

MULTICULTURAL COLLABORATION

A cultural models approach to improving multicultural collaboration focuses on the development of hybrid team cultures (Sieck & Mueller, 2009). Hybrid cultures consist of a simplified set of shared assumptions, rules, expectations, and procedures that permit multicultural teams to function effectively. Hybrid cultures develop naturally over time as teams converge on a common process of interacting. However, cultural models can be used to inform the design of tools and processes that speed up the natural process. To take a concrete example, consider American and British expert planners' cultural models of quality plans (Rasmussen, Sieck, & Smart, 2009). These cultural models guide shared expectations about what the collaborative work product should look like in general form. Rasmussen and colleagues found that British planners' ideas included concepts such as "plan complexity" and "flexible execution," as well as shared causal beliefs such as that complex plans decrease the ability to execute flexibly. Finally, the British planners described ideas about desired/undesired states reflecting value. A summary of the resulting British model is that detailed specification of the rationale for actions in a plan will improve the capability of executors to adapt the plan in order to meet changing conditions. The ability to adapt is an important value, so planning team members are expected to focus much of their effort developing those components of the plan.

In contrast, American planners were found to place a premium on synchronization in execution and so focus on developing detailed actions. They expressed frustration with their British counterparts for spending so much time talking about goals at the expense of fleshing out all the relevant details. Understanding cultural frictions like this is useful for suggesting strategies or tools to improve collaboration. Recommendations included cultural model-based division of roles/functions in combined planning teams, and technologies to enhance multinational collaborative planning performance.

CONCLUSION

We all have the ability to think and speculate about the behavior of objects, events, and other people. We do this naturally in a variety of domains. In the social domain, we are able to make guesses about other people's thoughts and therefore speculate about their intentions and their motives. Human interaction and communication relies heavily on our ability to anticipate each other's intentions and actions, but that ability is heavily dependent on a shared cultural background. In this chapter, a cultural network analysis method was described for explicitly mapping commonly understood decision making within a cultural group. Specifically, a cultural group's shared knowledge within a situation is analyzed and displayed using a network representation of consensus elements. We also illustrated the use of the method for several applications, including improvements to cultural training, cross-cultural communications, and multicultural collaboration. A core assumption of our program is that peoples' intuitive understandings of others' decisions are fundamental to many more complex domains of interest in cultural research and applications. Hence, investigations using cultural network analysis provide a useful starting point for addressing these more complex cultural domains.

REFERENCES

Atran, S., Medin, D. L., & Ross, N. O. (2005). The cultural mind: Environmental decision making and cultural modeling within and across populations. *Psychological Review,* 112(4), 744-776.

Garro, L. C. (2000). Remembering what one knows and the construction of the past: A comparison of cultural consensus theory and cultural schema theory. *Ethos,* 28, 275-319.

Hofstede, G. (2001) *Culture's consequences* (2 ed.). Sage, Thousand Oaks, CA.

Howard, R. A. (1989) Knowledge maps. *Management Science,* 35, 903-922.

Keeney, R. L. (1994). Creativity in decision making with value-focused thinking. *Sloan Management Review, 35*(4), 33-41

McFarquhar, N. (2006). It's Muslim boy meets girl, yes, but please don't call it dating. *The New York Times,* Sept. 19.

Mueller, S. T., & Veinott, E. S. (2008). Cultural mixture modeling: Identifying cultural consensus (and disagreement) using finite mixture modeling. *Proceedings of the Cognitive Science Society.* Washington, DC.

Rasmussen, L. J., Sieck, W. R., & Smart, P. (2009). What is a good plan? Cultural variations in expert planners' concepts of plan quality. *Journal of Cognitive Engineering & Decision Making,* 3, 228-249.

Sieck, W. R., & Mueller, S. T. (2009). Cultural variations in collaborative decision making: Driven by beliefs or social norms? In *Proceedings of the International Workshop on Intercultural Collaboration* (pp. 111-118). Palo Alto, CA.

Sieck, W. R., Rasmussen, L. J., & Smart, P. (2010). Cultural Network Analysis: A Cognitive Approach to Cultural Modeling. In D. Verma (Ed.), *Network Science for Military Coalition Operations: Information Extraction and Interactions,* IGI Global.

Sieck, W. R., Smith, J. L, & Rasmussen, L. J. (2008). Expertise in making sense of cultural surprises. *Interservice/Industry Training, Simulation, and Education Conference (I/ITSEC),* December 2008, Orlando, FL.

Sperber, D. (1996). *Explaining culture: A naturalistic approach.* Malden, MA: Blackwell.

CHAPTER 28

Target Audience Simulation Kit: Modeling Culture and Persuasion

Glenn Taylor, Keith Knudsen, Robert Marinier III,
Michael Quist, Steve Furtwangler

Soar Technology, Inc.
3600 Green Court Suite 600
Ann Arbor, MI 48105, USA

ABSTRACT

We describe a system called the Target Audience Simulation Kit (TASK) that is intended to help marketing departments better understand their target audiences and develop more effective persuasive messages. We describe the relevant literature related to persuasion and marketing, and how we have incorporated cognitive theories and related methods into a computational framework to model target audiences and their responses to different messages. We describe the current capabilities of the system, challenges in building such a system, and future work.

Keywords: Cultural Models, Influence, Persuasion

INTRODUCTION

"Know your audience" is a mantra that applies to everything from product advertising, to public service announcements, to military influence operations (Trent and Doty 2005). Knowing what motivates an audience is critical to estimating how well different messages will resonate with that audience, and how likely the audience will, for example, buy a product or otherwise be persuaded to change their behavior. Understanding international audiences adds yet another layer of complexity when creating persuasive messages abroad. A tendency is to "mirror" one's own beliefs and values onto people from other cultures, which fails to capture the unique ways in which those cultures perceive and think about the world. To help practitioners better understand audiences within their own and other cultures, we have been developing modeling and simulation tools to represent and express culture-specific behaviors.

These tools, collectively known as the Target Audience Simulation Kit (TASK), incorporate marketing best practices into a desktop application that can enable marketing departments to create more effective marketing messages. TASK makes this possible by helping marketers better understand the motivations and concerns of an audience and how those can translate to effective messages. TASK builds models from data collected from a target audience, and uses those models to estimate the kinds of messages that would most resonate with the represented audience. The core of this work is our Cultural Cognitive Architecture (CCA), which implements theories of culture and cognition to enable the development of computational models of cultural behavior. In this work, we have extended CCA specifically for building models of the audiences, and estimating the effects of marketing messages.

APPROACH: TARGET AUDIENCE SIMULATION KIT (TASK)

Our approach to the problem of modeling target audiences is not a single model, but rather a modeling environment (a "toolkit") that can be used to construct a range of computational models of target audiences. Models can be constructed in different ways, including directly from pre-segmented data, or by generating similar clusters algorithmically. Those models can be analyzed in a number of ways, including how they might respond to the content of persuasive messages.

THEORY-BASED CULTURAL COGNITIVE ARCHITECTURE

The computational basis for this work is a cognitive architecture, called the Cultural Cognitive Architecture (CCA) (Taylor et al. 2007), designed to represent the cultural values of a representative sample of a population and the cognitive processes related to cultural behavior. The architecture is based on a number of theories related to culture and cognition. The work described here extends CCA with theories related specifically to persuasion and attitudes. This section describes some of the relevant theories and how those theories are implemented in the TASK computational architecture.

Cultural Schema Theory

Cultural Schema Theory (D'Andrade 1992) is a specialization of schema theory from psychology (Bartlett 1932), and posits that much of culture is *knowledge*, specifically in the mental structures individuals construct, maintain, and use to understand and operate in the world around them. An individual may have schema for culture-specific ways of interacting with others in one's same culture versus other schema for how to interact with outsiders, schema for determining the salience of objects in the environment, or schema for determining in-groups and out-groups (Shore 1996). Schemas in TASK are represented as concept networks that link behaviors and the cultural values that motivate them.

Persuasion Theories

Numerous researchers have developed theories to help explain how people are persuaded to changed their behavior – whether to buy a product or change their habits. Various mechanisms are hypothesized as affecting, causing or even resisting persuasion. A core concept at the center of persuasion is that of attitude, which is typically defined as an evaluative measure of how a person feels about a particular concept (Petty and Cacioppo 1981). Some researchers have tried to identify various factors in how persuasion can affect a person's intent to change their behavior, such as attitude toward the behavior, behavioral norms and social influences (Ajzen and Fishbein 2005). Some researchers have taken a more cognitive approach to understanding what motivates people to change their behavior (Bagozzi et al. 2002). Others cite motivation and amount of processing (elaboration) of the message that determines if a message will be persuasive, and how durable that persuasion might be (Petty and Cacioppo 1984). Generally speaking, attitudes are taken to be indicators of intention to act, and understanding and affecting attitudes is a core aspect of persuasion. This idea shapes the basic framework of TASK.

Appraisal Theory

Appraisal Theories (Scherer et al. 2001) describe the process of assessing (appraising) a situation along a number of dimensions, including how well the situation fits with one's goals, how surprising the situation is, etc., to help determine an emotional response to that situation. Appraisal is a cognitive process that can be any combination of automatic and fast, or deliberate and slow, depending on the available knowledge. In addition to determining the emotional response, appraisals can also inform symbolic reasoning via the activation of appraisal-cued schema. In CCA, generated emotions are a concrete way to demonstrate when cues in the environment align or clash with desires or expectations as defined in cognitive schema. A model in TASK generates an aggregate appraisal in response to a message, and this appraisal is used as a proxy for attitude.

Means-Ends Chaining Theory and Hierarchical Value Maps

Means-End Chaining Theory (Gutman 1982; Reynolds and Olson 2001), often used within social psychology and marketing research, provides a framework for understanding human behavior specifically focused on the linkages between the behavior (the "means"), the consequences of that behavior and the personal or cultural values (the "ends") the consequences reinforce. The premise is that people and cultures learn to choose behaviors that are instrumental to achieving their desired consequences. Means-End Chaining Theory simply specifies the rationale underlying why consequences are important: namely, personal and cultural values. A Means-End Chain, as a knowledge representation form, relates very well to the concept of cultural schema, which is already the basis of the CCA. Indeed, (Bagozzi et al. 2003) has used Means-Ends Chains with Cultural Schema Theory to understand cultural differences in decision-making.

The core representational concept in Means-End Chaining is the Hierarchical Value Map (HVM), which organizes the prevalent concepts from the target audience in a way that captures their regularities, including how often the concept occurs among the target audience and how that concept is connected to other concepts. An HVM is a directed graph, where the nodes correspond to the concepts, and the edges between the nodes are weighted based on how many times those two concepts were mentioned together. A hypothetical example of an HVM is given in Figure 1. The concepts drawn higher in the graph are more abstract concepts, often representative of core cultural values. The concept drawn at the bottom of the HVM represents the target behavior that a persuasive message is attempting to affect.

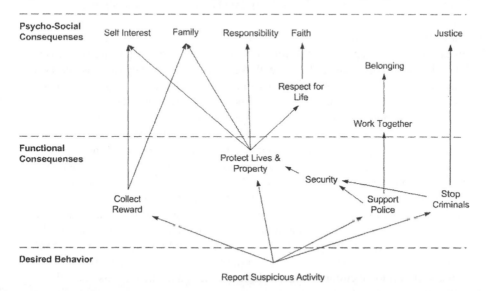

Figure 1. Hypothetical hierarchical value map (HVM) related to the behavior of "reporting suspicious activity".

TASK IMPLEMENTATION

A TASK model is essentially a Hierarchical Value Map, as described above, that captures the concepts from a target audience relative to a particular behavior. The concepts themselves, as well as the links between them, have weights associated with them, determined by the frequency in the source data. Using this core representation scheme, TASK applies a number of processes related to the theories described earlier, to generate an overall attitude toward the contents of a message.

TASK implements a concept activation algorithm that allows for related concepts to be activated by a message. Activation follows the strength of links between concepts: links and concepts with greater strength are more likely to be activated. For a concept with weak activation relationships, several related concepts may need to be active in order for it to activate. This activation process is meant to capture the process of associative memory: for example, when presented with the concept of "stopping criminals", concepts related to "justice" may be activated. The sub-network of concepts that are activated above a fixed threshold represents those concepts that are being actively thought about by the model. Activation relationships also capture the fact that an individual does not typically utilize their complete knowledge (i.e., set of concepts) when evaluating a message. Instead, people are typically only thinking about a small subset of concepts that affect their evaluation of a message. For example, a public service announcement to encourage people to report suspicious activity may explicitly draw a connection between reporting crime and protecting one's family to make the message more personal, and leave out other concepts on purpose. Activating these related concepts makes the audience think about them, while other related concepts may remain inactive.

Along with activation, the network captures the causal nature of the concepts. The direction on the link between nodes is causal in nature: the target audience believes that one concept is a cause for another concept (e.g., "reporting suspicious activity" leads to "collect reward" in the above example). These causal links carry a likelihood of how likely a person thinks the concept is (or will be) true. Causal relationships can also be negative—one concept can decrease the likelihood of another. For example, "reporting crime" may negatively affect how much crime occurs in the city (that is, cause "crime" to *not* happen). Causal relationships capture the fact that thinking about concepts together may make some seem more or less believable. For example, if there are positive causal relationships between "reporting suspicious activity" and "justice" (and these concepts are both active), then considering reporting to the police will increase the likelihood of "justice."

Some concepts are treated as goals or values – things that the target audience model wants to be true (if the importance is positive) or to not be true (if the importance is negative). In general, most concepts could be values, but TASK typically treats only the concepts with the highest importance (as identified in target audience surveys) to be cultural values. These are the "ends" in a means-ends chain: the ultimate motivation for behavior. Behaviors that relate to the ends, according to the theory, are more likely to be acted on.

With a given behavior (as suggested by a message), and with the activation process, we get an activated sub-network of strongly related concepts that comprise a causal chain from means (behavior) to ends (values). The activated sub-network is used to compute the appraisal of, and thus the model's attitude toward, the suggested behavior concept. Appraisals across multiple dimensions are combined to produce an overall attitude response. Example dimensions include Conduciveness (how good or bad a situation is for my goals), Outcome Probability (how likely a situation is to lead to progress towards my goals), and Standards Compatibility (how good or bad a situation is for my values). Appraisals are generated and combined to form an attitude. Generally, appraisal theories define a mapping from dimension values to emotions, which could be used to produce a categorical value (e.g., Anger or Joy). For our implementation here, we are not interested in emotional state per se, so the system just produces a valence value between -1 and 1, where 1 translates to a positive attitude about the concept, and -1 translates to a negative attitude about the concept. To produce a valence, simple appraisal theories tend to just multiply the absolute values of the dimension values together (Gratch and Marsella 2004). This works fine when there are only a small number of appraisal dimensions, but when there are more than a few, a more complex function for combining the values may be desirable (Marinier et al. 2009).

USING TASK

The use of TASK in support of planning a marketing campaign currently begins with information about a target audience. Models can be entered manually from expert knowledge, generated from coded respondent-level data from a survey of the population, or imported directly from a condensed form of that data called an incidence matrix (concepts and a frequency measure of each concept being mentioned together). Optionally, a user may request that multiple models be generated from coded responses from the sampled audience. In this case, TASK runs the data through a clustering algorithm to find related subsets of the population based on characteristics such as their responses and the demographic data. The algorithmically generated models represent sub-segments of the total sampled population.

Once a model has been generated, TASK allows a user to analyze the model's structure to determine which messages would resonate best with the population represented by the model. A "message" consists of a chain of concepts that is essentially an argument deliberately connecting the target behavior to an important cultural value, where the intermediate concepts provide the logical causal support between the two concepts. The idea of constructing a message as a causal chain of concepts is actually very general, and imposes no constraints on the actual medium of the message. In developing a particular message for a particular medium, a set of concepts can be communicated via text, speech, images, or any combination. From the model's perspective, the medium is abstracted away and only the core concepts are presented. A model takes in a message argument, and generates an attitude response, which is a measure of how good or bad the message argument fits with the target audience values and beliefs. Messages are scored relative to each other based on the attitude response generated (-1...+1).

For example, Figure 2 shows a TASK model built directly from data related to how people think about reducing their high blood pressure (Taylor et al. 2006). In this case, a message with concepts such as "prevent disease" and "family" generates a positive attitude from the model. While a message is a single thread through the network, the process of activation will cause a larger sub-network to be activated for appraisal. So, the selected message may in fact be connected to only slightly activated concepts, unimportant concepts, or even concepts related to negative consequences. Different messages, then, can result in different attitudes being generated.

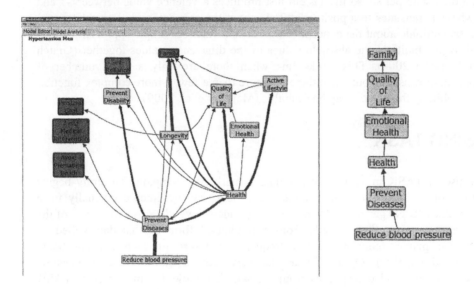

Figure 2. Snapshots of a model of a target audience in TASK (left) and a single message argument through the value map (right) that connects a behavior (reduce blood pressure) with a cultural value (family). Green colored concepts relate to important cultural values. Data from (Taylor et al. 2006).

The TASK system actually goes well beyond just evaluating a single message. It is able to perform a search across an entire network and identify and rank all messages that might be generated relative to the concepts in the network. The message set can be limited to only a subset of concepts specified by the user, or across all concepts in the network. Additionally, a user can apply messages to multiple models, to determine if the same message would be effective across multiple target populations. Highly resonant messages can be stored for later comparison.

To support the idea of a message campaign that lasts over a long period of time, models can be organized into a "project" that lets the user track multiple models and messages that are estimated as highly effective. A project might contain multiple models representing audience sub-segments, or models that capture different snapshots in time, and different messages that reach those different audiences.

CHALLENGES AND FUTURE WORK

One challenge is the sheer breadth of the psychology of persuasion. Literally hundreds of theories exist, covering aspects of how humans perceive the particular media used by an advertisement, or how they process the content of the message within an advertisement, to how social considerations might weigh into whether a behavior is adopted. There is currently no "unified theory of persuasion" that can be leveraged, so bounding the work to a narrow problem was essential in actually constructing a system, with acknowledgement that we cannot account for all aspects of persuasion in the models that are constructed.

Another challenge in building a system such as this is that a proper evaluation of the system requires running a cross-comparison study that effectively implements a marketing campaign, with and without the tool. This is a large effort that we have not yet undertaken. We have performed some micro-evaluations to determine that the appraisals generated by the system can be used as a proxy for attitudes, but we have not yet conducted a full system-level evaluation. Evaluating all aspects of TASK is a clear next step in the development of the system, from the underlying architecture to the usability aspects of the user interface.

A core challenge in marketing and persuasion studies is the availability of data about an audience, and the ability of practitioners to go out and collect new data. If circumstances are such that data cannot be collected, then the models that are constructed manually can only be best guesses by an expert. Many techniques exist for gathering data about audiences and their responses (surveys, interviews, focus groups), but the process itself is viewed as a barrier to many practitioners. Certainly cultural experts might be used to construct models, but empirical data will tend to produce more reliable assessments of target audiences.

Our next step in the development of TASK is to deal with the data issue head-on by incorporating into TASK some in-tool support for collecting and analyzing data. A technique that goes hand-in-hand with Means-Ends Chaining is called "laddering" (Miles and Rowe 2004; Taylor et al. 2006). The purpose of a laddering interview is to get people to describe how their low-level decisions are linked to their high-level motivations or values. In the laddering approach, respondents are asked to list reasons why they believe something or want to do something. For example, someone might be asked why he wants to lower his blood pressure. Then, for each reason listed, the respondent is asked to generate a reason why that is important. And then a reason for that reason is requested. Once raw data has been collected, the laddering technique includes coding and aggregating the data into an incidence matrix format. TASK can already directly import data in an incidence matrix format, or from the coded response data.

The laddering method is, by itself, quite straightforward, though can be time consuming with large amounts of data. Data collection and analysis are not free, but tools have been developed in the past to help make the process easier (Gengler 1995). Future work on TASK will include direct support for conducting laddering studies (e.g., developing the survey instruments) and assisting a user in coding and analyzing the resultant data.

SUMMARY AND CONCLUSIONS

We have described a simulation-based tool for modeling the behavior of target audiences, called the Target Audience Simulation Kit (TASK). TASK is aimed at putting simulation tools in the hands of marketing departments, public service centers, or influence operations planners. TASK can be used to understand the core concepts and values of a target audience, can help draw out the most resonant themes to use in persuasive messages, and then help estimate the effectiveness of those messages on different subsets of the audience. TASK incorporates theories from culture and persuasion, and effectively implements some best practices and methodologies from marketing science.

With TASK, we have taken cognitive approach to analyzing target audiences: understand concepts that resonate with those audiences, and construct messages with the greatest resonance related to a desired behavior. The current implantation does not capture all of the subtleties of human culture and cognition; however, our approach helps to answer the cognitive questions that other models of media and message reach do not touch.

ACKNOWLEDGEMENTS

Many thanks to our consultants on this work: Dr. Richard Bagozzi of the University of Michigan, Dr. Cliff Behrens of Telcordia Technologies, and Dr. Afzal Upal of the University of Toledo.

REFERENCES

Ajzen, I. and Fishbein, M. (2005). The influence of attitudes on behavior. The handbook of attitudes. D. Albarracín, B. T. Johnson and M. P. Zanna. Mahwah, NJ: Erlbaum: 173-221.

Bagozzi, R., Bergami, M. and Leone, L. (2003). "Hierarchical Representation of Motives in Goal-setting." Applied Psychology 88: 915-943.

Bagozzi, R., Gurhan-Canli, Z. and Priester, J. (2002). The Social Psychology of Consumer Behavior, Open University Press.

Bartlett, F. C. (1932). Remembering: An Experimental and Social Study. Cambridge, Cambridge University Press.

D'Andrade, R. G. (1992). Schemas and motivation. Human motives and cultural models. R. G. D'Andrade and C. Strauss. Cambridge, UK, Cambridge University Press: 23-44.

Gengler, C. (1995). "LADDERMAP-Version 4.0." Journal of Marketing Research 32(4): 494-496.

Gratch, J. and Marsella, S. (2004). "A Domain-independent Framework for Modeling Emotion." Journal of Cognitive Systems Research 5(4): 269-306.

Gutman, J. (1982). "A means-end chain model based on consumer categorization processes." Journal of Marketing 46: 60-72.

Marinier, R., Laird, J. and Lewis, R. (2009). "A Computational Unification of Cognitive Behavior and Emotion." Journal of Cognitive Systems Research 5(4).

Miles, S. and Rowe, G. (2004). The Laddering Technique. Doing Social Psychology Research. G. M. Breakwell, Wiley-Blackwell.

Petty, R. E. and Cacioppo, J. T. (1981). Attitudes and Persuasion: Classic and Contemporary Approaches. Dubuque, IA, Wm. C. Brown.

Petty, R. E. and Cacioppo, J. T. (1984). "The effects of involvement on response to argument quantity and quality: Central and peripheral routes to persuasion." Journal of Personality and Social Psychology 46: 69-81.

Reynolds, T. J. and Olson, J. C., Eds. (2001). Understanding Consumer Decision Making: The Means-End Approach to Marketing and Advertising Strategy. Mahwah, NJ, Lawrence Erlbaum Associates.

Scherer, K. R., Schorr, A. and Johnstone, T., Eds. (2001). Appraisal processes in emotion: Theory, Methods, Research. Appraisal processes in emotion: Theory, Methods, Research, Oxford University Press.

Shore, B. (1996). Culture in Mind. Oxford, UK, Oxford University Press.

Taylor, G., Quist, M., Furtwangler, S. and Knudsen, K. (2007). Toward a Hybrid Cultural Cognitive Architecture. CogSci Workshop on Culture and Cognition, Nashville, TN, Cognitive Science Society.

Taylor, S., Bagozzi, R., Gaither, C. and Jamerson, K. (2006). "The bases of goal setting in the self-regulation of hypertension." Journal of Health Psychology 11(1).

Trent, S. and Doty, J. L. (2005). "Marketing: An Overlooked Aspect of Information Operations." Military Review July-Aug.

CHAPTER 29

Data Problems for Cross-Cultural Decision Making

William J. Salter, Ian Yohai, Georgiy Levchuk, Bruce Skarin

Aptima, Inc.
Woburn, MA, USA

ABSTRACT

In recent years, there has been an increasing focus on developing socio-cultural models to assist military planners. The utility of all such models is constrained by the quality of available data. We discuss how problems with data can easily lead to false inferences, and provide real-world examples based on research regarding public opinion in Afghanistan and terrorist events in Iraq. In particular, we focus on problems caused by missing data, and discuss statistical methods to address them.

Keywords: Cross-Cultural Decision Making, Missing Data, Multiple Imputation

INTRODUCTION

The focus of American military operations has changed dramatically over the last two years. The emphasis has shifted from a standard, kinetic military approach to a much more complex mission, stressing counter-insurgency and similar operations. This shift places much greater emphasis on the importance of social and cultural issues in planning and implementing missions. Strategy, operational plans, and implementing procedures must reflect understanding of local cultural constraints to be effective, and soldiers must appropriately interact with local people whose culture, language, lifestyle, and beliefs are very different from their own. Gen. Stanley McChrystal, Commander of U.S. and NATO forces in Afghanistan, summarized this in a discussion with reporters on February 4, 2010: "This is all a war of perceptions. This is not a physical war in terms of how many people you kill or how much ground you capture, how many bridges you blow up. This is all in the minds of the participants" (Shanker, 2010).

Dealing with cultural issues is difficult for several reasons. First, the new missions are typically more complex than traditional combat missions. They tend to have multiple objectives; measures of effectiveness are often difficult to develop and apply; and paths to mission success are not well mapped in existing tactics, techniques and procedures, the standard form in which military actions are expressed for planning and training. Second, of course, culture itself is complex, affecting many aspects of human behavior and social organization, from norms for interacting with individuals of different genders and ages through how political and economic power are organized, expressed, and exercised.

Third, achieving mission success often involves complex tradeoffs. A simple, but frequently relevant, example concerns dealing with a local leader with whom one wants to establish an effective working relationship. This leader may not get along well with another local leader with whom it is also important to work effectively; such problems are difficult in one's own culture, but more so in a foreign war zone. Tradeoffs can also occur across geographic and temporal scales: fostering economic development in one village may harm a neighboring village; if those villages differ in ethnicity or tribal affiliation, the local problem may have larger-scale implications. Or building a well in a village may help farming in the near term but may cause more rapid depletion of the aquifer in the medium term, or may greatly benefit nearby farmers and disadvantage others over time. The importance of cultural issues cuts across all levels, from strategic through operational to direct tactical engagements, and affects mission planning, execution, review, and training.

The military research and development community has been responsive to this new requirement, of course, and a major human socio-cultural behavior (HSCB) program has begun: conferences (including the one for which this paper was written) have been held, attracting a lively mix of academic, military, other government, and industry participants; research and development grants and contracts have been awarded; a number of offices and task forces have been stood up across Services and Commands. Modeling is a major focus of HSCB activity: enhancing and integrating existing models; applying old models to new data and new challenges; building new models for a variety of purposes, from a variety of theoretical perspectives, and with a variety of technologies.

HSCB modeling is also complicated for a number of reasons. In this paper we consider several aspects of only one class of problem – those concerning data – and discuss approaches that modelers, and other users of such data, can take to reduce associated risks, to make the best uses of the data they have, and to get maximum value from additional data collection efforts.

Data are vital to applying models but they are also critical even in the absence of models. In a recent report, Major General Michael T. Flynn (Flynn et al., 2009), the senior military intelligence officer in Afghanistan, argues vigorously that "the vast intelligence apparatus is unable to answer fundamental questions about the environment in which U.S. and allied forces operate and the people they seek to persuade" (p. 7). His goal is to close that gap. Intelligence must "acquire and provide knowledge about the population, the economy, the government, and other aspects of the dynamic environment we are trying to shape, secure, and successfully leave behind" (p.10). His emphasis on answering "fundamental questions," on "provid[ing] knowledge," makes clear that he is primarily concerned with description, not prediction or sophisticated analysis of alternative scenarios.

Examples abound where data issues are relevant to social policy questions. For example, rates of domestic violence have remained relatively steady for the past three decades, despite major efforts to increase reporting, reduce stigma, and increase punishment for abusers and protection for victims. Does that mean that those efforts have been unsuccessful? That they have been helpful but underlying drivers of abuse have increased? That they have been helpful and the steady rates reflect the combination of reduced abuse and increased reporting? Similarly, the lively debate in some quarters about autism incidence includes discussion of increased reporting and broadening of the definition to incorporate more symptom patterns, as well as considerations of possible mechanisms of causation. Similar discussions are taking place about various other mental disorders, such as ADHD, depression, and social anxiety. Discussions of survey results, methods, and interpretations pervade media coverage of politics. Determining what counts and how to count it – that is, questions about the data – are central to all these issues.

Data problems are of various types. Many available datasets contain missing observations; how to deal with them can have major implications for resulting analyses. Many such methods involve discarding records for which some data are missing, but, in general, it is better to use all of the data than to throw some out. "It is critically important to address missing data, as it arises in almost all real-world investigations. Accounting for incomplete observations is particularly important for observational analyses with many predictors" (Horton and Kleinman, 2007, p. 89), which is often the case for real-world socio-cultural datasets. Indeed, how missing data are dealt with can affect not only confidence limits of conclusions or forecasts but the direction of detected relationships and the nature of predicted outcomes (Rubin, 1987; King et al., 2001; Little and Rubin, 2002; King and Honaker, 2010).

In other cases, datasets may have underlying sampling or similar problems, which can complicate analysis or, even worse, lead to erroneous conclusions. We briefly discuss these issues below, with real examples, and suggest approaches to addressing some important data problems.

ANALYZING PUBLIC OPINION IN AFGHANISTAN

Here, we briefly describe an example from research on the impact of ethnic identities on population attitudes toward coalition forces in Afghanistan. A survey conducted in August 2009 revealed that 40% of Pashtuns held a "low" or "very low" opinion of the International Security and Assistance Force (ISAF), compared with only 21% of Tajiks. Based on this information alone, one might conclude that Pashtuns as a group are hostile toward the coalition. Disaggregating the data, however, paints a fundamentally different picture. Indeed, there are wide regional variations in attitudes toward the coalition: respondents in the southern provinces were much more likely to express negative attitudes than in the northern ones. Since the South is overwhelming Pashtun, we have a classic case of a "confounding" variable. are Pashtuns as a group actually more hostile toward the coalition, or is it simply that more Pashtuns live in the South, an area where the Taliban is strongest and which has been most affected by the increase in violence?

One solution is to compare the attitudes of Pashtuns and Tajiks in provinces with significant numbers of both groups. Within a given province, we are able to hold environmental factors, such as the level of violence or degree of Taliban infiltration, relatively constant. Indeed, the data show that, in contrast to our initial finding, Tajiks and Pashtuns living in ethnically mixed provinces actually have quite similar attitudes toward ISAF. In short, then, a simple national level cross-tabulation – even with reasonably good survey data – could quickly have led us to a false inference.

TERRORIST EVENTS IN IRAQ

Another example comes from a dataset provided by JIEDDO (Joint IED Defeat Organization) dealing with terrorist-related "events" in Iraq in 2005 and 2006, taken from mainstream news sources and coded by contractors.

To make better predictions about when and where future terrorist attacks are likely to occur, it is useful to examine historical data on such events to search for relevant patterns. Obviously, if one terrorist organization is found to operate primarily in one geographic area, or tends to use the same attack mechanism, this information is quite important in helping to make accurate forecasts. The dataset provided to us by JIEDDO seemingly contained a wealth of variables along these lines related to 6590 terrorist "events" in Iraq during 2005 and 2006. These variables included the date of the attack, the geographic location, the organization thought to be responsible, the target, the attack mode (IED, gun, etc.), the number of causalities, and the news source from which these data were coded. Once we began to work with the dataset, however, we noticed several problems that, if left unaddressed, could have lead to seriously mistaken inferences. Some problems were seemingly trivial: there were misspellings of the names of the terrorist organizations. For example, "Al Qaeda" was spelled in a number of different ways, which might lead the casual observer to conclude that *different* groups were responsible. While an educated consumer of the data could go through each of the 6590 events and clean up these minor errors, it is a quite labor intensive process that it is difficult to automate. Moreover, such corrections must be made *before* using complicated modeling strategies such as pattern matching algorithms. How can an algorithm find important patterns if the data, as initially coded, indicate that several different groups are behind a number of attacks, when, in reality, the *same* group was responsible?

Upon further investigation, the dataset contained much more significant problems. The total number of attacks increased dramatically between the two years included in the sample. Whereas there were only 921 attacks reported for 2005, this figure ballooned to 5669 in 2006. Those familiar with the trajectory of the Iraq war would know that the level of violence likely did increase between the two years, but did the data really support a *six-fold* increase? More troubling, the purported increase in attacks was especially abrupt. Throughout the latter half of 2005, about 100 attacks per month were reported on average, which jumped to over 400 in the early months of 2006. While it was certainly possible that the number of attacks increased suddenly, the apparent spike caused us to dig deeper.

Our first step was to examine the news reports from which the terrorist event data were coded. We found that a wider array of news sources was used to identify the events in 2006 than had been used in 2005. In 2005, 61% of the events was coded from Reuters, while only 42% was coded from Reuters in 2006. Many more events were coded in 2006 from the Associated Press (AP), Agence-France Press (AFP), BBC, CNN, and others. Indeed, the difference in coverage from the AP was striking: reported attacks increased from a mere 70 in 2005 to over 1300 in 2006. We also found a substantial amount of month-to-month fluctuation when breaking out the number of attacks reported by each news source.

To help correct for this, we used regression smoothing to help estimate trend lines. The results, however, still showed substantial variation among the different news organizations. Reuters indicated an average of about 20 attacks a month in early 2005, increasing steadily to an average of nearly 400 attacks a month in late 2006. By contrast, the AP series indicated an average of about 2 attacks in early 2005, peaked at 100 attacks a month near the middle of 2006, before declining again to 83 attacks per month in late 2006. Looking solely at the AP data, one might have concluded that the level of violence had leveled off substantially, or even declined; the Reuters data supported exactly the opposite conclusion. That is, news source proved to be a key explanatory variable for the number (and types) of attacks.

Clearly, then, the dramatic increase in attacks between 2005 and 2006 could potentially be explained by the different mix of news sources consulted in construction of the dataset. To confirm this finding, we compared the number of attacks in our dataset to another dataset that we believe was constructed in a more consistent manner. The Worldwide Incident Tracking System (WITS), complied by the National Counterterrorism Center (NCTC), attempts to standardize the way in which terrorist events are counted and coded, across the entire world. To be sure, the WITS database suffers from its own limitations. In particular, the WITS website notes that it is difficult to obtain open source reporting of terrorist attacks from areas that see the most violence. As a result, the number of attacks is probably underestimated, particularly in the most violent regions of Iraq and Afghanistan. Nevertheless, it is useful benchmark against which to compare to our own data.

In Figure 1, we plot a histogram of the number of attacks in Iraq by month for both the WITS dataset and the JIEDDO dataset. There are many fewer attacks in the JIEDDO dataset than in the WITS dataset in 2005. This is not surprising given the limited number of news sources used to construct the former. In 2006, the two series track each other much more closely, although the JIEDDO dataset indicates a more significant drop-off in violence toward the end of that year. Still, though, the WITS dataset contains a greater number of attacks in *every* month, except in February of 2006. The apparent spike in the JIEDDO dataset during February is almost entirely the result of several AP news stories relating to the bombing of the Samarra mosque, which led to several sectarian reprisals. In short, our dataset seems to be heavily dependent on the particular mix of new organizations that were used at any one time to construct the list of events.

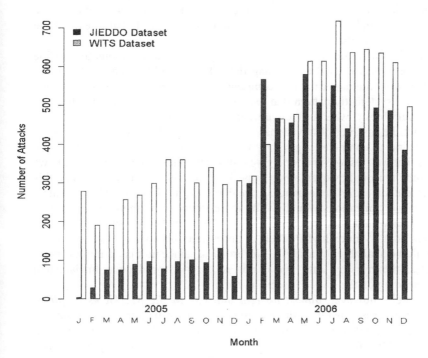

FIGURE 1. Comparison of the JIEDDO and WITS Datasets

While this discussion has not been overly technical, relying mainly on descriptive statistics and simple visualization plots such as histograms, we believe it illustrates a larger point. *Simply looking at one's data* before employing more sophisticated analysis or modeling strategies is an invaluable first step that will improve the results obtained later on. In this case, it was quite clear that the 2005 data significantly undercounted the number of attacks, and we were able to isolate the reason why. Certainly, it would be possible to conduct further pre-processing of the

286

data, including coding additional news sources from 2005, in order to better match the procedure that was evidently used in 2006.

A MISSING DATA FRAMEWORK

In addition to the coding problems outlined above, we also discovered another important limitation of our initial dataset: for many terrorist events, several variables were missing. For example, the organization responsible for the attack was not recorded in a large fraction of cases. This might be due to a number of factors: no group claimed responsibility or could otherwise be identified; the news report from which the event was coded failed to mention the group even if the information was available; or finally, the human coder simply failed to extract the name of the group from the news story.

The statistical literature describes three categories of assumptions that define the scope of problems caused by missing data (Rubin 1987, King et al. 2001).[1] Most optimistically, the data can be *missing completely at random* (MCAR). In this case, there is no systematic pattern that explains the reason why certain values are missing. No variables – whether they are contained in the dataset or not – can be used to predict the missing observations. An example of this process might be as follows: imagine a computer algorithm was used to extract the name of the organization that was responsible for a given terrorist attack from the relevant news story. For some reason, however, a bug caused the algorithm to randomly skip this step for one out of every three attacks. In this case, no bias would be created, because certain organizations would *not* be more likely to be missing. However, the missing data would still impose real costs. Any conclusions drawn from this dataset would contain much higher levels of uncertainty because only two-thirds of the data would be available for analysis. Nevertheless, since no bias is introduced in our estimates, the missing data would not necessarily produce a fundamental problem, depending on the level of precision required.

Unfortunately, however, *almost no real datasets fit the MCAR assumption.* Inevitably, the process generating the missing data cannot be attributed to random error. If a single variable – whether we capture it or not – helps predict the missing values, then the MCAR assumption is violated. This suggests an easy way to test whether the MCAR assumption is valid: for each variable in the dataset, simply compare the distribution of values when other variables are observed to when they are missing. Figure 2 depicts one such exercise based on the JIEDDO dataset. As the figure shows, shootings are the largest proportion of attack types when the name of the organization responsible is observed. Yet bombings are the largest proportion when the group behind the attack is unknown. Had we simply discarded all observations that contained at least one missing variable – a common practice known as *listwise* deletion – we would have concluded that bombings were more prevalent than shootings, a finding that obviously would have a large impact in

---

[1] The discussion below draws heavily from King et al. 2001.

devising future strategy. Clearly, then, the MCAR assumption is violated in this case. If the two distributions in Figure 2 had been the same, we would then proceed to the next variable and perform the same type of analysis.

Another possibility is that the data are *missing at random* (MAR; this standard terminology is confusing for historical reasons). Under this assumption, the missing values can be predicted by other *observed* variables. In the case, location, date, target, and type of attack, as well as number of causalities, might all be relevant in determining which terrorist organization was responsible. Unlike MCAR, however, missing observations *will* lead to bias if the data are generated by a MAR process, as in the example above regarding the common type of attack modes (bombings or shootings). As we discuss in greater detail below, statistical methods are available to mitigate the bias caused by missing data under the MAR assumption.

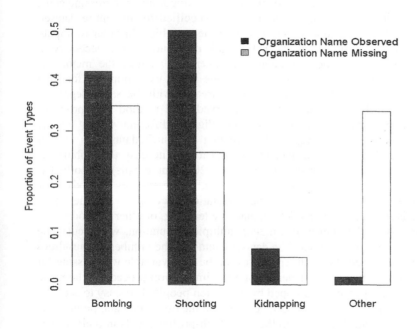

FIGURE 2. Comparison of Event Types with and without Missing Data

Finally, the data may be categorized as *non-ignorable* (NI). Here, the missing data are generated by a non-random process (unlike MCAR), but also cannot be fully predicted by other observed variables (unlike MAR). Even knowing several characteristics of a terrorist attack may not be enough to know which group was responsible. The missing values, then, depend on some *unobserved* set of variables. As a result, determining whether the data fall into the MAR or NI categories is not possible unless the additional variables become available for analysis. And, of course, we cannot use *unobserved* variables to help predict missing values. Thus, in order to make further progress, we must assume, as do the methods discussed below, that the data are MAR. How reasonable this assumption is obviously depends on the specific case, but empirical studies have shown that these methods can still generate accurate results even under mild violations of MAR (thus making the missing data technically NI; see King et al. 2001, Little and Rubin 2002).

METHODS FOR MISSING DATA: MULTIPLE IMPUTATION

Standard methods for addressing the problems caused by missing data rely on two key insights. First, building on the MAR assumption, the observed portion of the dataset contains information relevant for filling-in, or "imputing," the missing cells. Accordingly, a joint statistical distribution can be specified for the entire dataset, which formalizes the correlation structure among the variables. In order to impute missing values, repeated draws are taken from this distribution. It is necessary to take *repeated* draws to properly account for uncertainty; hence the method is referred to as *multiple* imputation. If only a single value were imputed for each missing cell, then the imputed values would be treated with the same degree of confidence as the cells that were actually observed. The method thus produces several "complete" datasets – each containing slightly different values for the missing cells. The literature suggests that creating five such datasets is usually enough to capture the underlying uncertainty, and it is quite easy to combine the outputs for substantive analysis (see King et al., 2001 for details). Importantly, multiple imputation does *not* constrain the range of ultimate analysis methods. *Any* type of analysis is possible following the imputation process --- whether it be descriptive statistics, regression modeling, machine learning, or other methods.

To illustrate the importance of utilizing multiple imputation, we discuss one example from the JIEDDO dataset. The dataset contained the number of causalities (both killed and wounded) for each terrorist event, but this variable was missing for several of the coded attacks. After performing the imputation procedure, the mean number of causalities per attack dropped by nearly ten people. It is easy to see why the reported number of causalities might be higher in the observed data: perhaps the news organizations tended to focus on the more high-profile events in their stories, thus painting a skewed portrait of the scale of the typical attack. And the different result would have implications for policy: defending against many smaller attacks might well require a different strategy than defending against fewer, larger ones.

While multiple imputation has become the "gold standard" in dealing with missing data, it is certainly not a cure-all. The method cannot be expected to produce accurate imputations if the missing values depend substantially on unobserved variables (i.e., the data fall into the NI case). But simply ignoring the missing data problem or using a method like listwise deletion is guaranteed to be substantially worse. A range of diagnostic procedures is available to test the sensitivity of the imputations to specific method employed (see King et al.,

2001 for details), and if the imputations tend to be highly variable from one draw of the distribution to another, this may be a sign that results should be interpreted with a heavy dose of caution. Naturally, if other datasets are available against which to compare results, they should be exploited to the fullest extent possible. Often, additional data will be available over time that can be used directly for validation purposes.

CONCLUSION

The quality of available data is a critical factor in drawing reasonable conclusions even from highly sophisticated models. Simple coding errors, differences in sampling coverage, and incomplete data all can easily lead to substantial bias and false inferences. Adequately addressing these problems often requires some degree of "handwork" – whether that involves simply visualizing data or interpreting the results of sensitivity analysis stemming from more advanced procedures like multiple imputation. Taking the time necessary to full understand the limitations of data will undoubtedly improve a wide variety of analysis methods and can be far more cost-effective than development of yet more complex models.

ACKNOWLEDGEMENTS

This work was partially supported by contract number FA8750-09-C-0095 from the Air Force Research Laboratory/RIEA and contract number FA8650-08-C-6922 from AFRL/RHCS. All opinions expressed herein are the authors', and do not reflect those of AFRL, the DoD, or the U.S. Government.

REFERENCES

Honaker, J. and King, G. (2010). "What to do About Missing Values in Time Series Cross-Sectional Data." *American Journal of Political Science*, 54(2): 561-581.
Horton, N. J. and Kleinman, K. P. (2007). "Much Ado About Nothing: A Comparison of Missing Data Methods and Software to Fit Incomplete Data Regression Models." *The American Statistician* 61(1): 79-90.
King, G., et al. (2001). "Analyzing Incomplete Political Science Data: An Alternative Algorithm for Multiple Imputation." *American Political Science Review*, 95(1): 49-69.
Little, R.J.A. and Rubin, D.B. (2002). *Statistical Analysis with Missing Data: Second Edition*. Hoboken, NJ: John Wiley & Sons.
Rubin, D.B. (1987). *Multiple Imputation for Nonresponse in Surveys*. New York: Wiley.

Shanker, T. (2010) U.S. General Offers Upbeat Views on Afghan War. *New York Times, February 4, 2010*

CHAPTER 30

MASON RebeLand and Data Aspects of Agent-Based Simulation Models

Claudio Cioffi-Revilla

Center for Social Complexity
George Mason University
Fairfax, VA 22030 USA

ABSTRACT

Empirically referenced agent-based social simulation models in general, and socio-natural systems models in particular, pose new challenges regarding data needs. Although some challenges are solved by drawing on previous social simulation experience, others require new solutions. This paper discusses data requirements and related issues in the contexts of model calibration (fitting) and validation (testing) with illustrations from the MASON RebeLand model of the Mason-HRAF Joint Project on Eastern Africa.

Keywords: Social simulation, agent-based modeling, RebeLand, MASON toolkit, model calibration, model verification, external validity, Eastern Africa, coupled socio-natural systems

INTRODUCTION

Social simulation models from computational social science are beginning to provide some advances in terms of implementing more complex social, human, and natural dynamics that are characteristic of how countries operate in the real world. Although social simulation was pioneered several decades ago through dynamical models (e.g., Sternman, 2000), it has not been until the generation of object-oriented programming languages that social simulations have gained in empirical fidelity and experimental potential (Gilbert and Troitzch, 2005).

In particular, increasingly realistic agent-based models (ABMs) can improve capacity for understanding and prediction. Examples of pioneering agent-based models are the Schelling "segregation model", Heatbugs, and Sugarscape (Gilbert, 2008). Although these are generally theoretical abstract models intended mostly as proof-of-concept, other more recent models have shown applicability in a number of areas of public policy and not strictly academic interest. These include areas such as transportation, urban growth, land use and cover change, biodefense, cyber security, and numerous other areas (North and Macal, 2007).

This paper focuses on data-related aspects illustrated by the recent MASON RebeLand agent-based model developed by the Mason-HRAF Joint Project on Inner Asia (Cioffi and Rouleau, 2010). Data-related aspects include issues pertaining to external validity, such as calibration and validation in a narrow sense. (By contrast, issues of internal validity concern model verification and are not the topic of this paper.)

DATA ASPECTS OF AGENT-BASED MODELS

The purpose of this section is twofold: to present an overview of data types that are common in agent-based models, and to outline aspects of calibration and validation with special reference to empirical data in ABMs. The section that follows illustrates some of these ideas in the context of the MASON RebeLand model.

TYPES OF DATA

This analysis on data requirements in ABMs focused on socio-natural or social-ecological systems (e.g., Liu et al., 2007). Empirical data for any kind of simulation model is selected as a function of the kinds of research questions being addressed by the model in question—all data requirements are ultimately driven by research questions. The data-related issues raised here therefore apply to a broad class of models in social simulation, especially those involving natural or technical environmental contexts or ecologies in interaction with human and social dynamics.

Empirically, and at a high level of formal modeling description (Taber and Timpone, 1996), spatial agent-based models require three classes of data, corresponding to ontologies concerning natural, artifactual, and social model components. Such an overall ontological architecture is somewhat similar to but distinct from the high-level data-layer structure that is characteristic of GIS (Geographic Information Systems). However, in ABMs the data files may or may be GIS-based (e.g., NDVI distribution; Cioffi, Rogers, and Hailegiorgis, 2009).

The three classes of data are as follows. *Natural data* is used for describing terrain (topography, hydrology, and other terrain features), land cover or use (e.g., as in LUCC models; Parker et al., 2002), and weather features (e.g., as in the

MASON HouseholdsWorld model; Cioffi, Rogers, et al., 2010; Rogers et al., 2009). The inclusion of weather data in social ABMs, or of other geophysical data, is still relatively new but essential for models in which weather phenomena are deemed relevant for answering research questions e.g., crucially, for modeling societal impacts of climate change or variability. Generally speaking, natural data refer to the physical and biological ecosystem of the target system or focus region, including all land features (terrain, cover, biota) included in the simulation system. (In the Mason-HRAF Joint Project on Eastern Africa we are interested in applying the new ecosystems classification being developed by the USGS; Sayre, 2009.)

Social data refers to people and human-related features, such as population settlement patterns and other demographic distributions and social relations. Human agents in social ABMs are typically heterogeneous (e.g., classes of agents such as sedentary or nomadic; buyers, sellers, and lenders; land-based, maritime, airborne), in proportion to human diversity found in the focus region.

Finally, *artifactual data* refers to all man-made artificial or engineered components of the target system or focus area, including physical and organizational entities along with their relevant features and dynamics; in the sense of Simon (1996). *Physical data* include all materially engineered systems or man-made structures, such as buildings (public, private; residential, commercial), roads, power and utility grids, transportation systems, parks, irrigation systems, port facilities, industrial facilities, bridges, and other infrastructure system components. *Organizational data* includes institutions, administrative hierarchies, and other decision making or policy making systems distributed over territory.

While not all models contain all of these data types, most socio-natural systems will contain all three classes.

CALIBRATION AND VALIDATION

Empirical data are associated with ABMs two distinct contexts: calibration and validation. The former concerns the proper quantitative and qualitative representation attained by the simulation model, or level of fidelity, with respect to a given target system and the specific research questions being asked. The latter concerns the extent to which simulation results or artificial outputs from the model are able to replicate the observed system being modeled. These are two distinct modeling contexts that are akin to "fitting" and "testing", respectively, in the more traditional context of mathematical modeling (Muncaster, 1987).

More specifically, data requirements for *calibration* purposes (i.e., for "fitting" the simulation model to the target system or focus area) involves proper tuning, parametrization, scaling, and similar model development activities beyond the abstraction of relevant entities, attributes, dynamics, and relations (i.e., after the essential ontology is designed). Examples of calibration-related items can include:

- Relevant empirical quantities (demographic parameters, seasonal patterns, cost quantities);

- Distributions of size or time (time between event onsets, durations);
- Interaction topology features and neighborhoods (triangular, square, hexagonal, or other sites), depending on spatial features; and
- Proper temporal units (from milliseconds associated with human decision making to years and longer associated with longer-term human and social dynamics).

Whereas many of these features are not viewed as important in purely formal or theoretical ABMs (e.g., Heatbugs, Conway's Life, or even Schelling's Segregation model), features such as basic empirical parameters, distributions, topologies, and temporal units are essential for proper calibration on socio-natural ABMs—especially those that may be intended for policy analysis applications. For example, a model containing settlement patterns of any arbitrary form (e.g., uniform, or Gaussian) would be socially rare or unnatural (because human settlement patterns are rarely uniform or bell-shaped), which would result in poor or inadequate model calibration. Similarly, an ABM designed with socially impossible or rare inter-event distributions (e.g., some highly skewed power law very far from exponential) would not provide for good calibration—unless, of course, such rare temporal distributions happen to be observed in the target system.

Data requirements for *validation* purposes (i.e., for "testing" the simulation model), on the other hand, involve obtaining relevant empirical data with which to compare "synthetic" output data generated by simulation runs. Examples of validation can include but are not limited to the following (which partially overlap with calibration data):

- Key quantitative proportions (agent populations, seasonal fluctuations, spatial patterns);
- Overall qualitative properties (tends, fluctuations, spatial organization);
- Distributions of size or time (time between event onsets, durations); and
- Time series features (stationarity or Markovian properties, such as ergodicity and steady-state).

Again, although these features may not be as important in purely formal ABMs, they are critical for testing social ABM simulations that have a clear empirical reference. Accordingly, a model producing patterns of human settlement must generate approximately Zipfian rank-size distributions, such as is common in many societies, not in any arbitrary form (e.g., Gaussian). Similarly, a model that produces conflict events should generate inter-event and duration distributions that are approximately exponential, not with some highly skewed power law distribution, whereas size distributions should approximate a power law and not an exponential distribution (again, unless such rare temporal distributions are observed in the target system).

There are many data features that can be used in both contexts, such as the case of size distributions and temporal distributions. Other important examples of dual-context data include *network structures* (e.g., degree distributions for distinguishing random, scale-free, and other kinds of networks), which can be used in both contexts, and *social institutions* (e.g., official government and alternative and/or competing subsidiarity organizations and public good provides; Ferguson and Mansbach, 1996; Mousseau, 2007; Cioffi, 2009). Sometimes the same data set is used, especially in the context of validation or testing, as in a classical within-sample and out-of-sample research design. Also, all three large classes of data—social, natural, and artifactual—are involved in both contexts.

CALIBRATION AND VALIDATION IN RebeLand

In this section I describe data requirements within the context of the Mason-HRAF Joint Project on Eastern Africa in the first subsection, and in the second subsection one of the models of the project is used to illustrate types of data and contexts just discussed. The Mason-HRAF project and model information provided is necessarily brief, due to space limitations, so the reader should use references provided to obtain further details.

THE MASON-HRAF JOINT PROJECT ON EASTERN AFRICA

This ONR-sponsored project is focused on the contemporary societies, polities, and ecosystems of Eastern Asia, defined as the traditional region of East Africa plus the Horn of Africa—namely Kenya, Uganda, Tanzania, Rwanda, Burundi, Somalia, Ethiopia, southern Sudan, and eastern Chad (Cioffi et al., 2009). These nine countries comprise numerous ecosystems, ranging from desert to high mountain, with grasslands, forests, and other ecosystems within a diverse spectrum (Sayre, 2009). Although the interest in this project concerns understanding and analysis of complex contemporary issues through modeling and simulation, the region is also know for being the cradle of human evolution as well as the location of significant geophysical activity in the Great Rift Valley.

The project seeks the following related research goals specifically pertaining to Eastern Africa and, from a more general cross-cultural comparative perspective, with respect to the global international system:

- Improving basic and applied scientific understanding of social dynamics such as conflict and cooperation, especially in the context of socio-ecological systems and multi-agent systems;
- Creating new datasets and cross-cultural methodologies that integrate and bridge the gap between classical statistical social science approaches (e.g., cross-cultural research) and more recent social simulation approaches, such as socio-natural agent-based computational modeling;
- Developing a suite of new computational ABMs for simulating a range of *in silico* experiments, including societal impacts of endogenous, exogenous, and combined stresses, including economic crises, insurgency,

climate change, and others, as well as innovative applications of evolutionary computation (de Jong, 2009; Luke et al., 2009).

The main products of the project consist of new data and models. Data are mostly in the form of ethnographies and cross-cultural datasets for comparative research. Models are primarily rendered as agent-based models of Eastern Africa on various spatial scales for addressing the set of research questions.

The MASON system (Luke et al., 2005) has been used to develop four main ABMs with increasing spatial (and ecological) scale:

1. HerderLand is a simulation model that captures local interactions among herders (i.e., herder-herder relations) and between herders and farmers, inspired by conditions in the Mandera Triangle of northeast Kenya (Kennedy et al., 2010). Experiments with this model have focused on resource access (e.g., watering holes for cattle) and resulting social relations (conflict) under various climate conditions pertaining to the incidence of drought in the region.

2. RebeLand represents a generic Eastern African country-level polity situated in several ecosystems with resources and climate (Cioffi and Rouleau, 2010). Experiments include the inducement of state failure conditions under various scenarios for political instability, insurgency, and climate change, among others

3. TurkanaLand is a regional model comprising national territory as well as borderlands, inspired by the Marsabit region of Turkana District in northwest Kenya and comprising the borderlands with Uganda, Sudan, and Ethiopia. This model is presently in the early stages of development. A prototype is expected by the end of 2010.

4. AfriLand is the largest area model, comprising the whole of Eastern Africa and its ecosystems (Cioffi and Rouleau, 2009). The early prototype of AfriLand was developed by aggregating a large region composed of ten neighboring RebeLand-like country models. Subsequent versions of AfriLand will be specialized to reflect actual Eastern Africa.

Each model in the project includes some of the three classes of data discussed earlier: Social data on agents, norms, beliefs, internal and international administrative boundaries; natural data for topography and hydrology, NDVI biomass distribution for land-cover, climate; and artifactual data pertaining to road networks, urban areas, borders, and others. For example, governance data includes local traditional institutions (e.g., McCabe, 2004; Mwangi and Ostrom, 2009). The extent of calibration and validation also varies across models, with HerderLand and TurkanaLand being closer to the empirical focus regions and the initial RebeLand and AfriLand prototypes being more abstract.

THE MASON REBELAND MODEL

The MASON RebeLand model (Cioffi and Rouleau, 2010) illustrates many of the issues raised by data requirements of calibration and validation in the Mason-Smithsonian project. A separate paper provides a more detailed comparative analysis of data aspects across the suite of models in the project (Cioffi, 2010b).

As mentioned earlier, the MASON RebeLand model is an agent-based social simulation model of a country-scale society. The model presents three innovations over earlier models: (1) an explicit polity model with politically complete structure and processes (i.e., a social system with governance recognizable to a political scientist); (2) social and natural model components within an integrated socionatural system; and (3) generative dynamics where insurgency and the state of the polity (stable, unstable, failing, failed, recovering) occur as emergent phenomena under a range of social and environmental conditions, not as exogenously imposed conditions. Three scenarios have thus far been demonstrated with this model, showing stable, unstable, and failing polity conditions. The MASON computational system for agent-based and network modeling (Luke et al., 2005) also permits additional experiments and extensions that are currently underway (Cioffi, De Jong, and Bassett, 2010; De Jong, Bassett, and Cioffi, 2010; based on De Jong, 2006, 2009; and Luke et al., 2010).

In the context of construction and *calibration*, the RebeLand model covers all three main data classes as follows:

1. *Natural data* inputs include terrain topography (mountains, valleys, coastal features, natural resources), land cover (biomass), and weather (ambient moisture/precipitation). Weather fluctuates with simple sinusoidal seasonal change, and with two unequal annual seasons (large and small) in subsequent versions to reflect actual fluctuating Eastern Africa weather patterns. Weather is used to force an additional source of stress on the polity, in addition to social stresses (e.g., inflation or terrorism), so weather patterns are calibrated to resemble prevailing patterns in Eastern Africa. Similarly, natural resources represent oil, diamonds, or precious metals (e.g., gold).

2. *Social data* consists primarily of the distribution of population in cities, including a capital city (the largest concentration of inhabitants), three provincial capitals, and a number of smaller towns. Calibration features of social features in the model include:
   a. A two-level system of governance with federal and provincial decision making for public administration of implemented policies;
   b. Realistic features, in the system of governance, such as levels of corruption, policy efficiency, and a range of tax rates;
   c. Citizen satisfaction dependent on basic needs and public sentiments in support of government and, under conditions of stress, support for armed opposition (insurgency);

        d.   Settlement sizes of cities and towns chosen from a Zipf-like harmonic distribution generated by the largest (capital) city;

        e.   Public issue onset rates and durations given by an exponential (Poisson-like) distribution, reflecting known patterns in political events data (e.g., coups and violence);

        f.   Public issue intensity is power-law distributed, again reflecting known empirical patterns;

        g.   Time units calibrated to the range of weeks to months for the main simulation loop, with years to decades in terms of simulation runs.

3. *Artifactual data* consists of both man-made systems data and organizational systems data. The former consists primarily of roads drawn by a greedy algorithm, linking population centers and natural resource locations. The latter consists of a public administration system headquartered at the capital and composed of three provinces, as previously highlighted.

These and other features make RebeLand a reasonably well-calibrated model, given the model objectives and research questions being investigated.

In the context of *validation*, thus far RebeLand has used the following strategies to match and test the correspondence between simulation and empirical data:

1. Lognormal distribution of household income, reflecting empirical patterns in many societies.
2. Emergence of state instability and failure through conditions resembling the emergence of corresponding processes throughout the political history of developing African polities (e.g., Bates, 2008).
3. Bursts of political violence and insurgency under a range of credible conditions when complex societal stresses occur, including state failure when nasty conjunctions of natural disasters (e.g., drought) occur in combination with social stresses (inflation, insurgency);
4. Delays of "hysterisis" and related dynamic effects emulating patterns of political metastability.

Additional validation tests are underway to provide further exploration and insights. For example, parameter sweeps provide an approach to uncover not only bugs in the model (verification) but, importantly, tests of validity by reference to the operation of actual Eastern African polities.

CONCLUSIONS

Data-related research in the context of calibration and validation has been a major activity in the development of ABMs of coupled socio-natural such as those discussed in this paper. Subsequent versions of RebeLand in the Mason-HRAF project will include additional data inputs across all classes of data and in both contexts. For example, the class of natural objects will be enhanced with data that approximates increasingly realistic patterns of natural hazards, including not just drought (already present in Rebel and v.1.0) but also severe storms, infestations, and floods. The social class will be enhanced to include heterogeneous agent identities, which can be co-located (spatially integrated) or segregated. The class of artifactual objects will be enhanced to include additional physical systems as well as organizational systems, such as rogue polities and organized opposition to the official government.

As is the case in every complex social simulation project (Cioffi, 2010a), the addition of entities and dynamics (model development) takes place according to a specific design determined by the main research questions in the project. RebeLand v.1.0 investigates the generic conditions under which a ("basic") country maintains stability, becomes unstable, or fails, given a variety of stresses and operating conditions. Subsequent versions will add more data classes to approximate specific countries in our region of interest (Eastern Africa).

ACKNOWLEDGEMENTS

Funding for this study was provided by ONR MURI grant no. N00014-08-1-0921 and by the Center for Social Complexity of George Mason University. Thanks to members of the Mason-HRAF Joint Project on Eastern Africa (MURI Team) for helpful criticisms and comments. The opinions, findings, and conclusions or recommendations expressed in this work are those of the author and do not necessarily reflect the views of the sponsors.

REFERENCES

Bates, R. H. (2008), *When Things Fell Apart*, Cambridge University Press.

Cioffi-Revilla, C. (2010a), "On the methodology of complex social simulations." *Journal of Artificial Societies and Social Simulations*, 13(1), 7. Online.

Cioffi-Revilla, C. (2010b), "Comparing agent-based computational simulation Models in Cross-Cultural Research," Annual Conference of the Society for Cross-Cultural Research, Albuquerque, NM, 17–21 February, 2010.

Cioffi-Revilla, C., Rogers, J.D., and Hailegiorgis, A.D., (2009), "GIS and spatial agent-based model simulations for sustainable development", AfricaGIS 2009 International Conference on Geo-Spatial Information and Sustainable Development, Kampala, Uganda, 26–30 October 2009.

Cioffi-Revilla, C., De Jong, K., and Bassett, J. (2010), *Evolutionary computation approach to analyzing polity dynamics in an agent-based modeling*, Working Paper, Mason-HRAF Joint Project on Eastern Africa, Center for Social Complexity, George Mason University.

Cioffi-Revilla, C., and Rouleau, M. (2009), "MASON AfriLand: A regional multi-country agent-based model with cultural and environmental dynamics", Human Behavior-Computational Modeling and Interoperability Conference 2009, Oak Ridge National Laboratory, Oak Ridge, TN, June 23–24, 2009.

Cioffi-Revilla, C., and Rouleau, M. (2010), "MASON RebeLand: An agent-based model of politics, environment, and insurgency." *International Studies Review*, 12(1), 31–46.

De Jong, K. (2006), *Evolutionary Computation*. MIT Press, Cambridge, MA.

De Jong, K. (2009), "Evolutionary computation". *Wiley Interdisciplinary Reviews (WIREs) Computational Statistics*, 1(1), 52–56.

De Jong, K., Bassett, J., and Cioffi-Revilla, C. (2010), *Evolutionary computation approach to analyzing social dynamics in an agent-based modeling*, Working Paper, Mason-HRAF Joint Project on Eastern Africa, Center for Social Complexity, George Mason University.

Ferguson, Y.H., and Mansbach, R.W. (1996), *Polities*. University of South Carolina Press, Columbia, SC.

Gilbert, N. (2008), *Agent-Based Models*. Sage Publishers, Thousand Oaks, CA.

Gilbert, N., and Troitzsch, K. (2005), *Simulation for the Social Scientist*, 2nd ed. Open University Press, Buckingham, UK and Philadelphia, PN.

Kennedy, W.G., Hailegiorgis, A.B., Rouleau, M., Bassett, J.K., Coletti, M., Balan, G.C., and Gulden, T. (2010), "An agent-based model of conflict in East Africa and the effect of watering holes". Annual Conference on Behavioral Representation in Modeling and Simulaiton (BRiMS 2010).

Liu, J., et al. (2007), "Complexity of coupled human and natural systems". *Science*, 317(5844), 1513–1516.

Luke, S., Cioffi-Revilla, C., Panait, L., and Sullivan, K. (2005), "MASON: A Java multi-agent simulation environment", *Simulation*, 81(7), 517–527.

Luke, S. et al. (2010), *ECJ: A Java-based Evolutionary Computation Research System*. Available online: http://cs.gmu.edu/~eclab/projects/ecj/

McCabe, T. (2004), *Cattle Bring Us to Our Enemies*. University of Michigan Press, Ann Arbor, MI.

Mousseau, M., and Mousseau, D.Y. (2007), "How the evolution of markets reduces the risk of civil war," Annual ECPR General Conference, 6–7 September 2007, University of Pisa, Italy.

Muncaster, R.G. (1987), *Fitting vs. Testing*. Working Paper. Merriam Lab for Analytical Political Research, University of Illinois, Urbana, IL.

Mwangi, E., and Ostrom, E. (2009), "Top-down solutions." *Environment*, 51(1), 34–44.

North, M.J., and Macal, C.M. (2007), *Managing Business Complexity*. Oxford University Press.

Parker, D.C., Berger, T., and Manson, S.M., (2002), *Agent-Based Models of Land-Use and Land-Cover Change*. Anthropological Center for Training and Research on Global Environmental Change, Bloomington, IN.

Sayre, R. (2009), "Mapping standardized terrestrial ecosystems for Africa". Africa Geospatial Science and Technology Workshop, 28 May 2009, Kennedy School of Government, Harvard University, Cambridge, MA.

Simon, H.A. (1996), *The Sciences of the Artificial*. MIT Press, Cambridge, MA.

Sterman, J.D. (2000), *Business Dynamics*. Irwin McGraw-Hill, Boston, MA.
Taber, C.S., and Timpone, R.J. (1996), *Computational Modeling*. Sage Publications, Thousand Oaks, London and New Dehli.

Chapter 31

Terrorist Profiles: From Their Own Words

P. M. "Pooch" Picucci

Institute for Defense Analyses
4850 Mark Center Drive
Alexandria, VA 22311-1882

ABSTRACT

Decision-making processes can be significantly colored by beliefs regarding the nature of the political universe. This is particularly true of organizations with relatively small leadership circles operating under conditions of high threat. The Operational Code approach provides a reproducible means (by way of computerized content analysis) of deriving estimates of these beliefs in a form suitable for quantitative comparison. This paper provides operational code analyses of two organizations (al-Qaeda and Hamas) and compares their operational codes to a control group and to each other in an effort to demonstrate both the strengths and weaknesses of this approach when applied beyond its usual subject matter of individual state leaders.

Keywords: Operational Code, Terrorism, al-Qaeda, Hamas

INTRODUCTION

The decision-making of violent extremist organizations is particularly amenable to influence by beliefs. The very nature of such an organization and its members, its position within international society resulting from the types of actions it takes, and the peculiar influence induced by that position, argue for a dominant collective identity that is expressed in a set of powerful communal beliefs. This belief system can be extracted via Operational Code analysis such that it can be compared across other international actors as well as other violent extremist groups. These comparisons can lead to valuable insight into the behaviors and motivations of violent extremist organizations.

This paper reports part of the findings of a doctoral dissertation conducted at the University of Kansas that performed an analysis of the operational codes of al-Qaeda and Hamas (Picucci 2008). In order to generate this analysis, statements for both al-Qaeda and Hamas spanning several years were collected. These statements were aggregated into groups based on proximity in time to each other and were then coded via computerized content analysis. The resulting indices were then compared to each other and to a set of state leader mean values (the norming group). These were evaluated against each other in order to determine the degree of distinctiveness of these two actors from state leaders. The belief systems were also compared against one another and differences were linked to observable structural, motivational and behavioral differences between the organizations. Additional operational codes based on divisions within the organizations were also performed and these were compared against each other as well as to the overall organizational operational code. While speculative in some instances the results of each of these analysis do indicate that not only is the Operational Code approach appropriate to the study of these kinds of actors but that it is capable of generating valuable insights into terrorist behaviors and motivations.

The remainder of this paper first briefly describes Operational Code analysis. Second, it discusses the rationale for expanding this form of analysis to violent extremist groups in general and al-Qaeda and Hamas in particular. Third it describes the belief systems of these two organizations, comparing them to a norming group of state leaders and then to each other. Fourth it discusses the value and implications of these analyses and identifies additional research directions utilizing Operational Code analysis on violent extremist organizations.

OPERATIONAL CODE ANALYSIS

Operational Code analysis is an approach to the study of political behaviors that focuses on a specific set of political beliefs embedded in the personality of a leader or arising from the shared identity of a collective. This set of beliefs acts as constraints upon an actor's rational decision-making processes, thus creating a bounded rationality that directly impacts the behavior of that actor. These beliefs are presumed to manifest in the verbal characteristics of the actor and are expressed as answers to a series of questions about the political world and the role of the actor within that world. Systemization of the answers to these questions is accomplished by means of numerical indices derived from the Verbs In Context System (VICS) coding scheme developed in 1998 by Walker, Schafer and Young.

The concept of the operational code has its origin with Nathan Leites' work on the Soviet Politburo. As originally formulated (Leites, 1951 & 1953), the operational code concept was intended as a measurement of the psychological impacts upon foreign policy decision–making of culture, ideology, and personality type. Alexander George (1969) categorized the results of Leites work into answers to a series of questions regarding philosophical and instrumental beliefs that related to the perceived state of the world, the role of the individual within that world, and attitudes toward the efficacy of various instrumental means. The philosophical beliefs referred to the assumptions and premises about the fundamental nature of the political universe while the instrumental beliefs related ends to means.

GEORGE'S QUESTIONS AND THEIR CORRESPONDING INDICES

Philosophical Questions

- What is the "essential" nature of political life? Is the political universe essentially one of harmony or conflict? What is the fundamental character of one's political opponents? [P-1]
- What are the prospects for the eventual realization of one's fundamental political values and aspirations? Can one be optimistic, or must one be pessimistic on this score; in what respects the one and/or the other? [P-2]
- Is the political future predictable? In what sense and to what extent? [P-3]
- How much "control" or "mastery" can one have over historical development? What is one's role in "moving" and "shaping" history in the desired direction? [P-4]
- What is the role of "chance" in human affairs and in historical development? [P-5]

Instrumental Questions

- What is the best approach for selecting goals or objectives for political action? [I-1]
- How are the goals of action pursued most effectively? [I-2]
- How are the risk of political actions calculated, controlled and accepted? [I-3]
- What is the best "timing" of action to advance one's interests? [I-4a corresponding to the flexibility between conflict and cooperation and I-4b corresponding the flexibility between words and deeds.]
- What is the utility and role of different means for advancing one's interests? [I-5a through I-5f corresponding to six categories of tactical options: Reward, Promise, Support, Approve, Oppose, Threaten and Punish respectively.]

In 1998 the operational code construct became systematized and reproducible via the construction of a set of numerical indices that directly related to the philosophical and instrumental questions of George (Walker, Schafer & Young, 1998). Each belief question corresponds to its own numerical index that is obtained from the coding of speech acts using the Verbs in Context System (VICS) coding scheme. The result of which is a set of quantitative indicators that allowed for "direct, meaningful comparisons across our subjects and conduct statistical analyses that allow for probabilistic generalizations" (Schafer & Walker, 2006b, p. 27).[1] The VICS coding scheme proved to be amenable to generation via computerized content analysis thus marking considerable improvement over the potentially less reliable qualitative operational code measures that had been in use prior to that time. Utilization of the VICS scheme for the generation of comparable operational codes has been substantial and has covered a wide range of subject and issue areas since that time.[2]

USING OPERATIONAL CODE ANALYSIS TO STUDY VIOLENT EXTREMIST ORGANIZATIONS

Primarily Operational Code analysis has been applied to specific individuals (usually state leaders) however limited work has been done developing operational codes based on shared or prevailing belief systems of groups. In theory the operational code of a group, necessarily based on shared communication of that belief system, could be more indicative of the behavior of said group than would the operational code of the group's leader (Picucci 2008, pgs 127-129). Violent extremist organizations are a subset political actors for which situational conditions are likely to empower the collective belief system and therefore make an operational code analysis of that group a particularly valuable instrument for explaining and possibly predicting behavior.[3] Two such organizations, al-Qaeda and Hamas are the subject of this report.

Both al-Qaeda and Hamas are currently designated by the US State Department as Foreign Terrorist Organizations that pose substantial threats to US interests; either directly threatening domestic security or threating the security of critical allies. They are both exemplars of modern networked organizations that lack a direct hierarchical structure extending from a single (or small set of) leader(s) down to individuals undertaking field operations. Both are relatively large and capable organizations (especially in comparison to past images of terrorist groups consisting of, at most, several dozen members). Both organizations have become active in the same time period (post-Cold War era) and have proven adept at operating in their highly interconnected, multi-polar operating environment. Although there are clearly differences in the cultural background of the primary leaders of these groups, they are, in a broad sense, drawn from a common greater Arabic culture.

---

[1] For a detailed look at the coding scheme and the content analysis process see Picucci 2008 pgs 135-153.

[2] For an overview of operational code analysis, its history and its varied uses see Schafer & Walker 2006a.

[3] See Picucci 2008 pgs 127-136.

306

Both organizations are also overtly religious in nature. This is not to claim that the religious views expressed by Hamas and al-Qaeda are identical or to minimize the differences between mainstream Sunni beliefs and those of the Salafi tradition but merely that the religious beliefs espoused by both organizations have common origins. The fact that both draw heavily upon the Islamic revivalist traditions espoused by al-Banna and Qutb is particularly indicative of this commonality. These factors provide a reasonable set of controls for the analysis while still allowing for meaningful distinctions of both structural and motivational considerations.

THE OPERATIONAL CODES OF AL-QAEDA AND HAMAS

The initial hypothesis of this project is that the operational code indices of Hamas and al–Qaeda will be noticeably distinct from the values obtained from the norming group of state leaders. The only other application of Operational Code analysis to terrorism (Lazarevska, Sholl, and Young, 2006) has indicated that terrorist leaders can be usefully distinguished from non–terrorist political leaders on the basis of seven indicators from Leadership Trait Analysis (LTA) and two from Operational Code analysis (the P–1 and I–1 indices). The alternative conceptualization of the operational code, as a measure of the belief system of the organization, utilized in this study makes it possible to determine if the results of Lazarevska, Sholl, and Young (2006) extend to the organization or are limited to specific terrorist leaders. A second hypothesis is that these two actors, despite a shared categorization as violent extremist organizations and significant shared characteristics, can be usefully distinguished from each other on the basis of their beliefs as indicated by differences in their operational codes. The degree of difference between their belief systems is not a trivial issue. Levitt (2007, p. 938) has indicated that al-Qaeda operatives are actively seeking recruitment of Hamas members.[4] The degree of similarity in belief systems between these two organizations can serve as a measure of potential for the success of these recruitment activities. Additionally, taken in concert with each other, the operational code indices can be used to achieve a measure of the degree of expressive or instrumental violence of the organization.

COMPARISON TO THE STATE LEADER NORMING GROUP

Do the operational codes of al–Qaeda and Hamas differ from the mean state leader values in such a way as to indicate the presence of a belief system specific to terrorist organizations? Primary evidence for this contention would be indicated by a highly conflictual worldview, a strategic and tactical reliance upon conflict, belief in a largely predictable international system, high levels of risk acceptance, and

---

[4] See also Cohen, 2009 for recent evidence of these behaviors.

tactical reliance upon the use and threat of violence for both organizations. The operational code indices corresponding to these beliefs (P–1, P–2, I–1, I–2, P–3, I–3, I–5e, and I–5f respectively) are generally supportive of this position.[5] Both al–Qaeda and Hamas view other political actors as being primarily and intensely motivated toward the use conflictual behaviors as is indicated by P–1 and P–2 scores. They also share a perception of a political universe that is highly predictable indicated by their P–3 index scores. Although the P-4 and P-5 indices do not indicate a uniform distinction between the two groups and the state leaders, they measure beliefs for which there is either no reason to presume a significant difference from other international actors or the evidence in favor of such differences is contradictory and therefore inconclusive.

It would seem consequently that there is good reason to believe that these two organizations are distinctly different from state leaders with regard to their philosophical beliefs. While this may seem no more than a common sense conclusion, independent and reproducible verification that these kinds of organizations have a largely different philosophical view of the political universe is an important factor in both the academic and policy–centric studying of these organizations. It calls into question the validity of applying a strict rational approach that subsumes belief and value systems consistent with other actors in the system. At the same time the commonality of their differences from the state leaders also lends credence to an approach that treats them as a single class of actor: violent extremist organizations. The differences between the organizations are significant enough however to avoid a simplistic reduction of all terrorist actors to a single form. While their differences may only be the degree to which they differ from the state leader norms, that degree of difference may be critical to understanding the differences in behavior of the organizations.

With regard to their instrumental beliefs there is still evidence to indicate Hamas and al–Qaeda distinctiveness from state leaders however these differences are much less uniform between the two organizations. The difference between the organizations and state leaders are only uniformly significantly different on the tactical means of goal pursuit (I–2) and the disutility of the reward tactic (I–5a). Only the risk acceptance index (I–3) does not differ significantly at 0.10 or better level although it is not far off the prediction at a 0.20 significance. Problematic however are the instrumental indices derived for Hamas. Only the tactical intensity (I–2) and risk acceptance (I–3) indices were significant in directions indicative of a coherent terrorist belief system. The other key instrumental indices did not differ significantly from the state leader means. Particularly troubling is the fact that the utility attached to the use of Punish tactics (I–5f) was indistinguishable from the state leader utility level.

---

[5] All significance tests were done using a difference of means t-test. For actual significance scores see Picucci 2008.

COMPARISON ACROSS AL-QAEDA AND HAMAS

As was previously indicated, despite a common portrayal as "Islamic terrorist organizations" there are significant differences between the Hamas and al–Qaeda organizations with regard to their structures, causes, types of operations, constituencies, and degree of social connection. Do these differences manifest as differences between their operational codes? It is clear from the preceding discussion that the operational codes for these two organizations differ significantly from those of the state leaders but is it equally clear that they differ significantly from each other in important ways? Aside from the risk acceptance index (I–3), the cooperation/conflict flexibility index (I–4a) and several of the tactical utility indices (I–5), all other philosophical and instrumental beliefs differ significantly between al–Qaeda and Hamas.[6] In most instances these represent varying degrees of difference from the state leader values and are therefore expressions of the extent to which the belief is held rather than substantive divisions between the two. This is particularly true of the philosophical indices. While both organizations display belief in a political universe that is highly conflictual (P–1 and P–2) and highly predictable (P–3), all three beliefs are more extreme for al–Qaeda than for Hamas.

The instrumental indices display similar traits although there are greater instances of substantive difference (defined as varying in direction from the state leader mean). In general, as noted previously, the instrumental indices for Hamas display a greater commonality with the state leader values than with those of al–Qaeda. Despite similar perceptions of their political universe, there appears to be a difference in both the strategic and tactical choices between al–Qaeda and Hamas. Hamas seems to have chosen to adopt a mixed strategy in the pursuit of its objectives and to this end employs a lower than expected level of conflict–oriented tactical options. This lends further plausibility to the presumption that we can differentiate between motivational types of organizations on the basis of their belief structures.

CONCLUSION & THE WAY FORWARD

The preceding examinations of the operational codes of al–Qaeda and Hamas, and their comparisons to each other and to the norming group of state leaders have largely supported the primary contentions of this research project. Both Hamas and al–Qaeda view the political universe, their role within it, and optimal behavior under those conditions, in a manner that differs remarkably from that of state leaders. Although this still does not imply a singular belief system that is generalizable to all terrorists, or even to religious terrorists, it does indicate that certain philosophical and instrumental beliefs may be useful in the distinguishing of violent extremist groups from other international actors. These differences appear to be more pronounced for al–Qaeda than for Hamas, whose instrumental beliefs often

---

[6] The P-5 index, indicating the perception of the role of chance, only differs weakly at a 0.13 significance level.

display strong similarities to those of the state leaders. Further, the results of this analysis indicate that the differences between these organizations are also significant and can be used to distinguish between them particularly with regard to the issue of instrumental and expressive motivations.[7] In general then the case for the utility of the Operational Code approach to the study of the belief systems of terrorist entities appears to be a compelling one.

THE WAY FORWARD

Beyond comparing measures of a static operational code at the group level, the project from which the preceding findings were taken also explored alterations in the belief systems over time,and of sub-groups within the al-Qaeda and Hamas organizations. The results of these analyses were similarly supportive of the utility of applying the Operational Code approach to violent extremist organizations While the results of this study certainly indicate the utility of Operational Code analysis, the preliminary nature of much of this analysis suggests that additional applications of this approach are warranted. The confining of this project to the al–Qaeda and Hamas organizations was necessitated by practical considerations. However, the intent of this project has always been to develop a database of operational codes for violent extremist organizations. Expansion beyond the two in this study is necessary for the exploration of a number of the tentative conclusions suggested by this research. In particular, the search for an operational code or operational code traits specific to these kinds organizations (whether shared across the group or specific to certain sub–elements) requires the study of additional groups. Similarly, while differences in the operational codes of al–Qaeda and Hamas have, in many cases, been linked to specific characteristics of those organizations, systematic correspondence of those operational code values and organizational traits cannot be accomplished without compilation of additional operational codes. Similarly expansion of this approach to groups that are presumed to be "at-a-risk" for turning to violent extremism is a potentially viable line of research particularly if it can be determined that specific changes in operational code presage such a shift.

This is not to say that expansion of this work can only come about through the study of additional groups. The operational code work on al–Qaeda and Hamas is far from being exhausted. So long as these groups continue to exist as relevant political entities additional statements will continue to be released and are increasingly available in full–text translated forms. Even within the time constraints of this study, additional statements occasionally become accessible. Documents are declassified or full–text translations become available of statements that were previously either un–translated or were unavailable in their full–text forms. The addition of these statements to the operational codes of al–Qaeda and Hamas increase the confidence in the accuracy of the values obtained, make possible the refinement of the previously conducted analyses, and provide the opportunity to explore research areas previously untouched. In the case of al–Qaeda additional

---

[7] See Picucci 2008, pgs. 236-242.

statements from Zawahiri, allowing for a more confident derivation of his operational code, could confirm the presence of a division between the rest of al–Qaeda and its primary leaders, as well as allowing for the exploration of differences in the operational codes between bin Laden and Zawahiri. The limited numbers of leadership statements from Hamas eliminated the possibility of a similar leader to leader comparison however access to additional statements by those leaders would resolve this issue. The subgroup analyses of Hamas demonstrated significant belief system differences between its leadership and the Qassam Military Brigades. With additional leader statements it ought to be possible to determine whether differences in belief structure also occur between the internal and external political wings of the organization.

Deliberately excluded from this project were statements from leadership sources of both organizations that could not be directly linked to being representative of the organization. Introduction of those statements offers additional opportunities to explore the belief systems of these groups. An assumption of this project has been that the collective belief system of the organization is dominant and that personal beliefs either are replaced by the beliefs of the collective or are repressed in light of extreme pressures against dissension. Comparison of statements reflecting personal beliefs to those reflecting the organizational belief system would provide one means of testing this assumption. High degrees of coherence between personal and group operational codes would indicate a dominant collective identity while low levels of coherence would indicate repression of personal beliefs.

FINAL REMARKS

Since the events of September 11, 2001 the motives and decision-making processes of violent extemist organizations have been approached in a myriad of ways. It seems clear that belief systems play a crucial role in their decision-making processes and that there is a definitive need for a strictly comparable quantitative measure of those beliefs systems and a means of determing those beliefs using an at-a-distance measure. The Operational Code approach is ideally suited to fill this need. As indicated by the preceding analysis it is capable of drawing significant insights into these kinds of organizations. It is generalizable at the group level but also capable of being applied to specific individuals and sub-sections within those groups. The application of this approach to al-Qaeda and Hamas argues convincingly that the belief systems of these actors differ in significant ways from those of state leaders and that it is sensitive enough to reflect relatively nuanced motivational differences in these organizations.

REFERENCES

Cohen, Yoram. (2009), *Jihadist Groups in Gaza: A Developing Threat.* Washington Institute for Near East Policy Policy Watch #1449 available at http://www.washingtoninstitute.org/templateC05.php?CID=2981.

George, Alexander. (1969), "The Operational Code: A Neglected Approach to the Study of Political Leaders and Decision Making." *International Studies Quarterly,* 13(2), 190-222.

Lazarevska, E., Sholl, Jayne M., and Young, Michael D. (2006), Links Among Beliefs and Personality Traits: The Distinctive Language of Terrorists. In Mark Schafer and Stephen G. Walker (eds.), *Beliefs and Leadership in World Politics: Methods and Applications of Operational Code Analysis.* Palgrave Macmillan, New York, 171-183.

Leites, Nathan. (1951), *The Operational Code of the Politburo.* McGraw-Hill, New York.

Leites, Nathan. (1953), *A Study of Bolshevism.* Free Press, New York.

Levitt, Matthew A. (2007), "Could Hamas Target the West?" *Studies in Conflict & Terrorism* 30(11), 925-945.

Picucci, Peter M. (2008), *Terrorism's Operational Code: An Examination of the Belief Systems of al-Qaeda and Hamas.* Doctoral Dissertation available at http://kuscholarworks.ku.edu/dspace/handle/1808/4051.

Schafer, Mark and Walker, Stephen G. (2006a), *Beliefs and Leadership in World Politics: Methods and Applications of Operational Code Analysis.* Palgrave Macmillan, New York.

Schafer, Mark, and Walker, Stephen G. (2006b), Operational Code Analysis at a Distance: The Verbs in Context System of Content Analysis. In Mark Schafer and Stephen G. Walker (eds.), *Beliefs and Leadership in World Politics: Methods and Applications of Operational Code Analysis.* Palgrave Macmillan, New York, 25-52.

Walker, Stephen G., Mark Schafer, and Michael D. Young. (1998), "Systematic Procedures for Operational Code Analysis: Measuring and Modeling Jimmy Carter's Operational Code." *International Studies Quarterly* 42, 175-190.

Chapter 32

Dynamic Decision Making Games and Conflict Resolution

Cleotilde Gonzalez, Christian Lebiere, Jolie Martin, Ion Juvina

Dynamic Decision Making Laboratory
Carnegie Mellon University
Porter Hall 208-15213, Pittsburgh, PA

ABSTRACT

This paper discusses an approach to study conflict resolution as a socio-cognitive dynamic decision making process. Our approach unifies computational and experimental methods to balance the realism of complex cross-cultural interaction with the experimental control of abstracted conflict from Game Theory. We offer an innovative approach to the study of conflict resolution as a dynamic decision making (DDM) process, in which we develop a socio-cognitive computational theory of conflict resolution. We present our use of commercial games that address particular cases of international conflict, such as *PeaceMaker* (ImpactGames, 2006). *PeaceMaker* is an interactive representation of the Israeli-Palestinian conflict that we use in controlled laboratory experiments to study the influence of diverse socio-cognitive variables. We also develop our own DDM games that address conflict more generically. We present our extension of the traditional Prisoner Dilemma 2x2 game, into an iterative group game, the Iterative Prisoner Dilemma-Squared (IPD^2), that enriches the possibilities to study the dynamics of group behavior through computational cognitive modeling.

Keywords: Dynamic Decision Making, Conflict Resolution, Adversarial Behavior, Culture, Social Factors, PeaceMaker, Iterative Prisoner Dilemma-Squared

INTRODUCTION

Conflict resolution is a socio-cognitive dynamic decision making (DDM) process in which leaders attempt to make decisions while having to respond to both their own internal cognitive and social limitations and to the external events in real-time with accompanying time pressure and stress (Geva, Redd, & Mintz, 1997; Kelman, 2008; Mintz, Geva, Redd, & Carnes, 1997). One goal of our research program is to develop a better understanding of DDM in adversarial situations; particularly in relation to the influence that socio-cultural variables have on the way people make decisions in these situations. Additionally, mathematical and computational representations of cognitive and social theories can help formalize the behavioral observations obtained from real-world or laboratory studies. Computational models may support the anticipation and prediction of the emergence of violence and terror in an adversarial situation. Historically, the computational representation of human behavior has tended to divide between approaches that attempt to provide high performance and scalability at the cost of cognitive plausibility, and cognitive psychology approaches that model human performance in high detail but only on relatively trivial tasks. Thus, a second goal of our research program is to apply cognitive modeling to model increasingly complex social interactions without sacrificing plausibility and insights into the human decision processes. We particularly seek to advance cognitive computational theories by formulating the mathematical representations of the effects and dynamics of socio-cultural variables in the models' explanations of DDM behavior.

In this paper we present the approaches we have followed to study DDM in conflict resolution. We argue that following a "bottom-up approach," where one draws conclusions from controlled experiments in very simplified versions of conflict situations, has multiple advantages. It allows us to identify clear strategies and measures of DDM, and thus make theoretical progress by drawing generic inferences about conflict. However, this approach has some limitations; notably, the applicability to the study of conflict in realistic situations. Simplified games abstract many important aspects of real-world conflict situations in a way that limits their ability to provide reliable explanations and predictions in realistic situations of conflict that are ill structured. On the other hand, the "top-down approach," where one draws conclusions from realistic cases of conflict and decision-making, is a very relevant to the study the socio-cultural aspects of conflict. However, it too has multiple limitations. Realistic studies demonstrate only particular examples of political and conflict situations from which general predictions and inferences are hard to derive. We present several ways in which we can increase the degree of convergence between these two approaches.

BACKGROUND

Traditional economic theory of conflict often makes predictions based on "rational theory," the simple and intuitive idea that people search for the optimal outcome (often monetary outcome). Common game theoretic paradigms such as the Prisoner's Dilemma (PD) (Rapoport & Chammah, 1965) demonstrate why two people might not cooperate, even when their payoffs would lead them to benefit from cooperation. In the PD, a game is played where each of the two players,

"Player1" and "Player2," has two choices that can be referred to as "cooperate" (C) and "defect" (D). The players make their "moves" simultaneously before they receive their payoffs, which are calculated according to a "payoff matrix" in which the *maximum total* outcome for the two players occurs when both decide to cooperate. If both players defect, they both loose, but if one defects while the other one cooperates, the player who defects gets more money than the player who cooperates. Rational choice leads the two players to both play D, even though each player's individual reward would be greater if they both played C. The challenge is therefore for the players to establish trust in each other to cooperate through iterative plays. It has been shown that rational players that interact repeatedly for an unknown number of games can sustain cooperative outcome in the long run (Aumann, 1959).

In an iterative PD where people make repeated choices, it is possible to study the dynamics of decision-making strategies (Axelrod, 1984). For example, a good yet simple strategy in the iterative PD game is *tit for tat* (Axelrod, 1984), which involves initial cooperation and then responding in agreement to the opponent's previous action. If the opponent previously was cooperative the player is cooperative, if not the player is not. After extended analyses of the conditions that lead to the best performance in this game, Axelrod determined that a successful strategy in the PD needs to be "nice" (do not play "D" before the opponent), "retaliating" (don't be nice all the time, sometimes punish defection), "forgiving" (if the opponent does not continue to defect, then cooperate), and "non-envious" (don't strive to score more than the opponent).

In political science, the PD is often used to illustrate situations of conflict between two parties for real or assumed cases of military supremacy, or for any competition between nations in which one attempts to stay ahead of the other. Governments sometimes follow a "deterrence strategy" by which a player threatens retaliation if attacked in a way that forces the other player to change its strategy if it does not wish to suffer great damage. Although important for illustration purposes, the PD as studied in the laboratory rarely applies directly to a real-world conflict resolution situation.

Those who study real-world affairs are confronted with serious challenges. They often study nations, large groups, societies as represented by a leader, but not the person alone. Real-life dilemmas often involve multiple players in the face of foreign policy making, including decisions on conflict resolution. Leaders typically confront many uncertainties about the available options, they have inadequate information about their options; they rarely know the likely costs and benefits, and the value trade-offs they entail. Although one could expect that leaders have clear goals and that their decisions would promote those goals, the reality is that decision-making in real world political situations is seldom rational, and in fact it is often hard to understand what constitutes rational choice under such conditions. For example, as explained in documents of realistic studies (Blight, Allyn, & Welch, 1993; Lebow & Stein, 1994), it is hard to interpret President Kennedy's actions in regard to the 1962 Cuban Missile Crisis in terms of his Soviet foreign policy or any other factors. Even with detailed records, it would not be possible to tell what inferences he made about the USSR's motives. Like this example, there are multiple discussions and in-depth case studies of particular decisions in international conflict. These studies provide powerful explanations of specific events, but do not offer concrete theoretical results that allow more general inferences.

As concluded by Stein & Welch (1997, p. 61): "Identifying the psychological processes involved in decision making is difficult enough in the laboratory; it is a daunting task in a naturalistic setting." We propose a novel approach that presents an effective compromise between the two extremes described above: laboratory experiments using both realistic conflict resolution scenarios and simplified DDM games (Gonzalez & Czlonka, 2010).

BRINGING THE TWO APPROACHES TOGETHER

There are several ways to improve the degree of isomorphism between the Bottom-Up and Top-Down approaches discussed above. The first way is the conversion of a realistic conflict into a DDM game. Here we present a case study of *PeaceMaker* (ImpactGames, 2006), a realistic video game that simulates Israeli-Palestinian interactions, with a player assuming the role of either the Israeli Prime Minister or the President of the Palestinian Authority. The second way is the extension of the traditional iterative PD to include some of the complicating factors described previously. Here we will discuss one particular paradigm we are developing called Iterative Prisoner's Dilemma - squared (*IPD^2*) consisting of two levels of competition and cooperation within and between groups of individuals.

Several researchers have characterized the major features of DDM environments to highlight the shortcomings of typical lab experiments and field studies for eliciting human decision processes. Edwards (1962) was one of the first to emphasize the interdependence of sequential decisions, and the frequent tradeoff between payoffs in one time period and information acquisition for future decisions. Optimal decision-making is especially difficult under such circumstances, given environmental changes over time – both autonomously and as a result of previous decisions – for individuals with inherently limited computational capacity. Others have illustrated the importance of task complexity and feedback in studies of DDM, and have suggested "microworlds" (here called DDM games) as tools for manipulating these factors and assessing behavioral outcomes (Brehmer and Allard, 1991; Brehmer, 1992). For instance, Sterman (1989) used an investment task to investigate misperceptions of feedback due to temporal delays between cause and effect. In a classic DDM inventory management task, Diehl and Sterman (1995) also found decreases in performance as causal relationships were ambiguated by side effects. In such situations, observing a participant's entire sequence of actions permits inferences about the underlying decision -making heuristics being employed.

Gonzalez, Vanyukov, and Martin (2005) supply a taxonomy of features that can be built into a DDM game to accommodate a researcher's specific aims. First, DDM games can capture the *dynamics* of path-dependence, meaning that the options (and their corresponding utilities) available to a decision maker at any time will be influenced by choices made in the past and/or external evolution of context. Second, *complexity* can be manipulated within DDM games by altering the number and type of relationships, such as contingency of individual outcomes on the simultaneous decisions of others. Third, increasing the *opaqueness* of information or action outcomes within a DDM game is a means to assess participants' adaptation to uncertainty. Lastly, DDM games can include *dynamic complexity*, or intricacies of causal relations such as nonlinearity or time lags, in the structure of feedback loops. Each of these elements can be implemented to reflect the key aspects of a DDM scenario that a researcher wishes to isolate, while omitting other sources of noise encountered in field studies.

These components of DDM tasks are difficult to explore via conventional laboratory methods and traditional research tools, and, at the same time, they are impossible to measure in the real world. DDM games offer a highly flexible intermediate solution.

LABORATORY STUDIES WITH PEACEMAKER

In the past years we have engaged in a set of laboratory studies intended to demonstrate the influence of individual identity variables such as religion, personality, and political affiliation on an individual's success in resolving the conflict represented in the *PeaceMaker* game. Often, people's desire to belong to a group and to feel attached to that group determines the way they perceive a conflict and address its resolution (Garzke & Gleditsch, 2006; Jackson, 2006; Levi, 2007; Lugo, 2007; Shamir & Sagiv-Schifter, 2006). Here we summarize the main features of the DDM game and some of our current findings from experimental studies using this game.

PeaceMaker was developed by ImpactGames (2006) as a commercial, educational video game. A player takes on the role of either the Israeli Prime Minister or the Palestinian President, and engages in a series of decisions with the goal of satisfying constituents on both sides of the conflict. PeaceMaker is currently available in three languages (Arabic, English, and Hebrew), and can be played at three difficulty levels (calm, tense, and violent) that differ in the frequency of stochastic inciting incidents that tend to disrupt stability. PeaceMaker encapsulates many interesting features of the conflict, such as the need to balance competing interests (Gonzalez, Czlonka, Saner, & Eisenberg, 2009). Furthermore, players must deal with events beyond their control, given uncertainty about precisely how different constituents will respond to actions taken (Gonzalez & Saner, in press). In this sense, PeaceMaker allows us to infer behavioral responses to psychological stimuli in a highly complex and dynamic environment. See Figure 1 for a screenshot of PeaceMaker with the major elements labeled.

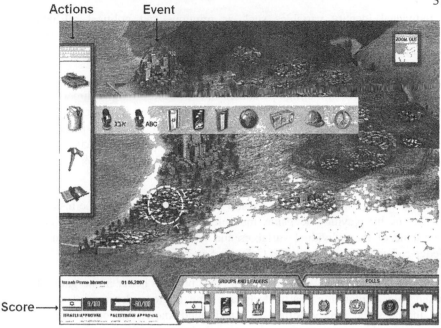

Figure 1. PeaceMaker Screenshot. © [2007] [Eric Brown]. Used with permission.

The underlying assumption is of a non-zero-sum game, since any stable outcome must take into account both Israeli and Palestinian concerns. Of greater interest to us is the manner with which a player works toward this end. In addition to choosing between numerous actions available in the game, players also encounter random inciting incidents modeling those of the real-world conflict (and including real storylines and footage from the news), and must decide how to appropriately respond. The three types of actions available on the main menu are construction, political, and security, and branching out within each of these are a variety of alternatives depending on the role being played. Both player actions and inciting incidents affect the approval of constituents (including factions within the region such as Fatah, as well as international parties such as the United Nations). These constituent views are aggregated into two main scores that represent the approval of stakeholders aligning with one's own role in the simulation and stakeholders aligning primarily with the other side. Players choose actions and accumulate points based on the effects those actions have on the approval ratings of the various interest groups. If the player reaches a balanced and highly positive score of 100 with both Israeli and the Palestinian groups, the player wins. If a player cannot sufficiently please his/her own people or the other side's constituents, the player loses. The scores are available to a player at all times and after each action taken, players can see how those actions affected the scores.

We have used PeaceMaker in several laboratory experiments including populations from the USA, Qatar, and Israel. Our results have started to reveal interesting patterns of behavior:

1) Individual identity variables such as religious and political affiliation, personal affiliation to the conflict, and general trust disposition relate to the scores obtained in the game (Gonzalez & Czlonka, 2010).

2) The players' decision -making personality is related to their performance in PeaceMaker. Players of "thinking" personality (as measured on one dimension of the Myers-Briggs Type Indicator) were more successful at reaching a conflict resolution in the game compared to the players with "feeling" personality, suggesting that those who are more assertive and impersonal, rather than affective and personal, are more successful in conflict resolution (Gonzalez & Saner, in press).

3) Scores improve in the game as a result of either knowledge of the history of the conflict or by playing the game multiple times. Further, the correlations between some of the individual variables and game performance decrease with knowledge and experience (Gonzalez, Czlonka, Saner, & Eisenberg, 2009).

4) Those who explore a variety of action types perform better in the game. For the Israeli role, exploration (particularly avoiding a preponderance of security actions) mediates the positive effect of background knowledge on winning the game. For both roles, non-winners are consistently less exploratory from beginning to end, with a disproportionate number of security actions serving as a strong predictor of poor performance (Gonzalez, Kampf, & Martin, 2010).

Our research is starting to elucidate some interesting patterns of individuals' identity variables such as religion, personality, and political affiliation as they relate to resolving the conflict represented in the *PeaceMaker* game.

MODELING AND EXPERIMENTING WITH IPD^2

IPD^2 is an extension of the well-known Iterated Prisoner's Dilemma (IPD) paradigm. The IPD has been studied widely, both behaviorally and computationally, particularly in relation to strategic behavior and the conditions needed to be successful in the game, as well as the cognitive mechanisms underpinning competitive behavior (Lebiere, Wallach and West, 2000; Gonzalez & Lebiere, 2005). Despite its popularity and success in accounting for many real-world situations, the IPD does not realistically represent the dynamics that one can find in situations involving more than two actors. Often, political and economic decision-makers have to take into account not only their interaction with other opposing decision-makers but also stakeholders in their own camp. These additional constraints can result in a narrower, distorted range of choices, which are in turn reflected in the higher-level dynamics between opposing camps.

In IPD^2, there are two teams of two players playing an inter-team IPD game. Each team can be seen as a player in Table 1.

Table 1. Payoff matrix used in Prisoner's dilemma (with entries in the table representing Player1 payoff / Player2 payoff)

		Player2	
		Cooperate	Defect
Player1	Cooperate	1/1	-4/4
	Defect	4/-4	-1/-1

Within a team, both players choose between Cooperation (C) and Defection (D), but only one player's choice counts as the choice of the team. What determines whether a player's choice counts as the team's choice is the player's "power". A player's "power" is a quantity assigned at the start of the game and increased or decreased after each round of the game depending on the player's choice, the teammate's choice, and the opponent team's choice. The sum of power within a team remains constant throughout the game. All players start the game with the same amount of power. A random value is added or subtracted from the initial level, which causes the two players of a team to have different powers. In the remainder of this description, the player with higher power in a team will be called "high-power" and the other player will be called "low-power".

After a round, if the two members of a team made the same decision (both played C or both played D), their powers do not change. If they made different decisions, their powers change in a way that depends on the outcome of the inter-team game (determined from the payoff matrix in Table 1). A fraction of the payoff received from the payoff matrix is added to the power of the high-power player and the same amount is subtracted from the power of the low-player player. Note that the values in the payoff matrix used in this study can be positive or negative Thus, if the team receives a positive payoff the power of the high-power player increases whereas the power of the low-power player decreases. If the team receives a negative payoff, the power of the high-power player decreases whereas the power of the low-power player increases. A player's power does not only allow a player's decision to count as its team's decision. It also determines each player's payoff. The payoff that the team makes is shared between the two teammates in direct proportion with their power. Again, since the team's payoff can be positive or negative, differences in individual payoff between teammates can be directly or indirectly proportional to differences in power. For example, if the team gets a negative payoff, then the individual payoff of the high-power player is decremented by a larger amount than the individual payoff of the low-power player.

The IPD^2 task has the potential to model real-world conflict situations in which human actors find themselves cooperating or competing with one another both directly and indirectly via the groups to which they belong. Players might need to go through the intermediate step of acquiring power within their own group before securing positive payoffs for themselves and for their groups. They might need to consider the interactions between power and payoff and possibly trade one for another. We have developed a cognitive model of IPD^2 that is able to elicit humanlike behavior through predicting the moves of its opponent, exploring various strategies and selecting those that yield positive outcomes, and correcting its own errors based on the perceived state of the environment (see Juvina, Martin, Lebiere, & Gonzalez, submitted). Currently, a series of empirical studies with both IPD and IPD^2 are in progress. The main goal of these studies is to bridge the gap between laboratory studies and simulations on one side and real-world conflicts on the other.

CONCLUSIONS

We have illustrated some of the challenges involved in the study of conflict resolution. On the one hand, traditional economic theory presents a well -known paradigm to study cooperation and conflict in simplified situations. This approach has multiple advantages including the possibility of measurement, isolation, and control over central variables of interest to the study of conflict, but unfortunately, they rarely apply directly to real cases of conflict resolution. On the other hand, the study of real-world affairs provides specific, in-depth examples that illustrate detailed processes of real-world conflict resolution. Obviously, these studies are extremely beneficial for documentation and deep understanding of particular solutions; however, it is hard to generate some general inferences about conflict resolution from these unique examples.

We provide a unique and innovative approach to studying conflict resolution, which represents a compromise between the two extremes illustrated in this paper. We have used DDM games and utilized behavioral and cognitive modeling methodologies as a way to improve the degree of isomorphism between these two approaches. The DDM games are of two types: computational representations of particular, realistic situations of conflict used in laboratory experiments; and computational extensions of traditional economic games that better reflect the real world conflict challenges. We are starting to demonstrate how this approach is leading us to concrete and interesting predictions of conflict resolution.

ACKNOWLEDGEMENTS

This research is supported by the Defense Threat Reduction Agency (DTRA) grant number: HDTRA1-09-1-0053 to Cleotilde Gonzalez and Christian Lebiere.

REFERENCES

Aumann, R. J. (1959). "Acceptable points in general cooperative n-person game." In A. W. Tucker & R. C. Luce (Eds.), *Contributions to the theory of games IV. Annals of mathematics studies* (Vol. 40, pp. 287-324). Princeton, NJ: Princeton University Press.
Axelrod, R. (1984). "*The evolution of cooperation*." New York, NY: Basic Books.
Blight, J. G., Allyn, B. J., & Welch, D. (1993). "*Cuba on the brink*." New York: Pantheon Books.
Brehmer, B. (1992). "Dynamic decision making: Human control of complex systems." *Acta Psychologica, 81*(3), 211-241.
Brehmer, B., & Allard, R. (1991). "Dynamic decision making: The effects of task complexity and feedback delay." In J. Rasmussen, B. Brehmer & J. Leplat (Eds.), *Distributed*

decision making: Cognitive models of cooperative work (pp. 319-334). Chichester: Wiley.

Diehl, E., & Sterman, J. D. (1995). "Effects of feedback complexity on dynamic decision making." *Organizational Behavior and Human Decision Processes, 62*(2), 198-215.

Edwards, W. (1962). "Dynamic decision theory and probabilistic information processing." *Human Factors, 4, 59-73.*

Geva, N., Redd, S. B., & Mintz, A. (1997). "Decisionmaking on war and peace: Challenges for future research." In N. Geva & A. Mintz (Eds.), *Decisionmaking on war and peace: The cognitive-rational debate* (pp. 215-222). Boulder, CO: Lynne Rienner Publishers, Inc.

Gonzalez, C., & Czlonka, L. (2010). "Games for peace: Empirical investigations with PeaceMaker." In J. Cannon-Bowers & C. Bowser (Eds.), *Serious game design and development: Technologies for training and learning.* Hershey, PA: IGI Global.

Gonzalez, C., Czlonka, L., Saner, L. D., & Eisenberg, L. (2009). "Learning to stand in the other's shoes: A computer video game experience of the Israeli-Palestinian conflict." Unpublished manuscript under review.

Gonzalez, C., Kampf, R., & Martin, J. (2010). "Homogenity of actions and success in conflict resolution." Unpublished manuscript under review.

Gonzalez, C., & Saner, L. D. (in press). "Thinking or feeling? Effects of decision making personality in conflict resolution." In J. V. Brinken, H. Konietzny & M. Meadows (Eds.), *Emotional gaming.*

Gonzalez, C., Vanyukov, P., & Martin, M. K. (2005). "The use of microworlds to study dynamic decision making." *Computers in Human Behavior, 21*(2), 273-286.

ImpactGames (2006). PeaceMaker. Pittsburgh, PA: ImpactGames.

Jackson, R. (2006). "Doctrinal war: Religion and ideology in international conflict." *The Monist, 89*(2), 274-300.

Juvina, I., Martin, J., Lebiere, C., & Gonzalez, C. (submitted). "A game paradigm to study the dynamics of power".

Kelman, H. C. (2008). "A social-psychological approach to conflict analysis and resolution." In D. Sandole, S. Byrne, I. Sandole-Staroste & J. Senehi (Eds.), *Handbook on conflict analysis and resolution* (pp. 170-183). London and New York: Routledge [Taylor & Francis].

Lebow, R. N., & Stein, J. G. (1994). "*We all lost the Cold War.*" Princeton: Princeton University Press.

Levi, I. (2007). "Identity and conflict." *Social Research, 74*(1), 25-50.

Lugo, L. (2007). "International obligations and the morality of war." *Society, 44*(6), 109-112.

Mintz, A. (1997). "Foreign policy decisionmaking: Bridging the gap between the cognitive psychology and rational actor "schools"." In N. Geva & A. Mintz (Eds.), *Decisionmaking on war and peace: The cognitive-rational debate* (pp. 1-7). Boulder, CO: Lynne Rienner Publishers.

Rapoport, A., & Chammah, A. M. (1965). "*Prisoner's dilemma: A study in conflict and cooperation.*" Ann Arbor: University of Michigan Press.

Shamir, M., & Sagiv-Schifter, T. (2006). "Conflict, identity, and tolerance: Israel in the Al-Aqsa Intifada." *Political Psychology, 27*(4), 569-595.

Stein, J. G., & Welch, D. A. (1997). "Rational and psychological approaches to the study of international conflict. Comparative strengths and weaknesses." In N. Geva & A. Mintz (Eds.), *Decisionmaking on war and peace: The cognitive-rational debate* (pp. 51-77). Boulder, CO: Lynne Rienner Publishers.

Sterman, J. D. (1989). "Misperceptions of feedback in dynamic decision making." *Organizational Behavior and Human Decision Processes, 43*(3), 301-335.

Chapter 33

Lethal Combinations: Studying the Structure of Terrorist Networks

Ian Anderson, R. Karl Rethemeyer, Victor Asal

Project on Violent Conflict
Rockefeller College of Public Affairs & Policy
University at Albany, SUNY
Albany, NY 12203

ABSTRACT

Why do terrorist organizations ally with other terrorist organizations? Are all connections created equal? To answer the first question, this paper looks to the literature on why terrorist organizations form alliances. To answer the latter question we argue that all connections are not created equal, but rather that there are three major types of terrorist network connections: ideological, business, and enemy of my enemy. The structure of each of these types of connections is observed throughout a global terrorist network of organizations built from data compiled in the Big Allied And Dangerous (BAAD) dataset. Finally a brief look at specific examples using egocentric networks of key groups is performed.

Keywords: Terrorism, Social Network Analysis, Alliance Structures

INTRODUCTION

Why do terrorist organizations form alliances? What forms of alliance are prevalent in the global network of terrorist organizations? Our analysis of data in the Big Allied and Dangerous Version 2.0 (BAAD2) dataset suggests that there are three broad types of alliances: ideological alliances, business alliances, and enemy of my enemy (EME) alliances. Ideological alliances form when organizations share a common ideology or see themselves as part of a larger movement. Most terrorist organizations share ideological network connections – see, for instances, the relationships between the Kashmiri jihadist groups Jaish-e-Mohammad, Lashkar-e-Taiba, and Harakat ul-Mudjahidin. Business alliances are forged by terrorist organizations that act as entrepreneurs, bringing together groups without a common goal for a material payoff. Usually this entails the trading of resources or services. Examples of terrorist business alliances include the Irish Republican Army's (IRA) relationship with the Revolutionary Armed Forces of Columbia (FARC) for drug smuggling and al-Qaeda relationship with the Revolutionary United Front (RUF) for diamond smuggling. EME alliances are formed between groups that seek to combine resources against a common enemy. Examples include the Moro Islamic Liberation Front's (MILF) alliance with the New People's Army (NPA) and the alliances formed by Pakistan's Inter-Services Intelligence (ISI) with Kashmiri Jihadists and ethnonationalists operating in the "Seven Sisters" region of India.

Structurally, ideological connections are expected to be the most common and the strongest, while EME and business connections would be more unique and tenuous. Ideological connections are also expected to subdivide into cliques and subnetworks, while EME and business connections are likely to form paths between subnetworks. Of particular note is how these three forms may be "mixed and matched" to create "new combinations [that] could produce new and deadlier adversaries" (Hoffman 2006).

DATA & METHODOLOGY

To create our typology of terrorist relationships, we rely on data from the Big Allied And Dangerous project. The BAAD project focuses on creation and maintenance of a comprehensive database of terrorist organizations. To date, this project has involved two datasets. Version 1.0 (BAAD1) contains a single snapshot of 395 terrorist organizations active between 1998-2005. Version 2 (BAAD2) improves upon BAAD1 by (1) extending coverage through 2007 (2) increasing the number of organizations covered to 971 (3) collecting yearly data for a total of 10 time slices, and (4) increasing the number and depth of variables collected and coded. BAAD2 is composed of two major components. The first is the data on organizational variables. These variables include: group name, aliases, homebase country, ideology, size, age, structure, financial support, electoral involvement, social service provision, leadership loss, territorial control, and CBRN pursuit or use. The number of incidents, injuries, and fatalities are recorded from the Global Terrorism Database (GTD). The second component is the social network data, which characterizes relations between terrorist organizations as well as between countries and terrorist organizations. Relationships are coded for categories such as: suspected ally, ally, founding group, faction, splinter group, rival, enemy, target, and state sponsor. The network data also captures material support flows such as weapons, money, and training. This data can then be used to create dynamic network visualizations to show the networks evolving over the 10 years included in the dataset.

The data used in this study is based on preliminary data from BAAD2. The BAAD2 data was gathered from academic journals, government publications, books, and news sources on Lexis Nexis. For the purpose of this study the network categories were the collapsed into three tie strength measures: weak (mainly suspected ties), normal (typical ally relations), and strong (familial or shared membership ties). This study excludes relationships of enmity.

Social network analysis and visualizations were done using UCINET 6 and NetDraw. Data is symmetrical for the purposes of this study, though the final BAAD2 dataset will contain directional data.

WHY DO TERRORIST ORGANIZATIONS ALLY?

Terrorist organizations, like licit organizations, have goals strategies and tactics to reach those goals. First and foremost "organizations enter partnerships when they perceive critical strategic interdependence with other organizations in their environment...in which one organization has resources or capabilities beneficial to but not possessed by the other" (Gulati 1998). By forming an alliance terrorist groups can import new resources and ideas. For instance, terrorist organizations may have expertise with certain tactics, allowing for diffusion of tactics through training. While tactical diffusion theories (Enders & Sandler 2006, Lia & Skolberg 2004) account for organizational learning through observation and imitation, training by a veteran organization can make the trainee more "successful." Al-Qaeda has become proficient at spreading "best practices" through training (Weitz & Neal 2007). Use of improvised explosive devices may also be spreading through networks (Brook 2009).

Asal and Rethemeyer have highlighted the power of networks in two areas. First, the more direct alliances a group creates the more lethal the group tends to be (Asal & Rethemeyer 2008). Given growing interest in perpetrating mass casualties incidents (see examples in Hoffman 2006 and Tucker 2001), alliance formation may help achieve this goal. Second, Asal, Rethemeyer, and Ackerman (2009) find that network connections are a key factor in attempts to acquire chemical, biological, radiological, or nuclear (CBRN) weapons While religion has been posited as the driving force in the desire to pursue CBRN weapons, alliance connections appear to be more powerful

Alliances can increase the expected lifespan of terrorist organizations. Alliances with other terrorist organizations or state sponsors has helped sustain terrorist organizations and terrorist campaigns across time (Byman 2005).

Critically, networks have built-in redundancy that allows for continued operations even if a key group or member is destroyed. Denser connectivity also helps to stave off fragmentation over time (Sageman 2004). Dense, redundant movements survive longer.

Terrorist groups may form alliances through brokerage by a third party. For instance, Granovetter (1973) hypothesized long ago that in most relationships a "friend of a friend" will also become friend. That is, if a strong connection exists between A and B and a strong connection exists between A and C, then there is a good chance at least some connection will exist between B and C (Granovetter 1973). Granovetter gives two major explanations for this. First, if A and B as well as A and C have frequent interaction, then it is likely that B and C must interact too. Since both B and C are friendly towards A, then it is also likely that B and C will desire to be on good terms so as not to cause a strain with A. Second, it is suggested that the stronger the connection that exists between two individuals, the more similar they are likely to be. This line of thought can be carried to a group level. So if A is similar to B and C, then logically B and C will also be similar as well.

Let us turn to a real example to illustrate. Al-Qaeda is allied with the GSPC in Algeria. Al-Qaeda is also allied with MILF in Southeast Asia. In 1999, three GSPC militants were arrested in the Philippines. According to reports, seasoned terrorist trainers from the GSPC were sent to the Philippines, "with the intention of providing further assistance to the MILF at the request of al-Qaeda" (Schanzer 2004). Interactions between al-Qaeda and MILF and al-Qaeda and the GSPC lead to the interaction of MILF and the GSPC. Here, a friend of a friend became a friend.

The formation of alliances decreases competition between organizations (Gulati 1998). Competition exists for the attention of target audiences and scarce resources, including weapons, money, and personnel (Enders & Sandler 2006; Hoffman 2006; Nacos 2006). Alliances can alleviate the strains of competition by sharing resources and personnel. Just such a structure of competition coupled with alliances and frequently shared memberships is observed in the network of Palestinian terrorist organizations.

Finally, technology, communication, and globalization are often cited as factors enabling increased terrorism (see Crenshaw 2001; Cronin 2003; Li & Schaub 2004). The same can be said specifically for networking of terrorist organizations. While terrorists of the past had to meet at clandestine safe houses or under the protection of a friendly government in desert camps, today's terrorists have a new easy and anonymous meeting place, the Internet. Gabriel Weimann discusses the Internet as the new arena for terrorists, where organizations spread their message, communicate with one annother, obtain training material, or actually plan attacks (Weimann 2006). The anonymous and inherently networked nature of the Internet alleviates the risks of exposure that more traditional alliances might have, such as infiltration by police.

GLOBAL NETWORK OF TERRORISM

In order to look for alliance patterns theorized earlier, the first step is to analyze the global network. The network pictured in Figure 1 contains 572 organizations (mainly terrorist) with 1483 ties active in the 1998-2007 time period. The first observation is that the network can be broken into fairly distinct subnetworks, which we will refer to as "syndicates." Syndicate form around location and ideology. In the diagram, the node colors indicate syndicate membership. Shades of red represent jihadist terrorist organizations; shades of green represent ethnonationalist/separatist terrorist organizations; and shades of blue represent communist/socialist terrorist organizations.

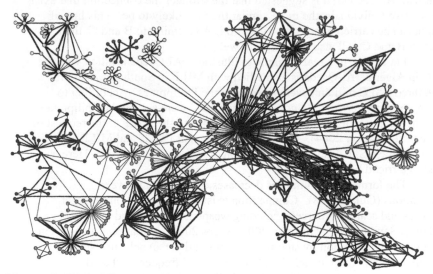

Figure 1: Global Terrorist Network broken down by syndicate.

The second observation concerns the strength and placement of ties. The thickness of the lines in the diagram signifies the strength of the relationship, classified as weak, normal, or strong. It appears that most of the connections across syndicates are weak or suspected ties, while within syndicate ties tend to be confirmed, strong ties. Where cross-syndicate ties are strong, shared ideology is the explanation see, for instance, the regional jihadist groups connected to the central al-Qaeda cluster.

In order to locate what appear to be the key actors in the dataset, we first ran a number of centrality measures. It is important to realize that while measures of centrality are all effectively measures of power based on a node's connection, each type of centrality measures a different type of power. Degree centrality focuses on direct alliances. Organizations central to the syndicates that form numerous ideological connections are likely to have high degree centrality.

For measures that consider the entire network, we must turn to betweenness and closeness centrality measures. Closeness centrality measures both direct and indirect connections to see how "short" the pathways are from a particular node to all others in the network. Closeness centrality can be conceptualized a measure of independence for a node. Betweenness centrality looks at organizations that lie on geodesics (or shortest paths) between actors over the entire network. Whereas closeness can be seen as measuring independence, betweenness can be said to measure control as these key actors can act as brokers. Groups that form cross-syndicate connections are hypothesized (a) to form business and EME connections as bridges and (b) to have higher measures on closeness and betweenness centrality. The top ten terrorist groups for each of these types of centrality are summarized in Table 1 below. Examining these lists it is clear that al-Qaeda plays a dominant role in the network, as it tops all three lists. Hezbollah is the only other group to appear on all three lists. This indicates that these groups may be thought of as "super" groups, as they are central to their own ideological syndicates as well as reaching into and bridging syndicates.

Table 1: Top Ten Terrorist Groups for Key Centrality Measures

Degree	Closeness	Betweenness
Al-Qaeda	Al-Qaeda	Al-Qaeda
Jemaah Islamiya (JI)	Hezbollah	Hezbollah
Revolutionary People's Struggle (ELA)	Hamas	Kurdistan Workers' Party (PKK)
Lashkar-e-Taiba	Palestinian Islamic Jihad (PIJ)	United Liberation Front of Assam (ULFA)
Popular Front for the Liberation of Palestine	Taliban	Revolutionary Organization 17 November (RON17)
Harakat ul-Mudjahidin (HuM)	Islamic Movement for Change	Revolutionary Armed Forces of Colombia (FARC)
Hezbollah	Al-Gama'a al-Islamiyya	Shining Path
Islamic Front for Iraqi Resistance	Egyptian Islamic Jihad	Basque Fatherland and Freedom (ETA)
Jaish-e-Mohammad (JeM)	Revolutionary United Front	International Solidarity
Al-Fatah	Libyan Islamic Fighting Group (LIFG)	Hamas

Another approach for identifying key actors it to use structural holes measures. Structural holes are the "separation between nonredundant contacts;" therefore, they provide a unique bridge between nonredundant contacts (Burt 1992). Structural holes are based on egocentric measures for each node in a network and can also be thought of as the location at gaps between clusters (Burt

2005). As Burt discusses, the measurement of structural holes closely relates to Granovetter's argument regarding the strength of weak ties. Logic dictates that subnetwork clusters (syndicates in this case), which are comprised of dense strong ties, will have actors that share some features – most likely valuable knowledge, communication, and material resources that can travel through syndicates at high velocity. The weak ties spanning structural holes between syndicates provide channels across which new information and resources may flow to syndicates. Terrorist organizations located at the "edge" of structural holes will therefore hold an important position in the overall structure of the global terrorist network. Table 2 contains the top ten terrorist organizations based on structural holes scores for the main network. Once again al-Qaeda and Hezbollah appear, providing further evidence of their prominence.

Table 2: Top Ten Terrorist Groups for Structural Holes Measure

Al-Qaeda
Revolutionary Organization 17 November (RON17)
Jemaah Islamiya (JI)
Hezbollah
Revolutionary People's Struggle
Inter-Services Intelligence (ISI) – Pakistan's Intelligence Agency
Islamic Front for Iraqi Resistance
United Liberation Front of Assam (ULFA)
Lashkar-e-Taiba (LeT)
Kurdistan Workers' Party (PKK)

Now we turn to a quick look at a few select cases from these lists that typify the three major alliance types.

A CLOSER LOOK AT ALLIANCE STRUCTURES

Three members of the top ten for degree centrality are members of the Pakistani-Kashmiri Jihadist syndicate: Lashkar-e-Taiba (LeT), Harakat ul-Mudjahidin (HuM), and Jaish-e-Mohammad (JeM). Figure 2 below is a visualization of this syndicate extracted from the main network, with LeT, HuM, and JeM represented by the larger nodes. Notice the large number of strong ties and the density of ties that typifies an ideological brotherhood of organizations.

Using our structural holes analysis, the ISI's role in an EME alliance structure becomes clearer. ISI views India as the most important strategic threat to Pakistan. As a buffer against India's intentions, ISI has helped to support jihadist elements in Kashmir as well as a collection of anti-Indian movements. Figure 3 below shows the egocentric network of the ISI (center) and how it brings together three syndicates: Pakistani-Kashmiri jihadists (top), South Asian Communists (right), and Anti-India Ethnonationalists (bottom). Notice that the strength of the connections in this network are much weaker than in the ideological network. ISI stands out as a unique node in this network.

Figure 2: Pakistani-Kashmiri Syndicate

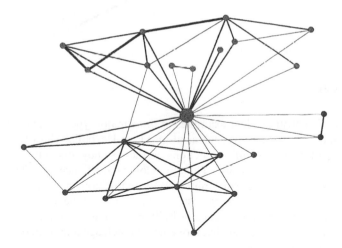

Figure 3: ISI Egocentric Network

No relationship embodies terrorist entrepreneurship like the Revolutionary Armed Forces of Colombia's (FARC) involvement in the drug trade as means to finance operations. One such business arrangement – between the Irish Republican Army (IRA) and FARC – came to light after the arrest of three IRA members in Bogata, Colombia in August 2001. Apparently, the IRA member were in Colombia to provide FARC with training in urban combat / bombings as well as a supply of explosives in exchange for drugs that the IRA could sell in Europe. Figure 4 shows the FARC-IRA nexus, with the FARC (enlarged) and its syndicate on the left and IRA (enlarged) and its syndicate on the right. Similar to EME alliances, connections of this type are weaker and less dense than within the syndicates.

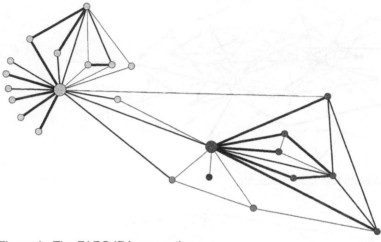

Figure 4: The FARC-IRA connection

CONCLUSION

The networks above demonstrate clear examples of three major types of terrorist organizational alliances. Other groups such as al-Qaeda or Hezbollah show the power of combining all three types of connections, as they occupy uniquely powerful positions in the global network. Though ideological alliances are the most numerous and strongest, the greatest potential for innovation and surprise comes from business and EME ties that span structural holes. It is the lethal combination of strong, fast-diffusing syndicates built on ideological connections coupled with weak cross-syndicate ties that move new knowledge and resources that create new, unexpected dangers.

REFERENCES

Asal, V., G. Ackerman, et al. (2009). Connections Can Be Toxic: Terrorist Organizational Factors and the Pursuit of CBRN Terrorism.
Asal, V. and K. Rethemeyer (2008). "The Nature of the Beast: Organizational

Structures and the Lethality of Terrorist Attacks." Journal of Politics 70(2): 437-449.

Drook, T. V. (2009). IED attacks soar in Afghanistan. USA Today.

Burt, R. (1992). Structural Holes: The Structure of Competition. Cambridge, Harvard University Press.

Burt, R. (2005). Brokerage & Closure: An Introduction to Social Capital. New York, Oxford University Press.

Byman, D. (2005). Deadly Connections: States that Sponsor Terrorism. New York, Cambridge University Press.

Crenshaw, M. (1998). The Logic of Terrorism: Terrorist Behavior as a Product of Strategic Choice. Terrorism and Counterterrorism: Understanding the New Security Environment. R. D. Howard and R. L. Sawyer. Guilford, McGraw-Hill: 54-66.

Cronin, A. K. (2004). "Terrorist Motivations for Chemical and Biological Weapons Use: Placing the Threat in Context." Defense and Security Analysis 20(4): 313-320.

Drake, C. J. M. (1998). Terrorist's Target Selection. New York, St. Martin's.

Enders, W. and T. Sandler (2006). The Political Economy of Terrorism. New York, Cambridge University Press.

Gibson, H., S. Brodzinsky, et al. (2001). Global Networking Trading in Terror? Time Europe. 158: 28-29.

Granovetter, M. (1973). "The Strength of Weak Ties." American Journal of Sociology 78(6): 1360-1380.

Gulati, R. (1998). "Alliances & Networks." Strategic Management Journal 19: 293-317.

Gunaratna, R. (2003). Inside Al-Qaeda. New York, Berkley.

Hoffman, B. (2006). Inside Terrorism. New York, Columbia University Press.

Li, Q. and Drew Schaub (2004). "Economic Globalization and Transnational Terrorism." Journal of Conflict Resolution 48(2): 230-258

Lia, B. and K. H. Skjolberg (2004). "Why Terrorism Occurs: A Survey of Theories and Hypotheses on the Causes of Terrorism." Journal of Conflict and Violence Research 6(1): 121-163.

Midlarsky, M. I., M. Crenshaw, et al. (1980). "Why Violence Spread: The Contagion of International Terrorism." International Studies Quarterly 24(2): 262-298.

Nacos, B. (2006). Terrorism and Counterterrorism: Understanding Threats and Responses in a Post-9/11 World. New York, Pearson Longman.

Sageman, M. (2004) Understanding Terror Networks. Philadelphia, University of Pennsylvania Press.

Schanzer, J. (2004). Al-Qaeda's Armies: Middle East Affiliate Groups and the Next Generation of Terror. New York, Specialist Press International.

Tucker, D. (2001). "What Is New about the New Terrorism and How Dangerous is It?" Terrorism and Political Violence 13(3): 1-14.

332

Weitz, R. and S. R. Neal (2007). Preventing Terrorist Best Practices from Going Mass Market: A Case Study of Suicide Attacks "Crossing the Chasm". Terrornomics. S. Costigan and D. Gold. London, Ashgate Publishing.

CHAPTER 34

Multi-Perspective, Multi-Future Modeling and Model Analysis

H. Van Dyke Parunak, Sven A. Brueckner

Vector Research Center, Division of TTGSI
3520 Green Court, Suite 250
Ann Arbor, MI 48105

ABSTRACT

Many simulation models are constructed to answer a specific question in a limited semantic domain (such as social interactions, geospatial movement, or progress toward accomplishment of a task). This approach limits the reusability of individual models, and makes the construction of large, complex models inefficient. We describe a system approach to modeling that integrates multiple perspectives, efficiently computes multiple futures, and supports rigorous analysis.

Keywords: Agent-based models, multi-future, multi-perspective, model analysis

INTRODUCTION

Many simulation models are ad hoc, constructed to answer a specific question with a main emphasis on a limited semantic domain (such as social interactions, geospatial movement, or progress toward accomplishment of a task). This approach limits the reusability of individual models, and makes the construction of large, complex models inefficient.

This paper discusses three characteristics of a systems approach to modeling.

Multi-future: We need to compute the envelope of possible behaviors with probability distributions over the relevant state variables, rather than generating only a single system trajectory with each run.

Multi-perspective: We must be able to integrate different model dimensions, including geospatial, social, process, financial, communications, and political, in a single model. For such a model to be computationally tractable and understandable, these different perspectives should be implementable as distinct modules that can be interconnected according to the needs of a specific model.

Analyzable: We need a set of mathematical tools that can express and test objective propositions about model behavior, instead of relying on subjective impressions formed from human observation of runs.

A modular **multi-perspective** modeling technology allows the straightforward combination and reuse of different specialized models into more complex systemic models. A **multi-future** approach guards us from unjustified generalizations based on limited numbers of runs that undersample the behavioral space of a model, and provide the basis for a statistically grounded **analytic** toolbox that can assess the significance and confidence to be associated with model outputs.

A novel agent-based modeling formalism, the polyagent, allows efficient sampling of multiple futures. This formalism supports multi-perspective modeling, and enables a formal analytic framework.

POLYAGENTS AND MULTI-FUTURE MODELING

THE PROBLEM

Imagine $n + 1$ entities in discrete time. At each step, each entity interacts with one of the other n. Thus at time t its interaction history $h(t)$ is a string in n^t. Its behavior is a function of $h(t)$. This toy model generalizes many domains, including predator-prey, combat, innovation, diffusion of ideas, and disease propagation.

It would be convenient if a few runs of such a system told us all we need to know, but this is not likely to be the case.

- We may have imperfect knowledge of the agents' internal states or details of the environment. If we change our assumptions about these unknown details, we can expect the agents' behaviors to change.
- The agents may behave non-deterministically, either because of noise in their perceptions, or because they use a stochastic decision algorithm.
- Even if agents' reasoning and interactions are deterministic and we have accurate knowledge of all state variables, nonlinear decision mechanisms or interactions can result in chaotic dynamics, so that tiny differences in individual state variables can lead to arbitrarily large divergences in agent behavior. A nonlinearity can be as simple as a predator's hunger threshold for eating a prey or a prey's energy threshold for mating.

An Equation-Based Model (EBM) typically deals with aggregate observables across the population (e.g., predator population, prey population, average predator energy level, or average prey energy level), as functions of time. No attempt is made to model the trajectory of an individual entity.

An Agent-Based Model (ABM) describes the trajectory of each agent. A single run of the model captures only a subset of possible interactions among the agents.

In our general model, during a run of length τ, each entity will experience one of n^τ possible histories (a worst-case estimate, since domain constraints may make many of these histories inaccessible). The population of $n + 1$ entities will sample $n + 1$ of these possible histories. It is often the case that the length of a run is orders of magnitude larger than the number of modeled entities ($\tau \gg n$).

Multiple runs with different random seeds are only a partial solution. Each run only samples one set of possible interactions. For large populations and scenarios that permit multiple interactions on the part of each agent, the number of runs needed to sample alternative interactions can quickly become prohibitive. In one recent application, $n \sim 50$ and $\tau \sim 10,000$, so the sample of the space of n^τ possible entity histories actually sampled by a single run is vanishingly small. We would need on the order of τ runs to generate a meaningful sample, and executing that many runs is out of the question. Previous multi-trajectory approaches (Gilmer and Sullivan 1998; Gilmer and Sullivan 2000) essentially replicate single trajectory runs, and the sampling approach they take to avoid explosion in the search space leads to an ergodicity failure (Gilmer and Sullivan 2001).

THE POLYAGENT MODELING CONSTRUCT

In the polyagent modeling construct (Parunak and Brueckner 2006), a persistent *avatar* manages a stream of transient *ghosts*, each of which explores an alternative future for the entity in a simulated world. These futures are executed in one or more virtual *environments*, such as a book of temporally successive geospatial maps or a task network, whose topology reflects that of the problem domain.

Ghosts are tropistic. Their behavior is determined by a set of fields in their environment (called "digital pheromones" after the insect mechanism that inspired them). Each field associates a scalar with each cell of the environment. Some fields are emitted by objects of interest (such as roads or buildings). Others are deposited by the ghosts. A ghost's behavior is determined by a weighted sum of the pheromones in its vicinity. The weights define the ghost's personality and can be either manually coded or learned by observing the entity that the ghost represents.

In the real world, one entity's behavior can depend on the presence or absence of another entity. A ghost's behavior depends on the fields of other entities, and thus reflects an average response across all of the locations of the other entities that their ghosts have explored (Figure 1.1).

Each ghost's state changes to reflect its interaction with the environment. For instance, its strength may change as a result of combat in a battlefield model. The ghost's strength can be interpreted as its degree of health, or more abstractly as the probability that the entity that it represents would be at full strength at the ghost's time and place. A log of each ghost's strength as a function of time is an additional

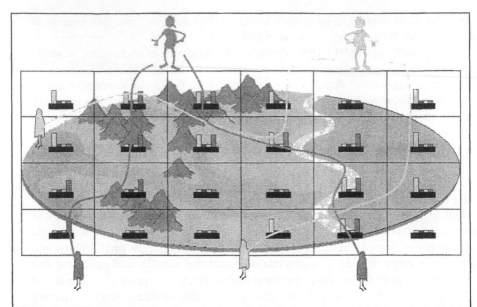

Figure 1.1 Ghosts explore alternate futures for their avatars, and interact via fields that they maintain in the environment.

resource, alongside the pheromone fields deposited by the ghost, for deriving fields. An entity's avatar can estimate the strength of its entity by taking the average of the strengths of its ghosts.

Each ghost increments the field corresponding to the entity it represents as it moves. The strength of this particular field at a location represents how frequently ghosts of that entity visit that location. The amount of the deposit depends on the ghost's strength, so its field takes into account the effects of attrition. A ghost can increment multiple fields. For example, one might correspond to its own avatar, one to the entire team to which its avatar belongs, and one to a unit within the team. A field modulated by strength yields an estimate of the probability of encountering a unit of force at each location. Fields can also be modulated by a ghost characteristic than strength, such as current preference for a given course of action, yielding a field with different semantics that may be useful in some applications.

Field strength depends not only on entity type and location, but also on time. In predictive applications, we maintain a set of field maps, one for each successive time step from a specified time in the past (the "insertion horizon") to a specified time in the future (the "prediction horizon"). Each page covers the entire area of interest (AOI). This set of maps is called the "book of maps." The number of pages is fixed, and as real time advances, we drop the oldest page and add a new page one time step further into the future. Pages are indexed by τ. $\tau = 0$ corresponds to "now." $\tau < 0$ indexes pages in the past (used to train the ghosts by evolution against observations (Parunak 2007)), and $\tau > 0$ indexes pages in the future. Thus if the current real time is t, the real time represented on a given page with index τ is $\tau + t$.

Pages in the book of maps for which $\tau \geq 0$ have real fields only for relatively persistent environmental features such as topography or clan territories. Otherwise, the fields to which ghosts respond on these pages are built up by the ghosts themselves as they traverse them. The first ghosts to visit each page do not see any ghost-generated fields, and their behavior is constrained only by persistent features. To enable ghosts to respond to one another, avatars release them in *shifts*. In one application, each avatar releases a total of 200 ghosts over 100 shifts, two per shift. The ghosts in each shift respond to the state of the fields as modified by the previous shifts. Ghosts in early shifts do not have well-defined fields to which to respond, so their movements are not as reliable an estimator of entity movement as those in later shifts, when the fields have converged. To accommodate this increase in accuracy over time, at each simulation step the field strengths on each page are attenuated by a constant factor E (a process inspired by pheromone evaporation in insect systems). The effect is to weight deposits by later shifts more strongly than those by earlier ones. When the shifts are complete, the avatars take their action in the real world based on the information from their ghosts, we index the book of maps, and start a new avatar cycle.

Often we are interested not in the location of an individual entity, but in the distribution of a group of entities (for example, all members of a team). In this case, all ghosts of entities on the same team increment the same field, which now reflects the probability of encountering any team entity over the area of interest.

MULTIPLE FUTURES FROM POLYAGENTS

How many futures does this approach explore? Each avatar sends $g*\sigma$ ghosts, where g is the number of ghosts per shift. In one application, each of $n = 5$ avatars sent $g = 2$ ghosts per shift over $\sigma = 100$ shifts into a book of 60 future pages. Each ghost could in principle follow a distinct path through the book of pheromones. Our probability distributions reflect the resulting distribution of trajectories.

A state of the world consists of the state of all avatars. Because we can capture multiple avatar states concurrently, we capture a number of states of the world equal to the product of the number of states visible for each avatar. Naively, we might estimate the number of possible futures as $(g*\sigma)^n \sim 3.2*10^{11}$. This is an overestimate, for two reasons:

- The number of ghosts that have visited a given page depends on the page. Pages further in the future see fewer ghosts from the current avatar cycle. (Ghosts from earlier cycles may still be on the page.)
- A ghost interacts with later ghosts through the field that it increments, and this field evaporates over time. So we should not count all ghosts equally.

Assume that we are at shift σ and page $\tau < \sigma$, so that the page in question has been visited. The oldest deposit on page τ was made by the g ghosts issued at $\sigma = \tau$, and a fraction $g*E^{\sigma-\tau}$ remains. The most recent deposit, made at σ, contributes g. So each avatar's "virtual presence" on the page is

$$g \sum_{i=0}^{\sigma-\tau} E^i = g \frac{1-E^{\sigma-\tau}}{1-E} \tag{1}$$

In the near future (pages 10 and lower), we explore nearly 40 alternative behaviors per avatar. Happily, the function is concave: the drop-off in parallelism is gradual until we get to more distant futures, where the prediction horizon effect (Parunak, Belding et al. 2007) (the increasingly random divergence of future trajectories under nonlinear iteration) makes predictions less reliable anyway.

The number of states explored on each page is this value raised to the power of the number of entities. Averaging this value over the 60 pages yields an average number of parallel futures of 8.4 x 10^7. This is several orders of magnitude lower than the 3.2 x 10^{11} estimate based on 200 ghosts per avatar, but still far more than a single-trajectory simulation can explore.

MULTI-PERSPECTIVE MODELS

THE PROBLEM

Modern analysis must consider multiple perspectives on the world, including geospatial, social, process, financial, communications, and political. Concurrent analysis of such a set of systems face three major challenges: combinatorics, nonlinearity, and model maintenance.

Combinatorics.—Let network i be $<N_i, E_i>$, where N is a set of nodes and E a set of edges. The set of places at which n networks can influence each other is the set product

$$\prod_{i=1}^{n} N_i \tag{2}$$

without even taking into account the types of edges (typically more than one type per network).

Nonlinearities.—The dynamics of processes on even a single network are highly non-linear. For example, even a random graph, an oversimplified model of almost any network, exhibits a phase transition as the probability of node connection increases (Erdös and Rényi 1960). These nonlinearities mean that network predictions are subject not only to state uncertainty (resulting from ignorance about the network), but also to dynamic uncertainty (resulting from the divergence of trajectories). Nonlinear systems are notoriously resistant to closed-form analysis, and are commonly studied using simulation, but dynamic uncertainty makes single simulation runs uninformative, and the space of possible interactions is so large that even replications of conventional simulations grossly undersample it.

Model Maintenance.—Conventional modeling techniques require close the various model perspectives. It is difficult to construct and test components independently, then easily reconfigure them into a variety of combinations.

HOW POLYAGENTS HELP

A given entity exists simultaneously in multiple environments. Conventional modeling focuses on the environment, but we focus on the agent, which as an Inter-Tier Entity (Figure 1.2) links the environments together (Parunak, Brueckner et al. 2010). The agent's experiences in one environment change its behavior in another.

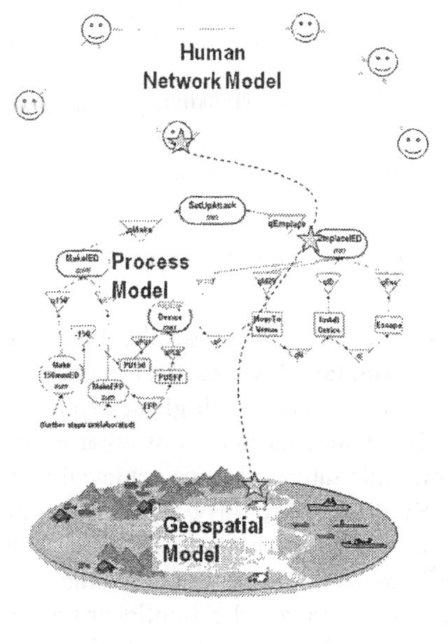

To achieve a **multi-perspective** model, an agent travels from one topology to another during its execution. For example, if a method in a process graph requires geospatial movement, an agent on a task node of the process graph detours through the geospatial topology to estimate the task's duration and success. Entities interact by means of the fields that their agents generate, effectively interacting with all of the futures explored by those agents.

Figure 1.3 shows the interaction of a process model (consisting of a series of actions with ordering constraints among them) and a geospatial model. The process model specifies that the agent purchases something at its initial

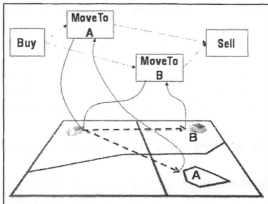

Figure 1.2   Inter-Tier Entities (ITES) coordinate the dynamics of multiple environments

location, then moves to one of two alternative locations, where it seeks to sell its purchase. In a pure process simulation, the duration and success of the "Move" actions would be sampled from a distribution. In our approach, when an agent reaches a "Move" action, it drops down into the geospatial environment and executes the move. It may encounter obstacles or other agents in that space that

affect the success or duration of its move. If it completes the move, it returns to the process model and advances to the next task.

Polyagents can explore multiple futures in a process graph just as they can in a geospatial map. In the example of Figure 1.3, some ghosts will sample one "Move" task, others will sample the other. This search can be influenced by the success of their venture by having the ghosts retrace their steps through process space, marking each task node with

Figure   1.3   Inter-Tier   Entities   concurrently explore multiple perspectives

digital pheromone indicating the value received, and weighting the choices of subsequent ghosts by this value pheromone. The result will be the emergence of the most profitable route.

Another paper (Parunak, Brueckner et al. 2010) walks through our multi-perspective approach in more detail, and (Parunak, Sauter et al. 2009) describes its application to a military problem.

SOLVING THE PROBLEM

This architecture addresses the three challenges of multi-perspective modeling.

Combinatorics.—Ghosts interact by means of pheromones in whatever space they explore. Thus each ghost explores the interactions of its particular future with the distribution of futures of other entities, providing the same efficient exploration of multiple futures that we achieved in geospatial modeling..

Nonlinearities.—By actually emulating agent activities in the appropriate space, we obtain a much more realistic distribution of outcomes than if we sampled artificial distributions. In our example, the agents executing moves in physical space are subject to all the nonlinearities of interaction with other agents, and their completion times reflect a realistic distribution of possible experiences.

Model Maintenance.—The various subspaces can be constructed and tested separately, then combined at run-time as needed.

ANALYZING POLYAGENT ARTIFACTS

THE PROBLEM

Simulation models are excellent stand-alone tools for analyzing complex situations. However, they can be difficult to interface with other technologies. One particularly powerful set of methods is based on Bayesian probability theory. These methods require distributions over the alternatives that they explore. Conventional applications sample from distributions that are computationally convenient (e.g., the use of conjugate priors), but not necessarily realistic. A simulation model generates realistic distributions. We would like to derive statistical artifacts from our simulations to feed other, statistically-based, reasoning tools.

PHEROMONES ARE PROBABILITY FIELDS

Up to a normalizing constant, the pheromone field incremented by an entity's ghosts on a page of the book of maps is a probability field estimating the entity's location at the time represented by that page over all possible futures explored by that entity's ghosts. This claim is supported by the dynamics of pheromone field strength. The strength of the field in a cell is augmented by a constant deposit D

each time a ghost visits the cell, and decremented by a constant fraction E each time step. The strength $\varphi(t)$ for a single cell with a single ghost has dynamics.

$$\varphi(\tau) = E\varphi(\tau - 1) + D = D\sum_{t=0}^{\tau-1} E^t = D\frac{1-E^t}{1-E} \tag{3}$$

In the continuous time limit, $\varphi(t)$ converges exponentially to $\frac{D}{1-E}$. This result has two important consequences.

First, the field converges if enough shifts of ghosts visit a page. We determine experimentally how many shifts are needed in each application.

Second, the evaporation rate E does not change over time. Thus the converged strength of a pheromone field is proportional to the amount of deposit, even in the presence of evaporation. If multiple ghosts visit a cell over time and deposit the same pheromone flavor, the converged strength of the field in the cell is proportional to the average number of deposits experienced by the cell per unit time. In other words, pheromone strength measures ghost traffic through a cell.

To normalize the field, observe that all of an avatar's ghosts must pass through some cell on a given page as they move through the time interval represented by the page. The proportion of ghosts that visit a cell is equal to the ratio between their pheromone in that cell, and the total amount of pheromone deposited on the entire page. But this ratio is just the probability that the avatar will visit that cell.

We can use the field to estimate the probability that the entity is in a given region of the page. Let A be the total amount of the entity's pheromone on the entire page, and B the amount in a region of interest. Then B/A estimates the probability that the entity is in the limited region. The entity's most likely location is given by the center of mass of the probability field.

Often we are interested not in the location of an individual entity, but in the distribution of a group of entities (for example, all members of a team). In this case, all ghosts of entities on the same team increment the same field, which now reflects the probability of encountering any team entity over the area of interest.

In interpreting these fields, we must understand that the ghosts are moving under the constraints of a range of environmental influences on earlier pages (represented as pheromones of other flavors). The probability field that they estimate is thus not P(Avatar at location (x, t)), but

$$P(\text{Avatar at location } (x, t)|\text{Other conditions at } t' < t)$$

That is, the movements of individual ghosts are not independent of one another. They are all subject to the same conditions. However, those conditions form a Markov blanket for the locations of the ghosts, so *given* those conditions, the ghosts' locations (and thus the pheromone field that they generate) *can* be treated as independent samples of the avatar's location in space-time.

Elsewhere (Parunak 2009), we describe how to generate useful distributions, not only from the pheromone fields that ghosts lay down as they execute, but also from efficient logs of ghost state that we collect as they move from page to page.

CONCLUSION

Modern analysis requires models that can deliver statistically meaningful results from complex scenarios in a form that can interact with other analytic tools. The technologies that we have outlined in this paper address this need.

- By considering multiple futures concurrently, they avoid the sampling error of considering only one or a few trajectories.
- By modeling the location of entities in multiple environments concurrently and managing the flow of execution, they can configure complex environments easily from a library of standard components.
- The computational artifacts they produce can be interpreted readily by other systems.

REFERENCES

Erdös, P. and Rényi, A. (1960), "On the Evolution of Random Graphs." *Magyar Tud. Akad. Mat. Kutató Int. Közl.* 5: 17-61.

Gilmer, J.B., Jr. and Sullivan, F.J. (1998), "Alternative implementations of multitrajectory simulation." *the 30th Winter Simulation Conference*, Washington, D.C., United States, IEEE Computer Society Press.

Gilmer, J.B., Jr. and Sullivan, F.J. (2000), "Recursive simulation to aid models of decision making." *the 32nd Winter Simulation Conference*, Orlando, Florida, Society for Computer Simulation International.

Gilmer, J.B., Jr. and Sullivan, F.J. (2001), "Study of an ergodicity pitfall in multitrajectory simulation." *the 33nd Winter Simulation Conference*, Arlington, Virginia, IEEE Computer Society.

Parunak, H.V.D. (2007), "Real-Time Agent Characterization and Prediction." *International Joint Conference on Autonomous Agents and Multi-Agent Systems (AAMAS'07), Industrial Track*, Honolulu, Hawaii, ACM.

Parunak, H.V.D. (2009), "Interpreting Digital Pheromones as Probability Fields." *the 2009 Winter Simulation Conference*, Austin, TX.

Parunak, H.V.D., Belding, T.C. and Brueckner, S. (2007), "Prediction Horizons in Polyagent Models." *Sixth International Joint Conference on Autonomous Agents and Multi-Agent Systems (AAMAS07)*, Honolulu, HI.

Parunak, H.V.D. and Brueckner, S. (2006), "Concurrent Modeling of Alternative Worlds with Polyagents." *the Seventh International Workshop on Multi-Agent-Based Simulation (MABS06, at AAMAS06)*, Hakodate, Japan, Springer.

Parunak, H.V.D., Brueckner, S.A. and Bisson, R. (2010), "Agent Interaction, Multiple Perspectives, and Swarming Simulation." *the International Joint Conference on Autonomous Agents and Multi-Agent Systems (AAMAS 2010)*, Toronto, Canada, IFAAMAS.

Parunak, H.V.D., Sauter, J. and Crossman, J. (2009), "Multi-Layer Simulation for Analyzing IED Threats." *IEEE International Conference on Technologies for Homeland Security (HST 2009)*, Waltham, MA, IEEE.

Chapter 35

Building Cross Cultural Trust and Change: How Do I Obtain and Implement Local Knowledge?

Tara Tetrault[1], Joseph Godfrey[2]

[1]Department of Social Sciences
Montgomery College
9700 Takoma Avenue
Takoma Park, Maryland 20903

[2]WinSet Group LLC
4031 University Dr. Suite 200
Fairfax, VA 22044

ABSTRACT

Experience in the field has told us that "…digging a well in villages where irrigation is communal (may present problems in the long run?). For example, a new well can shift the power from a community to one the landowner. This action may strengthen the group paid to dig it or create divisiveness between indigenous populations or give the impression that you …do not understand what is going on in the culture or that you have unknowingly sided with one element or another. Yet, all you were trying to do is to provide water to a community," recites General McChrystal (*The Economist* 2009: 31). Understanding cultures and applying assistance is a common problem among organization members who are trying to improve the developing world. How do we go about understanding local values? Why should we take the time to find this out? How do we make changes that indigenous communities want to maintain and take ownership of? What possible value could this add and is it worth my time? Cultural Anthropologists have addressed these very issues for decades and have built a reservoir of data on cultures *all over the world*. Learn how Anthropologists have grappled with these very issues in the past and in many cases, made changes to communities that members want to own in the *Building Cross Cultural Trust and Change: how do I obtain and implement local knowledge?* Anthropologists have been working with indigenous communities worldwide and analyzing cultures according to the local context for decades. This paper discusses anthropological concepts to establish a common ground. Ethnographic studies are discussed from Ghana, West Africa, and Guatemala. Applied anthropology in the form of medical anthropology is illustrated (Benedict 1934; Mead 1931; Boaz and Powell 1966; Peaples and Bailey 2008).

Keywords: Indigenous, emic, ethnography, anthropologist, syncretism

USING ANTHROPOLOGY

Establishing a mutual understanding of cultural concepts offers a starting point for discussion. Anthropology is the study of human behavior in the past and present, where Cultural Anthropology is one of the five fields of Anthropology which focuses on studying people in present day contexts. Culture is the knowledge, norms and values people learned from being brought up in their own society. Ethnographers research urban and rural cultures, and then observe, interact and participate with people in their daily lives. They gather ethnographic data on the village cultural landscape, house form and structure, politics, economics, agriculture, lineage and kinship, marriage, roles, religion, language and art. One of the turning points in anthropology may help to understand current ethnographic approaches and one of the analysis trends people use. Here, Franz Boaz (1858-1942) is known as the father of American Anthropology because he spawned much of the discussion on contextual analysis. He and his colleagues encouraged anthropologists to using scientific methods as a way of measuring and validating their work. Known for his work with Native American languages, he coined the phrase, cultural relativity as a doctrine that says we shall not approach and view other people ethnocentrically, but rather, recognize each culture with respect and realize the purpose is to interpret other cultures on their own terms. With this in mind, analysts gather social data, record their ethnographic notes, put this into datasets, validate feedback, quantify datasets, publish results and request pier review. This ensures a level of quality control (Peoples and Bailey 2008).

Following this process, anthropologists have built a reservoir of cultural data in each of the four subfields of Anthropology including cultural anthropology (i.e., ethnography). The idea of applying anthropology to real world situations was born in the days of Margaret Mead (1933), Ruth Benedict (1934), and Gregory Bateson 1948) using what they learned about cultures to provide assistance to the local community in the form of providing immunization, medical services and education.

Those unfamiliar with anthropology may be challenged by the "relativism" in the approach anthropologists take to studying another culture. An example from management consulting may be helpful here. When a management consultant is asked to effect change in an organization, the starting point of a wise and experienced consultant is to appreciate the existing organizational process and recommend obtainable amendments within the existing structure. These recommendations enjoy a greater chance of success because they make sense to the organization.

BACKGROUND RESEARCH

Previous peer-reviewed and gray literature of cultures in this area enables you to become familiar with the previous work that has been done and identifies what needs doing, enabling you to better frame your research. One of the more useful cultural databanks is the Human Relations Area Files (HRAF) which houses a plethora of anthropological literature gathered since the beginning of the 20th Century along with research of peer-reviewed journals and grey literature (www.yale.edu/HRAF). Adequate research will assist in understanding people on their own terms (i.e., emic or insider perspective). To do this you need to gather data on language, kinship, economics, religion, social norms, politics and marriage practices. Identifying the cultural, linguistic and religious groupings associated with people in the local area is mandatory and will help you to avoid making mistakes or misinterpreting what is going on (Spradley 1977).

LOCALISMS

Establish local contacts through the local universities and colleges, as well as museums to establish a right to be there. Understand the appropriate attire, hire interpreters, and identify which villages you need to go to and who you need to meet to gain access to the local community. Once in country it may be easier to obtain a driver, who knows the roads, the urban and rural villages you need to go to establish contacts. In rural Ghana, for example, meetings were held with Chiefs, Queen Mothers, Elders, and Linguists Bring an "appropriate gift" by first consulting with locals to determine what represents adequate value to be considered a "gift". Then through the available Liaison, or interpreter, arrange the meeting.

Plan to stay for a period of time so that you can get to know the people in the village or your informants. Ask open-ended questions to obtain the most information. Repeat some of your line of questioning to check your data and take time to unpack the traditions, meanings and norms of the village. Do not assume that your ideas are the best way to remedy a local problem in a country or culture you are not from. You will have a stronger positive impact on people if you can implement change using cultural relativity. Learn to incorporate the local knowledge. Otherwise the actions you take to "help" may not really being providing wanted change and as a result, may not be adopted and used by the local community. The following examples in Ghana and Guatemala may help provide some insight into cultural studies and health.

THE GHANA EXAMPLE

Background research on the Akan area suggests that the Akan people are matrilineal. Women own the land they farm on. They play a major role in the transmission of land and other property, and in the allocation of financial and labor assistance. Lineage elders also allocate considerable property and financial assistance through decisions on inheritance, loans or gifts, residence rules, and schooling, with women participating as elders and as recipients. Access to capital and labor affect women through their roles as sisters, mothers, and aunts. Principles of leadership and inheritance give women less access than men to lineage-controlled resources, but ensure women's access at a minimal subsistence level. While we need to understand politics and economics, many analysts are finding that in addition details on marriage and kinship are intimately connected to people's ability to be major players in politics and economics (Gillespie 2000; Tetrault 1997).

Owing to the shared power between genders, the Egaa Cultural Area was divided into four sections (i.e., Egaa I-IV). Each village was run by a Male Village Chief or female Elder who had voting rights in the Royal Family. Decisions were made by all four members using an egalitarian system. After requesting entry to the village, meetings were held, a ritual ceremony was performed, prayers were said at the door entry and libation was poured into the ritual clay vessel. A level of trust was established once village Chiefs and elders were confident the purpose of the visit was not exploitive. Subsequently, ethnographic data was gathered on clans, political roles, medicinal and religions roles, village, architecture, association with other neighboring villages-cultures and pottery use (Ethnographic Report, Egaa, Tetrault 1997).

Detailed ethnographic inquiries were used to locate active pottery making villages, to identify vessels made and used today and to assist in the identification of vessels from a local archaeology site. People in the area were from the Akan Ethnolinguistic group and local pottery reflected Fante, Ga, Asante, and Ewe cultures. Women who made pottery were from the Drummer clan. While other clans included the Royals, Warriors, and Linguists also had prescribed career paths. Women may marry outside of their village and cultural group, but they took their knowledge of pottery making with them to their new home and created pottery according to their own cultural identity.

Ethnographic inquiry found that Ashanti culture was present in other villages as indicated by their pottery. One such twentieth century example was made evident by the Asante vessels in Elmina. Potters had married into Elminan families generations ago but the tradition remained as mothers and daughters continued to make carinated (angled) vessels (Tetrault 1997). The following example illustrates how anthropology may be applied to understand dual cultures.

APPLYING ANTHROPOLGY: MEDICAL KNOWLEDGE

Medical Anthropology is a subfield of applied anthropology that draws upon social, cultural, biological, and linguistic anthropology to better understand what is meant by health and the health practitioner in a given culture. Medical anthropology captures the local terms and definitions for diseases and identifies the religious and/or medical practices used to prevent or treat the illness. Medical anthropologists define the healing processes, the social relations with practitioners and the utilization of pluralistic medical systems.

Medical services if combined with local cultural knowledge are more apt to be used by communities. Anthropologists apply their skills to a range of factors driving health, nutrition and health care transitions. For example, anthropologists have worked with local hospitals and the Hmong American community to combine healing rituals with hospital regulations (Doherty 2000; Fadisman 1991). Analysts gather data on ethnomedicine, pluralistic healing modalities, and healing processes and apply this new knowledge to help the indigenous population use the latest medical practices

Historically, religion has been linked to medicine in many cultures worldwide. This may come out when practioners or family members choose to combine herbal medicines with prayers or chants. In the 1950s, medical anthropologists such as Richard N. Adams, Benjamin D. Paul, and Lois Paul wrote monographs dedicated to the Maya medical beliefs and practices. Richard N. Adams, described the relationship between Maya medical beliefs and practices and Western science, illustrating why Mayans rejected projects applied by the Institute of Nutrition for Central America and Panama (INCAP). His work is seen as setting the stage for four decades of medical anthropology in Guatemala by diagnosing the communication breakdown caused by "ignorance of local beliefs and practices." Many of those once affiliated with INCAP have since published works on various topics of interest to medical anthropology in Guatemala.

Several things came to undermine the indigenous way of practicing medicine. For example, religious persecution was first administered by the Catholics by the Spanish, and later, by the Protestant evangelical religions. This resulted in the prohibition of their members from consulting traditional healers. Beginning in the 1980s, the Guatemalan national health care system relied heavily on Western medicine and began to suppress traditional healers by banning them from practicing. The health care system began making efforts to train local midwives (Gonzalez 1988). Regardless of public suppression, religious and healing practices did not die out but continued to be practiced in the home. People combine religious faiths and may continue to perform practices associated with health, childbirth, and to ensure recovery from illness.

While medical facilities are available in modern cities, millions of people in rural areas lack the same level of health care. Today, within established medical facilities in Solola, Guatemala, Nicole Berry's anthropological assessment found that patients had a lack of trust in the quality of care they received. The outcome is often cited as a major factor promoting reluctance to seek biomedical help for obstetric emergencies. The indigenous community had poor perceptions of the quality of care received when seeking obstetric care in the hospital. Using data collected over two years, Berry set out to understand why interviewees repeatedly complain that hospital staff "do not attend to you." This lack of communication and attention to the patient's specific needs (in the form of a healer or combination of herbal remedies) points to the need for any health organization to incorporate the indigenous perspective into their treatment plan. This may include providing interpreters or allowing healers practice along side physicians as seen in multiple examples from the Hmong-American community (Fadisman 1991; Doherty 2000; Berry 2009).

MAKING CHANGES INDIGENOUS POPULATIONS WANT TO ADOPT?

The examples discussed call to question, how do we make changes that indigenous communities want to maintain and take ownership of? First, we must provide assistance through the existing network which may be an economic, political or kinship system, if not a combination. Gathering ethnographic data to understand values and the underlying cultural system at work may mean the difference between spending a small amount of money to implement AID projects versus over spending for development projects. Then, ask someone from within the culture for feedback on your plan. Involve the indigenous community in testing out models of how possible changes would work if implemented. Assistance is *for* the local population, so make sure the indigenous population *want*s to use it. This is key.

FOREIGN POLICY

The current interest by the Department of Defense, and the various Services, to involve anthropologists and other social scientists in support of military objectives is not new. Well known anthropologists were involved in military tasks during World War I and II, as well as Viet Nam (McFate 2005). Project Camelot (circa 1964), for example, has been extensively studied to understand why it failed, and the ethical questions it raised concerning the relationship of academic and government institutions, particularly the deployment of academic resources in pursuit of military interests (Horowitz,1974). We cannot in this paper revisit this complex and difficult history. But one dimension in the failure of Project Camelot is immediately relevant: trust. We have noted in our Ghana example that establishing a basic level of trust is essential in in collaborating with another culture. Despite the fact that all the elements of Project Camelot were open and no participant thought of himself as "spying" for the United States, the perception by Latin American countries was that the endeavor was to deploy social scientists to discover how to more effectively suppress leftist and left-centrist revolts. This was exacerbated by the already established suspicion held by many citizens of Latin American countries concerning their own militaries and the coincidental but contemporaneous intervention in the Dominican Republic.

The American Anthropological Association (AAA) has expressed concern over using anthropology for military purposes. The AAA Responsibility to the Discipline states that Anthropologists should undertake no secret research or any research whose results cannot be freely derived and publicly reported (http://www.aaanet.org/stmts/ethstmnt.htm). Given the new role of DOD, in being tasked and financed to both implement war _and_ change in the form of development projects, it is natural for DOD to turn to anthropologists for guidance and help. Anthropologists have the expertise to provide a depth of knowledge on cultures and assist in aid related projects. This kind of knowledge _can_ be used to implement constructive change. However, by assisting the government in providing development to help a community obtain food and water resources, the data collected may also be used for other defense related purposes with or without the anthropologist's knowledge. The end result of combining defense and aid projects together means that anthropologists run the risk of acting against the AAA recommendations & ethics.

Resolving this impediment to the contribution of anthropologists would lessen the problems experienced by McChrystal (2009) and increase reports like the recommendations that Dehgan and Palmer-Moloney published in their article on Water Security (2009). According to Dehgan and Palmer-Moloney, three of six waterways connect Afghanistan to Iran and therefore, are used by people in both places. Iran is dependent on using this water. Any action taken on the part of Afghanistan's waterways will inevitably affect Iran and its ability to feed its people. These are useful recommendations in rethinking county budgets and development projects (2009).

RECOMMENDATIONS:

1) Insert Cultural Anthropology concepts and methods into the professional training non-profit and for profit organizations providing any assistance cross-culturally, in the USA or abroad. Anthropological methods should appear in organization guidebook and mandate cultural anthropology as an official training class given by an anthropologist such as: Introduction to Anthropology Methods & Interpreting cultures.

2) Many women have a pivotal role in bringing their families and community forward in the future worldwide. If you want to assist areas in need, provide education and family financial assistance to women who bare the brunt of the responsibility of raising children.

3) To help another community include the user in the decision making process to identify what changes to make (if any) and how to implement and provide assistance that will be used.

REFERENCES

Adams, Richard N (1988) Texas Papers in Latin America: Prepublished Papers of the Institute of Latin American Studies. University of Texas: Autsin.

Agar, Michael (1994) Language Shock: understanding the culture of Conversation. Morrow: New York.

American Anthropological Association Ethics Statement. http://www.aaanet.org/stmts/ethstmnt.htm .

Benedict, Ruth ((1934) Patterns of Culture. Houflin, Miffton Co.: NY.

Berry, Nicole. Who's Judging the Quality of care? Indigenous Maya and the problem of "not being attended" Simon Frasier University: Burnaby, BC, Canada.

Boaz, Franz and J. N. Powell (1966) Introduction to the North American Indian Languages: Indian Linguistic Families of the American North of Mexico. University of Nebraska Press: Lincoln.

Dohery, Steven (2000) Political Behavior and Candidate Emergence in the Hmong-American Community. Hmong Studies Jourbnal:8: 1-25.

Dehgan, Alex and Jean Palmer-Molony (2008). Water Security and Scarcity: Potential Destabilization in Western Afghanistan. United States Corps of Engineers: Washington, D.C.

Emerson, Robert M., Rachel Fretz, and Linda Shaw (1995) Writing Ethnographic Fieldnotes. University of Chicago: Chicago.

Fadisman, Anne (1997) The Spirit catches You and then You Fall Down: A Hmong Child, Her American doctors and the collision of two cultures. Farrar, Strauss and Giroux: New York.

Freidel, David and Linda Schele and Joy Parker. Maya Cosmos: Three Thousand Years on the Shaman's Path. William Morrow and Company: New York.

Horowitz, Irving L. (1974), The Rise and Fall of Project Camelot, MIT Press: Cambridge

Gonzalez, Nancie and Carolyn S. McCommon (1989) Conflict, Migration and the Expression of Ethnicity. Westview press: Boulder.

Kleiner, Carolyn F. and H. Christian Brewer (2008) "The Importance of Cultural Knowledge for Today's Warrior-Diplomats". Carlilse, PA.

Kroeber, Alfred L. and T. T. Waterman (1920) A Sourcebook in Anthropology. University of California Press: Berkley.

McChrystal, W. (2009) The Economist: 31. London.

McFate, Montgomery J.D. (2005) Military Review, March-April:24-38.

Mead, Margaret (1977) An Anthropologist at Work: the writings of Ruth Benedict. Greenwood Press: Westport.

Mead, Margaret (1931). Growing up in New Guinea: a comparative study of primitive society. W. Morrow and Co. : New York.

Mead, Margaret (1978) The Mead Collection. www.libraryofcongress.com

Peoples, James and Garrick Bailey (2008) Humanity: an Introduction to Cultural Anthropology Thomson Wadsworth: Belmont, California

Spradley, James. (1979) The Ethnographic Interview. Holt Rinehart: NY

Spradley, James. (1979) Participant Observation. Holt Rinehart: NY

Tetrault, Tara L. (1997) Continuity and Innovation: Pottery Manufacture and Use among the Coastal Akan. Masters Thesis on file, University of Maryland, College Park, Maryland, (2007 Study, forthcoming).

Tetrault, Tara L. (1997) "Ethnographic Reports on Akan Villages", on file, University of Maryland, College Park

Warren, Dennis M. (n.d.) The Role of the Emic Analysis in Medical Anthropology: the case of the Bono of Ghana. Iowa State University: Des Moines.

www.yale.edu/HRA

Chapter 36

Applying Epidemiological Modeling to Idea Spread

William J. Salter, Robert McCormack

Aptima, Inc.
12 Gill Street
Woburn, MA   01801, USA

ABSTRACT

Epidemiological models have been extremely effective in public health for both analysis and policy making. They formalize insights into biological and social processes of disease propagation into (low parameter) mathematical models. The work reported here applied relatively simple epidemiological models to the spread of ideas extracted by applying statistical language processing methods to real-world data. These models reproduced observed patterns of information spread, demonstrating the applicability and utility of epidemiological models. Both epidemiology and statistical language processing offer methods that can enhance our results. Such approaches can be applied to track, to evaluate, and ultimately to predict, spread of ideas across multiple temporal and geographic scales, in a variety of media, with widespread military, intelligence, and other applications.

Keywords: Information Diffusion, War of Ideas, Epidemiological Modeling, Language Analysis

INTRODUCTION

In early February 2010, in the run up to the Marja operation, General Stanley McChrystal, Commander of U.S. and NATO troops in Afghanistan, explained our objectives in Afghanistan: "This is all a war of perceptions. This is not a physical war in terms of how many people you kill or how much ground you capture, how many bridges you blow up. This is all in the minds of the participants." (Shanker, 2010) This focus on "the minds of the participants" makes clear how the military's emphasis has shifted from standard, kinetic approaches to more complex missions stressing counter-insurgency and similar operations.

A recent report by Major General Michael Flynn, senior military intelligence officer in Afghanistan, demands a dramatic change in the organization and focus of intelligence operations (Flynn *et al.*, 2009). He argues that "the vast intelligence apparatus is unable to answer fundamental questions about the environment in which U.S. and allied forces operate and the people they seek to persuade" (p. 7). Intelligence must "acquire and provide knowledge about the population, the economy, the government, and other aspects of the dynamic environment we are trying to shape, secure, and successfully leave behind" (p.10).

Methods for fighting wars of ideas are not well developed. They require changes in goals, strategy, tactics, techniques, and procedures across all levels, from strategic planning to targeting air strikes and how small units conduct cordon and search operations. The war of ideas is fought at multiple geographic and temporal scales, from immediate reactions of Marja's population to longer-term effects on local, regional, and national politics. And progress in that war is manifested in many ways, from blog posts and even Twitter feeds through reporting in the media, from attendance at local *shuras* to responses to survey questionnaires, from the content of Friday sermons through the composition of national political coalitions.

Assessment is also a concern: How can we evaluate the progress, the successes and failures, of the war of ideas? Of smaller battles in that larger war? How can we adapt our actions to increase our chances of success? Given the diversity of ways the war of ideas must be fought, and in which its progress can be manifested, a diversity of measurement and tracking methods must be developed.

The work reported here contributes to the set of methodological tools available. Our goal was to assess an intuitively appealing idea: powerful epidemiological methods can fruitfully be applied to modeling the spread of ideas across internet media. Some parallels are obvious: ideas spread by being "transmitted" from "carriers" to "uninfected" people; "transmission" requires "exposure" – the person being "infected" must encounter the idea – some people may have "resistance" to some ideas, while others may be much more "susceptible." We wanted to determine if these and other parallels were more than metaphorical – if mathematical methods derived from epidemiology could capture patterns observed in real-world data.

Epidemiological methods cannot fully address the requirements for measurement and assessment in fighting the war of ideas, of course, just as public health and clinical medicine use many methods for measurement and assessment, from sophisticated assays of specific manifestations of detailed biochemical or physiological processes to large outcome studies of overall effects on morbidity and mortality for particular prevention and treatment interventions. Epidemiological models focus on the spread of disease; similarly, epidemiologically-inspired models may be of particular utility in investigating and understanding the spread of ideas. Clearly, such understanding can be critical to prevailing in the war of ideas.

METHOD

MATHEMATICAL EPIDEMIOLOGY BACKGROUND

Disease, and explanations for it, have been present throughout human history. Early humanity attributed sickness to evil spirits, curses, and witchcraft. Through religious rites one could hope to stave off illness. Not until the Greeks did people begin to take a naturalistic view of disease. Hippocrates of Cos (of the Hippocratic Oath) was possibly the first to attempt to explain disease rationally. He believed that environmental factors, lifestyles, and habits were keys to disease, which was caused by physical imbalances in the four basic "humors." While simplistic and wrong in detail, this theory set the stage for scientific medicine. It was supplanted by the "Miasma Theory" in the Middle Ages, which held that disease was caused by poisonous vapors from decaying matter or stagnant water, thus making a fundamental connection between cleanliness and disease and suggesting that sickness can be "caught" from the environment.

Germ Theory, which began to take hold in the 19th century, built on the notion of communication by contact, and postulated microorganisms as the fundamental causes of contagious disease. With its introduction, formal mathematics could be used to help understand disease spread. Kermack and McKendrick (1927) introduced some of the first epidemiological models early in the 20th century, focusing on the rapid rise and subsequent decline in the number of infected patients often seen in epidemics. They formulated a nonlinear model which formally captured the ideas of infection and recovery. This very simple model reproduced the overall dynamics of epidemics, such as The Black Plague and cholera.

Over the past few decades, mathematical epidemiology has become one of the key drivers in setting public health policy. Epidemiological models take many shapes and sizes, from simple to complex; completely deterministic or stochastic; from individual to global. No one exemplar model sums up mathematical epidemiology. Such models provide both insight and predictive capabilities because they proceed from an understanding of the underlying biological dynamics and often make very few assumptions about the system. The original Kermack-McKendrick model, for example, contains only two parameters – infection and recovery rates – yet reproduces the common phenomenon of an initial large rise of infection, followed by a subsequent drop to endemic levels.

The type and scope of epidemiological models varies based on available information. If one only knows average rates of contact and recovery, a two-parameter model can provide insight. If one knows those rates for heterogeneous subpopulations, more complicated models can be used. Data on demographics, geography, and movement rates can also be factored into the models.

FORMALIZATION USED

To develop epidemiologically-inspired models of the spread of ideas, which we call "memes," we had to formalize the notion, find a suitable corpus to investigate, extract the memes, analyze them, select and parameterize models, and test their results against the data. We collected data from several real-world sources and extracted memes using natural language processing algorithms. We formatted the data into a set of time-series which show how the prevalence of each meme changes over time within each data source. We used cross-correlation analysis to compare the meme time-series between data sources. This produced two critical pieces of information: the maximum similarity of the meme signals and the time-lag between the data sources that produced the maximum correlation. We estimated the period of each meme time-series to determine how long memes remain prevalent in each data source. We used the results of these analyses to formalize and parameterize an epidemiological model, which was then simulated to produce approximations of the time-series, which were compared with the original time series.

We obtained data from five English-language blogs and news websites in the Islamic world, primarily Pakistan. We focused on two to analyze initially: a news website and a blog. We obtained approximately 4500 articles from The Pakistan Times (www.pak-times.com) between May 2007 and May 2009 and 4400 articles from Baithak Blog (baithak.blogspot.com) between August 2005 and May 2009. News websites and blogs, in general, operate in different ways. News websites are (often original) sources of information, while blogs tend to repeat or react to information. We chose these two data sources because of their chronological overlap and number of articles, so we could test if our models could find the publication of information and subsequent repetition or reaction to that information.

We operationalized memes as "topics" derived with natural language processing techniques. In particular, we used PLSA (Hofmann, 1999), one of a class of methods for data analysis called latent variable methods. Simply put, such methods find words that tend to be associated across documents, and extract such groupings without requiring that all the grouped words always co-occur. For example, if "bombing" often occurs with "terrorist," "insurgent," and "IED" in some documents, and "explosion" (but not "bombing") occurs with those same three words in another document, a latent variable method would correctly group "bombing" with "explosion," even though they never co-occur.

The input consists of a collection of documents from which a word-by-document matrix of frequency counts of each word in each document is constructed. This matrix is very sparse, since most documents use just a small portion of the total vocabulary. (We note that more restricted vocabularies, like those used in medical records or chat or Twitter posts, create smaller matrices than larger vocabularies, like those in newspaper articles. This can make latent variable methods more discriminating because they will find narrower topics.) Latent variable methods ignore word order and perform massive dimensionality reduction via latent (hidden) variables. For PLSA, these are the "topics," characterized by a (weighted) mix of words; documents then are characterized by a (weighted) mix of topics. By finding these latent variables to explain how ideas in a document are expressed in words, these techniques approach human judgments of meaning and conceptual understanding (Griffiths, Steyvers, & Tenenbaum, 2007). They are language-independent since word order is thrown out, requiring no grammars, dictionaries, or thesauruses, and no ontologies are used. (However, since words and source documents remain in the original language, language expertise is of course required to know what the topics mean.) The top words in a topic tend to be interpretable, as shown in Table 1 for topic 66.

Table 1: Top 10 words for topic 66.

Word	Probability		Word	Probability
Iran	0.0337		Iraq	0.0090
Iranian	0.0182		Military	0.0075
Nuclear	0.0145		Country	0.0074
Missile	0.0139		Tehran	0.0069
Israel	0.0134		Egypt	0.0053

We applied PLSA to both sources and extracted a 100-topic model. We then built time series for each topic for each source, which yielded 100 "waveforms" for each source, where each waveform represents the prevalence of one topic over time in that source. We then performed a standard cross-correlation analysis on the waveforms that computed similarity of the two waveforms for all possible temporal offsets, which yielded a distribution of similarity measurements as a function of time-lag for each topic. From this, we extracted two potentially important pieces of epidemiological information: the maximum correlation of the signals and the time-lag between sources that yields that maximum; these varied across topics.

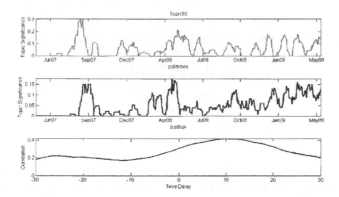

Figure 1: Topic 66 time-series for Pakistan Times (top) and Baithak Blog (middle) and correlations between them for different time lags (bottom)

Figure 1 shows the time-series of Topic 66 for the Pakistan Times on top, for the Baithak Blog in the middle, and their correlations for different time lags below. The two signals appear to show interesting dynamics, with a large spike in activity around July-August 2007 in the Pakistan Times followed by a similar increase in the Baithak Blog. In late 2008 and early 2009, there appear to be periodic dynamics in both signals. Thus, upon inspection, the data from the two sources appear related. The maximum correlation is 0.4, between about a 9- and 12-day lag, with a positive lag interpreted as the Baithak Blog lagging behind the Pakistan Times.

We also estimated the period of each meme, by correlating each individual time-series to a variety of sine curves with different periods. The period of the sine curve which best correlated to the data was our initial estimate of meme periodicity. This provided a third parameter for our epidemiological models. (In future efforts these estimates of periodicity can be enhanced by using Fourier transforms, which transform a function of time, such as the topic time-series, into a function of frequency. This allows extraction of the dominate frequency or frequencies, which will yield a higher resolution estimate of the periodicity of memes.)

As a preliminary epidemic model, we attempted to capture the relationship between the two sources, for individual topics, using three parameters extracted from our analysis. Our goal in this preliminary research was to develop an epidemiological model that can reproduce the empirical relationships we discovered; *we are not seeking to make any claims about the nature (or, indeed, existence) of causal mechanisms*. If epidemiological models do capture empirical relationships, further research can then be designed to investigate and exploit underlying mechanisms.

We used a standard stochastic epidemiological modeling technique based on an unrestricted random walk: a Markov process in which a random variable moves between a set of states (here, positive integers) based on certain probabilities. In this simple one-dimensional random walk, we are concerned only with the probabilities of the random variable increasing or decreasing. (This is the type of model initially used to provide insight into how the number of infections changes over time.) Our model used three parameters – maximum correlation, time lag for that maximum, and period of the meme – to predict the dynamics of the "influenced" source from the observed dynamics of the topic from the influencing source. The model is:

$$P\left\{I_{n+1} = i+1 \middle| I_n = i\right\} = \beta X_{n-d}$$

$$P\left\{I_{n+1} = i-1 \middle| I_n = i\right\} = \gamma i$$

$$P\left\{I_{n+1} = i \middle| I_n = i\right\} = 1 - \gamma i - \beta \hat{I}_{n-d}$$

where β is the maximum correlation based on the historical cross-correlation analysis, d is the time-lag of that maximum correlation, γ is the period estimated from the historical time-series signal, and X is the time-series signal from the influencing source. This model defines the probabilities of an increase, a decrease, and no change (one minus the other probabilities). We assume that the time step of the model is small enough that only one change occurs in that interval. The first line of the model gives the probability the topic occurring on the website in the next time step (tomorrow) based on the probability of the topic occurring in the current time step (today). It is defined as the maximum correlation times the probability that the topic occurred in the first website d days ago. This simple equation is intended to mimic the time-lagged influence between the data sources. The probability of a decrease in the next time step is given by γi, and the probability of no change is one minus the other probabilities.

Thus, this model estimates the probability of a topic increasing or decreasing on the website based on the influence of the other website and the specific parameters capturing the dynamics of the relationship between them. We simulated this model by choosing a random number from a uniform distribution at each time step. If it falls within the threshold of increase we add one to the current state, if it falls within the threshold of decrease we subtract one, otherwise we do nothing. In order to get a general understanding of the dynamics produced by the model we simulated 100 sample paths of the random walk and find the mean of those paths.

RESULTS

We conducted cross correlations at varying time lags between the time series of topic prevalence for each of the 100 topics for the two sources we focused on. The mean maximum correlation was 0.31, with a standard deviation of 0.1. Approximately one quarter of the correlations were above 0.4, and another 20% were below 0.2. The fact that some topics do *not* show correlations across sources suggests that there are real phenomena underlying the relationships, since always finding correlations would mean that we might be modeling noise. (Similar analyses between a third data source and the two discussed in this paper yielded a mean maximum correlations of topic time series in that third source with either of the other two of 0.14, with a standard deviation of 0.11, supporting the idea that we are not modeling noise.)

We used our three-parameter epidemiological model to attempt to reproduce the empirical relationships discovered between topics for the topics with maximum correlations above the median maximum correlation. The mean actual maximum correlation across those 50 topics was 0.4. (Thus, topic 66, as shown in Figure 1, was typical.) The epidemiological model produces a predicted time series for each topic in the "target" data source – the one being influenced, by hypothesis, after some time lag. The mean correlation between actual and predicted time series for the "target" news source was 0.46 . We take this as strongly encouraging support for the idea that epidemiological models can fruitfully be applied to look at the spread of ideas across news sources, and, more broadly, in mapping and understanding the dynamics of the relevant information space.

CONCLUSIONS

Our results are encouraging, showing that simple epidemiological models can capture empirical relationships in noisy real-world data. Our multi-step process – from downloading website data through statistical manipulations to parameterizing and running epidemiological models to assess how well they reproduce empirical relationships – can probably be improved at each step and, indeed, by including other procedures reflecting other insights. Perhaps the most significant implication of this work is the strong argument it makes for such further development.

Less simple models, less noisy data, more data sources and types, and more flexible approaches may well substantially enhance descriptive and predictive accuracy, and can potentially deliver real utility to operational users, as well as advancing theory and practice in modeling information diffusion more broadly. This work was performed under a Phase I SBIR contract, which imposes some temporal and other resource constraints, so we were unable to explore and apply a variety of possible enhancements and extensions. We briefly outline some possible applications, implications, and extensions for this work below.

Methods to assess ambient ideas and their patterns of diffusion will be valuable to multiple levels and for multiple uses by the military and intelligence communities. Such methods can greatly enhance their ability to measure progress in the "war of ideas." A range of potential metrics can be developed, grounded in solid theory and, perhaps even more important, embodied in well-defined, replicable procedures. In addition, insight into the dynamics of information diffusion, particularly when broken down by subpopulations (a standard approach in epidemiology), can significantly enhance our ability to design strategies and tactics in skirmishes, battles, and the overall "war of perceptions," as General McChrystal put it.

We have identified several ways in which this work can be extended and enhanced, and we are confident that there are others we have not considered. In general, more effective preprocessing of input data could enhance performance; incorporating robust methods for entity extraction and developing some language transformation rules to incorporate domain knowledge can almost certainly help as well. For example, Lawrence Hunter and his collaborators have developed powerful methods to extract meaningful information from the published biochemistry and pharmacology literature using a combination of formal quantitative methods and rules (embodied in regular expressions) that capture relevant domain knowledge. (See, for example, Baumgartner *et al*, 2008; Cohen *et al*, 2009; Hunter *et al*, 2008.)

A combination of epidemiological approaches and other methods – such as population surveys, economic and other statistics, studies by area experts – can potentially contribute substantially "to answer fundamental questions about the environment in which U.S. and allied forces operate and the people [we] seek to persuade," as General Flynn requires (Flynn *et al*, 2009). Social media, such as Twitter, are particularly appealing sources of data. Their strict size limits suggest they may be most valuable for narrowly defined temporal and geographic scopes, a potentially valuable supplement to slower, less fine-grained sources, such as news sources, blogs and the like. We believe that social media can be particularly useful when their analysis is embedded in a characterization of the larger "meme space."

The network topologies of the relevant information space can be considered and modeled in more detail. By "topology" we mean the structure of the networks, such as the link structure among news sources, blogs, and the like; the follower/following structure among users of Twitter and similar services; the geographical distribution and structure of tweets and, if possible, of other meme sources.

Attributes of meme content should also be considered, such as sentiment or valence; our pragmatic operationalization using PLSA does not distinguish between, for example, an article praising Israeli expansion of settlements and one criticizing it – both are "about" expansion of settlements – although such a distinction is clearly important. A subtler aspect of content concerns what may be termed "rhetoric": for example, some Muslim websites (both Jihadi-oriented and conservative, but non-violent, sites) use the high rhetoric of classical Arabic or Persian poetry. This rhetoric will appeal to some audiences and may alienate others. Determining the rhetorical approaches of sites and understanding the mapping of rhetorical styles to audiences can obviously sharpen analyses of meme spread.

Another aspect of content concerns "markers," like genetic markers – small fragments (for memes, sequences of words) that can help to track meme paths, just as small fragments of DNA can help in characterizing biological origins. For example, Leskovec and collaborators (Leskovec *et al*, 2009) investigated the spread of memes associated with "lipstick on a pig" in the 2008 American Presidential campaign, finding several distinct variants, which they traced to different sources.

More broadly, and moving beyond epidemiological concepts, the structure of "meme space" can be characterized. Simply put, some memes tend to co-occur, while others rarely appear together, and memes can certainly be viewed as having a similarity structure. A number of well-understood mathematical methods can be applied to the matrix structures that are used for meme extraction, and we believe that they can be quite useful in some applications and extensions of epidemiological modeling to investigate message spread.

ACKNOWLEDGEMENTS

This paper reports on work performed on a Phase I SBIR project entitled E-MEME (Epidemiological Modeling of the Evolution of Messages), contract N00014-09-M-0189, sponsored by the Office of Naval Research. All opinions expressed herein are the authors', and do not reflect those of ONR, the DoD, or the U.S. Government.

REFERENCES

Baumgartner WA Jr, Lu Z, Johnson HL, Caporaso JG, Paquette J, Lindemann A, White EK, Medvedeva O, Cohen KB, Hunter L. Concept recognition for extracting protein interaction relations from biomedical text. *Genome Biol.* 2008;9 Suppl 2:S9. Epub 2008 Sep 1.

Cohen, KB, Verspoor, K., Johnson, HL., Roeder, C., Ogren, PV., Baumgartner, WA. Jr., White, E., Tipney, H., Hunter, L. High-precision biological event extraction with a concept recognizer. *Proceedings of the Workshop on BioNLP: Shared Task*, 2008, pp 50-58, Association for Computational Linguistics.

Flynn, M.T., Pottinger, P., & Batchelor, P.D. (2009). Fixing Intel: A Blueprint for Making Intelligence Relevant in Afghanistan. The Center for a New American Security, Washington, DC.

362

Griffiths, T. L., Steyvers, M., & Tenenbaum, J. B. (2007). Topics in semantic representation. *Psychological Review , 114* (2), 211-244.

Hofmann, T. (1999). Proabilistic latent semantic indexing. *Proceedings of the 22nd annual international ACM SIGIR conference on Research and development in information retrieval*, (pp. 50-57). Berkeley, California, United States.

Hunter, L., Lu, Z., Firby, J., Baumgartner, WA. Jr., Johnson, HL, Ogren, PV, Cohen, KB, OpenDMAP: An open-source, ontology-driven concept analysis engine, with applications to capturing knowledge regarding protein transport, protein interactions and cell-specific gene expression. *BMC Bioinformatics*, 2008, Jan 31;9(1):78.

Kermack, W. O., & McKendrick, A. G. (1927). A Contribution to the Mathematical Theory of Epidemics. *Proc. Roy. Soc. Lond. , 115A*, 700-721.

Leskovec, J., Backstrom, L., & Kleinberg, J. (2009). Meme-tracking and the Dynamics of the News Cycle. *ACM SIGKDD Intl. Conf. on Knowledge Discovery and Data Mining*.

Shanker, T. (2010) U.S. General Offers Upbeat Views on Afghan War. *New York Times, February 4, 2010*

CHAPTER 37

Capturing Culture and Effects Variables Using Structured Argumentation

*Ken Murray[1], John Lowrance[1], Ken Sharpe[1], Doug Williams[2],
Keith Gremban[2], Kim Holloman[3], Clarke Speed[4]*

[1]Artificial Intelligence Center, SRI International
Menlo Park, CA 90425-3493, USA

[2]SET Corporation, an SAIC Company
Greenwood Village, CO 80111-3025, USA

[3]Applied Transformational Initiatives, SAIC
Sterling, VA 20164-7114, USA

[4]African Studies Program, University of Washington
Seattle, WA 98195-3650, USA

ABSTRACT

The Socio-Cultural Analysis Tool (S-CAT) is being developed to help decision makers better understand the plausible effects of actions taken in situations where the impact of culture is both significant and subtle. We describe the intended use of S-CAT on an illustrative use case, and discuss our use of *structured argumentation* as a representation technique to capture both culture variables and effects variables. Benefits of this approach include capturing multiple cultural theories and aggregating the forecasts of effects from multiple sources.

Keywords: Modeling Culture, Effects-Based Modeling, Structured Argumentation

INTRODUCTION

The United States and other countries are engaged in Security, Stability, Transition, and Reconstruction (SSTR) operations in many parts of the world. Successful outcomes for such operations require understanding the societies and cultures within which they are conducted (Ng et al, 2005). Misunderstandings due to cultural differences can yield unforeseen and potentially adverse consequences, delaying progress in the best cases and producing instability and tragedy in the worst cases (Hearing, 2004; Sharpe et al, 2007).

To improve effectiveness, SSTR operations must be conducted with improved understanding of the human socio-cultural behavior (HSCB) of the host nation, and they must pursue courses of action (COAs) that avoid adverse consequences and exploit favorable opportunities. Selection among candidate COAs must be informed by their plausible consequences and requires understanding how they may impact the attitudes, activities and circumstances of the host nation people. However, not every SSTR planning analyst can become a social scientist. Instead, technology can gather, organize, and represent HSCB knowledge from diverse sources, including social scientists and regional experts, to capture culturally informed action-effect relationships that forecast the plausible consequences of candidate COAs (Sharpe et al, 2007).

We are developing the Socio-Cultural Analysis Tool (S-CAT) to help analysts and decision makers better understand the plausible effects of actions taken in situations where the impact of culture is both significant and subtle. Although its application is potentially much broader, it is specifically being developed to support planning analysts and decision makers in conducting SSTR operations.

EXAMPLE SCENARIO: GUINEA COAST PIRACY

The use of S-CAT is illustrated with the hypothetical SSTR scenario[1] in Figure 1.

S-CAT contributes to COA selection by identifying their plausible effects. In the example described, S-CAT exploits the following types of knowledge: **Ontology:** an extensible set of classes, relation types and attribute variables used to describe the situation where actions will be performed, the actions performed, and the effects of the actions. **Site model:** a set of the entities (e.g., people, groups, places) and activities described with relations and attribute profiles to collectively represent the situation where actions will be performed. **Plan:** a set of actions that compose a candidate COA. **Effects rules**: a set of rules that forecast plausible effects of actions, including rules to capture both direct effects and indirect effects. S-CAT uses these types of knowledge to forecast plausible effects of a candidate plan for elements (i.e., entities and activities) of the site model.

---

[1] The people and events described are not real. The COAs are for purposes of illustration only and do not represent the real or intended actions of any government.

Following the international crackdown on and resultant reduction in the level of piracy in Sudan, the international shipping community is recognizing an increase in piracy operations off the coast of Sierra Leone. Most raiding vessels seem to come from the southern region of Bonthe. Violence, so far, has been low; the raiders use stealth and board a vessel at night, take command, and either off-load select cargo or sail the vessel to the shoreline, where cargo is removed.

Jonathan (Pa Yangi-Yangi) Tucker, a descendent of powerful traders and merchants, is the leader of the fishermen and farmers on and around Kita Island. He is proud of his family's historic prominence in the region, and he has an intense desire to see his people continue their traditions into perpetuity, preserving their way of life on the land he has always known. History has deprived his village of goods to trade, and his people are being forced to live as little more than subsistence farmers. The humiliation and poverty have become intolerable to Pa Yangi-Yangi. From his point of view, the Westerners and their lackeys in Freetown stole his birthright, and he is determined to get it back. His forefathers seized trade goods (e.g., slaves, food, gold) from other tribes and sold the goods to traders on ships. He wants to simply reverse the process: seizing goods from ships for trading and selling to other tribes and villages. For Pa Yangi-Yangi, piracy is nothing more than an entrepreneurial opportunity, a moral path to sustain his people and their traditions.

Pa Yangi-Yangi leads a group of his village fisherman on a raid of a large trawler off the coast of Turtle Island. He brings back food, goods, and money to the village. His bold and successful act immediately expands his power and standing within the tribal group. He is now living his birthright: he is a Tucker, and like his ancestors he is leading his people, ensuring that they and their traditions will endure.

The SSTR operation must decide how best to curtail the emerging piracy on the Bonthe coast. Two candidate COAs are

(1) Provide monetary aid to the local governmental leader (a paramount chief) to discourage piracy in his jurisdiction.

(2) Provide fishing and agriculture equipment (e.g., fish-finding sonars and rice irrigation systems) to the local fishing and rice-farming enterprises.

FIGURE 1: Hypothetical SSTR scenario

To generate a forecast of plausible effects for a candidate COA, S-CAT uses two distinct reasoning engines, a rule-based probative reasoner and an agent-based simulator. S-CAT aggregates the results from each into a single forecast of plausible effects presented to the user. The following sections describe how S-CAT uses structured argumentation to represent culture and effects variables and reason about their values while identifying plausible effects resulting from the second candidate COA in the scenario (Figure 1).

CAPTURING VARIABLES IN S-CAT

The ontology includes several kinds of attribute variables. *Profile* variables are used to capture the attribute (i.e., non-relational) descriptions of the site elements, and they include *culture* variables, used to capture aspects of culture considered during reasoning. *Effects* variables are used to capture the plausible consequences of actions. S-CAT currently reasons about the political, military, economic, social, infrastructure and information (PMESII) effects of candidate COAs. In S-CAT these variables and their values are captured using the techniques of template-based structured argumentation (Lowrance et al, 2007).

Variables are captured using structured argument *templates*. Each template is specified as hierarchy of questions about an implicit object of interest. Structurally, a template includes primitive (i.e., leaf) questions, which can be directly answered, and derivative (i.e., non-leaf) questions, whose answers are derived from the answers to more specialized (e.g., primitive) questions. For the example scenario, a simple template is used to capture the PMESII effects (see Figure 2).

---

1. Political: what are the political effects?
   1.1 Leadership: what is the effect on political leadership?
   1.2 Stability: what is the effect on political stability?
   1.3 Secular: what is the effect on secular political influence?
   1.4 Religious: what is the effect on religious political influence?
   1.5 Ethnic: what is the effect on ethnic political influence?
   1.6 External: what is the effect on external political influence?

3. Economic: what are the economic effects?
   3.1 Wealth: what is the effect on available wealth?
   3.2 Macroeconomic: what is the effect on macroeconomic stability?
   3.3 Microeconomic: what is the effect on microeconomic stability?
   3.4 Formal: what is the effect on formal economic activity?
   3.5 Informal: what is the effect on informal economic activity?
   3.6 External: what is the effect on external economic influences?

4. Social: what are the social effects?
   4.1 Stability: what is the effect on social stability?
   4.2 Welfare: what is the effect on the welfare of society?
   4.3 Public Security: what is the effect on public security?
   4.4 Public Contentment: what is the effect on public contentment?
   4.5 Formal: what is the effect on approval by formal leaders?
   4.6 Informal: what is the effect on approval by informal leaders?
   4.7 External: what is the effect on external approval?
   4.8 Hearts and Minds: what is the effect on appeal to the populace?

5. Infrastructure: what are the infrastructure effects?
   5.1 Equipment: What is the effect on the available equipment?
   5.2 Facilities: What is the effect on the available facilities and venues?

...

FIGURE 2: Partial contents of a simple PMESII effects template

Aspects of culture relevant to the scenario and appropriate for the region are captured in the Speed[2] culture template (see Figure 3). The example also makes use of a very simple profile template for describing selected aspects of affluence relevant to the scenario (see Figure 4).

---

1. Power Attainment: is influence achieved vs. ascribed?

2. Effective Power Site: is power located in formal or informal power structures?

3. Allocative Power: is political power allocative (relying on material support or the promise of material support)?

4. Transparency: are the behaviors and motivations apparent (open) or secret?

5. Causation: is causation perceived as being material or mystical?

6. Lineage: is preserving family and traditions a strong or weak priority?

7. Time Perception: is the passage of time perceived as linear or circular?

8. Resource Perception: is the availability of resources (opportunities, food) perceived to be something that can grow or fixed (zero-sum)?

FIGURE 3: Speed culture template

---

1. Education: What is the achieved level of formal education?

2. Wealth

   2.1 Aspiration: Is the primary aspiration quality of life or survival?

   2.2 Perception: Is the perceived level of wealth abundance or scarcity?

   2.3 Health: Is health care available whenever needed or unavailable?

3. Technology

   3.1 Tools: are the available tools modern or primitive?

   3.2 Activities: Are activities enabled more by technology or by labor

---

FIGURE 4: Simple affluence profile template

Each question in a structured argument template defines a variable, and the answer to that question for a given site element identifies the value of the variable for that element. A *structured argument* is created using a template by answering one or more of the primitive questions for a particular site element. A PMESII argument for an element of the site model describes effects forecast for that element; S-CAT uses automated inference to forecast plausible effects for elements of the site model by inferring answers to the PMESII template primitive questions for those elements. A Speed argument for a site element is a cultural description of that element. Profile arguments (or simply profiles), including culture, for site-model elements are provided to S-CAT as part of the site-model specification. Figure 5 provides the Speed profile (the answers to the Speed template) for Pa Yangi-Yangi.

---

[2] This template is named after Clarke Speed, its principal author.

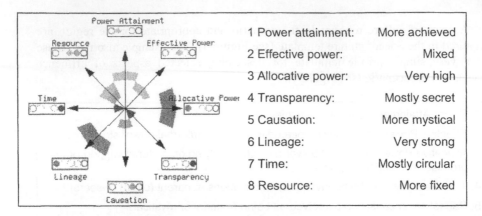

1 Power attainment:	More achieved
2 Effective power:	Mixed
3 Allocative power:	Very high
4 Transparency:	Mostly secret
5 Causation:	More mystical
6 Lineage:	Very strong
7 Time:	Mostly circular
8 Resource:	More fixed

FIGURE 5: Speed profile for Pa Yangi-Yangi

Answers to template questions are multiple choices of discrete scalar values: each candidate answer corresponds to a discrete point or interval on a continuous scale. The answers to the questions of effects templates capture change. For example, the candidate answer set for question 3.1 (and several other questions) of Figure 2 includes *"Significant increase," "Moderate increase," "Little change," "Moderate decrease"* and *"Significant decrease."* The answers to the questions of profile templates capture descriptions at a given time. For example, the candidate answer set for question 3.1 of Figure 4 includes *"Mostly modern," "More modern," "Mixed," "More primitive,"* and *"Mostly primitive."* Answers can be either single valued or intervals (e.g., *"Little change to significant decrease"*). Each answer to a primitive template question is annotated with text that explains the given value. For example, the explanation of the answer *"Very strong"* to question 6 (*"Lineage: is preserving family and traditions a strong or weak priority?"*) in the Speed profile for Pa Yangi-Yangi (Figure 5) is *"Pa Yangi-Yangi's greatest desire is the long-term preservation of his family, his people, and his traditions on the land he has known all his life."*

In S-CAT, structured arguments are used in several ways. Profiles are used in two ways. First, profiles are provided for the elements in the site model. Second, profiles are used as constraints in the antecedents of effects rules. For example, the antecedent of an effects rule can constrain the values of a culture variable so that the rule applies only to site elements having select cultural properties. This enables effects rules to capture plausible, culture-specific consequences of actions. Effects arguments are used in similar ways. First, effects arguments capture forecast effects: they are used in the consequences of effects rules to capture the plausible effects that each rule infers, and they are used to forecast effects for a candidate COA reported to the user. Second, effects arguments are used as constraints in the antecedents of indirect-effects rules in order to capture the propagation of effects from one entity or activity to another. The next section provides example uses of profile and effects arguments as S-CAT forecasts plausible effects of the second COA in the example scenario.

FORECASTING PLAUSIBLE EFFECTS

S-CAT infers the plausible effects of a candidate COA by first computing the direct effects using simple direct-effects rules that infer values of effect variables for those elements in the site model directly impacted by the COA actions. For the second COA in the example scenario, gifting modern fishing equipment and agricultural equipment to the Kita Island fishing and rice-farming operations has the direct effect of improving the equipment used by these two enterprises. This inference is made by the direct-effects rule: *"if new equipment is provided to a technologically primitive enterprise, then a plausible direct effect is that the equipment available to the enterprise improves."* This direct-effects rule requires that the given enterprise is technologically primitive (i.e., gifting equipment to a modern company may save a few expenses but probably would not make a fundamental improvement). This condition is captured with a constraint on the affluence profile for the enterprise that requires question 3.1 of Figure 4 to be *"More primitive"* or *"Mostly primitive"* (or the interval that includes both values). Because *"Kita fishing"* and *"Kita rice farming"* both have *"More primitive"* for this question, the constraint is satisfied and the rule applies to each of them and infers as direct effects that their access to equipment (question 5.1 in Figure 2) improves.

Having identified the direct effects for a candidate COA, S-CAT next identifies indirect effects that plausibly result from the COA actions. This involves applying all the indirect-effects rules whose antecedents are satisfied after establishing the direct effects. Indirect-effects rules capture how effects for one site element propagate to related elements, or how values for one effect variable determine values of other effect variables for a given site element. For example, a plausible consequence of improved equipment for an economic enterprise is improved wealth creation by that enterprise; therefore S-CAT infers that the Kita fishing and rice-farming operations provide more accessible wealth (e.g., food). Having established the initial, or *first order,* indirect effects, S-CAT iteratively identifies additional (*second order, third order,* etc.) indirect effects. The indirect-effects rules applied in each round have effect constraints that became satisfied by the inferences made during the immediately prior round; each round that establishes new effects may enable new applications of indirect-effects rules in the following iteration. This enables indirect-effects rules to chain, revealing how the initial direct effects of an action propagate among elements of the site model. The user initiates each iteration and vets the resulting effects forecasts.

In the example, the additional wealth provided by improved fishing and rice farming is controlled by the leader of the local fishermen and farmers. This inference is culturally specific: it is not presumed to be the case that in all circumstances the entire collective proceeds from an enterprise are controlled personally by the leader of those performing the enterprise. The people in this region generally adopt a social contract between informal leaders (e.g., Pa Yangi-Yangi) and their followers (e.g., Kita fishermen and farmers) where the leader controls all labor and proceeds but in return assumes the obligation of providing material support to the laborers and their families. This *"allocative power"* relationship between leaders and their followers is captured as question 3 of the Speed template (Figure 3).

The rules for computing indirect effects are typically more complicated than those for direct effects; a set of relations over typed, site-element variables expresses part of each rule's antecedent. Indirect-effects rules can have profile constraints in their antecedents, similar to the affluence-profile requirement for the direct-effects rule described above. Two indirect-effects rules constrained by *"allocative-power"* and one constrained by *"lineage"* (question 6 of Figure 3), are described in Figure 6. These rules establish that Pa Yangi-Yangi has improving wealth, improving political strength, and decreasing support for piracy. Because Pa Yangi-Yangi controls the local labor force, his loss of support for piracy causes a decrease of piracy in the Kita area, and the intended impact of the COA is confirmed and explained.

Rule 1:   **IF** an enterprise has improving wealth creation
                 AND its performers embrace allocative power
        THEN the performers' leader has improving wealth
    BECAUSE allocative leaders control labor and enterprise proceeds

Rule 2:   **IF** an allocative leader has improving wealth
        THEN the leader's followers have moderately improving wealth
                 AND the leader's wealth declines moderately
                 AND the leader's political strength improves
    BECAUSE allocative leaders invest in their power by supporting followers

Rule 3:   **IF** a leader is strongly motivated by lineage preservation
                 AND the leader controls both safe and dangerous enterprises
                 AND the safe enterprise has increasing wealth creation
        THEN there is reduced support for the dangerous activity by the leader
    BECAUSE the leader doesn't want to risk losing productive followers

---

FIGURE 6: Example culture-specific indirect-effects rules

Figure 7 presents part of the structured-argument display of the plausible effects forecast by S-CAT for Pa Yangi-Yangi after four rounds of computing indirect effects. Similar forecast displays are created for other elements of the site model impacted by the COA. As illustrated in Figure 7, a traffic-light metaphor is used to succinctly present the forecast values of effects variables *("Significant increase"* is green, *"Moderate increase"* is yellow-green, *"Little change"* is yellow, *"Moderate decrease"* is orange, and *"Significant decrease"* is red). Each forecast value is explained and sibling values are aggregated into values for the parent variables. Question 3.1 of Figure 7 provides an example of aggregating multiple values: the two independent forecast values of increase (i.e., the interval including both *"Significant increase"* and *"Moderate increase"* values) and two forecast values *"Moderate decrease"* yield the aggregate value *"Moderate increase."*

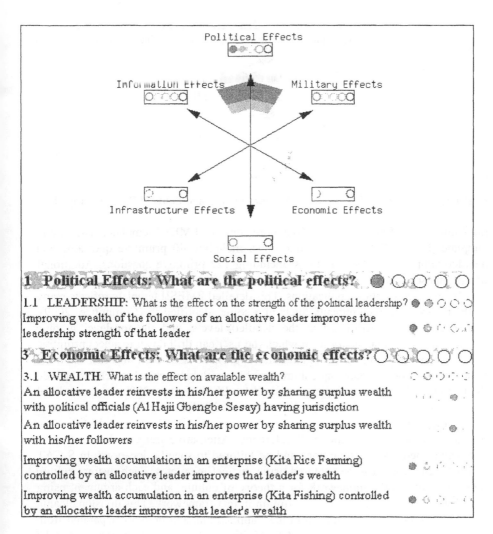

FIGURE 7: Forecast effects argument for Pa Yangi-Yangi

BENEFITS OF STRUCTURED ARGUMENTATION

The example scenario illustrates the use of structured argumentation in S-CAT as it forecasts plausible effects of a candidate COA. Some benefits of using templates to capture culture and effects variables and using structured arguments that instantiate the templates to reason about their values include:

Structure: Templates are organized hierarchically with top-level questions decomposed into more specific questions. Figure 8 contrasts the graphical structures of only the economic-effects argument forecast by a direct-effects rule using either

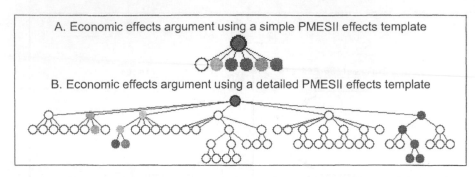

FIGURE 8: Economic effects arguments from simple and detailed PMESII templates

(a) a simple PMESII template, or (b) a more complex PMESII template. The simple template (Figure 2) has 46 total questions, including 40 primitive questions; the detailed template has 180 questions, including 122 primitive questions. Additional detail imposes a greater modeling burden on the site model and effects rules, but it may be required by some applications. This detailed PMESII template reflects many PMESII modeling requirements, including some identified in (Committee, 2009). Reasoning can be performed at the detailed levels, and the top-level graphic provides a compact summary from which the user can drill down to select primitive questions to review the explanations and aggregations of their forecast answers. Other approaches to organizing culture variables with hierarchical structures include the CONTACT taxonomy (Salazar et al, 2010; Shuffler et al, 2009) and multilevel cultural models (Eriz and Gati, 2004; House et al, 2004).

Multiple frameworks: The *framework* of a cultural theory identifies the essential variables that theory will reference. Alternative frameworks can reflect competing theoretical assumptions and be tailored to specific locations. In S-CAT distinct frameworks can be captured using distinct templates. Each distinct culture template captures the framework for a theory of culture, and each argument using the template provides a cultural description in that framework. This division cleanly separates theoretic frameworks from their application, and both are separated from the rest of the knowledge in S-CAT. Multiple, alternative frameworks can be captured: S-CAT currently includes culture templates for both Speed (Figure 3) and Hofstede (Hofstede, 2001; Hofstede and Hofstede, 2004) frameworks; others could be added. Multiple cultural profiles can be provided to describe site-model elements or to constrain effects rules. S-CAT does not commit to a single culture framework: it accommodates alternatives that can be used selectively, collectively or individually.

Aggregating effects: As illustrated in question 3.1 of Figure 7, multiple effects can be forecast for a primitive question for a particular site-model element. The forecasts can come from different effects rules within a reasoning engine or from different reasoning engines. The rule or reasoner supporting each specific forecast effect provides evidence (e.g., the explanatory text) for that effect. The structured argumentation techniques enable maintaining the evidence supporting each effect inferred for a question while producing an aggregate effect answer for the question.

Explaining and vetting: Each effects rule includes text that explains the effects it forecasts. All effects forecast for a candidate plan are explained to the user by the rules used in the forecast. Furthermore, users can review and modify forecast effects; for example, when users know that in the given circumstances a forecast effect is not possible, they can edit the forecast to reflect that impossibility. S-CAT is intended to complement human analysis, not replace it.

ACKNOWLEDGEMENTS

This work has benefited from contributions by Ian Harrison, Janet Murdoch, Eric Yeh, Robert Tynes and David Anhalt, and it has been supported in part through a contract with the Combating Terrorism Technical Support Office with funding from the Department of Defense, Office of the Director for Defense Research and Engineering, Human Social Culture Behavior Modeling Program.

REFERENCES

Committee (2009). *Requirements for a Government Owned DIME/PMESII Model Suit,* Office of the Secretary of Defense Modeling & Simulation Steering Committee.

Erez, M. and Gati, E. (2004). A Dynamic, Multi-Level Model of Culture: From the Micro Level of the Individual to the Macro Level of a Global Culture. *Applied Psychology: An International Review, 53*(4), pp. 583-598.

Hearing (2004). *Army Transformation Hearing before the Committee on Armed Services,* United States House of Representatives, Second Session, Washington, D.C. July 15-24.

Hofstede, G. (2001). *Culture's Consequences: Comparing Values, Behaviors, Institutions and Organizations Across Nations,* 2nd Edition. Thousand Oaks, CA: Sage Publications.

Hofstede, G. and Hofstede, G.J. (2004). *Cultures and Organizations: Software of the Mind,* Revised 2nd Edition. New York: McGraw-Hill.

House R. J., P. J. Hanges, M. Javidan, P. W. Dorfman and V. Gupta. Eds (2004). *Culture, Leadership, and Organizations - The GLOBE study of 62 societies.* Thousand Oaks, CA: Sage Publications.

Lowrance, J., Harrison, I., Rodriguez, A., Yeh, E., Boyce, T., Murdock, J. and Murray, K. (2008). Template-Based Structured Argumentation, in *Knowledge Cartography: Software Tools and Mapping Techniques,* Springer.

Ng, K.Y., Ramaya, R., Teo, T.M.S. and Wong, S.F. (2005) Cultural Intelligence: Its Potential for Military Leadership Development, in *Proceedings of the 47th International Military Testing Association Conference (IMTA),* Singapore.

Salazar, M., Shuffler, M., Salas, E. (2010) Conceptualizing our Nations Together: A Cultural Taxonomy (CONTACT). (UCF, Institute for Simulation & Training Working Paper).

Sharpe, K., Gremban, K., and Holloman, K. (2007), Evaluating the Impact of Culture on Planning and Executing Multinational Joint Force Stability, Security, Transition and Reconstruction Operations, in *Proceedings of the 4th International Conference on Knowledge Systems for Coalition Operations (KSCO),* Waltham, MA.

Shuffler, M., Salazar, M., Bedwell, W., Salas, E. and Burke, C.S. (2009). Culture on the Front Line: Dimensions that Matter, in *Proceedings of the Interservice/Industry Training, Simulation & Education Conference (I/ITSEC),* Orlando, FL.

CHAPTER 38

Evaluating Human, Social, Cultural, and Behavioral (HSCB) Models for Operational Use

Anne V. Russell[1], Mark A. Clark[1], Richard W. La Valley[1], William C. Hardy[1], I. William Zartman[2]

[1]Science Applications International Corporation
4001 N. Fairfax Drive, Suite 725
Arlington, VA 22203

[2]School of Advanced International Studies
Johns Hopkins University
1740 Massachusetts Avenue, NW
Washington, DC 20036

ABSTRACT

While interest in using human, social, cultural, and behavioral (HSCB) models in operational, field, and business environments grows, the challenge of developing metrics that can verify, validate, scale and replicate the complexity of human driven events on computational platforms continues to confound. For HSCB models to make the jump from academic exercises into operational decision-making processes, metrics that can help decision-makers determine which model to trust and which results matter are essential. In this paper, the authors pull from traditional social science and mathematical approaches to provide an overview of the authors' approach for verifying and validating HSCB models for decision-making. Using a non-computational HSCB model of conflict theory, the authors walk through each stage of the evaluation process to identify verification and validation methodologies from the hard and soft science that can be applied to measure HSCB models on a consistent basis.

Keywords: HSCD model evaluation, computational social science, human factors, human terrain modeling, intelligence analysis, forecasting events, face validity, content validity, construct validity, criterion validity, net utility assessment

INTRODUCTION

Operational decision-makers show increasing interest in formalized human, social, cultural, and behavioral (HSCB) models and the potential they hold for use in their business process. Still lacking, however is a means to effectively measure HSCB model outputs such that operational users can determine a model's relevance to their mission. Demand for metrics that scale and account for the complexity and uncertainty in human events and comprehensively characterize the performance of the models is growing (Klein 2009). Government agencies, academics, and operational end users have spent years attempting to determine the metrics but have been confounded by their inability to apply common measures of performance and effectiveness. Yet modelers are still challenged to build and implement HSCB models in the absence of meaningful metrics. Part of the issue lays in the embryonic nature of the HSCB sciences. Still in its infancy, HSCB modelers are in the process of agreeing on basic definitions, scales of importance, and verifiable and repeatable processes. Part of the challenge also lies in the complexity and reality of modeling HSCB events. Many events are sparse and even when they are not, the data necessary to assess and anticipate those events can be. The HSCB field has minimal scientific laws and truths. Given the state of the art and the operational focus the models aim to serve, traditional scientific modeling and simulation evaluation techniques alone cannot provide sufficient metrics for evaluating disparate HSCB models.

By their nature, HSCB models must capture and replicate how experts in varying disciplines postulate and prove that events occurred, their significances, and the causal change that led to the event. Ultimately, validation of HSCB models should depend on how they approximate the drivers and events in the real world (Hanson 2009). However, with such disagreement over basic definition of events, HSCB modelers should set a threshold requiring HSCB models to approximate the end user's decision-making process (Avenhaus 2009). As such, HSCB modeling experts and model consumers need metrics that allow them to measure the utility of the model output along with the correctness of the social science theory the model is based upon.

HSCB EVALUATION IN THE CURRENT CONTEXT

In the void of comprehensive HSCB-specific metrics, four broad categories of models have tended to dominate. The first consist of models that are presented with no means of independently evaluating outputs. Models within this category have been dominated by the social and behavioral sciences. Within the confines of traditional social science methods for validation, the premises of these models may be considered sound. Take for instance the majority of political science-based forecasting models (Armstrong 2005). In these, evidence presented as "common sense" or "reasoned wisdom" is an acceptable measure of the usefulness of research methods common to the social sciences, such as empirical observations and comparative case studies, among others. Given the scientific method that the HSCB sciences seek to apply, however, hypothetical theories need the proof of hard, undeniable evidence. The confusion over whether social research, scientific, or a combination of methods apply in the HSCB sciences has led to several cases where modelers suggested that evaluators trust and accept their modeling results without scientifically measurable evidence. In some cases, it has also led HSCB modelers to confuse intellectual property with evaluation methods and to refuse to reveal either. With no means to evaluate the theoretical HSCB foundation of these models or their outputs, decision-makers have no process to determine whether an approach should receive further consideration (Frank 2005, Popp 2006, Shakarian 2008).

A second category of current HSCB models are those capable of providing scientifically measurable results produced in the lab with artificially produced input data. When tested independently with different data, the models did not perform as well. In part, this is a result of the relative rarity of distinct HSCB phenomena and the paucity of the accompanying availability of data to train, test and validate results. For example, there are less than 3,500 instances of armed conflict worldwide from 1946 to 2006 (Russell 2009). Each of the instances occurred over various geographic and cultural regions and under so many varying causal conditions that the numbers of truly similar events are in the single digits (Zartman 2005). Some data can be found to populate models of conflict events, but the data rarely exist in the quantity or form that formal modelers expect. In one example, a modeler built a model to measure the impact of population density, flows, and distribution by culture group on state stability in several countries. But after six months of searching, the modeler could not find census data on the regional, city-, or neighborhood-level demographics for those countries. Like many HSCB modelers, the best he could find was annualized data at the nation-state level, none of which provided evidence that could populate his model (Popp 2006).

A third group emerges from data-driven, statistical approaches where the modeler empirically derives the HSCB model from patterns identified in the data (Zachariah 2008). These inductive approaches can be insightful but are prone to spurious estimates of correlation and causality. While they more easily produce verified results through common metrics such as accuracy, precision and recall (APR), in the absence of a true theory, the modelers are hard pressed to explain the HSCB relevance of their results (Armstrong 2005).

The fourth and largest category consists of modeling and simulation approaches. This category encompasses the majority of computational social science attempts to model HSCB events to date. Running the gamut of various agent-based approaches that seek to use system dynamics, social network analysis, probabilistic reasoning, and game theory, initial experiments have shown the models' potential to enable better decision-making in the future (Zachariah 2008).

However, such approaches require large amounts of data and esoteric expertise in cleaning and segmenting the input data to fit the specific approach. Additionally, the results can be hard to explain and even harder to trace to dynamics in the input data. These methods have shown particular promise for implementing HSCB models where data is abundant and the societal norms underlying conditions are stable. However, the models fail to provide the adaptability and learning they have aimed to satisfy.

A WAY FORWARD

Each of the above classes of models shows promise in being able to capture specific aspects of HSCB modeling problem sets. Expert-based HSCB models show promise of formalization into computational models if done in such a way that enables independent validation. Data-driven approaches show the promise of capturing evidence. And formalized modeling processes such as those exemplified in agent-based approaches can show the dynamics behind HSCB events. Yet none of these approaches in and of themselves are sufficient to resolve the types of issues that HSCB modelers have encountered in their attempts to show the relevance of HSCB sciences to broader communities. What are lacking are objective performance and effectiveness measures that identify and optimize models' relative strengths and weakness in the context of the operational mission to which it will be applied.

Without standardized measures, it is impossible to contrast, compare, combine or otherwise judge which model or models work best and under what temporal, spatial, or other conditions. However, modelers must be careful about what can and cannot be validated at this juncture in the HSCB sciences. Take for instance Zartman's model on protest and revolt. In his work, Zartman posits that all cycles of protest and revolt are a reflection of societal grievances and go through a series of phases (Zartman 1995). Phase 1 is *petition,* in which various expressions of grievance are addressed to the government in an exercise of "normal politics." In Phase 2, *consolidation,* entrepreneurs work to consolidate the movement under a single leadership, bringing together the multiple expressions of grievance from the previous phase into a single force. Phase 3 marks the movement of protest into *confrontation* mode, where the outcome can take the form of: a) a victory of one side or the other; b) a mutually-hurting stalemate; c) a soft, stable self-service stalemate; or d) something idiosyncratic. Zartman's work clarifies how to identify and distinguish between protest phases and the forms it can take. Even on paper, it provides insight into how it can be constructed onto a computational platform to capture the trends and conditions of protest (Arnson and Zartman 2005).

Within Zartman's model, capturing and presenting data on demonstrations as one form of protest can be used as one of a number of measures to calculate the presence or potential of different phases of the model. But, it is highly unlikely that the model will ever be able to predict if a demonstration will happen exactly on a specific date for a specific reason with specific numbers and that demonstrators will take a specific path. It can, however, if fully formalized and automated, alert to indicators of impending phases and events, allowing for the constant possibility of free human choice. To date, the kind of scientifically measurable precision that predicts to the former level of specificity has not been achieved prospectively. Experiments on measuring precision retrospectively have shown promise but only under highly controlled conditions that do not match the real world (Shakarian 2008, Zacharias 2008).

The constant of uncertainty that typifies HSCB modeling and the embryonic nature of the science suggest that at this juncture, the goals for measuring HSCB models should focus on ensuring their operational relevance. That means that the model outputs should aid the understanding and consideration of HSCB events and fit into the operational workflow. In short, the authors propose metrics that capture the net utility of a model. Such metric should measure whether the model is internally consistent when applied to real data in the operation, has explainable outputs that match the operational mission, produces results with the spatial and geographical granularity required, and provide model outputs that can be used to formulate action or responses.

ROADMAP FOR VALIDATING HSCB MODELS

In the following section, the authors outline principles for formalizing and implementing HSCB models that enable measuring net utility. Over the course of the last decade, the authors have been involved in the majority of government-funded programs geared towards growing the HSCB sciences into its current embryonic state, including the Defense Advanced Research Projects Agency's (DARPA) Pre-Conflict Management Tools (PCMT), Pre-Conflict Assessment and Shaping (PCAS), and Integrated Crisis Early Warning System (ICEWS) programs. These principles are compilations of experiences and observations of what tends to enable clearer methods for validating models. The authors believe these principles provide a roadmap for formalizing and implementing HSCB models that will ensure the application of scientific methods to the HSCB modeling art irrespective of the model's intended application.

FORMALIZING HSCB THEORIES INTO COMPUTATIONAL MODELS

In order to overcome the paucity of hard scientific evidence inherent in expert-based paper models, the authors have identified four validation criteria to ensure that their computational formulations produce automated results that reflect the HSCB model's internal consistency. First, models must meet face validity criteria and determine the level of accuracy with which the computational model captures events related to the modeling hypotheses. For example, in Marshall's stability model, violence is a result of heightened levels of group factionalism within a society, not a measure of it. Concurrently, Zartman's protest and revolt model could use factionalism as one of many factors that can lead to violent conflict. Concepts overlap between the two models, a very common phenomenon in HSCB modeling where one model may feed into another model's higher-level aggregation scheme. However, each model assesses indicators and produces outputs that are consistent with the modeler's respective theoretical constructs, accompanying definitions and terms, and the types of evidence needed to separately determine whether protest and revolt and/or factionalism is in evidence. Using face validity as a key measure also enables modelers to build ontologies related to specific concepts expressed within their models. Since the HSCB expert defines the terms, they can in turn be used to build bottom-up, domain-independent libraries of synonymous terms linked to each other by their use of different models. This kind of construct is particularly important in the HSCB sciences where perception of events and competing definitions of very like events hinder top-down ontological development (Dyer 2005).

Second, modelers must ensure that the formalized model is content sufficient such that the model's implementation correlates with the domain expert's intuitive modeling and/or judgment process. At a minimum, the model must ensure the appropriate number of indicators or nodes and aggregation schemes represented by the modeler's hypothesis. For example, in Zartman's protest model, the Level of Grievance indicator should at minimum be able to capture different forms of grievance from basic needs such as food, water, shelter, and medical care. As a weighted aggregate reflecting a cultural norm, the model must also be able to determine whether the accumulated evidence meets the threshold of grievance as Phase 1 "normal politics", Phase 2 "consolidation", or Phase 3 "confrontation". For instance, are there certain indicators or nodes that are more or less important than others? Are there indicator aggregations that establish patterns of greater importance to the overall judgment call? Are there interdependencies between indicators and atomic elements that will change aggregated indicator levels? If Zartman's model posits that mass and multiple demonstrations in France constitute "normal politics", does the formalized, implemented model produce results to capture that evidence?

380

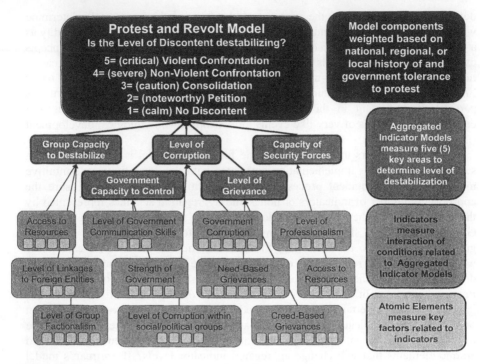

FIGURE 1. A hierarchical view of Zartman's Protest and Revolt Model that is used to automatically ingest and analyze HSCB events in accordance with Zartman's theoretical principles. By identifying interdependencies between and across indicators and atomic elements, the same model can be displayed as trend lines or as a system dynamics model to show the ebbs and flows of events over time, or mapped to geo-spatial context. Patterns emerging from trends also fuel agent-based, probabilistic reasoning, and/or Bayesian or HMM forecasting engines, among others.

Third, models must have construct validity such that assessments of the measures of an event aggregate consistently and logically. If the model's assessment of events of interest changes without corresponding movement in the model's measures, the model does not have construct validity. As such, a model's construct validity is essential in helping users understand how the dynamics in the measures affects the event of interest.

Finally, models must have some degree of criterion validity. The model outputs must have some consistency with the observed dynamics in the society. For example, a model that repeatedly forecasts societal unrest when societal conditions remain calm has little criterion validity. It may take some time for all to agree on basic definitions such that it is reasonable to mark a model's forecast of a riot wrong when the event was classified as a massive march. However, for a model to meet criterion validity, its assessments and forecast must have a significant correlation to manifestations in a dynamic populace.

These four criteria enable models to achieve a minimum standard for formalizing, instantiating and testing their models on automated platforms. The four criteria can further help to ensure confidence that models are reliable and credible, enabling the measurement of their outputs against ground truth data and against measures of net utility.

MEASURING OPERATIONAL USEFULNESS

Given the real world environment in which these online models and judgments will be used, the authors also propose measuring a model's net utility to ensure various HSCB modeling implementations are of operational value to end users. To measure net utility, implemented models should at minimum have: 1) the capacity to be decomposed such that their underlying structure can be evaluated for face, content, criteria and construct validity; 2) understandable and documented data sensitivity measures so that confidence measures can be computed; 3) clear, simple, and realistic data requirements that ensure the model can ingest available real world data; 4) reliably observable indices that provide rapid comprehension of events of interest when they occur; 5) spatially granular analytic capacity that retains fidelity when applied to different geographic or administrative levels; 6) timely analysis capacity that can synch with the mission/stakeholder's information ingest cycle; and 7) analytic results which fit with mission/stakeholder areas of interest. Meeting criteria such as these further enables the decision-maker to determine whether or not a particular model is operationally useful.

The principal measure the authors would propose for HSCB models is a measure of utility with regards to the mission of the decision-maker or stakeholder. Utility can be thought of as the probability that the process will result in *timely* actions for which the stakeholder was comfortable with the decision at the time it was needed. The classical metrics for any information provided to decision-makers or stakeholders in support of action are: timeliness, completeness, accuracy, relevancy and presentation. These notions are attributes of what is provided to decision-makers, and expected to be commonly and readily understood within the context of the problem the model is addressing.

Table 1. Proposed Metrics and User Perception for HSCB Models

HSCB Metric	User Perception Shaped by:
Timeliness	**Timeliness**
Probability that the HSCB model outputs will be generated with acceptable completeness, accuracy, relevancy quickly enough to satisfy decision-makers' needs.	Ability to obtain and use HSCB model outputs soon enough to arrive at a comfortable decision before any feasibility action is pre-empted
Completeness	**Actionability**
The model produces outputs that are based on all relevant data and information readily available to the decision-maker.	Confidence to act on estimates and inferences without excessive risk of erroneous evaluation of alternatives.

Accuracy		Cogency	
Model output accurately reflects model's construct, factual content, representation of uncertainty and/or durability, and the integrity of production.		Ability to readily understand and intuitively test model's underlying rationale, inferencing methodologies, and inferences.	
Relevancy		Information Content	
Model provides definitive and actionable output directly relevant to its internal construct and the decision-maker's mission.		Ability to use the model's outputs without further analysis or interpretation.	
Presentation		Information Form	
Model outputs presented in form decision-maker finds most useful to mission.		Ability to use the model's outputs without further processing to facilitate its assimilation into current business processes.	

As illustrated in Table 1, the extent to which these requirements will be met will depend on how well an HSCB model has recognized and addressed five basic concerns of the decision-makers who would use them:

o **Completeness:** Is the information from the model based on consideration of all available data that is likely to affect the conclusion?

o **Accuracy:** Are the data from which the information has been derived reliable? Have the assumptions and uncertainties attendant the conclusion been clearly recognized? Has the answer been derived in a way that would assure: (a) the accuracy of the data on which it is based; and (b) the validity of the reasoning applied in interpreting that data?

o **Relevance:** Does the report contain the information sufficient to represent the basis for a direct and definitive answer to the question(s) that stimulated its production?

o **Accessibility:** How readily can I interpret the information in the form which it has been received to achieve the desired reduction in the uncertainty that stimulated the need for the information?

o **Credibility:** How readily can I understand and endorse the methodology by which the information was produced? How readily can I convince myself that the analytical methodology used to generate the answer was sound?

o **Timeliness:** Will I receive an answer to my question in time to make use of it? Can the answer as received be used or made usable in time to support the action selection problem that generated the need for it? Are the accuracy and completeness that can be achieved with time constraints adequate for my comfort?

With these ideas in mind, an overall measure of model utility with respect to the decision-makers mission can be developed. It should have various component parts which should be different measures of effectiveness as described in the five basic concerns above and each could have multiple measures of performance.

THE CHALLENGES OF VALIDATION - FUTURE AREAS FOR GROWTH

In the current state of the art, each HSCB modeling project must develop its own set of ground truth data against which to validate findings. HSCB ground truth is neither readily available nor universally agreed upon. Without a standardized ground truth dataset in the spatial and temporal granularity necessary, modelers and model customers can not use traditional verification and validation methods. With the need to include HSCB assessments in operations in a more scalable manner beyond a core group of SMEs, using metrics of net utility can assist the modeler to improve and measure their models and the consumer to evaluate and select the appropriate HSCB models for their use. The use of net utility metrics also enables a foundation of common standards from which to compare and contrast HSCB models. Such standards will help to ensure that consistent scientific methods are applied and enable the development of more robust measures that capture the full capacity of HSCB models in computational environment, their application to decision-makers, and their future capacity to incorporate and measure the effect of emerging advanced computational techniques on improving performance.

REFERENCES

Arnson, C. and Zartman, I.W. eds, (2005) *Rethinking the Economics of War: The Intersection of Need, Creed and Greed.* Woodrow Wilson Center and Johns Hopkins Presses.

Armstrong, J.S., ed., (2005), *Principles of Forecasting*, Norwell, MA: Kluwer.

Avenhaus, R. and Zartman, I W (2009). *Diplomacy Games: Formal* Models and International Negotiation. New York: Springer.

Dyer, D., Cross, S., Knoblock, C., Minton, S., and Tate, A. (2005) "Planning with Templates," *Institute for Electrical and Electronic Engineers (IEEE)*

Frank, A. (2005) "Pre-Conflict Management Tools: Winning the Peace," *Center for Technology and National Security Policy, National Defense University.*

Hanson, L. and Russell, A. (2009) "The Challenge of the Black Box: Techno-social Predictive Analytics in the Real World." *Association for the Advancement of Artificial Intelligence (AAAI), Spring Symposium.*

Klein, G. (2009) "Summary Evaluation Proceedings," *HSCB Focus2010 Conference.* Office of the Secretary of Defense.

Popp, R. Kaisler, S.H. Allen, D. Cioffi-Revilla, C. Carley, K.M. Azam, M. Russell, A. Choucri, N. and Kugler, J. (2006) "Assessing Nation State Instability and Failure" *Aerospace Conference, Institute for Electrical and Electronic Engineers (IEEE)*, 10.1109/AERO.2006.1656054.

Russell, A. and Clark, M. (2009) "Modeling Human, Social, Cultural or Behavioral Events for Real World Application: Results and Implications from the State Stability Project.*" International Conference on Computational Science and*

Engineering, Institute for Electrical and Electronic Engineers (IEEE), cse, vol. 4, 683-690.

Shakarian, P. "The Future of Analytic Tools: Prediction in a Counterinsurgency Fight." (2008) *Military Intelligence Professional Bulletin (MIPB)*

Zacharias, G., MacMillan, J., and Van Hemel, S., (2008) *Behavioral Modeling and Simulation: From Individuals to Societies.* National Academies Press.

Zartman, I.W. ed, (1995) *Elusive Peace: Negotiating to End Civil Wars* Brookings 1995

Zartman, I.W. (2005). "Comparative Case Studies," *International Negotiation*, Vol. 10, 1:3-15

Chapter 39

Cross-Cultural Decision Making Training Using Behavioral Game-Theoretic Framework

Azad M. Madni, Assad Moini, Carla C. Madni

Intelligent Systems Technology, Inc.
12122 Victoria Avenue
Los Angeles, California 90066

ABSTRACT

The recent surge in interest in cross-cultural decision making training stems primarily from the need in the commercial world to improve global business relations and the need for military commanders to conduct non-kinetic operations in theaters where non-western cultures predominate. Existing approaches to cross-cultural decision making training are grounded in multi-agent simulations and classical game theory. The former tend to be ad hoc while the latter takes a limited view of strategic decision making. Empirical research in strategic economic games has clearly shown that humans respond to more than just monetary incentives and material gains. Furthermore, research has shown that cultural norms play a key role in human decision making behavior. This paper presents an innovative and generalizable game-based simulation approach for cross-cultural decision making training. The approach combines findings from behavioral game theory research with classical game theory and agent-based modeling. An illustrative game for cross-cultural decision making training is presented.

Keywords: Cross-cultural decision making, game theory, behavioral game theory, game-based simulation, non-kinetic operations

INTRODUCTION

Cross-cultural decision making training is a multi-faceted problem in which the decision making behavior of individuals is tempered by the culture they live in and the implicit need to conform to the norms of that culture. Cross-cultural training has begun to take center stage over the past decade with international businesses beginning to tap into a variety of "intercultural training programs" for their employees. The value proposition of cross-cultural decision making training in the commercial world is achieving harmony and warding off cultural shock in cross-cultural business relations in general and negotiations in particular. Similarly, the value proposition of cross-cultural decision making training in the military is being able to conduct effective negotiations with indigenous populations when engaged in non-kinetic operations such as civil reconstruction and humanitarian assistance. Cross-cultural training seeks to make people aware of socio-cultural factors that are key to effective communications in a wide variety of cultural contexts. Cross-cultural training also seeks to instill in people the right motivations needed for strategic negotiations.

Existing approaches to cross-cultural decision making training are grounded in one of two approaches: *multi-agent simulations* and *classical game theory*. Multi-agent simulations tend to be ad hoc (i.e., lack theoretical underpinnings). While they have been successfully used in depicting representative behaviors of people from different cultures, they are inadequate when it comes to guiding individuals in cross-cultural decision making contexts. Classical game theory formulations, on the other hand, tend to take a limited view of strategic decision making in that they assume that humans are self-regarding maximizers of material gains. Empirical studies, especially over the last decade, have repeatedly shown that humans respond to more than just monetary gains in strategic contexts (Gintis, 2000, Henrich et al, 2005). In fact, humans are capable of showing altruism and reciprocity in negotiations and, based on their cultural roots, reflect their socio-cultural norms and decision making style when making decisions.

This paper presents key socio-cultural considerations that come into play when dealing with non-Western cultures, and presents an innovative game-based simulation approach based on behavioral game theory, classical game theory, and multi-agent simulation.

SOCIO-CULTURAL CONSIDERATIONS

Socio-cultural considerations can have a profound impact on decision making. This effect is especially stark when comparing Western cultures to non-Western cultures (Klein, 2008). Some of the more striking differences between Western cultures and non-Western cultures such as Iraq and Afghanistan are discussed below.

- **Socio-cultural Norms**: The perspectives and beliefs of families and tribes in non-Western cultures are rooted in their cultural norms. The emphasis on honor is part of their culture.
- **Change Propensity**: Where Westerner's feel comfortable with and sometimes embrace change, people from non-Western countries such as Afghanistan and Iraq do not. They tend to be influenced by fatalism and an abiding belief in the limits to human power.
- **Power Distance**: Power distance in Western societies tends to be small, unlike in Middle Eastern and Asian societies where it tends to be large. This disparity can lead to conflict and confusion during negotiations and the conduct of joint operations.
- **Thinking Style**: Thinking style is key to effective coordination. People from Middle Eastern and Asian cultures tend to think holistically, whereas those from Western cultures tend to think analytically. This difference influences the planning process and the plans that are created. These differences need to be taken into account when making judgments about how people from other cultures are likely to behave in a particular context.
- **Communication Style**: Westerners prefer and employ direct communication whereas people from Middle Eastern and Asian cultures use and respect non-verbal communications.
- **Reasoning Style**: Westerners tend to be at ease when engaging in hypothetical and abstract reasoning, whereas non-Westerners prefer concrete reasoning. These differences often lead to different choices when evaluating courses of action and can, in fact, produce disagreement when conducting joint operations.

PROMISE OF BEHAVIORAL GAME THEORY

Behavioral game theory (BGT) is an empirical science. Its findings are a direct result of empirical research. Behavioral game theorists employ economic games to conduct experiments. For example, Heinrich et al (2005) employ three different economic games as the basis for their economic experiments. BGT can be viewed as extending classical game theory with concepts from behavioral economics, sociology, and cognitive psychology. To appreciate this distinction, we first discuss classical game theory and its implications and then discuss the extensions embodied in BGT. Classical game theory is a mathematical approach to strategic decision making. It provides a language and framework for determining the strategic responses of people based on anticipating what they are likely to do and choosing moves based on expectations of what others (e.g., opponents) will do (Buena de Mesquita, 2009). Game theory makes a few rather simple assumptions about people that many view as problematic. Game theory assumes that people are *rational*, i.e., they try to do what they believe is in their best interests. Rationality does not mean that people have perfect foresight, or even good foresight. It does not mean exhaustively evaluating available options. Indeed, one might argue that attempting to consider everything one might be able to do is more than likely behaving irrationally (Buena de Mesquita, 2009), because the cost of continuing to search for the best option is eventually going to exceed the expected gain from the continued search! Clearly, when the cost exceeds the expected gain, it makes little sense to continue with the search. Thus, rationality simply means that people do what they believe are in their best interests. The latter implies that people have values and beliefs. *Values* are shaped by their families, communities, experiences, and their cultural environment. *Beliefs* pertain to how they perceive others would react to them and the types of people they perceive that others are. Thus, belief is based on not having certain knowledge about how others will respond in specific situations. Therefore, decision makers invariably face uncertainty in negotiations and consensus building activities. Finally, decision makers make choices under *constraints*. The decision making problem ultimately boils down to determining the optimal way to behave given one's values, beliefs, and constraints (Buena de Mesquita, 2009) and recognizing that the environment is competitive but with potential opportunities to cooperate.

BGT is concerned with how people actually behave based on experimentation with real economic situations and people's decision making behavior (Gintis, 2000, Henrich et al, 2005). While preserving game theory principles in its regime of applicability, BGT attempts to go beyond by fitting a model to actual observations, not narrow self-interests. The three key tenets of BGT are that: a) humans use only a *limited amount of backward induction* and purely logical reasoning because they do not have unlimited computation and reasoning capacity; b) humans exhibit *inequality aversion* in that they often deviate from maximizing purely monetary gains towards "fairness" and reciprocity, and other non-monetary considerations; and c) humans need to be *incentivized* to behave in desired ways.

In our ongoing research, we are combining the findings from BGT with findings from cross-cultural research to develop a game-based simulation for analyzing cross-cultural effects in decision making and for developing a cross-cultural decision making training game.

GAME-BASED SIMULATION BASED ON BGT

Our game-based simulation which exploits BGT findings is a multi-person, cross-cultural decision making game that allows players to learn how to negotiate, bargain and co-opt/neutralize various stakeholders in non-Western cultural settings. The game is intended to allow players to: understand how people from different cultures behave under different situations; learn what are appropriate forms of social interactions under different situations; and predict within a bounded behavior and risk envelope the likely behaviors of non-Western actors when pursuing non-kinetic operations such a civil reconstruction, humanitarian assistance, and other diplomatic missions. The key assumptions and characteristics of the behavioral game are presented in Table 1.

Table 1: Behavioral game characteristics

- Focus
    - multi-person, cross-cultural decision making training
- Assumptions
    - humans use a limited amount of backward induction
    - humans pursue more that purely monetary gains and often exhibit a sense of fairness and reciprocity
    - humans need to be incentivized to behave in desired ways
- Key Features
    - turn-based – later players have partial knowledge about action of previous players
    - increasing difficulty level (information gaps, time-stress)
    - player has incomplete/imperfect information about beliefs and payoff functions of other players – needs to learn through interaction
    - dynamic game trajectory (players' decisions, decision order, and prevailing uncertainty)
    - player learns to apply a variety of monetary and non-monetary incentives (including social capital) based on player's reading of different stakeholders
    - multiple contributors to game outcome – prevailing uncertainties, information not known or determined by actions of other players, decisions of other players, order of their decisions, differential weighting of decisions by players
    - terminating criteria – stakeholders co-opted/neutralized, or resources expended
- What Players Learn
    - how to characterize opponent (coercive, inflexible, cooperative, demanding, ...)
    - what strategy to pursue based on opponent characteristics, resources available, and environmental factors (population's view, political climate, NGO/USAID goals, ...)
    - how to "read" opponent's (i.e., stakeholders) responses
    - how to adapt strategy based on opponent/stakeholders' responses
    - how to modify strategy if time element/resource availability becomes a concern

ILLUSTRATIVE GAME AND TESTBED

We are applying our overall game-theoretic framework to the development of a cross-cultural decision making training game and experimentation testbed. The purpose of the game is to allow the player (e.g. unit commander in Afghanistan) to learn how to conduct effective negotiations and consensus building with individuals primarily from non-Western cultures. Achieving a high score in this game requires the learners to understand socio-cultural norms, beliefs and non-material motivations of people.

The purpose of the game-based simulation is two-fold: (a) allow social scientists and cognitive psychologists to experiment with different human, social, cultural, behavioral (HSCB) models; (b) train military planners and unit commanders in culturally appropriate behaviors when conducting non-kinetic operations such as civil reconstruction or disaster relief. Specifically, social scientists and cognitive psychologists can explore possible outcomes with different HSCB models and changes to their parameters. They can also assess the sensitivity and impact of monetary and non-monetary incentives (including social capital) on mission outcomes. In the training mode, military planners and commanders get to experience first-hand the challenges in negotiating and building consensus with individuals from non-Western societies and cultures. They learn how to effectively employ incentives (monetary, non-monetary, social capital) during negotiations and consensus building operations. They learn strategies that go into co-opting, placating, or neutralizing ambivalent and hostile actors who are key to mission accomplishment. The game-based simulation is a convenient tool to cost-effectively explore possible futures for a wide range of scenarios through game play. A snapshot of an illustrative game for a school reconstruction project in Kandahar province of Afghanistan is presented in Table 2.

Table 2: Key assumptions and features of behavioral game

- **Player Goal**: Achieve desired mission outcome by employing appropriate socio-culturally-aware strategies and incentives
- **Desired Mission Outcome**: Achieve consensus among stakeholders on school reconstruction project in Kandun with available resources
- **Stakeholders**: Commander, Head Cleric, Warlord, Village Elder, Friendly Citizen, Girls of school-going age
- **Resources**: Financial, food and supplies, trained personnel (security, infrastructure, reconstruction), social capital
- **Gameplay**: Employ socio-culturally appropriate strategies and incentives to achieve desired outcomes
- **Strategies**: Determine proper allocation of resources, and types and timing of messages to be delivered to various stakeholders
- **Messages**: Convey intent, objections, incentives, disincentives, ...
- **Scoring**: Based on time and cost to reach consensus and resources remaining

The gameplay that unfolds within the context of this scenario is as follows. The player assumes the role of military commander in the game. The player engages in simulated conversation with the non-player characters (NPCs) in the "story." The player faces negative, occasionally threatening, responses from NPCs. The player responds to these using free-form text input. The player draws on helpful NPCs for support. The player expends resources and employs a variety of incentives to co-opt/placate the various stakeholders while attempting to make steady progress towards the desired outcome. The overall score achieved by the player is a function of the number of stakeholders who are co-opted/placated, the correct characterization of the various stakeholders (e.g., in terms of their predisposition to negotiate), the correct use of incentives based on stakeholder characterization, the time and cost to reach consensus, and resources expended.

The instructional strategies used during and upon completion of gameplay are grounded in sound pedagogy. Exemplar instructional strategies are presented in Table 3.

Table 3: Exemplar instructional strategies

- **Demonstration** of appropriate behavior (by key NPCs in tutorial phase)
- **Simulated conversations** of increasing difficulty (i e , severity of objection) with player
- **Cultural cues** embedded in what an NPC says and how an NPC behaves
- **Coaching and prompting** during key interactions (NPC, voice of mentor)
- **Variable "stress interventions"** that increase tension ("stress inoculation")
- **After Action Review** at conclusion of the game (post hoc analysis, demo)

GAME IMPLEMENTATION

Our interactive agent-based simulation environment (Figure 1) employs a game-theoretic framework to represent and generate strategic behavior of NPCs (i.e., agents) responding to actions/decisions of other agents.

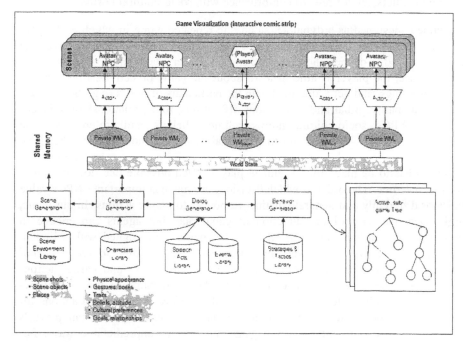

Figure 1. Conceptual architecture of game environment

Within this framework (Madni et al, 2008, Madni, 2009), self-interested agents form beliefs based on what they expect others might do (i.e., engage in strategic thinking) and choose their *best* response in light of those beliefs. They continue to adjust their respective strategies until they reach a level of mutual consistency between their beliefs, desires and intentions. When reaching consistency is not feasible, an agent may violate rationality and consistency constraints. This is in sharp contrast with the classical game theory assumption that all agents are rational.

Recognizing that strict adherence to rationality in classical game theory rarely holds true in complex, cross-cultural interactions, we have chosen a more realistic approach inspired by recent advances in BGT (Madni et al, 2008; Madni 2009) and tempered by socio-cultural considerations (Klein, 2008). Another characteristic of classical game theory is that it takes a fundamentally static view of agent interactions. This view disregards the evolving and adaptive nature of agent choices and preferences in strategic, adversarial interactions. Our game-based simulation framework does not impose this constraint. Rather, it offers a formal, predictive model of agent decision making by assuming that: (a) each agent maximizes a *social welfare utility function* by evaluating the consequences of its actions, plans ahead, and forms beliefs about other agents' likely actions; (b) motivated by non-material factors (e.g., altruism, reciprocity, sense of duty to God, tribe or country), an agent may willingly forsake personal gains and the pursuit of self-interest; (c) bound by a sense of duty and honor, individual agents may act to punish violations of religious edicts, and/or reward conformance with socio-cultural norms.

The game is designed as a dynamic, stochastic repeated game in which each agent has a repertoire of tactics and techniques. At the game outset, the human player has limited knowledge of the other agents (i.e., non-player character). Using the game user interface, the player is able to learn about the game objectives, rules, setting and surroundings, and get acquainted with the particulars of NPCs (e.g. their names, gender, nationality, tribal and/or religious affiliations, roles, relationships/alliances with other NPCs, socio-cultural norms, religious beliefs, and preferences). The human player begins by choosing to play against one or more NPCs. The game progresses in accord with a branching tree structure where at every branching point a decision has to be made by the human player. Each possible decision leads to a different outcome. The order in which decisions are made is important. NPCs value decisions differently and generate distinct responses to player actions. In our game, single agent can control the direction of game play but agents can engage in coalition-formation behavior. The NPCs in our game are designed with a limited capacity for *remembrance* in the sense that each NPC is able to retain *partial* knowledge of previous game states and the interactions with and earlier actions taken by the human player and other agents. The remembrance function allows NPCs to selectively collect and retain a record of *relevant* events and actions for future recall. Uncertainties, which represent factors outside the human player's control, are allowed to influence the game outcome in unpredictable ways. For instance, the game outcome can be partly based on information not yet known and not determined by other NPCs' actions. Each NPC has a set of initial beliefs about the world state (including that of the human player) and other NPCs. These beliefs are updated with world/game state changes (e.g., based on actions of NPCs). In our current approach, we employ a probabilistic approach to representing beliefs (i.e., a belief system assigns probabilities to information about events, actions or facts known to an NPC). We are currently exploring the use of Bayesian Belief Networks to model the actions of NPCs that drive the stochastic behavior of the branching game tree. Combining remembrance with a probabilistic belief system transforms the game of *incomplete information* to a game of *imperfect information* in which recent game history need not be available to all NPCs. In addition to its system of beliefs, each NPC has a *type* (described in terms of biology, traits, socio-cultural biases and preferences, goals, etc.) fully characterizing its social welfare utility function. Each NPC is aware of its own type and has limited knowledge, in a probabilistic sense, of the types of the other NPCs. Figure 2 presents a screenshot of the character authoring tool. The game user interface is implemented as an interactive comic book, consisting of a series of scenes. The scenes are automatically generated in layers (i.e., background, character location, assignment of character behavior).

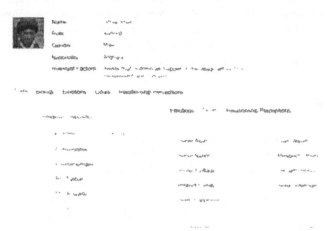

Figure 2. Character authoring tool user interface

Figure 3 provides a screenshot of the scene authoring tool. The dialogs and responses of NPCs are generated from speech-acts associated with each character.

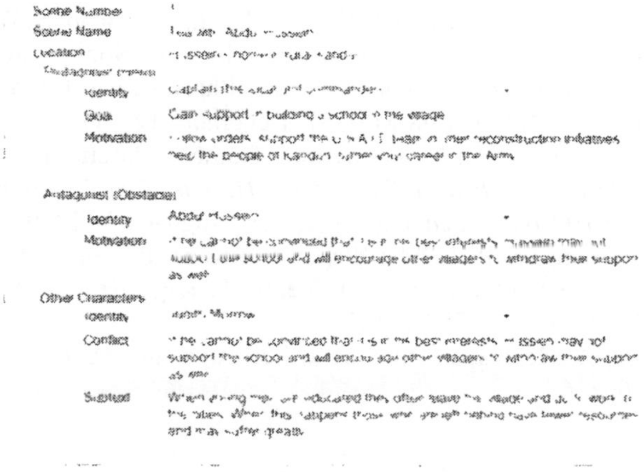

Figure 3. Scene description authoring tool user interface

CONCLUDING REMARKS

Cross-cultural decision making training continues to gain in importance as global businesses seek to improve business relations worldwide and the U.S. military seeks to increase it awareness of non-Western cultural norms to conduct effective non-kinetic operations. Existing methods tend to be ad hoc or impose stringent constraints that preclude the ability to incorporate cultural factors and socio-cultural norms. In this paper, we have presented a game-based simulation approach for cross-cultural decision making training which combines behavioral game theory findings, classical game theory, and multi-agent simulation. This approach overcomes the limitations of existing methods while making learning enjoyable, memorable, and fun. The game-based simulation framework can be employed to teach other soft skills (e.g., leadership skills) as well as skills that are hazardous or too expensive to teach through in the realworld through live exercises.

REFERENCES

Buena de Mesquita, B. (2009) *The Predictioner's Game: Using the Logic of Brazen Self-interest to See and Shape the Future*, Random House.

Camerer, C (2003), *Behavioral Game Theory*, Princeton: Princeton University Press.

Gintis, H. (2000) *Game Theory Evolving*. Princeton, NJ: Princeton University Press.

Henrich, J. et al. (2005). 'Economic Man' in cross-cultural perspective: Behavioral experiments in 15 small-scale societies. *Behavioral and Brain Sciences*, 28: 795-855.

Klein, H.A. (2008) Cognition in Natural Settings: The Cultural Lens Model. In Michael Kaplan's (Ed.) *Cultural Ergonomics: Advances in Human Performance and Cognitive Engineering Research*, Vol. 4, Chapter 9, pp. 249-280, Elsevier Press Ltd.

Madni, A.M., Madni, C.C., MacGregor, D., Riedl, M., and Rouse, W.B. (2008) "GameSim™: Game-based Simulation for Human, Social, and Cultural Training," Intelligent Systems Technology, Inc. Phase I Final Report, ISTI-FR-601-12/08, Contract # N00014-08-M-0337, December 12, 2008.

Madni, A.M. (2009) "GameSim™: Behavioral Game-Theoretic Framework for Socio-Cultural Training," *Proceedings of the Human Social Culture Behavior (HSCB) Focus 2010 Conference,* Chantilly, VA, August 5-7, 2009.

Pillutla, M.M. and Chen, X. P. (1999) Social Norms and Cooperation in Social Dilemmas: The Effects of Context and Feedback. *Organizational Behavior & Human Decision Processes*, 78:81–103.

Simon, H. A. (1990) A Mechanism for Social Selection and Successful Altruism. *Science*, 250:1665–67.

Chapter 40

Language Understanding Technology for Cross-Cultural Decision Making

Marjorie Freedman, Alex Baron, Ralph Weischedel

Raytheon BBN Technologies
10 Moulton St
Cambridge Ma, 02138, USA

ABSTRACT

This paper summarizes research in progress on extracting socio-cultural content from language. Unlike prior work in information extraction from text, where only explicitly stated facts are the goal, in this paper the goal is to infer properties and facts from language use, even though the properties and facts may never be stated explicitly. Further, the focus is on interactions among informal groups; prior work on information extraction has focused on third party reporting, e.g., news. The paper covers corpus collection, human assessments, and evaluation methodology. The study is underway in English and in Arabic.

Keywords Socio-cultural content, extraction, natural language, informal communication

INTRODUCTION

The past generation of language understanding has focused on extracting explicitly stated information from text. The technology tested in the MUC and ACE evaluations detected explicitly stated relations between entities. For example, upon encountering the text '*Bob Dole, former leader of the Republican Party,*'an ACE system would extract the relation LeaderOf(Bob_Dole, Reuplican_Party);, '*Bob's daughter Sally'* leads to the relation ParentOf(Bob, Sally). The question answering systems evaluated in TREC were developed for questions like *'When was Mozart born?'* or *'Who is Vladimir Putin?'*; the expectation is that the answer is contained in one or a set of extracted text phrases. Language, however, encodes much more than what is explicitly stated. Work on sentiment analysis has started down the path of making predictions that are implicit rather than explicit, but that work has been focused primarily on product reviews and delves into only one aspect of the implicit meaning of a message.

In this paper, we describe techniques for using statistical language understanding to extract a broad range of implicit information, for example, the author's purpose when making a post to an online discussion forum (e.g., to agree, to disagree, to introduce a new topic, to ask a question, or to answer a question). Since our hypothesis is that language use conveying such implicit information is culturally dependent, our initial study will focus on online discussion forums in English and in Arabic. Our goal is to conduct a parallel study in Mandarin Chinese a year from now. In addition to the author's purpose in posting a message to a newsgroup, we also describe several implicit labels for a user's tone.

First, we describe the importance of developing systems that understand online communications and how language understanding technology can aid in that process. Since the statistical learning algorithms used by our language understanding technology require training data, we then describe the set of communications (corpus) selected; the human annotation process (including an iterative process of defining the labels can allow a naïve user to consistently annotate difficult judgment calls); and a case study in human annotation. We describe alternative techniques for evaluation and then describe our future plans for developing predictive models that will enable better understanding of online communication. .

CULTURAL INDICATORS

Internet communication has dramatically changed the nature, availability and quantity of textual inter-personal communication. Wikipedia suggests that over 25,000 messages are posted to newsgroups each day. Furthermore, new kinds of media seem to appear each year. Consider Twitter; in the beginning of 2009, there were 2.5 million tweets per day, by January 2010, this jumped to 50 million per day (Ionescu, 2010). These rapidly growing resources provide rich potential for understanding human socio-cultural behavior. They provide insights into the opinions and beliefs of their users and into how these users interact within a group.

Much prior research in extracting information from informal, inter-personal communications has focused on utilizing the metadata present in both email and blogs. Naturally, the meta-data associated with a message on a newsgroup, allows one to, for example, use the frequency with which a user posts to provide some information about that person's attachment to the group. When research involving online communication examines the content of the messages, typically only shallow features are used: Fortuna (2007) incorporate word overlap between a message and responses to the message in a model that predicts agreement between newsgroup posters; Chung and Pennebaker (2007) show that a broad range of characteristics of an author including gender, age and status are correlated with different uses of function words. However, researchers in natural language processing can provide a much richer understanding of the text than what is available through such simple lexical techniques. By automatically providing insight into the meaning of the text (predicate argument structure: *who did what to whom* and discourse features such as the links between pronouns and their antecedents), NLP systems can provide a more comprehensive understanding of the content of a message. By harnessing the richer content information, predictive models will be able to be more accurate, more specific, and provide more intuitively understandable explanations. For example, a feature that indicated that the tone of an author's message displayed anger could be used to understand the author's role in a group even without the prior knowledge of group topic assumed by Fisher (2006). Furthermore, being able to identify the angry post can lead to more explicit justification of a system's predictions to help target information (e.g. the angry post) that would be particularly useful for understanding the cultural atmosphere in which group communication occurs.

Unfortunately though the quantity of information is overwhelming, actual, relevant, cultural content can be quite sparse. In order to develop predictive models that utilize this stream of information, we need to be able to identify relevant social cues such as tone, persuasion, and purpose automatically. There is simply too much content for any individual to review, and because the cues themselves are sparse any person tasked with identifying these cues in text would spend most of their time reviewing messages that would not contribute to the eventual predictions.

CORPUS DEVELOPMENT

SELECTION OF RELEVANT MATERIAL

The content of online forums varies significantly; some message threads consist primarily of spam or reposting of news articles, others have little dialogue between members. For our purposes, we are most interested in message threads which contain active discussion between multiple participants. Limiting the corpus that we annotate to message threads that are likely to contain content of interest makes our annotation more efficient—annotators spend less time reviewing messages that are unlikely to provide value. We started with a very general harvesting principle: selecting only threads that that consisted of 20 or more messages and contained messages from at least six distinct posters. Those two selection criteria are language independent and have been applied to both the Arabic and English collections. We also implemented two more language specific filters in English. English online-posts, like e-mail, frequently contain metadata that provides the exact text that appeared in the message to which the writer is responding. Use of such direct quotation can suggest more involvement on the part of the writer. And for a portion of our corpus we ensure that this quoting behavior appears in at least one of the messages in the thread. Finally, we looked at a small set of shallow-linguistic features of the message and ensured that for a portion of our corpus, at least 30% of the messages in a thread used first-person pronouns or made use of punctuation or capitalization to show emotion.

Because our selection is at the level of threads (and not at the level of messages), even with the more restrictive set of selection heuristics, annotators evaluate a combination of messages that have the selected characteristics and messages that do not. To further ensure that our corpus is not overly tailored to aspects of online communication that we already understand, we include some threads randomly sampled from the initial set in which the only requirement is the presence of at least 20 messages and at least 6 distinct posters. .

The final step in our corpus selection procedure is to have a single annotator manually review the automatically selected threads. This first pass annotation is designed to be much less detailed and thus much faster than full annotation. The annotator is asked to judge whether or not the message thread contains 'meaningful conversation,' but not to look for any specific characteristics of the conversation or language use. This final step will provide us with a general corpus that can be used for an evolving set of more specific annotation targets.

ANNOTATION PROCESS

ACHIEVING HIGHLY CONSISTENT ANNOTATION

As described earlier, the development of statistically trained predictive models requires annotation of the concepts that will be predicted. Typically, the accuracy of a predictive system depends at least partially on the consistency of its training material. Inconsistent training material provides the system with conflicting information which can be difficult to interpret. Furthermore, the agreement between people on the concepts that are being labeled provides an implicit ceiling on system accuracy. If agreement is very low, then any prediction that the system makes is likely to be judged unsuitable by some users of the system.

Word sense disambiguation (WSD) is an example of a case where modifying the label-set to ensure that people agree on the correct label led to significant gains in accuracy. WSD is the problem of determining which of several possible meanings to assign to a specific instance of a word. For example the two sentences "*I called her last night*" and "*They called for immediate action*" use different senses of the word '*called*'. In the first sense, *called* most likely involves using a telephone. In the second sentence, *called* refers to a public statement to act. Initial experiments marking word sense usage in context had human agreement rates of 70% (Veronis, 1998); when trained on such data, systems perform in the mid-60s. As a part of the OntoNotes project (Hovy, et al. 2006), both the sense inventory and the guidelines for annotation were revised. Under OntoNotes, agreement on the annotation of word sense reached 90%. Systems trained on the OntoNotes annotation reach accuracies in the mid-80s.

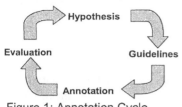

Figure 1: Annotation Cycle

Annotating socio-culturally relevant concepts , like annotating word-sense, presents the challenge of developing a set of labels and guidelines that can be agreed upon. Decisions about the correct granularity and definition of concepts present challenges that must be addressed by looking at data. Here, we describe the approach that has been shown to be successful for Word Sense disambiguation and describe how we are adapting it to the annotation of socio-culturally relevant language uses. To ensure that the annotation is highly consistent, our annotation process will follow an iterative cycle (illustrated in Figure 1). The first step in this process is developing a set of properties of interest and definitions/guidelines for the properties. The label set is designed in consultation with experts in the area to ensure that the end result is theoretically grounded and represents a useful characterization of the data. The guidelines are developed with examples from real data so that the task is grounded in the types of examples the annotators are likely to see. The next step in the process is to perform an initial round of annotation. We then measure agreement on each property between the annotators. If agreement is below the desired threshold (e.g. 90% for the word-sense-disambiguation task), then the guidelines and property set are refined to reflect the disagreements. For word-sense-disambiguation, in some cases this involved collapsing two senses because the distinctions were too difficult for an annotator to make, In other cases, refinement is simply a matter of clarifying ambiguous cases in the guidelines for annotation.

The language use properties we define for predicting socio-cultural content in online communication will be designed to be cross-lingual. For example, we have begun an investigation of annotating *'persuasion to believe'* as a language use in English and in Arabic. While we believe that the high level language use will be extant in communications in both languages, the characteristics that indicate the presence of the property will be language specific. As a result, guidelines development necessarily involves the adaptation of guidelines in one language to other languages. Because the guidelines development process is a collaboration between experts in the domain of the properties being annotated, the developers of the statistical models, and the annotators themselves, our work has started with English. We develop an initial set of guidelines for English annotation, and do an initial round of annotation. Questions that arise during this initial round of annotation are discussed so that the guidelines are more precise. Before starting annotation in a second language, the English-speaking and Arabic-speaking annotation staff review the guidelines together. The Arabic-speaking annotation staff extends the guidelines to include Arabic examples and only then do they begin annotation in Arabic. Arabic annotation, like English annotation begins with multiple people annotating each document. As with the English annotation, if agreement is not sufficiently high after the first round, guidelines will be revised and the property set may even be changed to provide a better characterization of communication. We have started this process in English and in Arabic, in the future we intend to extend it to Chinese as well.

ACTIVE LEARNING FOR SPARSE PROPERTIES

As mentioned in the section on corpus development, a challenge for annotating a socio-culturally relevant property set is that while there is a large amount of data available, the properties of interest may appear in only a fraction of the messages. Techniques in active learning and unsupervised learning can increase the value and impact of messages that are annotated and mitigate the challenge of relatively sparse properties. In our discussion of corpus selection, we described how a simple filtering mechanism can increase the yield. With active learning, the statistical model selects high value instances in which the model is most unsure from a large volume of unannotated data. By tuning models to over-select particularly rare phenomena, we will be able to identify some cases that would be missed by a more conservative model, thereby adding to the instances of the rare phenomena. We will further address the challenge of rare phenomena by investigating unsupervised techniques such as clustering to identity documents similar to a rare instance on key dimensions. Active learning has been shown to significantly reduce the cost of annotation for tasks such as name extraction (Miller, 2004).

ANNOTATION TO ENABLE AUTOMATIC EVALUATION

Annotation is a necessary part of developing statistical models; we have described a process for efficient, effective annotation. However, training models is not the only use for the annotated material. Having a fully annotated test set also allows us to perform automatic evaluation of our predictive models. Our end goal is to develop models that will make predictions on new human communications. To

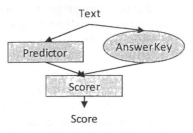

Figure 2: Scheme for Evaluation

simulate this situation, we will divide the annotated data into three disjoint (non-overlapping) sets: a training set, which the algorithms use to estimate the parameters of the statistical model; a development set, which is used to measure performance as often as we make a change in the model, e.g., weekly; and a blind test set, which is used rarely (perhaps quarterly) to estimate performance when the algorithm would be deployed and process new, current data. As we discussed above, ideally, humans exhibit high agreement in their assessments, and as a result the human assessments for the development and blind test set can be considered a 'gold standard', a reference, or answer key for automatic scoring (see Figure 2). This allows evaluation of different configurations of a system without the repeated cost of annotation. Separating the development data from the blind test data allows us to have a set on which we run many experiments (the development data) and also a set that is more removed from algorithm development (the blind test). Repeatedly testing on any data, even if the data is never examined by a person, is very likely to lead to misleading results; features or parameters that increase scores on that data will be adopted; those that do not will be discarded. The adoption of these parameters, which may or may not be reflective of the best parameters for a different test set results in over-fitting.

ANNOTATION CASE STUDY

Our current annotation targets language use constituents, that is, the reason the author is using language in the manner in which he or she writes. Language use constituents may be annotated on a single message, or they may be a reflection of how that message fits into the message thread. We are currently targeting 3 broad categories: persuasion, establishing status/credibility, and topic control. Within each broad category, annotation is broken into more specific constructs: for example persuasion is broken into persuasion to act and persuasion to believe. Establishing credibility is broken into declaring authority, using directives, and answering questions. Our selection of these categories has been informed by those characteristics of groups that have been found to be useful by social network analysis tools to both identify the type of discussion forum and the roles that individuals play in the forum, for example the inventory suggested by Fisher et al (2006). These categories are not specifically tied to any particular type of language, the annotators are instead instructed to read the message and evaluate whether or not the property appears to be a characteristic of the message.

Q Why does our head of state live overseas? A Because that is where she has her main residence Many Canadians happen to do so also

O Why is our head of state not a Canadian? A Who says she is not? The leader of the opposition is a Russian Count The last two Governors-General have been from Haiti and Hong Kong

Q Why is our head of state decided by birth rather than merit? A It is as good as any other method of selection, far less costly and isn't partisan What's best is that the person who gets the job isn't a politician

No head of state is decided by merit The person who is president of Italy or France isn't there because he passed any examinations and obtained a qualification for the job or went through a rigourous job interview Indeed, most of the elected heads of state would probably fail they are there because they are party hacks, with different degrees of slime trailing behind them

Actually we should cut all ties with the British Monarch we as an independant country have every right to decide who our head of state will be Some one whose permanent residence is IN CANADA , not in some foreign country Also we do not need any dumb snooty Brits presuming that they know better how to run our country ,topically when it comes to selecting our Head of State If we don't want the Queen of England as our head of state(of Canada) then that is final ' None of your business CJ	☑ Persuade Persuading to act ☐ Persuade Persuading to believe ☐ Persuade Persuasion (other) ☐ Status Answering questions / Explaining ☐ Status Attempt to establish status ☐ Status Declaring authority ☐ Status Using directives ☐ Topic Agreeing (with the earlier message) ☐ Topic Agreeing (with the original message) ☐ Topic Attempt to control a topic (other) ☐ Topic Attempt to gain information ☐ Topic Broadening topic focus ☐ Topic Changing topic focus ☑ Topic Disagreeing (with the earlier message)	☑ Persuade Persuading to act ☑ Persuade Persuading to believe ☐ Persuade Persuasion (other) ☐ Status Answering questions / Explaining ☐ Status Attempt to establish status ☑ Status Declaring authority ☐ Status Using directives ☐ Topic Agreeing (with the earlier message) ☑ Topic Agreeing (with the original message) ☐ Topic Attempt to control a topic (other) ☐ Topic Attempt to gain information ☐ Topic Broadening topic focus ☐ Topic Changing topic focus ☑ Topic Disagreeing (with the earlier message)

Figure 3: Sample Language Use Constituent Annotation

Figure 3 illustrates annotation by two annotators on a message. Because understanding any given message and how language is used within that message frequently requires the context of the thread in which a message appears, the annotator sees the message to be annotated (bottom left) as well as earlier messages. On the bottom right of the message, we see that the two annotators agree about some of the labels (persuading to act and disagreeing), but disagree about others. The points of disagreement indicate places where annotation guidelines need refinement.

In addition to annotating higher level concepts described above, we intend to annotate some aspects of language use that are more closely tied to the text. Specifically, we intend to extend our annotation to include general socio-cultural language indicators that reflect the tone of the message, for example annotating anger, politeness, familiarity, and formality. Examples of these indicators appear in **Table 1**. While less closely tied to a specific purpose, we believe that this property set, which is more directly tied to specific words in a message, will be easier for people to agree on and also easier for our models to predict. These more robust predictions will provide supporting evidence for the more challenging properties. .

Indicator	Example
Familiar	*I am sorry,.......I know how much you cared for her*
Polite/Formal	*Please follow the general guidelines of this group. Offending posts will be removed.*
Anger	*'There is NO NEGOTIATING with them. There is only one solution to stop them - KILL THEM ALL, *NOW*!!!!!!!'* is very angry text.

Table 1: Examples of Indicators

The language use constituents in turn will be useful for providing a characterization of the groups and the roles in which individuals play in the group. Work such as Fisher et al (2006) and Kelly et al (2006) has shown that the level of disagreement within a group can be used to classify whether the group is ideological or information seeking.

ALTERNATIVES FOR ANNOTATING SOCIO-CULTURAL CONTENT

The focus of our current effort is to develop an annotation standard with which we can achieve a high-level of agreement both to improve the accuracy of our statistical models and also to enable automatic evaluation. While we believe that this goal is achievable for the property set we describe below, there are alternatives if high agreement annotation proves to be impossible. For some language understanding tasks, developing a single, correct gold-standard has proved impossible. One example of such a task is automatic summarization. When generating summaries for an article, there are many potentially correct answers. These correct answers will vary in both the facts they include and the language that is used. Nenkova et al.

(2007) present the Pyramid Method as a technique for allowing automatic evaluation from a diverse set of references. Until recently, a challenge for Pyramid annotation was the cost involved in including multiple references. However, the availability of crowd-sourced annotation through services such as Amazon's Mechanical Turk significantly reduces the cost of annotation (Snow, et al, 2008).

While these techniques provide interesting options, they do not achieve the goal of a system that assigns labels from a property set to a message or set of messages. Therefore, we will continue in our effort to develop clear guidelines and consistent annotation

FUTURE DIRECTIONS

The work described here is in its early stages. We have recently completed our first round of annotation and are beginning the process of revising our guidelines and property set to improve agreement between annotators. As we build up sufficient annotation to develop predictive models for our property set, we will begin to investigate which features of language are predictive of labels in our property set: for example investigating if complex predicate argument structure is predictive of more formal text, or if sentences without subjects (imperative sentences) appear more frequently in text designed to persuade others to act. As we develop better predictive models of both the indicators that are more directly tied to specific language (anger, formality) and those that are more tied to the author's social goals (e.g. persuasion to believe, establishing credibility), we will explore how these predictions can be incorporated into models that predict the roles that each poster plays in the group.

While we have focused on the collection of material from online discussion forums, they are only one of many potential sources of online conversation. Certain desirable media, like e-mail, are difficult to acquire because of privacy concerns. However an organized collection effort (similar to the Linguistic Data Consortium's collection of telephone speech) would be highly beneficial to the research community. Other resources, such as comments posted in response to news articles, public twitter feeds, and potentially public chat-rooms are much more available. The challenge faced by researchers who use these resources is the rapidly changing nature of online resources. Typically, researchers collect their own corpora, e.g. Honeycutt & Herring (2006) Boyd et al (2010), which may or may not be comparable to a corpus collected by a different research group. Further complicating the situation, the content may cease to be available on-line at any time. As researchers explore adding richer, annotations to these new-media corpora, having a shared data collection would allow synergy in the annotation efforts.

Acknowledgements

This work has been supported by the Intelligence Advanced Research Projects Activity (IARPA) via Army Research Laboratory (ARL) contract number W911NF-09-C-0136. The U.S. Government is authorized to reproduce and distribute reprints for Governmental purposes notwithstanding any copyright annotation thereon.

Disclaimer The views and conclusions contained herein are those of the authors and should not be interpreted as necessarily representing the official policies or endorsements, either expressed or implied, of IARPA, ARL, or the U.S. Government.

REFERENCES

Boyd, D, Golder, S, and Lotan, G. (2010), *Tweet, Tweet, Retweet: Conversational Aspects of Retweeting on Twitter*. HICSS-43. IEEE: Kauai, HI.

Chung, C.K., and Pennebaker, J.W. (2007). "The psychological functions of function words". In K. Fiedler (Ed.), Social communication, pp. 343-359. New York: Psychology Press.

Fisher D., Smith M., Welser H. (2006), *You Are Who You Talk To: Detecting Roles in Usenet Newsgroups*. Proceedings of the 39th Annual Hawaii International Conference on System Sciences, pp.59.2.

Fortuna, B., Rodrigues, E. M., and Milic-Frayling, N. (2007,. *Improving the classification of newsgroup messages through social network analysis*. In Proceedings of the Sixteenth ACM Conference on Conference on information and Knowledge Management.

Kelly, J., Fisher, D., Smith, D. (2006), *Friends, foes, and fringe: norms and structure in political discussion networks*, Proceedings of the 2006 international conference on Digital government research.

Honeycutt, C & Herring, S. (2009), *Beyond Microblogging: Conversation and Collaboration via Twitter*, hicss, pp.1-10, 42nd Hawaii International Conference on System Sciences.

Hovy E., Marcus M., Palmer M., Ramshaw L., and Weischedel R. (2006). *Ontonotes: The 90% solution*. Proceedings of the Human Language Technology Conference of the NAACL, Companion Volume: Short Papers, pp. 57–60. Association for Computational Linguistics, New York City, USA.

Ionescu, D. (2010), *Twitter Use Explodes, Hits 50 Million Tweets Per Day." PC World*. http://www.pcworld.com/article/190026/ twitter_use_explodes_hits_50_million_tweets_per_day.html. Febuary23.

Nenkova, A, Passonneau, R & McKeown, K. (2007). *The Pyramid Method: Incorporating human content selection variation in summarization evaluation*, ACM Transactions on Speech and Language Processing (TSLP).,

NIST Speech Group. (2008), *The ACE 2008 evaluation plan: Assessment of Detection and Recognition of Entities and Relations Within and Across Documents.* http://www.nist.gov/speech/tests/ace/2008/doc/ace08 - evalplan.v1.2d.pdf

Snow, R., O'Connor, Jurafsky, D., and Ng, A. (2008). *Cheap and fast-but is it good? Evaluating Non-Expert Annotations for Natural Language Tasks.* Proceedings of EMNLP-08.

Voorhees, E. & Tice, D. (2000), *Building a Question Answering Test Collection,* Proceedings of SIGIR, pp. 200-207.

Chapter 41

Multi-Culture Interaction Design

Javed Anjum Sheikh, Bob Fields, Elke Duncker

Institute for School of Engineering and Information Sciences
Middlesex University
London, UK

ABSTRACT

This research leads towards culturally-based design and shows how user concepts can be organized. In this context, we study cultural differences in categorization and classification by means of card sorting experiments in combination with observations and interviews. The analysis of data collected in Pakistan and UK reveals a number of differences between Pakistani and British participants as to how they classify every-day objects. The differences found suggest a number of design solutions for cultural inclusion. Therefore, the possible solution is to represent these differences by developing an interaction design obtained from local knowledge. It will allow user to explore effectively in comparison to a non-cultural based interface.

Keywords: Cross-cultural design; Cultural Information Access; Human-computer Interaction, Human Factors

INTRODUCTION

People always disown and resist against the things which do not relate to their culture norm, tradition, sentiments and values. The existing Interactive Information

Systems face the same problem. They have several challenges, including effective organization and efficient retrieval of information in cultural context. The limited extent of culture and unfamiliar terminology/classification create the demand of cultural based Interactive Information system. Therefore, this problem creates the need of a new mechanism to solve this problem. This would be based on cultural categorization. This research explores this matter extensively to find the way and techniques to overcome this problem to enhance the user access. For example during humanitarian crisis, volunteer teams need to obtain information about affected area's culture, languages, tradition etc.

There is no empirical study to design cultural based IIS. The use of cultural models in IIS is rare and few researchers used business oriented cultural models. These models neither fulfill the user's need nor compatible with IIS. Developers overlook cultural considerations to meet deadline or do not feel necessity to do research on cultures. IISs have followings issues (see Figure. 1)

- While some Interactive Information systems are global, the users are always local;
- Interactive Information systems heavily depend on western classification;
- All users are not familiar with Interactive Information Systems' organization;
- There is lack of culturally specific terminology for IIS;
- Due to global aspects of Interactive Information Systems, most users fail to get the desired result;

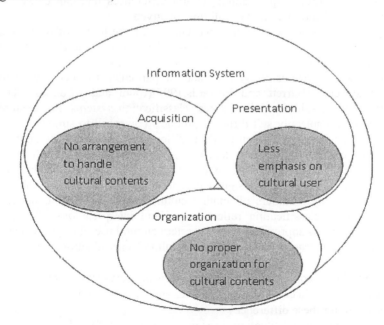

Figure 1. Existing Information System

In the light of above-mentioned figure 1, one main problem emerge that locally specific contents are not appropriately represented and classified in an Interactive design. The lack of representation and cultural differences increases the user problem (Sheikh et al2009a). Therefore, empirical study can provide a better understanding of cultural specific IIS to reduce the cultural gap.

In this paper we outline the approach in more detail and present results of a cross-cultural study that has allowed us to begin to investigate differences in classification. We reviews related work on culture and design. Than introduces the research methodology, explaining how the study has been done. After that, presents and discusses the results of the analysis. Finely, 0our conclusions and directions for future work.

BACKGROUND

The role of internet as a tool for global access to knowledge is undisputed. However, this role cannot be fully realised, as long as information and services of are less accessible to every culture. Problems do not only arise from obvious matters such as language translation, currency translation, formats of numbers and dates, etc. but also from deeply rooted cultural differences that can cause non-understanding and misinterpretation of information given.

A wide range of online classification schemes can be found, of which some seem to have a wider applicability and acceptance than others. The Dewy decimal system is used world-wide and yet, it classifies books differently to the German library classification systems in general and in particular to specialty related classification systems (Barrett and Edwards,1995)(Caidi and Komlodi, 2003). Similar things can be said for online stores' classification systems. For example, UK online stores do not only sell different products compared to similar German online stores, they often classify their products differently[1].This means that not only the content but also the way this content is organised and classified reflects the values and interpretive practices of the culture in which it was produced. Therefore, problems can arise, when content designed, organised and classified by members of one culture is used by members of another culture. Typically, web content, its organisation and its classification reflect values and interpretations of western cultures rendering it less appropriate to non-western cultural user groups. As part of a larger study, this research focuses on cross-cultural classification practices. It examines

- the way how people classify representations of every-day objects
- the differences in classification practices and classifications
- the cause for these differences

The above mentioned situation gives us an assumption for this research, that no scheme for organising information is likely to be equally effective for every cultural group. The current research aims to make a contribution in this area, not by finding a universal way of classifying information, but by providing a method for investigating classification in a locale in order to generate localised interface designs. The expected solution will be based on local user access needs and capability of the local users. Therefore it is need to integrate cultural differences with interaction capabilities (Khoja and Sheikh, 2002)

METHOD

In order to investigate the question of how different cultures organise their knowledge differently (see Figure 2), we used mixed methods. We used card sorting technique to get cultural knowledge and used cluster analysis which lead to knowledge representation, which help use to develop cultural based interaction design.

Figure 2. Methodology

CARD SORT

The card sorting technique help to develop and identify concepts, models, attitudes, trends, patterns and values for capturing information from the mental model of the participants. In a card sorting experiments participants are asked to arrange cards into groups. On these cards one finds pictures or the names of objects. Card-sorting experiments can reveal different ways in which participants organise their understanding of the world. Card sorting is widely used in the field of Human Computer Interaction, psychology, and knowledge engineering for knowledge elicitation. It helps to evoke participants' domain knowledge (Sackmary and Scalia, 1999) distinguish the level of the problem (Barrett and Edwards, 1996), and reflects ideas about knowledge (Zimmerman and Akerelrea, 2002) and knowledge acquisition (Russo and Boor, 1993). Furthermore, card sorting is often used to gather data about personal constructs, for instance menu structure specifications and to understand users' perceptions of relationships among items. The card sorting experiment was conducted in this way: Thirty-nine cards[1]

---

[1] Apple, bacon, banana, beans, beef, beer, bread, butter, cheese, chicken, coffee, cream, dessert, doughnut, egg, fish, garlic, ginger, grape, lamb, melon, milk, mushroom, naan,

of food items were used. As the study was a cross-cultural ne, the food items were translated into the participant's first language. Participants[2] were asked to group these cards. Subsequently, they were asked to label each group. Then they were asked for each group of cards, if they would like to subdivide the group. The participants labelled the subgroups as well. The process was repeated until participants no longer wanted to subdivide any groups. While the participants were grouping cards and labelling the groups the researcher recorded the emerging tree structures. We allow the participants to use as many layers as they find adequate, so that groups can be subdivided into lower level groups, in contrast to the above card sorting experiments, which always use one layer of grouping cards. The multi layered approach is closer to people's every day use of classification, but also poses quite a challenge for the analysis, particularly for large data sets. For this reason we automated part of the analysis, i.e. the measuring of the difference between two classifications as edit distance. Other differences were observed and analysed manually, such as the width and the depth of the classification. Furthermore, we employed cluster analysis (K-means) to determine, whether the cultural backgrounds of participants are a potential explanation for the observed differences.

The data collection generated hierarchical tree structures representing the classifications that the participants revealed by grouping the cards. The analysis of the data revolves around the discovery of similarities and differences between the hierarchies, and whether those similarities and differences are aligned with cultural identity. The investigation proceeded informally at first; looking for patterns in the data that were suggestive of culturally aligned classificatory practices. The initial analysis pointed the way for a more systematic analysis that lent itself to automated support.

ANALYSIS: INFORMAL

The results were conducted for patterns of similarity and difference between the hierarchies produced by members of the different cultures. Pakistani participant's categorisation is relatively flat, where as the British participant added an extra layer

---

onion, orange, pasta, peas, pizza, pork, potato, prawn, sweet potato, tea, tomato, turkey, wine, wraps and yogurt.

[2] A total of 160 (PK n=80 and UK n=80) subjects participated in this study and were selected based on the their ethnicity. They were literate, over 18 years of age and were familiar with all the items on the cards. Pakistani participants are from Karachi, Lahore, Islamabad, and Bahawa-l Pur. UK participants live in London and their grandparents are also UK born.

of categorisation (see Figure 3). The results showed that Pakistani and British participants differed in their categorisation judgments. However, they shared a common representation structure in some categories. The differences are also noticed within each culture. The results suggest clear differences between the categorisations produced by Pakistani participants and those produced by their British counterparts. For instance, fragments of typical categorisations produced by a Pakistani and British participant. In other cases, participants produced structurally identical classifications, but used different terminology to refer to parts of the classification tree. However, a fragment of the categorisation that was generated often being common to many participants, irrespective of their cultural background.

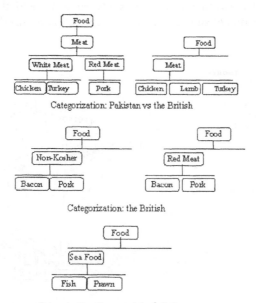

Categorization: Pakistan vs the British

Categorization: the British

Categorization: Common in both Cultures

Figure 3. Cauterization

The informal result of the study indicates a cultural difference in food categorisation among people belonging to different cultures that appears to be greater than the differences between people within the same culture. The studies suggest that both the 'national culture' and the 'belief system' of a participant shape the way they categorise items. By 'belief system' here, we refer roughly to religious background as this is a highly significant factor in the way people understand food and the various domestic practices that surround it. It seems likely that other elements of culture, such as professional cultures or membership of communities of practice would gain greater significance.

ANALYSIS: FORMAL

A first step towards conducting analysis in a more rigorous manner was to formalize the complimentary notions of 'similarity' and 'difference' that are at work. A number of possible formulations are possible, but the one that proved to be most promising was the notion of 'edit distance'. This measurement of distance was implemented in software based on a freely available framework called SimPack.[3] The algorithm for computing the 'edit distance' between trees facilitated the construction of a 'distance matrix' that encodes the edit distance between the hierarchies produced between all pairs of study subjects, and the discovery of structure in the population of subjects entails an exploration of this distance matrix.

```
Final assignments
Centroids
          Person7        Person2=
Person7    0.0           34.0=
Person2    34.0          0.0=
-----------Cluster 0---------
Cluster name: Cluster0
Sum dist: 51.0 Average 17.0
Centroid Person7
Points
    Person5
    Person7
    Person6
-----------Cluster 1---------
Cluster name: Cluster1
Sum dist: 69.0 Average 9.857142857142858
Centroid Person2
Points
    Person2
    Person4
    Person8
    Person10
    Person1
    Person9
    Person3
```

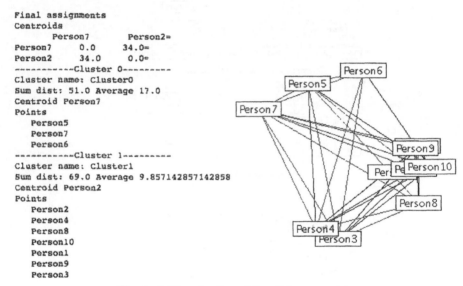

Figure 4. Visualization of the distance matrix

Two approaches to this exploratory task were employed. A more traditional statistically-based approach was implemented using a variant of the k-means cluster analysis algorithm to discover clusters of subjects who were 'close together' in that they produced similar hierarchies. This formal style of analysis was complemented with a more exploratory tool that produces a visualization of the distance matrix, (see figure 4) based around the physical analogy of data points

---

[3] SimPack is an open source collection of software tools for investigating the similarity between 'ontologies'. Available from http://www.ifi.uzh.ch/ddis/simpack.html.

joined by a collection of springs whose length is determined by the edit distances[4].
A simulation of such a system yields a dynamic network that tends to settle in a
'low energy' configuration. The latter technique provides a useful visual way of
seeing how a structure emerges from the confusion, as similarly similar trees tend to
gravitate towards one another.

Interaction Design

As earlier mentioned that there is no scheme for organising information is likely to
be equally effective for a range of cultural groups. Apart from this there is no
cultural based system (Sheikh at al 2009). The current research aims to make a
contribution in this area, not by finding a universal way of classifying information,
but by providing a method for investigating classification in a locale in order to
generate localised interface designs. The analysis has given a way of identifying
clusters of related structuring of a set of objects. Therefore two interface designs
developed by the result of cluster analysis and edit distance. One interface based on
Pakistani result and other based on UK (See figure 5).

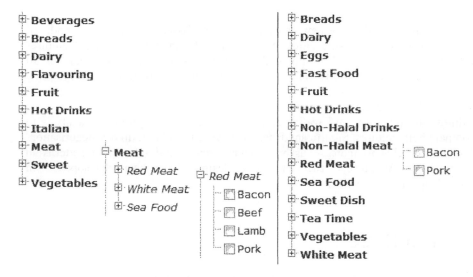

Figure 5. Multi-culture Interface

---

[4] The tool is based on the Graph demonstration program that is part of the Java Software
Development Kit available fromjhttp://java.sun.com.

414

The prototype cultural based interface obtained from local knowledge. Our interface shows a common concept which is result of hierarchical clustering analysis of multi cultural representation. It allow user to explore effectively in comparison to a non-cultural based interface. The interface is user perspective, which will help the user to interact effectively and close to human to human interaction. The paper presents a concept of cultural representation for interaction design. The important features of interaction design here are

- Sorting of cultural concepts
- Integration of concepts into categorization
- Interaction design for cultural representation.

Participants were asked to pick 15 items from 39 given items. Time and error noted. The valuation of both designs is in progress. Informal analysis tells that the British participants spend less time but do many mistake which fins items from Pakistani Interface.

CONCLUSIONS

This research introduces a new analysis technique based on SimPack's modified classes to discover/measure similarity and to understand how different cultures view similar concepts. The aim of this research is to propose a design for all cultures to increase usability enhancement and interaction patterns in categorizing that lead to browser design.

We used mixed methods to interpret the result. Our studies explored cultural difference by card sorting and result analysis through cluster analysis to compare both cultures. Significant differences were found in term of categorisation. The result helps to increase usability enhancement and interaction patterns in categorising. At present, two interactive designs have been developed. Our immediate aim is to evaluate them.

.

REFERENCES

Barrett, A. R. and Edwards, J. S. (1995). "Knowledge Elicitation and Knowledge Representation in a Large Domain with Multiple Experts", Expert Systems with Applications, vol. 8, pp. 169-176

Caidi, N., and Komlodi, A. 2003. Digital Libraries across Cultures: Design and Usability Issues. SIGIR Forum, 37(2), 62-64.

Khoja, S. and Sheikh, J. A. (2002) KIIT Digital Library: An open hypermedia Application. In: International Conference on Use of Information Technology in Teaching of Science, 12 - 14 March 2002, Islamabad, Pakistan

Russo, P. & Boor, S. (1993): 'How fluent is your interface? Designing for international users'. Human Factors and Computer Systems. Conference Proceedings of the conference on Human factors in computing systems. Boston: Addison-Wesley Longman Publishing. p342-347.

Sackmary, B., & Scalia L. M.(1999): Cultural patterns of World Wide Web business sites: A comparison of Mexican and U.S. companies. Paper presented at the 7th Cross-Cultural Consumer and Business Studies Research Conference, Cancun, Mexico..

Sheikh, J.A; Fields, B and Duncker,E. (2009) Cultural Representation for Multi-culture Interaction Design. In *Proceedings of the 3rd international Conference on internationalization, Design and Global Development: Held As Part of HCI international 2009* (San Diego, CA, July 19 - 24, . N. Aykin, Ed. Springer-Verlag, Berlin, Heidelberg, 99-107

Sheikh, J.A; Fields, B and Duncker,E. (2009a) Cultural Representation for Interactive Information system. *In: Proceedings of the 2009 International Conference on the Current Trends in Information Technology, Dubai.*

Zimmerman, D. E. and Akerelrea, C(2002).: "A Group Card Sorting Methodology for Developing Informational Web Sites", in the Proceedings of International Professional Communication Conference (IPCC).

Chapter 42

Generic Message Propagation Simulator: The Role of Cultural, Geographic and Demographic Factors

Tony van Vliet, Elja Huibregtse, Dianne van Hemert

TNO Defence & Security
POBox 23, 3769 ZG, Soesterberg, the Netherlands
in cooperation with FOI the Swedish Defence Research Agency

ABSTRACT

European defense organizations have to deal with a changed focus in military missions towards asymmetric and irregular warfare and Stability, Security, Transition and Reconstruction (SSTR) operations. A range of operations (e.g., psychological operations, humanitarian missions) is performed in a large variety of locations and cultures (e.g., Africa, Asia). In this type of mission, influencing target audiences is one of the (many) methods available to achieve the desired effects. One of these methods is to influence the behaviors and attitudes of target audiences with psychological messages, as is done in Psychological Operations (PsyOps).

There are many aspects to performing PsyOps such as target audience analyses, which channel to use, what the nature of the message should be and who the identified sender should be. For a PsyOps officer assistance in this activity is highly appreciated. PsyOps officers are particularly interested in being able to approximate what proportion of their target audience they reach and within what timeframe.

In order to comply with this need of our Armed Forces, TNO Human Factors is developing a generic message propagation simulation with the working title SHOUT (Simulating How Our Utterances Transmit) which should fulfill this need.

This work is done in cooperation with the Swedish Defence Research Agency (FOI). Because armed forces are operationally active in more than one theatre of operations this generic message propagation simulation needs to be culturally, geographically and demographically sensitive. In other words, the simulator needs parameters that account for these specific situations. On the other hand, the PsyOps officers are not helped with a simulation that needs extensive expert training to be used. Our challenge was to develop a simulator which can be used in specific settings without the overhead of a new expert to be added to a team.

The following solution was adopted to simulate the propagation of messages in the target audience. SHOUT explicitly incorporates five channels of communication (print, radio, TV, internet & word-of-mouth). The first four channels are relatively simple to simulate, and largely dependent on the availability and access to these media by the target audience. More difficult is the word-of-mouth propagation; however, for this channel we can make use of advances in epidemiological research. Within this body of knowledge, simulations have been developed that estimate the spread and infection rate of viruses. We have adapted an epidemiological simulation to simulate the propagation of messages by word-of-mouth. In this effort, we have included a host of cultural, geographical and demographical factors so that the generic simulator can be "tweaked" to a local situation. Included factors, are amongst others, literacy, household size, social network structure, rural/urban environment, ethnic diversity, and collectivism.

The goal of SHOUT is not to have an exact prediction of how a message propagates through a local population but to enhance the understanding by the PsyOps officer of their area of operations. SHOUT will enable the formulation of working hypotheses on which channel to use and what retention rates can be expected. By testing these hypotheses through probing actions, more insight can be gained about the target audience. This insight can then be used to adjust the parameter setting and in this manner, SHOUT becomes more sensitive.

INTRODUCTION

Defence organizations have to deal with a changed focus in military missions towards asymmetric and irregular warfare and Stability, Security, Transition and Reconstruction (SSTR) operations. A range of operations (e.g., psychological operations, humanitarian missions) is performed in a large variety of locations and cultures (e.g., Africa, Asia). In this type of mission, influencing target audiences is one of the (many) methods available to achieve the desired effects. One of these methods is to influence the behaviors and attitudes of target audiences with psychological messages, as is done in the activity called Psychological Operations (PsyOps).

There are many aspects to performing PsyOps such as target audience analyses, which channel to use, what the nature of the message should be and who the identified sender should be. For a PsyOps officer assistance in this activity is highly appreciated. PsyOps officers are particularly interested in being able to approximate what proportion of their target audience they reach and within what timeframe. An example of this type of message reads: "Do not choose to follow the enemies of the Islamic Republic of Afghanistan; choose peace and return home to your elders (ISAF)".

In order to comply with this need of the Armed Forces, TNO Human Factors is developing a generic message propagation simulation with the working title SHOUT (Simulating How Our Utterances Transmit), in cooperation with the Swedish Defence Research Agency (FOI) which should fulfill this need. Because our armed forces are operationally active in more than one theatre of operations this generic message propagation simulation needs to be culturally, geographically and demographically sensitive. In other words, SHOUT needs parameters that account for these specific situations. On the other hand, the PsyOps officers are not helped with a simulation that needs extensive expert training to be used. Our challenge was to develop a simulator which can be used in specific settings without the overhead of a new expert to be added to a team.

BACKGROUND OF THE MODEL

In general, PsyOps activities are guided by the Laswell formula (1948; see **Error! Reference source not found.**). The following five aspects are relevant in (persuasive) communication: sender, message, channel, receiver, and effect.

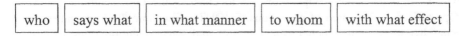

| who | says what | in what manner | to whom | with what effect |

Figure 1 Communication model of Laswell (1948)

The simulation we are developing also makes use of these aspects of communication. We are not distinguishing types of who because the tool is developed for our Armed Forces, and as such the sender is a known constant. The *characteristics of the message* (says what), the *channel* used (in what manner), *characteristics of the receiver* (to whom) and *how long it takes to arrive* (with what effect) are the focus of our attention.

In the following sections, these aspects are discussed in light of our main objective, developing a generic message propagation simulator. In our efforts, we are focusing on a simulator that:

- can be used by a PsyOps officer without extensive expert training;
- runs on a regular personal computer;
- requires minimum data; we know from experience that a high data input requirement will result in a tool which will not be used in military operational settings;
- is not used as a deterministic predictive tool, but as a hypothesis generation tool that can be used in support of sense making within this domain. SHOUT should facilitate "what if" analyses and help generate understanding of the social network the PsyOps officer is dealing with.

These constraints were directive in our choice for parameters and the level of resolution. We have chosen geographic, ethnographic and demographic manipulable parameters. We do not model individuals but use a social entity as unit of analysis; these entities represent subsets of the total population. The central assumption is that messages can be transmitted by five channels: word-of-mouth, print, TV, radio and internet. The channels print, TV, radio and internet are considered to be broadcasts (one sender, multiple receivers), whereas the channel word-of-mouth is considered as a one-on-one communication. We assume in our model that messages are broadcast initially at a targeted audience and are subsequently transmitted through the word-of-mouth channel (one-on-one). The model for broadcasting messages is relatively simple: Receiving these messages depends on the characteristics of the entity, such as the percentage of available TV and radio receivers, the percentage of illiterates, the percentage of newspaper readers and the like. For the word-of-mouth (one-on-one) propagation, we make use of algorithms developed within the field of epidemiology (Keeling & Rohani, 2008). Our assumption is that the spreading of a virus through a population is analogous to the spreading of a message, which requires one-on-one contact, either vis-à-vis or by phone or email. As said previously, entities are not individuals but subpopulations, which have one geographical location, belong to one ethnic group, are situated in one realm of power and have a predetermined number of individuals.

In the next section we will provide arguments for the choice of parameters in SHOUT with the Laswell formula as guiding principle.

SENDER CHARACTERISTICS

In the development of SHOUT we have not yet included variation in the characteristics of the sender. One factor authors agree upon is the trustworthiness of the sender. We are quite aware of the importance of this factor (Petty & Cacioppo, 1986) but for our initial development, we have decided that because the simulator is being developed for our armed forces in missions abroad, the sender is invariant, thus variation in this factor at this point in time is not of the highest priority.

MESSAGE CHARACTERISTICS

Research on information dissemination has been done since WW1. Until recently, the focus was on characteristics of the message that facilitate understanding, believing, and disseminating a message. For example, Clark (2009) summarizes these characteristics into ''The 8½ Laws of Rumor Spread'', mentioning that successful rumors tap into people's anxieties, are surprising but still fit existing biases, reflect the zeitgeist, are simple and concrete, and are difficult to disprove. Further, credibility of a message is also important, in addition to ambiguity, importance and anxiety (Rosnow, 1991). It must be noted that these are attributes of messages, but the attribution finds its source in the receiver.

In a chapter on the influence of rumors, DiFonzo and Bordia (2007) describe how rumors have been used to spread propaganda in the domain of politics. Two strategies are highlighted: innuendo (negative rumor about a candidate damages that candidate's reputation) and projection (accusing another person of the same things the accuser is guilty of leads to higher evaluations of the accuser and lower evaluations of the accused). They mention a few individual-level mechanisms of rumor spread: uncertainty and anxiety stimulate rumor spread, more outcome relevant involvement is related to more spread (when the topic is important to someone the likelihood of spreading the rumor increases), when people believe the rumor more they will spread it more, and motivation (relationship enhancement and self-enhancement). Finally, Knapp (1944) argued that negative rumors are more likely to be spread than positive rumors.

Taking the above considerations into account, we think that an estimation of how much impact a message can have on a population is of importance. We will have to formulate a mechanism that deals with this aspect. We need to have some indicator that estimates and captures the nature of the message. In our simulation, this aspect of the message is judged by the PsyOps officer along one dimension, importance of the message. The importance of the message is established by the officer on an ordinal scale (trivial, informative, and crucial).

CHANNEL CHARACTERISTICS

The distinction between broadcast-type channels and one-on-one channels is a useful and quite a distinctive one. Broadcasts are one-way disseminations, whereas one-on-one communication are interactions. Kimmel (2004, p102) writes on rumors that it "is not simply a message that is received and then automatically passed on to one other person"; instead it is "shared and evaluated as part of a two-way interaction between the transmitter and the recipient".

Broadcast
To be able to capture the variation in susceptibility to broadcasting we make use of the following aspects of the targeted population:
- ethnicity (to which group the entity belongs)
- population size (number of individuals that are part of the entity)
- initial susceptibility (literacy, access to TV, radio, internet, phones and newspapers)
- geographical location (where in the physical world the entity is located, how physical barriers impede propagation).

One-on-one propagation
This type of propagation, in our understanding, is analogous to the spreading of viruses within a population. We have made use of an already existing simulation on virus spreading (Keeling & Rohani, 2008) and adapted this to the propagation of messages. This simulation is based on proximity of entities in social networks. In SHOUT we have expanded the concept of proximity with geographic, ethnographic and demographic parameters. This brings us to the concept of social networks and the importance thereof in the simulation of message propagation.

Network characteristics

Watts, Dodds, and Newman (2002) state that group membership is a primary basis for social interaction; the probability of two people knowing each other decreases with decreasing similarities of the groups to which they belong. Consequently, boundaries between ethnic groups in social networks have been reported by many authors (see Baerveldt, Van Duijn, Vermeij, & Van Hemert, 2004). These boundaries can be explained in terms of the in-group versus out-group distinction.

Lazarsfeld, Berelson, and Gaudet (1948) proposed a two-step model of communication in which persons who have a central location in a social network serve as efficient agents in transmitting messages from media to mass. These persons are called opinion leaders. In later research it was suggested that dissemination within groups depends to a large extent on central persons, who have strong ties with group members, whereas dissemination between groups is mainly done by marginal persons, who are weakly tied to the group (Katz, 1987; Weimann, 1982). Lin (1986) suggests that information that is relevant to the preservation of a group's interest is more likely to be released within the group than across groups (that is, through strong ties). Onnela et al. (2007) studied communication patterns of millions of phone users in an unidentified Western country. They found that the removal of the weak ties leads to a sudden collapse of the entire network. In contrast, the removal of the strong ties results in gradual shrinking of the network, but no collapse. This finding illustrates that strong ties are mostly important within communities, whereas weak ties connect the different communities. Finally, the small-world principle (Watts, 2004) assumes that networks can be characterized by a few very well-connected hubs. This would explain why rumors can travel very fast, transmission episodes consist of a few well-connected persons spreading the rumor to many persons who do not disperse it further. According to small world theory the hubs are very important.

The above sketch of the literature on rumor spread in social networks suggests that the nature of the nodes (central or peripheral) in the network and the number of paths between nodes is of great importance to the propagation of messages. At a more generic level we assume that the propagation within and between entities differs. How to parameterize these differences, without having to simulate the interaction between each and every individual is the main challenge of our effort. In this, we rely on the work initiated by Hofstede (2001) on cultural differences.

RECEIVER CHARACTERISTICS

Cultural differences in networks

Cultural groups have been found to differ with respect to the nature of their social networks. A study in Hong Kong showed that people tend to share information with strong ties, or people with whom they perceive to have good relations (Lai & Wong, 2002). However, information transmitted via kin ties tends to arrive at the respondent faster than via non-kin ties or other communication channels. It is suggested that this preference for strong ties and kin ties is culturally influenced. In an overview of cross-cultural studies from the US, Marsden (1987) concluded that European Americans have the largest networks, followed by Hispanic Americans and African Americans, respectively.

A cross-cultural comparison of network density was conducted in Israel and the United States (Fischer & Shavit, 1995). Networks in the two countries were similar in terms of social contexts, felt intimacy, nature of social support, etc. Still, the Israeli networks were more tightly interconnected (dense) than American networks. They explained this in terms of the proportion of kin in the network: Israelis have more kin in their networks and this accounts for half of the national differences in density of the network. Also, the duration of relationships determines density; perhaps this also accounts for national differences (Israelis have known their contacts for a longer time, probably due to US mobility). A third reason for the denser networks of Israelis might be structural and cultural; Israel seems to be a smaller society with greater group cohesion, with as focus on the group instead of autonomy of the individual (like in the US).

Cultural dimensions

Culture-comparative studies on information dissemination are scarce. However, some inferences can be drawn from studies on cultural characteristics and information dissemination or communication channels in general.

Hofstede (1980, 2001) introduced five dimensions that describe national cultures: individualism-collectivism (focus on the individual versus the group), power distance (acceptance of hierarchy), uncertainty avoidance (need for rules), femininity-masculinity (focus on nurturing versus achievement), and long-term orientation (focus on long-term versus short-term view on life). The most obvious dimension that has relevance for message propagation is uncertainty avoidance. In his 2001 book, Hofstede argues that one would expect higher uncertainty avoidance to be related to seeking more information through the network. However, he continues to report that research has shown that the opposite is true, and convincingly so. For people in highly certainty seeking countries, new information is apparently more threatening than having less information.

Khalil and Seleim (2009) tested nine hypotheses on the impact of national culture dimensions on information dissemination capacity across 61 countries. They reported more information dissemination in countries with lower power distance (as information is controlled, few people have access to resources, higher uncertainty avoidance higher, more future orientation, higher institutional collectivism and lower in-group collectivism. Apparently, the role of collectivism is ambiguous.

Bagchi, Hart and Peterson (2004) studied the impact of culture information technology product adoption. They found more adoption in more individualist societies, less adoption in high power distance societies (as these are more centralized), less adoption in highly uncertainty avoidant countries (as highly uncertainty avoidant countries have less interpersonal trust, more written rules, and less use of internet and teletext), less adoption in more masculine countries, and less use of communication tools in countries with higher ethnic heterogeneity and in countries with more income inequality, and finally more use of tools in wealthier countries.

Consequences for the model parameters

In order to achieve a working simulation, different parameters need to be estimated by the used. First of all, ethnic groups have to be identified and geographically located and geographical areas attributed to power blocks. SHOUT allows for many groups, however this increases the data requirement factorially. Consequently an operator has to rate the ethnic groups on the five Hofstede dimensions (collectivism, power distance, femininity, uncertainty avoidance, long-term orientation) on 5-point Likert scales. These ratings result in a matrix of the within entity proximity. Secondly, the operator needs to create a distance matrix on eight difference scales that captures the cultural difference between the ethic entities (conflict, habits, attire, language, social-economic status, family size, political participation, and collectivism) by means of pair-wise comparisons. With these matrixes the proximity within and between entities is parameterized and the network is described in generic terms. Finally, the stability of the society assessment is done by choosing which of the following states best fits with the region of operations (Dziedzic, Sotirin, & Agoglia, 2008).

> *State Zero* (Externally Stabilized): Focused on immediate implementation tasks and altering most critical dynamics necessary for change. In State Zero, drivers of violent conflict persist, requiring the active and robust presence of external military forces in partnership with a sizable international civilian presence to perform vital functions such as imposing order, reducing violence, delivering essential services, moderating political conflict, and instituting an acceptable political framework pursuant to a peace accord.

> *State One* (Assisted Stability): Focused on the completion of mission tasks and transitioning responsibility to host nation authorities. State One is characterized by the drawdown of foreign intervention forces, and the increasingly visible presence of indigenous security and civilian institutions.

> *State Two* (Self-Sustaining Peace): Focused on autonomous, host nation sustaining efforts to expand and improve initial efforts. In State Two, international presence takes a truly background role as indigenous institutions prove increasingly capable of standing on their own. State Two is characterized by a shift from donor transition-oriented programs to long-term development and sustainability efforts, both in military-related (Security) and civilian law making and enforcement (Rule of Law), and traditional development sectors such as health, education and land resource management.

EFFECTS

The effect we simulate is the propagation of a single message through a population in terms of time and the number of persons reached. A very important effect PsyOps officers are interested in is to what extent the messages actually change a targeted behavior. This is not included in this simulation; we can only simulate an indication of the time and reach of the message.

MODEL

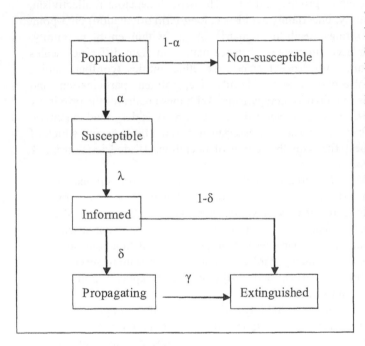

The model we have developed is schematically represented in Figure 2. The central premise is that only a part of the population is susceptible to messages (because they can read, etc). The susceptible proportion thus can be informed and are then to some extent capable of propagating the message further. Finally, the propagation stops because most people are informed or the message is not salient any more.

Figure 2   SHOUT model

After having initialized the simulation (set the parameters for the entities) the first step is to reduce the proportion of susceptibles (non-susceptibles are for instance individuals who cannot be informed, young children etc), this is based on the parameters which fall under process α (see section α). In the next step process λ is executed, this calculates the proportion of susceptible that become informed (see section λ for the relevant parameters).

Consequently a proportion of the informed start propagating the message, this is calculated in process δ. Finally the message ceases to be propagated which is calculated in process γ.

α *Characteristics of population, entities are defined with respect to:*
- ethnicity (group a, b, etc)
- population size (number of people)
- initial susceptibility (illiteracy, access to TV, radio, internet, phones and newspapers)
- geographical location
- realm of power these are located

λ *Characteristics of ethnicity within entities:*
- Collectivism
- Power distance
- Femininity
- Uncertainty avoidance
- Long-term orientation

λ *Comparison between entities:*
- Conflict
- Habits
- Attire
- Language
- Social economic status
- Family size
- Political participation
- Collectivism (Hofstede)

λ *Importance of message, Physical distance between entities, Instability community*

δ *Importance of message, Instability community*

γ *percentage of extinguished depends on:*
> % informed
> % propagating
> % extinguished
> if the sum of the above exceeds 85% entities stop propagating.

EXAMPLE RESULTS

INITIAL SETTINGS

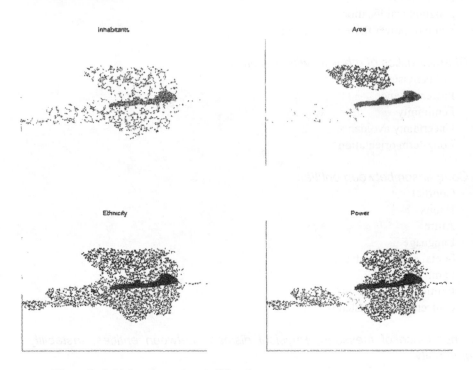

Figure 3   Initial settings scenario XLand

OUTPUT

The simulation is designed to generate all sorts of outputs. In this section we will present one output, the proportion of individuals that stop propagating the initial targeted message per entity over time (see Figure 4). In this run of SHOUT a small subset of entities was targeted with a broadcast message.

% Extinguished

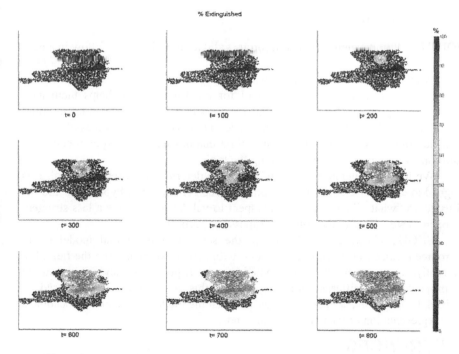

Figure 4   Propagation of one targeted message

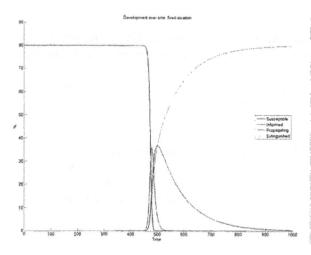

Figure 5   Propagation over time for one entity

In the same run we can also examine how for each entity the proportions of susceptible, informed, propagating and extinguished individuals change over time. In Figure 5, we see that the message informs the entity after 450 time units, after that propagation by this entity takes place, finally the susceptible group become extinguished with respect to this message.

DISCUSSION

SHOUT is still in development, and rigorous validation will have to follow in a next phase of our project.

What we would like to make clear is that combining insights from different fields can be useful in developing a tool for practitioners that helps them gain insight in their problem area. What we have experienced is that developing an initial demonstrator such as SHOUT is a prerequisite to understand what is feasible from a scientific point of view, but at the same time enables eliciting requirements with practitioners for whom we are developing the tool.

We have to reiterate that SHOUT is not a deterministic predictive tool. It is designed to enhance sense making, in other words generation of hypotheses and the ability to do "what if" analyses. With respect to validation, we take a less stringent approach as we adhere to the notion of fitness for purpose.

SHOUT allows for evolution, in the sense that the initial model of an environment can be changed to correspond with actual experiences in the field. Put differently, messages can be disseminated in a local population and by measuring the propagation at different point in reality; this information can be used as feedback to adapt the parameter settings. As such the model can adapt and become a better representation of the population at hand.

REFERENCES

Baerveldt, C., Van Duijn, M., Vermeij, L., & Van Hemert, D. A. (2004). Ethnic preferences in class rooms? Assessing the influence of ethnic composition of school populations on the distribution of social relationships. *Social Networks, 26,* 55.

Bagchi, K., Hart, P., & Peterson, M. F. (2004). National culture and information technology product adoption. *Journal of Global Information Technology Management, 7,* 29-46.

Clark, T. (2009). The 8½ laws of rumor spread. *Psychology Today online,* November 01, 2008 - last reviewed on September 16, 2009.

Di Fonzo, N. & Bordia, P. (2007). Rumor, gossip and urban legends. *Diogenes, 54,* 19-35.

Dziedzic, M, Sotirin, B, & Agoglia, J. (2008) Measuring progress in conflict environments (MPICE) - A metrics framework for assessing conflict transformation and stabilization. Version 1. http://www.usip.org/resources/measuring-progress-conflict-environments-mpice

Fischer, C. S., & Shavit, Y. (1995). National differences in network density: Israel and the United States.

Hofstede, G. (1980). *Culture's consequences.* Beverly Hills, CA: Sage.

Hofstede, G. (2001). *Culture's consequences. Comparing values, behaviors, institutions, and organizations across nations.* Thousand Oaks, CA: Sage.

Katz, E. (1987). Communication research since Lazarsfeld. *Public Opinion Quarterly, 51,* S25–S45.

Keeling, M. J. & Rohani, P. (2008). *Modeling infectious diseases in humans and animals*. Princeton: Princeton University Press.

Khalil, O. E. M. & Seleim, A. (2009) National Culture Practices and Societal Information Dissemination Capacity. *Proceedings of the 2009 conference on Information Science, Technology and Applications* (pp. 104-113). Kuwait: Kuwait.

Kimmel, A. J. (2004). *Rumors and rumor control: A manager's guide to understanding and combatting rumors*. Mahwah, NJ: Lawrence Erlbaum Publishers.

Knapp, R. H. (1944). A psychology of rumor. *Public Opinion Quarterly*, 8, 22-37.

Lai, G. & Wong, O. (2002). The tie effect on information dissemination: The spread of a commercial rumor in Hong Kong. *Social Networks, 24*, 2002, 49–75.

Laswell, H.D. (1948). The structure and function of communication in society. In L. Bryson, (Ed.), *The communication of ideas*. New York: Harper.

Lazarsfeld, P.F., Berelson, B., & Gaudet, H. (1948). *The people's choice*. New York, NY: Columbia University Press.

Lin, N. (1986). Conceptualizing social support. In: Lin, N., Dean, A., & Ensel, W.M. (Eds.), *Social support, life events and depression* (pp. 17-30). Orlando, FL: Academic Press.

Marsden, P. (1987). Core discussion networks of American. *American Sociological Review*, 52,122-131.

Onnela, J.-P., Saramäki, J., Hyvönen, J., Szabó, G., Lazer, D., Kaski, K., Kertész, J., Barabási, A.-L. (...). Structure and tie strengths in mobile communication networks, *PNAS,* 104, 7332–7336.

Petty, R. E., & Cacioppo, J. T. (1986). *Communication and Persuasion: Central and Peripheral Routes to Attitude Change*. New York: Springer-Verlag.

Rosnow, RL (1991). Inside rumor: A personal journey. *American Psychologist, 46*, 484–496.

Social Networks, 17, 129-145.

Watts, D. J. (2004). The "new" science of networks. *Annual Review of Sociology, 30,* 243–70.

Watts, D. J., Dodds, P. S., & Newman, M. E. J. (2002). Identity and search in social networks. *Science, 296,* 1302-1305.

Weimann G. (1982) On the importance of marginality: One more step into the two-step flow of communication. *American Sociological Review 47.*

Chapter 43

Using Conscript™ to Train Cross-Cultural Decision-Making in a Serious Game

Marjorie Zielke, Frank Dufour, Brent Friedman,
Daniel Hurd, Erin Jennings, Michael Kaiser

The University of Texas at Dallas
800 West Campbell Road
Richardson, Texas 75080

ABSTRACT

The First Person Cultural Trainer (FPCT), funded by TRADOC G2 Intelligence Support Activity, is a high-fidelity game-based simulation that stands alone as a way to train for unpredictable, non-linear cultural and behavioral situations and as a member of the Hybrid Irregular Warfare Improvised Explosive Device Network Defeat Toolkit (HI^{2}NT) program. The game inserts the player into Middle-Eastern settings where the goals are to explore the social and political settings through conversation, establish a presence within the area, provide humanitarian aid and eventually use the social status established to gather mission critical information or "golden nuggets." This paper focuses on Conscript™ a highly developed conversation system within FPCT that responds to unscripted gameplay. The game unfolds in the narrative, which uses Conscript™ to tie together written and spoken dialogue, NPC emotional states, sound elements and culturally-accurate motion-captured animations to present a high-fidelity virtual environment for the player to negotiate. In-game communication patterns cannot simply be memorized to progress to the goal. To successfully uncover the golden nuggets, the player must treat the non-player characters (NPCs) with respect, listening to their needs and evaluating their moods, while communicating and accomplishing objectives in a culturally acceptable manner.

Keywords: Simulation, Serious Games, Branching Narrative

INTRODUCTION

The First Person Cultural Trainer (FPCT) is a high-fidelity, game-based simulation, which trains cross-cultural decision-making within a 3D representation of a Middle-Eastern society. FPCT is part of the Hybrid Irregular Warfare Improvised Explosive Device Network Defeat Toolkit (HI^{2}NT) sponsored by TRADOC G2 Intelligence Support Activity. HI^{2}NT is a federation of virtual, constructive and gaming models, providing a combat training center (CTC)-like experience for individual through brigade-staff training audiences. HI^{2}NT is in its second year of development. FPCT functions as a stand-alone cultural trainer, or as part of the Federation.

In FPCT, the story is the game. The game unfolds in the base story or narrative, and the player is tasked with uncovering mission-critical information or "golden nuggets" from the population in a culturally aware and responsible manner. Much of the gameplay in FPCT unfolds within the interactions the player has with the non-player characters (NPCs). From the opening moments of the game, where the player is concerned with talking to the people in charge, to the latter stages where the player must identify golden nuggets, successful communication with the various NPCs is essential. The player essentially wins the game by having effective conversations with the NPCs to uncover information.

These gameplay conversations are processed as branching unscripted and scripted sequences that can be influenced positively or negatively by the NPCs' current emotional states and the player's choice of topics. Further, these conversations can, in turn, influence the behaviors, emotional states, and actions of other NPCs. Through these conversations the player will discuss and record important information. Among the cross-cultural decision-making that unfolds in the game are appropriate ways to engage indigenous populations; proper choice, timing and sequence of conversation topics; character and populace mood determination; and ultimately, skill development to use these cultural insights to accomplish missions.

Our model utilizes visual, auditory, cultural and behavioral components for immersive cultural training using the living-world construct. Living worlds offer nonlinear, unscripted processes for experiencing and safely learning the cognitive complexity and nuance of culture through emergent high-fidelity simulation (Zielke, Evans, Dufour et al. 2009). FPCT, as a living world, has several core components to facilitate cross-cultural decision-making training to include formally researched location typical visual and auditory environments and high fidelity and culturally accurate non-player characters, which emote both verbally and non-verbally—i.e., with gestures and facial expressions. Behavioral, psychological and cultural models, managed by our culture design tool, are also part of the development.

A key component of transmitting and practicing appropriate cultural decision-making within FPCT is Conscript^{TM}—a dialogue system through which the player can have conversations with the NPCs to discover community issues, resolve them and ultimately obtain mission-critical information. As illustrated by the diagram below, the Conscript™ scripting language shares data with the culture design tool, the NPC moods and attitudes model, the FPCT animation database and sound elements, and then calls the correct dialogue banks as the conversation progresses. Golden nuggets are embedded in the dialogues. The process of developing narrative stories through Conscript™, which drives the game-based cultural training tool, will be discussed in this paper.

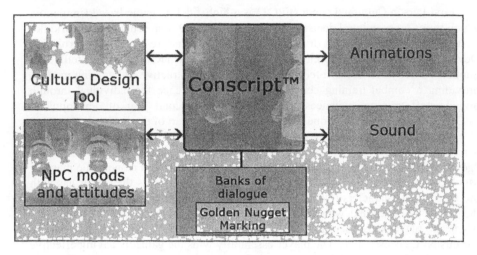

FIGURE 1. The Conscript™ scripting language shares data with the culture design tool, the NPC moods and attitude model, the FPCT animation database, and sound elements and then calls the correct dialogue banks as the conversation progresses. Golden nuggets are embedded in the dialogues.

The ability to be able to translate detailed narrative passages highlights the power of cross-cultural training tools such as Conscript™ to import fine translation of story into a virtual learning environment.

WHY CONSCRIPT™?

Within FPCT, the primary media of storytelling and gameplay are the conversations that the player has with the game characters. In order to implement the living-world construct defined above, these gameplay conversations must be culturally accurate, detailed, and context-sensitive. Given that dynamically generating stories is beyond the scope of our project, we choose to capture nuanced personalities and complex behaviors through writers individually crafting the narrative game conversations.

Conscript™ allows individuals with little or no programming experience to quickly develop context-sensitive dialogue for culturally accurate non-player characters. Writers develop conversations using any common text editing software and then attach the conversation to characters, or to classes of randomly generated characters, through our culture design tool. Individually crafting these multi-faceted conversations is no small undertaking, so several of the Conscript™ features are specifically aimed at reducing this workload. For example, conversation inheritance allows writers to reuse and prototype sections of dialogue across multiple interactions and characters in the game. Features like this have made development in Conscript™ easier than what might be found in other tools.

Conscript™ commands are executed sequentially and can branch at critical points in the conversation based upon context, and variations of context, generated by the player's actions. Any property of any character can be accessed through Conscript™, though, in practice, we focused on emotional state of the NPCs, gender, age, and dialogue text. Exactly what properties exist for a character is implementation defined. This design flexibility of Conscript™ allows for nuanced and dynamic story telling supporting the living world construct.

GAMEPLAY IN FPCT

As per our client's requests, we have developed the story within FPCT to align with five "prologues" or pre-narrative settings, which depict different geographic, cultural, and recent event data sets. Prologues are set within rural and semi-rural Afghanistan and rural, semi-rural, and urban Iraq. Gameplay within FPCT aligns with each of the four steps outlined by our client that represent the process of entering a village and securing information. Step one of this process is concerned with physically entering the village, meeting with the elders, and assessing the situation. Step two deals with ascertaining the humanitarian aid problems within the village while becoming more familiar with the village's way of life. Step three focuses on developing solutions to the problems discovered, such as facilitating construction of a new well. In step four, the player leverages the goodwill and trust he or she has developed with population to acquire mission-critical "golden nuggets."

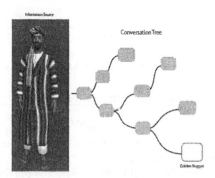

FIGURE 2. Conversations with NPCs through Conscript™ lead to golden nuggets and successful gameplay in FPCT.

USING CONSCRIPT™ TO TELL A STORY

As outlined above, the Conscript™ scripting language shares data with the culture design tool, the NPCs' moods/attitudes, FPCT's animation database and sound elements, and then calls the correct dialogue banks as the conversation progresses. The following is an example of the many ways that Conscript™ can direct the in-game flow of narrative. In one of the prologues or stories in FPCT, the player is directed to or may have heard about an elder NPC named Khan Agha. Depending on the choices that the player makes the conversations could turn out in many different ways. Upon first meeting the NPC Khan Agha, Conscript™ polls the culture design tool to determine what emotional and cultural variables have already been set for this character. Among the variables that can be represented are the NPC's gender, general age designation (elder, adult, or child), prologue variables (such as general population attitudes and cultural effects of recent good or bad events), and any NPC attitude adjustment from the NPC's perception of the player. Such information sharing is important, as it allows the NPC to respond to the player within the correct cultural and situational context.

If Conscript™ determines that Khan Agha's mood is not conducive to talking and the player doesn't introduce him/herself appropriately (which also notifies Conscript™), then the conversation may go like this:

Player: "I have a few questions for you."
Khan Agha: "I don't know anything. Leave me alone."
The NPC would then break off communication.

A player may be able to get help from other NPCs, but overall gameplay still affects ultimate success in obtaining information. In this next example, the player has spoken with Abdullah, who has offered some information. However, if the gameplay up to that point has not been culturally appropriate, this may not help acquire important information leading to the golden nugget. This is illustrated by the dialogue below:

Player: "Abdullah told me that you may have some important information for me."
Kahn Agha: "Then he was mistaken. I know nothing of use to you. Goodbye."
The NPC would then break off conversation.

The prior examples show that Conscript™ allows conversations to succeed and fail in different ways, and it illustrates that no conversation's outcome can be simply memorized to progress in the story.

Should Khan Agha be inclined to speak with the player, he may invite the player to have tea and food before discussing business, which can be accepted or declined. Declining the offer and attempting to talk about business makes the NPC uncomfortable, as this is impolite behavior and it puts Khan Agha in an awkward position.

> Khan Agha: "A man needs his strength. You should eat something."
> Player: "No thank you."
> Khan Agha: "You are not hungry? No matter. What do you want?"

Conscript™ influences the NPC's emotional model based on these player choices, and from this point onward, negotiation may be more difficult for the player, as Khan Agha now might view him or her as someone who is too focused on asking pointed questions rather than engaging in small talk and hearing about the village. This reaction is not set in stone, however, as Khan Agha's current mood and previous knowledge of the player may predisposition him towards forgiveness of small gaffes. This conversation could also take a more positive slant, however, if the player is convinced to take some time to share food and drink.

In this case, Conscript™ updates the NPC's emotional model with the player's choices, and it is enough to convince the NPC to continue the conversation. From this point, the player's choices will continue to affect the dialogue options available, as well as the potential to receive mission-critical information. Once the conversation is completed, Conscript™ lets the NPC "remember" the interaction, and that NPC may pass on impressions of the player to other family members or acquaintances, which could in turn alter their opinions of the player. Conscript™ receives NPC schema data from the culture design tool's behavioral and psychological modeling system and melds it with dialogue and NPC animation data, creating a conversation system that is dependent not simply on the player selecting the "right" choice, but also on the player's perception of the attitudes and situations of the people to whom they are talking, and their ability to adapt their conversations to each individual NPC.

REPRESENTING DIALOGUE WITH CONSCRIPT™

When the player activates an NPC conversation, the scene displayed on the screen changes from an environment shot to a close-up on the selected agent. The player engages the agent in dialogue by selecting one of the introductions generated by Conscript™ from a textual menu according to the context of the encounter. The answer selected by Conscript™ in the conversation tree is displayed in English text, while a recorded sound file in an appropriate language such as Pashto is played back. This sound file corresponds with synchronized animation, providing a realistic audiovisual representation. Facial expressions, body language and voice inflections allow for accurate understanding of the agent's mood and intentions.

This combination of full text segments of dialogue with audiovisual representations is meant to provide rich, complex, and in depth textual development of dialogues and highly realistic emotions and cultural traits without having to create all the audio recordings and facial motion captures associated with every segment of written dialogue expressed in any possible emotional state. Conscript™ dynamically associates every segment of written dialogue with an auditory sequence in line with the general meaning of this segment and with the emotional state of the

NPC. This strategy allows for the development of multiple new stories and dialogues without imposing the generation of additional audiovisual resources.

FIGURE 3. The player engages the agent in the dialogue by selecting one of the introductions generated by Conscript™ from a textual menu according to the context of this encounter.

ASSESSING GAMEPLAY: AFTER ACTION REVIEW

The After Action Review (AAR) tool assesses the player's in-game choices and provides feedback to improve future performance. The AAR relies on data passed from Conscript™ to assess three gameplay criteria. The first criterion, ground truth versus perceived truth, determines the accuracy of the player's understanding of the simulated environment. When conversing with the player, an NPC's body language, facial expression, and vocal inflection change in accordance with their mood. These cues, which are driven by Conscript™, enable the player to evaluate the NPC's emotional state. In the AAR, the ratings the player assigns to each NPC are compared to the actual values to determine how adept the player is at reading emotional cues.

The second assessment criterion, intelligence collection, compares the intelligence sources exploited by the player to the total number of sources available in each stage. Conscript™ is responsible for relaying to the AAR whether the player uncovered the intelligence "golden nuggets" accessible through conversation with NPCs.

In the final assessment criterion, *cause and effect*, the change in the NPC population's opinion of the player over time is analyzed. The population's opinion of the player is influenced by the player's actions and dialogue options selected during conversation, as well as the viral spread of information from one NPC to another.

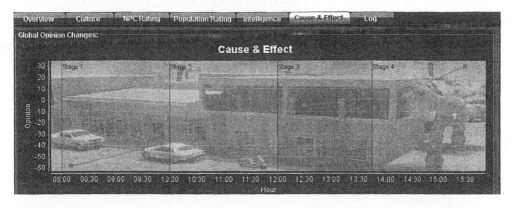

FIGURE 4. The after action review (AAR) allows the player to trace back the cause and effect of the overall attitude of the populace based on the result of gameplay that occurs through conversations developed with Conscript[TM].

DEVELOPMENT CHALLENGES/ OPPORTUNITIES FOR FUTURE DEVELOPMENT

DATA FROM OTHER FEDERATES

Currently, most of the information on geographical, environmental, cultural, behavioral and psychological influences is contained in the prologues outlined above. The information in these prologues directly affects the conversations the player can have and the information that will be available to them. In future iterations of FPCT, the prologues will be replaced with data provided by other HI[2]NT Federate simulations.

For example, data generated by interactions in FPCT and operational activities in the HI[2]NT ground maneuver model (OneSAF) are adjudicated by the HI[2]NT relationship model, the Joint Non-kinetic Effects Model (JNEM). This adjudication reflects civilian reactions and concerns toward military actions and other social factors that will be represented in FPCT through NPC attitudes and conversations and will hinder or help the player's ability to garner information from the NPC. This shift from scripted prologues to dynamic information will provide further opportunities to expand our living-world construct and effectively reduce the amount of time required to prepare FPCT for play.

THE SEMANTIC WEB AND OTHER WAYS TO DEVELOP THE GENERATIVE STORY PROCESS

To date, Conscript™ focuses on providing a scripting language for writers to support story telling in a living-world, game-based simulation. However, we are exploring other ways to generate complex stories that would incorporate other datasets and methodologies, to complement our current processes. Several other components are in the pipeline and will be evaluated for their ability to enhance the narrative aspects of the FPCT game-based simulation platform. These potential new developments include:

- Enhanced integration of gameplay dynamics.
- Dynamic prologue development through integration with other HI^{2}NT federates.
- Social web story development and data mining.
- Greater utilization of current development such as character schemas, and the development of families or social networks.
- Semantic Web-like ontology and annotation from appropriate sources.

Further, while dialogue creation in Conscript™ is geared toward the non-programmer, additional development is under consideration to add a graphical UI layer to facilitate creating flowchart-like dialogue instances, allowing users to better plan dialogues.

SUMMARY

Conscript™ is a multi-faceted design tool that allows writers to imbed detail and nuance on cross-cultural game-based simulations. Critical components in the development allow for complex, goal-driven storytelling as well as the integration of animations, sound and behavioral/psychological/cultural models. Conscript™ allows for multi-media human-based writing, and sets the foundation for rich-narrative, data-base driven platforms that could expedite training on cultural issues within a game-based simulation.

ACKNOWLEDGEMENTS

This project is possible from a contract from TRADOC G2 Intelligence Support Activity, and we appreciate their ongoing support. We also thank our colleagues at the University of Texas at Dallas for their contributions to this paper and the First Person Cultural Trainer program.

REFERENCES

Zielke, M., Evans, M. & Dufour, F., Christopher, T., Donahue, J., Johnson, P., Jennings, E., Friedman, B.,Ounekeo, P. & Flores, R. (2009). Serious games for immersive cultural training: Creating a living world. IEEE Computer Graphics and Applications. Vol. 29(2) p. 49.

Zielke, M., Linehan, T. (2009). The First Person Cultural Trainer. Proceedings Interservice/Industry Training, Simulation & Education Conference (IITSEC).

Chapter 44

Identifying Similarities and Differences of Pictorial Symbol Design and Evaluation of Two Culturally Different Groups

Yoon Suk Lee

The Grado Department of Industrial and Systems Engineering
Virginia Polytechnic Institute and State University
Blacksburg, VA 24061, USA

ABSTRACT

Pictorial symbols are widely used to convey warning messages. Despite various advantages of using pictorial symbols, studies have reported that people from different cultures may not comprehend the meaning of a pictorial in the same manner. In this work, two culturally different groups were recruited to perform a pictorial symbol design and evaluation. Mixed-methods design was conducted to identify the similarities and differences of pictorial symbol design and evaluation. Practical implications of the results are discussed.

Keywords: Cultural difference, Pictorial symbol, Safety symbol, Design and Evaluation

INTRODUCTION

Pictorial symbols have been widely recognized as a cost-effective medium to illustrate safety related messages (Davies, et al., 1998). Using pictorial symbols may also increase text comprehension when used together and can be more rapidly perceived than text (Rogers, Lamson, & Rousseau, 2000). Low literacy people may comprehend better and non-English speakers may also correctly interpret the symbolic information (Handcock, et al., 2004). Conveying meanings independent of language has gained significant amount of attention due to globalization. However, study results revealed that only very few pictorial symbols are universally understood (Davies, et al., 1998).

Accidents due to misinterpretation of warning symbols can be identified in various industries as well as in our daily lives, such as pharmaceutical industry (Ringseis & Caird, 1995), transportation industry (Bazire & Tijus, 2009), and consumer products (Davies, et al., 1998), just to list a few. One of the primary difficulties involved in designing a universally accepted symbol is due to cultural differences. Cultural differences may influence the level of understanding of certain concepts (Nakamura, et al., 1998; Nigam, 2003). To overcome the cultural barriers, The American National Standards Institute has established guidelines and procedures for symbol comprehension evaluation (ANSI, 2002). However, studies have demonstrated that the accuracy rate of correct interpretation for common ANSI symbols was rated below a satisfactory level (Hancock, Rogers, & Fisk, 1999).

Therefore, the study aims to identify cultural differences between the citizens of Korea and the United States to examine what aspects are similar or different in relation to the design and comprehension of a safety related pictorial symbol.

LITERATURE REVIEW

Previous studies have examined the mechanisms of how people perceive and interpret warning signals. The Communication-Human Information Processing (C-HIP) model, explains the phases as "source", "channel", "attention", "comprehension", "attitudes" / "beliefs", "motivation", and "behavior" (Wogalter, DeJoy, & Laughery, 1999). Similarly, Rogers and colleagues (2000) described the mechanism as a sequential process of "notice", "encode", "comprehend", and "comply". All of the phases are critical. However, the major problem addressed in the previous section can be associated with the "comprehension" phase of the models. People from different culture may perceive the same incident differently. The cause of such difference can be explained by the concept of mental model, which is individual's worldview that is used to describe, interpret, and predict the environments (Johnson-Laird, 1983). Multiple mental models form a collective worldview, also known as meta-schema, and culture is considered as a component of meta-schema (Smith-Jackson & Wogalter, 2004). Thus, cultural factor accounts a large portion in understanding the reason behind why people comprehend the same pictorial symbol in a different manner.

To overcome the cultural barriers, studies have examined different interpretation of pictorial symbols across different cultured groups. A recent cross cultural study conducted by Chan et al. (2009), demonstrated that Americans interpreted the American security safety symbols more accurately than Hong Kong-Chinese and Korean people. Likewise, most empirical studies have focused on the evaluation of pictorial symbols, which warrants more research on cultural influence on the design of pictorial symbols. Little knowledge is known about people's visual representation of a risk situation. These visualizations can be compared across different cultured groups to determine commonalities that can be further utilized as a universally acceptable pictorial symbol.

PURPOSE OF THE STUDY

The purpose of this study was to identify similarities and differences between the citizens of Korea and the U.S. in designing and evaluating safety pictorial symbols. Therefore, following research questions and hypotheses were posed.

Research question 1. What are the similarities and differences of safety symbol design between and within different ethnic groups?

- *Hypothesis 1: different patterns of pictorial symbol will emerge between different ethnic groups.*
- *Hypothesis 2: similar patterns of pictorial symbol will emerge within the same ethnic groups.*

Research question 2. How does the level of comprehension vary based on the symbol designers' cultural background?

- *Hypothesis 3: participants having similar cultural background with the symbol designers will rate the symbols higher.*

EXPERIMENTAL DESIGN

A mixed-method design was conducted to answer the research questions. A laboratory-based experiment was conducted specifically to answer the first research question. Participants were randomly assigned into three different dyads (Korean, U.S., and mixed-culture). Participants first designed a 'low overhead clearance' symbol individually. Then they were asked to collaboratively design the symbol. After the design tasks, the researcher conducted a semi-structured interview to understand the rationale and the meanings of their sketches. The entire session was video recorded to identify the similarities and differences within and between the two groups. Qualitative analysis was conducted to test the hypotheses.

An online survey was conducted to answer the second research question. Total 18 symbols designed from the lab-based experiment were used in addition to two ANSI Z535 approved symbols that represented 'low overhead clearance' (12 individual design, 6 collaborative design, and 2 ANSI approved symbols). The Questionnaire for the Comprehension Estimation (ANSI, 2002, p. 37) was used to examine the level of comprehension. The participants were asked to rate from 0% (no one will be able to understand this symbol) to 100% (everyone will be able to understand this symbol). Quantitative analyses were conducted to test the hypothesis.

PARTICIPANTS

For the laboratory-based experiment, 12 college students (six Korean, six United States) were recruited. Participants' mean age and standard deviation was 23 and 6.3 respectively. Students who have design experience or majoring design related disciplines were excluded from the study. In addition, the Korean participants must have stayed less than a year in the United States to reduce the acculturation biases. Likewise, the U.S. participants who lived abroad were excluded from the study. For the online survey, 68 participants responded. However, participants who lived outside of their country for more than half of their lives were excluded from the study, resulting 50 responses (25 Korean, 25 United States).

TASK

A written description, 'low overhead clearance', was provided to the participants in their own language. This was to control the language priming effect (Lechuga, 2008). The participants were first asked to individually design a pictorial symbol that best illustrates the written description for two minutes. The reason for such time restriction was to capture the first images that participants associated with the concept of 'low overhead clearance.' Then, the participants collaboratively designed a pictorial symbol for 15 minutes. The two minutes and 15 minutes constraints were set based on the three pilot sessions.

PROCEDURE

Upon arrival of the participants, an overview of the study protocol was provided. After obtaining consent for participation, the researcher explained about the tasks. Then, the participants designed pictorial symbols that illustrated the provided description, individually and then collaboratively. After completing both tasks, a short semi-structured interview session was conducted. Example questions were "why did you draw this part?" "What does this mean?"

RESULTS

The objectives of the study were to identify the similarities and differences of a pictorial symbol design and evaluation between two different ethnic groups. For the first research question, the design outcomes from the laboratory experiment were compared based on participants' nationality. Content analysis of the semi-structured interview was conducted to identify emerging themes, which further explained the reasons for such similarities and differences. To answer the second research question, quantitative analyses were conducted for the online survey responses using a statistic analysis software, JMP 7.0, with an alpha (α) value of .05. Normality was tested using Shapiro-Wilk W test. However, none of the responses were normally distributed. Therefore, the Wilcoxon signed-rank test was conducted to identify the differences between the Korean and U.S. participants' responses.

The similarity between the Korean and the U.S. participants were how they represented the hazardous component of 'low overhead clearance.' A rectangular object at the level of person's head [Figure 1-(a)] or a line indicating low ceiling [Figure 1-(b)] were the two primary hazardous components illustrated by the Korean and U.S. participants.

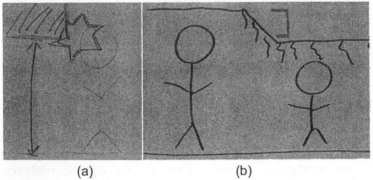

(a)   (b)

FIGURE 1. Example symbols that represent (a) a rectangular object, and (b) a low ceiling as the hazardous component.

The differences between the Korean and the U.S. participants were how they represented the accidents due to low overhead clearance. All of the Korean participants illustrated the accident as a star-shaped figure; whereas all of their U.S. counterparts represented the accident as squiggling lines (Figure 2).

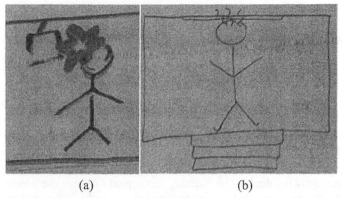

(a)   (b)

FIGURE 2. Example symbols representing (a) star-shaped figure, and (b) squiggling lines as accidents due to low overhead clearance.

Based on the theme analysis of the semi-structured interview, all of the Korean participants who illustrated the star-shaped figure intended to convey the 'moment of impact.' All of the U.S. participants who illustrated the squiggling lines intended to convey 'injury' or 'pain.'

The quantitative analysis of the online survey responses supported the third hypothesis. Significant differences between the Korean and the U.S. responses were identified for five out of 20 symbols ($p < .05$). Specifically, Korean participants rated significantly higher than their U.S. counterparts, meaning that more people will understand these symbols as 'low overhead clearance.' The Korean dyads designed three symbols, and mixed-culture dyads designed two symbols.

Additionally, a distinctive pattern was identified within the five symbols. The Korean participants rated significantly higher for the symbols that contained rectangular objects as hazardous materials, with a person at the moment of collision. The online survey responses were ranked in an ascending order to examine any further patterns within each group. Similar patterns were identified for the Korean groups. However, no pattern was found for the U.S. group.

DISCUSSION

Similarities and differences were identified between the Korean and the U.S. participants in designing and evaluating pictorial symbols that represent 'low overhead clearance.' The similarities identified from the design outcomes between the Korean and the U.S. participants were the hazardous components associated with 'low overhead clearance.' Both groups illustrated rectangular objects and low ceilings, meaning that these two types of illustrations were one of the first images that participants associated with 'low overhead clearance.' Previous studies have reported that graphic representation of nouns can be more culturally sensitive than verbs (Nigam, 2003). In case of 'low overhead clearance', the participants intended to sketch a hazardous object (noun) to effectively convey the concept. The two representations of hazardous objects (rectangular object and low ceiling) emerged in both groups, meaning that noun variation in the phrase 'low overhead clearance' are minor between the Korean and the U.S. groups.

The difference between the Korean and the U.S. participants' design was the representation of the outcomes due to failure to avoid the hazardous situation. All of the Korean participants illustrated a star-shaped figure to convey the 'moment of impact'; whereas, all of the U.S. participants sketched squiggling lines to convey 'pain or injury.' These two meanings explain two differing perspectives of perceiving a hazardous incident. The Korean participants were more objective to the event itself; whereas, the U.S. participants were emphasizing the negative outcome that may influence a person who failed to avoid the hazardous incident. In general, individuals from Western cultures are more egocentric than individuals from Eastern cultures (Heine & Lehman, 1997; Markus & Kitayama, 1991; Rose, et al., 2008). Egocentrism can be defined as regarding oneself to be the center of all things (Markus & Kitayama, 1991). When the researcher probed the meaning of squiggling lines, all of the U.S. participants mentioned either injury or pain. However, none of the Korean participants mentioned such words.

The online survey results indicated that the Korean respondents rated five symbols significantly higher than those of the U.S. respondents. The five symbols were either designed by the Korean or mixed-culture dyads, indicating that the symbol designers' nationality may influence the level of symbol comprehension. This result highlights the importance of cultural attributes associated with symbol design and comprehension. Heur (2000), in her empirical study of four differing cultural groups has also concluded that individuals with different language and experiences may not comprehend pictorial symbols in the same manner.

The ranked order of the online survey responses revealed a pattern for the Korean group. Korean participants rated higher for the symbols that contained a person at the moment of collision with a rectangular object. This can be explained by the different cognitive styles between Easterners and Westerners. Westerners have tendency to think more abstract and imaginative; whereas, Easterners are more likely to have a concrete thinking style (Rau, Choong, & Salvendy, 2004). The symbols that the Korean respondents rated higher were all simple and straightforward. The semi-structured interview results also supported this explanation. The participants who sketched low ceilings wanted to lower the height of the ceiling to make it more obvious. On the other hand, participants who sketched rectangular objects justified their drawings by explaining that they intended to make the accident as clear as possible.

To examine the reason for no emerging pattern for the U.S. group, a post-hoc analysis of within group variance was conducted. The responses by the Korean participants displayed larger variances than those of the U.S. group. In other words, it is likely that the U.S. participants did not have a distinctive preference of symbols that illustrated 'low overhead clearance.'

CONCLUSIONS

The findings from the study may yield practical guidelines and recommendations for safety related pictorial symbol design.

Design guidelines and suggestions for 'low overhead clearance':
- The hazardous component of 'low overhead clearance' can be illustrated as a rectangular object or a more obvious drawing of a low ceiling.
- The egocentric design may influence the level of comprehension as well as behavior outcomes. Easterners may interpret the squiggling lines incorrectly, which may lead to accidents. However, the squiggling lines may persuade Westerners to comply with the symbol to avoid injury or pain.
- The different cognitive styles must be incorporated in the design. Pictorial symbols are considered as abstract illustrations. However, to reduce the cultural barriers between Easterners and Westerners, more simple and concrete illustrations are suggested.

The general procedure for safety symbol design described in the ANSI Z535 does not consider cultural attributes that may influence the level of comprehension. Involving users in the early design and evaluation phases of a product life cycle is a basic Human Factors design principle that can be applied to overcome cultural barriers associated with symbol comprehension (Piamonte, Abeysekera, & Ohlsson, 2001) The empirical results from this study may contribute in developing recommendations for safety symbol design procedures. The steps for current safety symbol design described in the ANSI Z535 are as listed below.

- Identify hazard data
- Collect all existing symbols
    a. Do symbols exist?
    b. If no, artist designs new candidate symbols
- Make symbols comparable
- Identify target audience, message, etc.
- Evaluate (refer to ANSI Z535 for detailed information regarding evaluation procedures).

Based on the results from this study, following modifications are recommended.

- Identify hazard data
- Identify target audience, message, etc.
- Examine cultural attributes of the target audience
- Conduct participatory design with active user involvement (users must represent the sample population of the target audience)
- Inform the artists (symbol designers) of the preliminary design results

- Make symbols comparable
- Evaluate

The modified procedure identifies the target audience at the earlier design phase. This is to identify cultural attributes associated with the target audience, so that user profiles can be established for participatory design. For the participatory design, design tasks similar to the laboratory experiment conducted in this study can be utilized, and then the results can be used as guidelines for the symbol designers. As Young and Wogalter (2001) states in their study, the extent to which a symbol is comprehensible to the target population determines the symbol's utility of conveying information. This study is meaningful in a way that it not only empirically tested the influence of cultural attributes in symbol design and evaluation, but also developed practical guidelines and recommendations for design procedures that may enhance the level of comprehension by mitigating the cultural barriers. Future empirical studies utilizing the modified procedure needs to be tested.

ACKNOWLEDGEMENTS

This research was approved by the Virginia Tech Institutional Review Board, and was conducted as a course assignment. The author would like to express his appreciation for the supervision and inputs provided by Dr. Tonya Smith-Jackson. The author would like to also thank Doo Young Choi for his support on recruiting Korean participants.

REFERENCES

ANSI (2002). American National Standard Criteria for Safety Symbols. 1300 North 17th Street, Rossyln, VA 22209: National Electrical Manufactureres Association.

Bazire, M., & Tijus, C. (2009). Understanding road signs. *Safety Science, 47*(9), 1232-1240.

Chan, A., Han, S., Ng, A., & Park, W. (2009). Hong Kong Chinese and Korean comprehension of American security safety symbols. *International Journal of Industrial Ergonomics*.

Davies, S., Haines, H., Norris, B., & Wilson, R. J. (1998). Safety pictograms: are they getting the message across? *Applied Ergonomics, 29*(1), 15-23.

Hancock, H., Rogers, W., & Fisk, A. (1999). *Understanding age-related differences in the perception and comprehension of symbolic warning information.*

Handcock, H., Rogers, W., Schroeder, D., & Fisk, A. (2004). Safety symbol comprehension: Effects of symbol type, familiarity, and age. *Human Factors: The Journal of the Human Factors and Ergonomics Society, 46*(2), 183.

Heine, S., & Lehman, D. (1997). Culture, dissonance, and self-affirmation. *Personality and Social Psychology Bulletin, 23*, 389-400.

Huer, M. B. (2000). Examining perceptions of graphic symbols across cultures: Preliminary study of the impact of culture/ethnicity. *Augmentative and Alternative Communication, 16*(3), 180-185.

Johnson-Laird, P. N. (1983). *Mental Models: Towards a Cognitive Science of Language, Inference, and Consciousness*. Cambridge, MA: Harvard University Press.

Lechuga, J. (2008). Is Acculturation a Dynamic Construct?: The Influence of Method of Priming Culture on Acculturation. *Hispanic Journal of Behavioral Sciences, 30*(3), 324.

Markus, H., & Kitayama, S. (1991). Culture and the self: Implications for cognition, emotion, and motivation. *Psychological review, 98*(2), 224-253.

Nakamura, K., Newell, A., Alm, N., & Waller, A. (1998). How do members of different language communities compose sentences with a picture-based communication system?óa crossñcultural study of pictureñbased sentences constructed by English and Japanese speakers. *Augmentative and Alternative Communication, 14*(2), 71-80.

Nigam, R. (2003). Do Individuals from Diverse Cultural and Ethnic Backgrounds Perceive Graphic Symbols Differently? *Augmentative and Alternative Communication, 19*(2), 135-136.

Piamonte, D., Abeysekera, J., & Ohlsson, K. (2001). Understanding small graphical symbols: a cross-cultural study. *International Journal of Industrial Ergonomics, 27*(6), 399-404.

Rau, P., Choong, Y., & Salvendy, G. (2004). A cross cultural study on knowledge representation and structure in human computer interfaces. *International Journal of Industrial Ergonomics, 34*(2), 117-129.

Ringseis, E., & Caird, J. (1995). *The Comprehensibility and Legibility of Twenty Pharmaceutical Warning Pictograms*.

Rogers, W., Lamson, N., & Rousseau, G. (2000). Warning research: An integrative perspective. *Human Factors, 42*(1), 102.

Rose, J., Endo, Y., Windschitl, P., & Suls, J. (2008). Cultural differences in unrealistic optimism and pessimism: The role of egocentrism and direct versus indirect comparison measures. *Personality and Social Psychology Bulletin, 34*(9), 1236.

Smith-Jackson, T., & Wogalter, M. (2004). Potential uses of technology to communicate risk in manufacturing. *Human Factors and Ergonomics in Manufacturing, 14*(1).

Wogalter, M., DeJoy, D., & Laughery, K. (1999). *Warnings and risk communication*: CRC.

Young, S., & Wogalter, M. (2001). Predictors of pictorial symbol comprehension. *Information Design Journal, 10*(2), 124-132.

CHAPTER 45

Challenges and Approaches for Automating HSCB Decision-Making

Michael Gosnell, David Spurlock, Warren Noll

21st Century Systems, Inc.
6825 Pine Street Suite 141
Omaha, NE 68106, USA

ABSTRACT

This paper describes major existing challenges to developing and applying useful Human, Social, Cultural, and Behavioral (HSCB) models and implementations within decision-making applications. In addition, we describe progress we have made addressing some of these challenges and highlight areas where further research is needed to better capitalize on that progress.

Keywords: Human, Social, Cultural, Behavioral, Modeling, Decision-Making

INTRODUCTION

The effects of culture are only partially understood in many important contexts. Even when well-understood by experts, that understanding is not necessarily shared by non-experts who are engaged in intercultural or multicultural interactions. Moreover, the definitions of cultural constructs (and even the definition of culture itself) are often ambiguous or misunderstood across disciplines. However, it is generally accepted that culture plays a role that differentiates individual and group behaviors even when other situational variables remain unchanged. The desire to understand culture and its impact has existed for quite some time. However, it has only recently been emphasized in the military domain through DoD directives (2008 Directive 3000.07 on Irregular Warfare), doctrine updates (2008 Army Field

Manual 3-0; Chapter 3: Full Spectrum Operations), the 2006 Quadrennial Defense Review and increasing focus on the HSCB modeling domain.

We briefly outline some key challenges preventing widespread implementation and integration within existing tools and analyses, and present mechanisms which address these challenges. We first highlight HSCB model challenges discussing a few illustrative models relating to individualism and collectivism. Following discussion of these challenges, we briefly outline elements of a previous cultural response framework (Gosnell et al., 2010) and how this fits in the context of a proposed automated model selection framework. Presented next is a primitive baseline ontology which can capture general characteristics of HSCB models. These features are incorporated within the proposed model selection framework which is outlined and examined within the context of the existing cultural response framework toward increased usability.

MODEL BACKGROUND

Although many useful models in the behavioral and social sciences date back decades, new models or revised versions of existing models emerge often as both theory and measurement techniques improve. New social phenomena emerge as well due to advances in technology and changes in economic, political, legal, and demographic landscapes. Our concern here is with cultural models that might be candidates for predicting behavioral responses (actions) in contexts or situations where the potential for misunderstandings is significant. For a more comprehensive examination of HSCB models, theories, and data sources, see Hartley (2008).

To better appreciate the issues involved in related research, we present a brief summary of a variety of modeling considerations that pertain to the HSCB domain. We sketch typical approaches used to generate HSCB models and the very large number of possible variations in the nature of the models and accompanying data sets. This limited overview serves as a foundation for understanding the potential value of the associated frameworks, and the benefit for automated reasoning.

Models in the behavioral and social sciences are typically developed with some combination of theory-driven or data-driven processes that ideally complement one another. In some cases, a strictly empiricist investigation produces a model primarily or even exclusively from applying statistical techniques such as linear regression or factor analysis to data sets without any prior theoretical frameworks. In other words, a model is constructed from the data directly. Subsequently, a theory that explains the relations among the variables may be formulated. However, if the purpose of the model is merely to describe or, perhaps, predict relations among variables, no attempts to actually explain the phenomena via theory development may be attempted. This approach is usefully employed when analyzing preexisting data from archives or other sources originally collected without regard to a researcher's current investigation.

In other cases, a particular theoretical perspective is adopted first and that theory is employed to determine the variables that will be measured, how they will be measured, and how the empirical results are to be interpreted. In this approach, researchers may be attempting to provide support for (or extend) an existing theory or they may be trying to evaluate the relative merits of one or more theories. Usually the concern here goes beyond simple description or prediction and efforts are focused on explaining the phenomena of interest. Hybrid approaches that use theory for guidance (to varying intermediate degrees) are also common.

In addition to the vast array of actual HSCB models of specific phenomena, quantitative behavioral and social scientists have produced an enormous amount of literature on an extremely wide range of modeling approaches, complementing and incorporating models of statisticians, economists, computer scientists, and a number of other disciplines. It is beyond the scope of this paper to review even a tiny subset of these. However, it is important to note that over many decades, much effort has been devoted to exploring the critical aspects of model building, model testing, exploratory and confirmatory approaches, measures of "goodness of fit," appropriate levels of analysis (e.g., individual, group, societal), appropriate scales of measurement (e.g., nominal, ordinal, interval, ratio), reliability of measures, techniques for validating and generalizing models across samples, populations, and contexts, and a host of other methodological issues. Approaches or techniques, regardless of their complexity, that are implemented without consideration for limitations widely discussed in the literature and in textbooks should elicit skepticism from the audience evaluating the results obtained from such cavalier usage. Even when used properly, many methodological judgment calls may be required during investigations. Thus, because different judgments may be made by different investigators and because any given method has its weaknesses, the use of multiple methods as well as the replication of studies is usually encouraged to confidently answer any research question.

As part of the model development process for many common cultural models, researchers rely on techniques such as factor analysis, path analysis, and the combination of the two known as Structural Equation Modeling (SEM). Briefly, factor analysis tries to statistically ascertain the relationship among measured indicators (e.g., questionnaire items) to an abstract explanatory concept known as a factor (e.g., individualism-collectivism). Path analysis is used to describe the causal relations among the actual measures. Often these measures are scale scores aggregated across items that "load" on the underlying factor. SEM is used to model the relations among the underlying factors by combining the factor analyses and path analyses. In the process of constructing these models, researchers must make judgments from statistical information about the relative importance of particular measured indicators as well as the relative impact of particular factors. During the process, researchers will decide to include or exclude particular indicators or factors based upon both the statistical findings and justifiable theoretical considerations. Both types of decisions often require subtle judgments of meaning and precise conceptual discriminations.

In light of the foregoing, it is important to recognize that cultural theories and models are neither trivial nor arbitrary, even though each model is necessarily limited by the number of variables it can accommodate and the contexts in which it can usefully be used. No model in the behavioral and social sciences can make correct predictions of the behavior of a specific individual or small group every time in every situation. Most models perform better predicting the distributions of tendencies of large numbers of people in relatively circumscribed situations (e.g., percentage of voters who will vote for the Republican candidate in a particular state or district in a particular election). The probability that a specific behavioral prediction will be accurate in a specific instance will rise as the amount of relevant data pertaining to the prediction increases and as the relations among the variables in the model are accurately customized to reflect the particular circumstances in the situation of interest. Moreover, as the number of plausible action (response) options to be predicted is reduced, the more successful the predictions are likely to be, all other things being equal.

ILLUSTRATIVE CULTURAL MODELS

One of the most widely studied characteristics of culture is the relation of the individual to the social group. Although many particular aspects of this relation are worthy of study, the predominant concept that has emerged in recent decades is known as individualism-collectivism. Broadly, individualism refers to an emphasis on the primacy of the individual's independent self-identity, goals, desires, and personal attributes while collectivism subordinates those individual concerns to those of the social groups (e.g., family, community, society) in which the individual is embedded. Certainly individual concerns are often important in collectivist cultures just as community and family issues matter in individualist cultures, but the relative balance has been shown to be systematically different in significant ways in different cultures.

A full review of the literature on individualism-collectivism and related cultural concepts is impossible in this paper; we recommend excellent treatments by Triandis (1995), Matsumoto (1996), and Oyserman, Coon, & Kemmelmeier (2002). We will only briefly sketch some of these key ideas below to highlight that even models designed around a single root concept can pose challenges.

Geert Hofstede produced pioneering cross-cultural research and analyses, based upon results from surveys of IBM employees around the world beginning in the late 1960's and early 1970's. From Hofstede's early work and many subsequent refinements (2006), he defined and documented five cultural dimensions including a measure of individualism versus collectivism. Many have worked to refine measurement scales and develop insights into specific contributing values and behavioral implications. Singelis et al., (1995), among others, have produced and tested instruments that can be used to quickly measure individualism-collectivism with short scales. Triandis (1995) reviews much of the work on this topic emphasizing many of the measurement techniques and challenges as well as conceptual and practical issues. For instance, he explores a theoretical refinement of the original concept to differentiate horizontal and vertical subtypes of individualism and collectivism using the theory of self-construal proposed by Markus and Kitayama (1991). (The vertical aspect entails notions of rank and inequality and the horizontal aspect entails notions of similarity and group cohesion.) In addition, Triandis usefully cautions against confusing analyses of individuals and analyses of cultures. Hofstede has also emphasized that, within culture, individual differences cannot be overlooked.

The GLOBE study (House et al., 2004) examined a variety of factors for cultural distinction. House et al. expanded Hofstede's five dimensions to nine, adding additional categorical classifications such as splitting Collectivism into Institutional Collectivism and In-Group Collectivism. Recent scholarly literature includes critical commentary on the GLOBE study from Hofstede, among others, and it seems likely that these dimensions and their measurements will evolve as refinements in models and instruments are proposed and evaluated.

In work by Mary Douglas (1970) and others (Gross & Rayner 1985, Thompson et al., 1990), concepts not unrelated to individualism and collectivism were independently invented and developed to explain cultural characteristics. These concepts became known as "grid" and "group." Grid refers to a generalized form of individual relationships like a parent with a child or a manager with a subordinate worker and may be seen as vaguely corresponding to aspects of individualism. Group refers to degree of membership in and identification with a defined collectivity and may be viewed as containing aspects of collectivism. While the concepts of grid and group are then used in the model in ways that do not permit a simple transformation into individualism and collectivism, they do share enough conceptual territory to be intellectual neighbors, so to speak. Various methods provide quantitative measures of the Grid and Group dimensions (Chai et al., 2009; Kahan et al., 2007; Marris et al., 1998). Mamadouh (1999) provides a comprehensive background and overview of Douglas's Cultural Theory.

MODEL DIFFICULTIES

The illustrative cultural models described above have sufficient theoretical and empirical justification to warrant examination for predictive purposes in contexts where they are relevant. Among the many difficulties encountered using them is that they included somewhat overlapping concepts and measures. Advocates of each model may disagree about which concepts are more fundamental or which factors explain variations in other factors. Moreover, the proportion of variance in observed values of a particular predicted variable is often only a fraction of the total variation even under optimal conditions. While these models do not necessarily directly relate to behavioral predictions, individualism-collectivism measures have been correlated to behaviors which may be of interest in certain contexts, for example in Nisbett and Cohen's work on violence (1996).

FRAMEWORK OVERVIEW

Gosnell et al. (2010) have previously proposed a cultural response framework, which incorporates HSCB modeling techniques along with unique machine learning algorithms to provide analyses without bias toward any model, and while in the presence of incomplete input information. Although this framework tackles some of the aforementioned problems, there is still difficulty in selecting models within the specific requirements of the desired analyses.

At a high-level overview of the combined solution, users interact primarily with the Automated Model Selection (AMS) framework. The AMS framework can interact directly with HSCB models and associated data, or with HSCB models enhanced through the preexisting cultural response framework. Inference engine capabilities in the cultural response framework are used to supplement model analyses in the presence of incomplete information, extending the scope of the framework into the input domain. Multiple HSCB models performing identical reasoning are incorporated into a single output analysis without prejudice to any model, extending the response framework scope into the output domain.

The AMS framework, presented here, extends existing capabilities by adding HSCB model intelligence which allows users an intuitive way to interact with the models. The bulk of the intelligence is available due to the HSCB model metadata, formulated in concert with a fully developed ontology which provides the foundation for understanding models' capabilities and usefulness. As there is yet no standard HSCB ontology or generally accepted practices across disciplines, we provide a rudimentary ontology structure which will serve to illustrate the implementation potentials of the AMS framework. Following the discussion of the sample ontological specification, the framework is detailed along with suggested implementation considerations.

Even though our existing framework begins to tackle the data problems of source data availability and incorporation of uncertainty within the analyses, additional research and development is required for judging the quality and relevance of a given set of candidate models. The AMS framework, in concert or implemented separately from the response framework, begins to provide automated feedback on determining the best models at the right time for smarter HSCB integration.

PRELIMINARY ONTOLOGY

For acceptance and widespread HSCB usage to be successful, it is paramount to obtain standardization across the included domains. While the HSCB community as a whole will work toward common terminology and an acceptable model ontology which can help drive this standardization, we introduce a preliminary ontology structure which serves to illustrate how these standardized capabilities may be utilized within an automated model selection framework.

Of foremost importance is an understanding of a model's limitations. While there may be a variety of ways to enumerate limitations, a single measure of uncertainty may suffice. A single measure of uncertainty would provide a quick, clear measure of appropriateness of resulting solutions or predictions, but may risk discarding underlying information – such as if the uncertainty was due to inherent model unpredictability or characteristics of input data, etc. Conversely, benefits of a single measure include simplicity of a single value which could easily incorporate the error characteristics of reasoning under incomplete information, provided in our previous cultural response framework.

In addition to general model uncertainty, applicability needs to be captured. Each model is assumed to be applicable under certain conditions per its creator's expertise, being influenced by, or measuring influence on specific characteristics. However, multiple models can overlap applicability in part or in whole. Characteristics driving applicability need to be enumerated and we propose organized in a hierarchical manner when possible. This organizational feature will provide a robust way to reason about the data as will be discussed later.

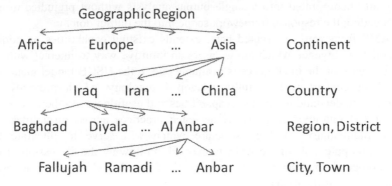

FIGURE 1. Example hierarchical depiction of Geographic Region.

Figure 1 shows an example of how a hierarchical approach could be constructed for geographic region. Within the hierarchical structure, numeric codes could represent each level similar to how the North American Industry Classification System identifies the type of activities a business performs. In such a structure, the first digit may indicate the continent, perhaps with 7 indicating the continent of Asia. Following this guide using two digits for country, region/district, and city/town (from Figure 1), one might end up with Fallujah being represented as 7110521 where 7 represents Asia, 11 identifies Iraq, 05 indicates Al Anbar, and 21 represents Fallujah itself. This type of hierarchical implementation can easily represent larger regions by using 00 as a "don't care" specification. In other words, Al Anbar would be represented as 7110500, and Iraq would equate to 7110000.

In addition to geographic region, political region, religious region or division, and model discipline are all areas of example which seem to fit within the *applicability* facet of our sample ontology. Finally, bringing everything together is some sort of measure of model scope such as an individual, organizational, and societal as Zacharias et al. discuss (2008).

The baseline ontology is then composed of the components of model uncertainty and applicability through hierarchical and generalizable divisions of geographical, political, and religious regions, discipline, and scope. Important characteristics of the ontology include the uncertainty, representation of components within the ontology, and the ability to know the appropriate scope at which the model applies.

AUTOMATED MODEL SELECTION

With a rudimentary ontology in place, models can be reasoned about within machine intelligence algorithms once an appropriate representation is available. Our automated model selection framework seeks to provide model selection, ranking, and recommendations based on the desired model application specified through the model metadata encapsulated within the ontology.

 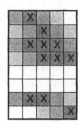 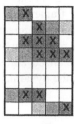

FIGURE 2. Ontological representations: explicitly, following a binomial distribution, and as an intelligent distribution.

Selection, ranking, and recommendations all stem from an understanding of model applicability, which was an underlying factor considered while creating the baseline ontology. Figure 2 represents an example of how the areas of applicability can be visualized with each applicability aspect appearing hierarchically within a row with the most general case on the left and the most specific classification on the right. A hierarchical tree structure (see Figure 1) does not map ideally into the representation of Figure 2, but each possible branch could be interpreted as a row. Using the ontological representation structure of Figure 2, each model could be identified in a number of ways from a simplistic representation to a more robust, tailored representation using probabilistic distributions, able to capture additional subject matter knowledge and uncertainty.

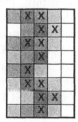 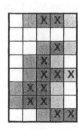 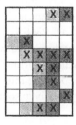

FIGURE 3. Illustrative Societal, Organizational, and Individual scoped ontological representations with intelligent distributions.

The fundamental features of the automated model selection framework include incorporation of model scope and uncertainty within each model's description. In the AMS framework, the ontology is used to provide the standard method of enumerating and processing model metadata, focused on the model's applicable characteristics. With this foundation, reasoning mechanisms can be easily incorporated to facilitate the selection and comparison mechanisms. For example, a simplistic difference measure comparing the intelligent distribution of Figure 2 to the illustrative societal, organizational, and individual scoped models of Figure 3 would show the most similarity (least difference) with the organizational model, followed by societal and individual.

Additional learning mechanisms could be incorporated within the reasoning components of our framework. In addition to the initial model selection and analyses, observing user interaction could allow additional learning, tailored to individual users, to provide appropriate selections based on user behavior. This would allow individualized adjustments based on both the known *a priori* model metadata as well as the dynamic interaction as users work to achieve their desired objectives.

FUTURE EFFORTS

There is clearly room for improvement in the HSCB domain, from providing appropriate initial models to incorporating HSCB reasoning within decision-making processes. We have highlighted some underlying difficulties and provided discussion points on methodologies and implementations which provide a stronger base for HSCB model implementation and use. The existing cultural response framework provides model input and output adaptability, allowing reasoning under incomplete information and providing capabilities to incorporate multiple models in output reasoning. Our automated model selection framework, presented here, can extend that work (or work independently with HSCB models) to provide the base capabilities necessary for enhancing user interaction. The AMS framework was presented around a primitive base ontology structure which incorporates uncertainty and model scope within a hierarchical approach, allowing subject matter experts to quickly and intuitively align the impact of models within the designated applicability – which can include characteristics of what the model measures influence on (such as PMESII factors – Political, Military, Economic, Social, Information, Infrastructure) and what influences the model (such as DIME factors – Diplomatic, Information, Military, Economic). Features of these frameworks can help propel HSCB modeling into mainstream decision-making by eliminating much of the user burden typically associated with navigating the sea of HSCB models.

REFERENCES

Chai, S., Liu, M., & Kim, M. (2009). Cultural Comparisons of Beliefs and Values: Applying the Grid-Group Approach to the World Values Survey. *Beliefs and Values*, *1*(2), 193–208.

Douglas, M. (1970). *Natural Symbols: Explorations in Cosmology*. London: Routledge.

Gosnell, M., Barker, J., Walenz, B., Woodley, R., & Noll, W. (2010). Adaptable Framework for Cultural Response Measurement. *International Journal of Computational Intelligence: Theory and Practice*, 5(1).

Gross, J., & Rayner, S. (1985). *Measuring Culture*. New York: Columbia University Press.

Hartley, D. (2008). *Human, Social, Cultural Behavior (HSCB) Modeling Workshop I: Characterizing the Capability Needs for HSCB Modeling* (Final Report No. ADA489736). National Defense University.

Hofstede, G. (2006). Dimensionalizing Cultures: The Hofstede Model in Context. In Online Readings in Psychology and Culture. Bellingham WA: Center for Cross-Cultural Research, Western Washington University.

House, R., Hanges, P., Javidan, M., Dorfman, P., & Gputa, V. (2004). *Culture, Leadership, and Organizations: The GLOBE Study of 62 Societies*. Sage Publications.

Kahan, D. M., Braman, D., Gastil, J., Slovic, P., & Mertz, C. K. (2007). Culture and Identity-Protective Cognition: Explaining the White Male Effect in Risk Perception. *Journal of Empirical Legal Studies*, 4(3), 465–505.

Mamadouh, V. (1999). Grid-Group Cultural Theory: An Introduction. *GeoJournal*, 47, 395–409.

Markus, H. R., & Kitayama, S. (1991). Culture and the self: Implications for cognition, emotion, and motivation. *Psychological Review*, 98(2), 224-253.

Marris, C., Langford, I. H., & O'Riordan, T. (1998). A Quantitative Test of the Cultural Theory of Risk Perceptions: Comparison with the Psychometric Paradigm. *Risk Analysis*, 18(5), 635–647.

Matsumoto, D. (1996). *Culture and Psychology*. Pacific Grove, CA: Brooks/Cole.

Nisbett, R., & Cohen, D. (1996). *Culture of Honor: The Psychology of Violence in the South*. Boulder, CO: Westview Press.

Oyserman, D., Coon, H., & Kemmelmeier, M. (2002). Rethinking Individualism and Collectivism: Evaluation of Theoretical Assumptions and Meta-Analyses. *Psychological Bulletin*, 128(1), 3–72.

Singelis, T. M., Triandis, H. C., Bhawuk, D. P. S., & Gelfand, M. J. (1995). Horizontal and Vertical Dimensions of Individualism and Collectivism: A Theoretical and Measurement Refinement. *Cross-Cultural Research*, 29(3), 240–275.

Thompson, M., Ellis, R., & Wildavsky, A. (1990). *Cultural Theory*. Boulder, CO: Westview Press.

Triandis, H. C. (1995). *Individualism and Collectivism*. Westview Press.

Zacharias, G., MacMillan, J., & Hemel, S. V. (Eds.). (2008). *Behavioral Modeling and Simulation: From Individuals to Societies*. National Research Council Board on Behavioral, Cognitive, and Sensory Sciences, Division of Behavioral and Social Sciences and Education. Washington, DC: The National Academic Press.

CHAPTER 46

Geospatial Campaign Management for Complex Operations

Laura Stroh, Alper Caglayan, Tarek Rashed, Dustin Burke, Gerry Eaton

Milcord, LLC
1050 Winter Street
Waltham, MA 02451, USA

ABSTRACT

Here we report initial findings from a research effort to understand the complexity of modern day insurgencies and the effects of counterinsurgency measures. This research integrates data-driven models, such as Bayesian Belief Networks, and goal-driven models, including multi-criteria decision making analysis, into a geospatial modeling environment in support of decision making for campaign management.

Keywords: multi-criteria decision making, Bayesian Belief Networks, campaign management, insurgency

INTRODUCTION

In the current global security environment threats are of a transnational nature, where cultural, social, and technological changes can drive instability across borders empowering hostile non-state actors and jeopardizing U.S. strategic interests. While conventional warfare is about destroying the enemy's military capabilities, complex operations, operations according to the 2009 National Defense Authorization Act that involve irregular warfare, counterinsurgency, and stability, security, transition, and reconstruction operations, require a focus on the human dimensions of conflict In essence, complex operations are campaigns to influence the support of the population Understanding the complexities of the local environment and adopting a geospatial analytical reasoning approach to monitor these population dynamics aids in the development and management of a campaign focusing on a mix of non-lethal and lethal activities.

In the following pages we report initial findings from a research effort to build a business intelligence platform that incorporates data-driven models, such as Bayesian belief networks, and goal-driven models, including multi-criteria decision analysis (MCDA), into a geospatial environment to support decision making for campaign management. This effort is part of Milcord's Predictive Societal Indicators of Radicalism (PSIR) project through Air Force Research Laboratory that involves developing a model of ethnically-based insurgencies for predictive analytics relating to low-intensity conflict. Our development approach supports tactical level commanders at the brigade, battalion, and company level, providing operationally relevant information on the relationships between factors driving the insurgency and leverage points for planning and executing more effective operations. In the following sections, we demonstrate some examples of the features of our tool. First we outline our use of MCDA templates for planning and managing operations. Then, we explore our development of business intelligence tools for military application. Lastly, we reveal insights from our use-case involving Bayesian analysis of data from Southern Afghanistan. Overall, these various components are integrated in our campaign management tool for complex operations.

MULTI-CRITERIA DECISION ANALYSIS

MCDA is a discipline for solving problems involving a set of alternatives that are evaluated on the basis of conflicting criteria (Ehrgott and Gandibleux, 2002). An MCDA framework involves breaking the problem structure down into several components that can include: (1) a goal or set of goals; (2) a decision maker and their preferences with respect to evaluation criteria; (3) a set of evaluation criteria to evaluate courses of action; (4) a set of decision alternatives; (5) the set of uncontrollable variables (decision environment); (6) and the set of outcomes or consequences associated with the alternative actions (Malczewski, 1999). In MCDA, a decision is evaluated on a set of measurable criteria that can either be quantitative or qualitative on a nominal scale. Military doctrine loosely follows this process (Bullock, 2009) in that during initial mission analysis the commander and staff ensure they understand the tactical end states and supporting objectives and develop a set of desired effects to meet the stated goals (see Figure 1.1) (Headquarters, Department of the Army, 2006).

462

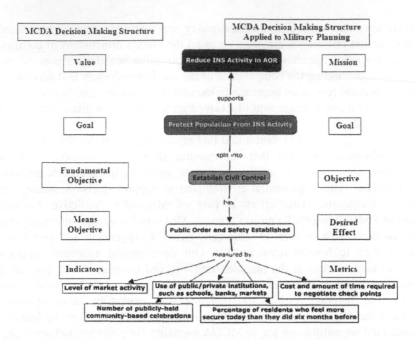

FIGURE 1.1 MCDA Structure Applied to Civil Control Line of Effort

Following the format listed above, the dark blue box at the top represents the mission, as articulated by the commander's guidance and intent. Below that, the blue box states the goal, followed by an objective along a line of effort. The green box lists a desired effect, and the yellow boxes are the metrics to gauge progress. The components of the lines of effort and metrics are part of our tool's library for user reuse. The tool provides collaborative concept mapping capabilities (Novak and Canas, 2006) to enable users to visually brainstorm about problems and their underlying factors. By linking metrics in a shared space, users can track progress across lines of effort and development initiatives, building a repository of knowledge for units to pass on as they transition out of the area of operations.

To support decision making we use Bayesian belief networks to find causal relationships, allowing the user to allocate resources and adjust campaign focus where necessary to achieve the desired end states. A user can track development along a line of effort with the opportunity to drill down and monitor progress for particular desired effects. The decision modeling capabilities enables users to integrate and optimize the MCDA models, taking into account user preferences by rating criteria for courses of action. The scenario can then be played out to determine the optimal location and project focus for various courses of action and can be displayed in a geospatial environment.

BUSINESS INTELLIGENCE PLATFORMS FOR MILITARY APPLICATIONS

In addition to our MCDA templates, we have developed a business intelligence suite of applications for our tool to support complex operations management. In business intelligence, the goal is to leverage existing data, determining new insights and relationships to assist in decision making. Our tool uses an open-source online analytical processing (OLAP) approach that provides a multidimensional conceptual analysis of the data. At its core, OLAP is based on cubic structure, consisting of facts that are categorized by dimensions. The cube metadata is created from a star schema, a data warehouse schema consisting of fact tables referencing any number of dimension tables in a relational database (see Figure 1.3). The cube structure of the data allows the analyst to easily navigate through the cube using various OLAP operations such as drilling down/up, slicing, dicing, and pivoting (see Figure 1.5). OLAP provides significant benefits for analyzing data because it is designed to convert data into usable information by allowing a user to break down data into various levels to determine interesting characteristics and relationships.

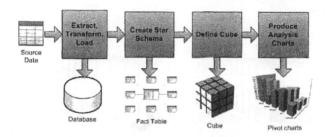

FIGURE 1.2 OLAP Process

To demonstrate the utility of our approach, we will show the power of OLAP analytics through our Tactical Conflict and Assessment Planning Framework (TCAPF) work with USAID Office of Military Affairs. TCAPF is a standardized diagnostic tool used by both military units and civilians operating in conflict environments to collect information from local inhabitants to understand the underlying causes of instability in the area and inform stabilization efforts. It should be noted that the analysis in this chapter is meant to demonstrate our technical approach and is based on a very small sample of data.

In demonstrating the analytical capabilities of our tool we used data sanitation techniques to preserve operational security. Our methodology involved a combination of data substitution, redaction, and data shuffling. We preserved the analytical value of the following fields: occupation, question 1, question 1-why, question 2, question 2-why, question 3 and question 4.

Following the process listed in Figure 1.3, we begin with the initial spreadsheets that are collected by units in the field. The next step in the process is to create a full Star Schema with TCAPF responses as the fact table and each element being a dimension of the OLAP cube. The dimensions are:

- Unit (that collected the data)
- Province/District/Village (where respondent was from)
- Tribe (respondent's tribal affiliation)
- Occupation (respondent's occupation)
- Question 1 – Has there been a change in village population in the last 12 months?
- Question 1 – Why?
- Question 2 – What is major problem facing the village?
- Question 2 – Why?
- Question 3 – Who can best solve the problem?
- Question 4 – What should be done first in village?
- Time – Year/Quarter/Month/Week/Day (when the response was collected)

From the schema, it is possible to define the various OLAP cube combinations (see Figure 1.4).

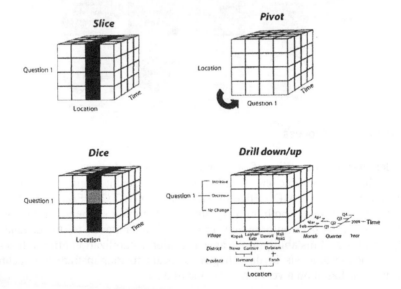

FIGURE 1.3 OLAP Cube Operations

As an example of the types of analysis possible with OLAP, using the slicing function the user can drill down to comparing responses to the question "has the population changed in the last 12 months" by province (see Figure 1.5). This type of analysis can be viewed in several formats including pie charts, 3-D bar charts, line graphs, or as shape files in a map environment.

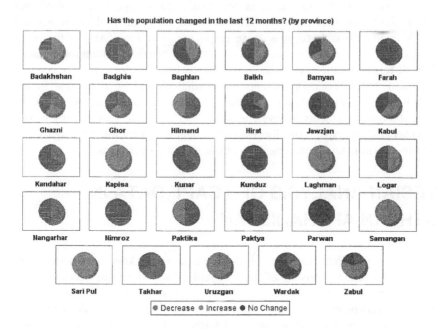

Has the population changed in the last 12 months? (by province)

⬤ Decrease ⬤ Increase ⬤ No Change

FIGURE 1.4 TCAPF Population Change Responses by Province

In the next example using the slicing function, we can view the responses for population change for Kabul Province on a monthly basis (see Figure 1.6).

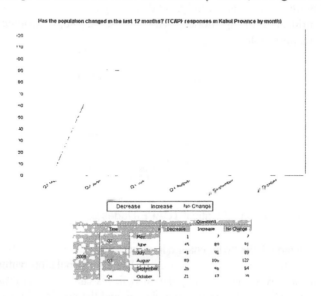

FIGURE 1.5 Population Change Responses by Month in Kabul Province

We can also view the responses at the district or village level listing who respondents think can solve problems based on what the major problem is. For the analysis, it is also possible to view these responses over time. Any of the other dimensions not involved in the cube can also serve as filters. For instance, the user can filter all responses for any of the above graphs but only look at the responses that identified "Paid Work Opportunities" as what should be done first in the village. It is also possible to pivot (swap the axes) to understand based on the major problem, who do respondents believe can solve it. The OLAP functionality of our tool provides analytic capabilities for leveraging data and determining insights into the operating environment, powering data-driven stability operations.

BAYESIAN ANALYSIS OF TCAPF DATA IN SOUTHERN AFGHANISTAN

Using the Bayesian belief network and influence diagram package GeNIe/Smile, we performed a Bayesian analysis of the TCAPF data. Bayesian analysis is a probabilistic graphical model that represents data as nodes, determining the conditional interdependencies and influences. The "Greedy Thick Thinning" algorithm with K2 priors method was used throughout the analysis for structural learning of the Bayesian belief networks, which has been shown to outperform other Bayesian network structural learning algorithms when the training set is sufficiently large and the maximal in-degree is sufficiently small (Dash and Cooper, 2004). The edges show the strength of influence using "Average Normalized Euclidean" method (see Figure 1.7). Interestingly, and perhaps counterintuitively, when examining the influences of the nodes, who can solve major problems is not connected to the major problem.

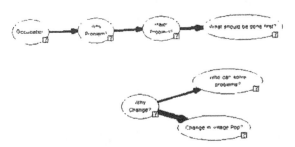

FIGURE 1.6 Influence Diagram of Critical TCAPF Nodes

In examining the model, the "why change in village population" and "who can solve the problems" nodes show that what has caused a change in population within the area conditionally influences who the local population believes can solve the problem. For example, corruption leads to a decrease in population, and the provincial government is the entity that is thought to address problems best (see Figure 1.8).

467

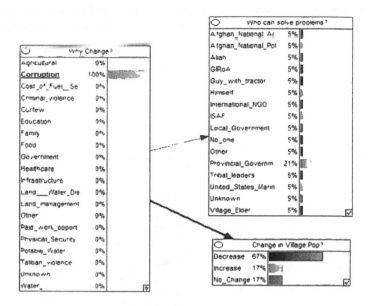

FIGURE 1.7 Influence of Corruption on Change in Village Population

Taliban violence leads to a decrease in village population and respondents believe the village elder can address their problems best (see Figure 1.9).

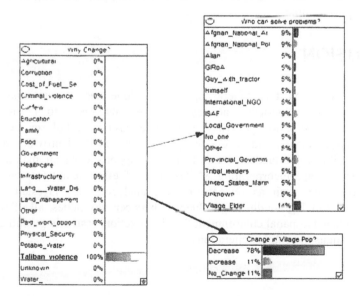

FIGURE 1.8 Influence of Taliban Violence on Change in Village Population

Paid work opportunities show an increase in population, and some of populace believes the Marines and ISAF forces can solve village problems (see Figure 1.10).

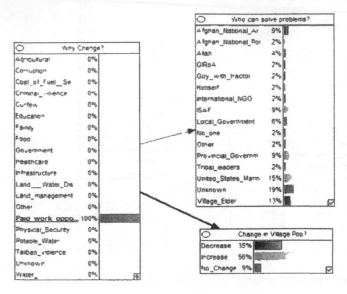

FIGURE 1.9 Influence of Paid Work Opportunities on Change in Village Population

Bayesian analysis of tactical data aids in understanding population sentiments to inform development initiatives and is an important element in our tool.

CONCLUSION

It is widely understood that kinetic operations alone cannot win against a determined adversary. Non-lethal actions must be undertaken to bolster the legitimacy of the host-nation government, provide basic essential services and infrastructure requirements, and secure the population. Our tool provides robust analytical capabilities for tactical level data that supports decision making for campaign management. Incorporating MCDA line of effort templates, business intelligence platforms such as OLAP for data analysis, and Bayesian belief networks for determining conditional influences between key nodes, all of which can be mapped to a geospatial environment for visualization, assists in planning and managing campaigns in complex operations.

ACKNOWLEDGEMENTS

The analysis discussed in this paper is based on an unclassified sample of TCAPF data provided by USAID and was given for demonstration purposes. The original sample was collected by the United Kingdom's 52nd Infantry Brigade.

REFERENCES

Bullock, X. L. (2009), *Engineering Design Theory: Applying the Success of the Modern World to Campaign Design*. School of Advanced Military Studies Fort Leavenworth, Kansas.

Cote, R. L. (2007), *Data-Driven Stabilization: The Process of Selecting Reconstruction & Development Efforts*. Civil Affairs Association, Columbia, Maryland.

Dash, D., and Cooper, G. (2004), "Model Averaging for Prediction with Discrete Bayesian Networks." *Journal of Machine Learning Research*, 5, 1177-1203.

Edgar, D. (2004), *Data Sanitization Techniques*, White Paper, Net 2000 Ltd.

Ehrgott, M., and Gandibleux, X. (Eds.) (2002), *Multiple Criteria Optimization: State of the Art Annotated Bibliographic Surveys*. Kluwer Academic Publishers, Massachusetts.

Headquarters, Department of the Army (2006), *Civil Affairs Operations*. Field Manual 3-05.40 (FM 41-10).

Headquarters, Department of the Army (2008), *Stability Operations*. Field Manual 3-07.

Malczewski, J. (1999), *GIS and Multicriteria Decision Analysis*. John Wiley and Sons, New York.

Muralidhar, K. (2008), *A Primer on Data Masking Techniques for Numerical Data*, Gatton College of Business & Economics, University of Kentucky, Lexington, Kentucky.

Muralidhar, K., and Sarathy, R. (2006), "Data Shuffling—A New Masking Approach for Numerical Data." *Management Science*, 52(5), 658-670.

Novak, J. D., and Canas, A. J. (2006), *The Theory Underlying Concept Maps and How to Construct Them*, Technical Report, Institute for Human and Machine Cognition, Pensacola, Florida.

Offices of Military Affairs and Conflict Management & Mitigation (2007), *Tactical Conflict Assessment Framework (TCAF)*. U.S. Army Peacekeeping and Stability Operations Institute.

United States Agency for International Development (2009), *TCAPF in Lashkar Gah*.

CHAPTER 47

A Metamodel Description Language for HSCB Modeling

Scott Neal Reilly, Avi Pfeffer, John Barnett

Charles River Analytics, Inc.
Cambridge, MA 02138, USA

ABSTRACT

Accurate models of social and cultural behavior have the potential to dramatically improve strategic planning, but in order to be incorporated into decision-making, models must be validated. It is important to develop computational tools to support the efficient, deep, validation of theoretical social science models. An effective computational tool requires a representation that is sufficiently formal to be validated but not a specific computational implementation of the model. We call these model descriptions *qualitative models* or *metamodels* and the language they are described in as a *metamodel description language*. We present the motivating goals of our representation, its key elements, and examples of its use.

Keywords: HSCB modeling, validation, metamodels, qualitative models.

INTRODUCTION

Accurate models of social and cultural behavior have the potential to dramatically improve strategic planning by providing decision makers with insights into the behavior of populations. For example, computational implementations of such models could form the basis of tools that forecast likely crises early and support planning for heading those crises off. However, to be incorporated into strategic decision making, the underlying theoretical models that such tools are built on, however, need to be validated. That is, we need to ensure that they accurately capture the behavior of real-world populations.

The most effective method of validating a model, therefore, is to compare its assumptions and forecasts against real-world data. Current approaches to validation rely on statistical analyses of what data is available, and while there are some computational tools available to support this analysis (e.g., MATLAB, R, Excel), they are not designed specifically for supporting validation and so tend to significantly limit the depth and efficiency of the validation process. For instance, they tend to provide poor support for temporal models or for validating causal claims made by most models. Therefore, we believe it is important to develop computational tools to support the efficient, deep, validation of theoretical social science models.

For such an approach to be practical, we need some form of the model to be expressed formally and computationally. We do not, however, want social scientists to have to build a working computational model of their theory, just a formal description at roughly the same level of detail as would appear in a standard social science paper. Furtherrmore, we want these model descriptions to be expressible apart from any particular computational modeling approach that might be used to implement the social science model as part of a deployable system (e.g., Bayesian networks or systems dynamics approached). We call these model descriptions *qualitative models* or *metamodels* and the language they are described in as a *metamodel description language.*

In designing our metamodel representation language, we have a number of objectives:

1. It should be natural and intuitive for modelers without mathematical or computational training to use.

2. It should capture the important aspects of social science models.

3. It should be independent of any specific computational modeling framework.

4. It should lend itself to exploration and validation, including by specifying the key assumptions of the model so they can be independently validated.

In this paper, we present the main elements of our language and examples of its use.

LANGUAGE ELEMENTS

HIGH LEVEL DESCRIPTION

Our metamodel language is graphical. Nodes represent either *factors*, which are aspects of the world that change over time, or *processes*, which take place in time and have a beginning and an end time. An *event* is a process in which the beginning and end are the same time. Edges may be either undirected (associational) or directed (causal). For example, there may be an edge from the factor *Unemployment* to the process *Strikes* labeled by "increases," to indicate that high unemployment increases the probability of strikes. Edges may be annotated to indicate additional features of the relationship, such as to specify that one node is a necessary or sufficient condition for another node. It is also possible to specify that a link between two nodes is dependent on a third node, for example that the link from *Unemployment* to *Strikes* is influenced by *Income Inequality*.

Our metamodel description language also supports *derived nodes*. Values of derived nodes are determined from other nodes in defined ways. These include mathematical functions, thresholds, aggregates, conjunctions, disjunctions, changes, and delays. For example, we can capture the statement that *Unemployment* causes *Social Unrest* if it remains at a level of more than 10% for two years or that *Unemployment* will lag other economic indices during an *Economic Upswing*.

FORMAL DESCRIPTION

A **metamodel** is like a directed graph, in that it has nodes and directed edges between them, but it also allows edges from nodes to edges.

There are two kinds of nodes: **factors**, and **processes**. For each kind of node, there are **basic nodes** and **derived nodes**.

A basic factor is an aspect of the world that changes over time. A factor has a **type**, that may be, for example, Boolean, ordinal, or continuous. A factor represents a function from time to its type. The type must be something that can be put on a continuous scale. Thus, Boolean can be mapped to 0 and 1, and ordinal nodes can be similarly mapped. However, categorical nodes cannot be so mapped, so they are not allowed. A categorical node must be represented as multiple Boolean nodes, one for every category. Such nodes can be marked as being mutually exclusive, or MUTEX. (In addition, though we will not focus on the user experience in this paper, the user interface (UI) for building these faux-categorical models will make this level of complexity largely invisible to the modeler.)

A basic process is something that takes place over time. A process has a **beginning time** and an **ending time**. An **event** is a process in which the beginning time and ending time are equal. There may be multiple instances of an event or process in a given scenario. For example, suicide bombing may be an event. A given scenario may contain many suicide bombings, but there will only be a single suicide bombing node in the metamodel.

A derived node represents a function of its inputs. Examples are threshold, product, aggregate, change and delay. These are defined as follows:

• A *threshold* node is a Boolean factor node that has a continuous node as a parent. It is defined to be true if its parent is above the given threshold.

• A *product* node takes two parents, and is a factor node defined to be the product of the values of its parents. Other mathematical functions can be defined similarly.

• *Aggregate* nodes take a single parent, and aggregate the values of the parent over a period of time. If the parent is a factor, the aggregate is the integral of the factor. If the parent is a process node, the aggregate is the number of such processes taking place in the given period of time. In either case, the aggregate node is a factor node.

- *Change* takes a factor node as parent. Its value is the change in the level of the function over a given period of time. A change node is a factor node.
- *Delay* takes any kind of node as parent. Its value is the value of its parent at an earlier time point, delayed by the given delay. If the parent is a factor node, the delay node is a factor node, otherwise it is a process node.

There are two basic types of edges: causal edges and mutual exclusion edges. Mutual exclusion edges indicate that two factors are not compatible, such as being in a state of war and a state of peace. Causal edges indicate some sort of causal relationship between the nodes.

Causal edges are further distinguished by their source and target. If the source of an edge is a factor node, its meaning is that the level of the factor node produces the effect described by the edge. If the source is a process node, it means that every time the process takes place, the effect is produced. If the target of an edge is a derived node, the effect is mathematically specified by the definition of the derived node. If the target is a basic process, a positive effect means that the probability of the target increases given the source. Similarly, a negative effect means that the probability decreases given the source. Causal edges can be further annotated in three ways.

1.   The direction of the effect. For instance, does the effect being modeled make something more likely or less likely?

2.   Whether the effect is necessary, sufficient, or both. A necessary condition means that, in the absence of the cause, we would not expect to see the effect. A sufficient condition means that if the cause is present, we would definitely expect to see the effect.

3.   The strength of the effect. If the target is a basic factor, for instance, a strong positive effect means that we would expect to see the target increase as the source increases. However, the strength of this relationship can manifest itself in different ways, either separately or simultaneously. A strong relationship might mean that the probability of seeing the effect is high, but it is not necessarily large. Or it might mean that the effect is large, but we don't see it with high probability. We support an additional annotation to differentiate these states.

4.   Whether there is no effect between factors. Some models explicitly posit that one factor does *not* affect another. We have added an annotation to capture this, which makes it possible to explicitly attempt to validate that non-effect, whereas if there is no link between nodes we would typically not attempt to prove or disprove a relationship between them.

There may also be an edge from a node X to an edge E between Y and Z. The meaning of such an edge is that the relationship specified by E is more likely or stronger in the presence of X than otherwise. For example, if E says that Y makes Z more likely, we would expect to see the difference between the frequency of Z given Y and the frequency of Z given not Y to be stronger if X holds. Such an edge can be annotated in the same way as edges into basic factor nodes.

EXAMPLES

The design just described was based on a rather lengthy analysis of existing social science literature and the types of models described there. Our goal was to ensure that our representation language was useful for representing a large number of useful social science models. To further confirm that our language can be so used, we have developed a number of qualitative models based on published social science models. We present two of these models here.

The first example comes from a study on fertility in Catholic Europe (Berman, Iannaccone, & Ragusa, 2007). This paper used regression models to attempt to explain the decline in fertility seen in Catholic countries in southern Europe after the Second Vatican Council. The paper did not state a formal theory, but we were able to construct a putative theory based on the discussion and equations in the paper, and represent it in our metamodel language.

The metamodel is shown in Figure 1.1. The figure is a screenshot from Connect^{TM}, Charles River Analytics' network-modeling tool. Stronger effects are shown by thicker arrows. The salient features of the model are that the Second Vatican Council had two effects: to decrease the number of clergy, which in turn decreased the amount of social service provision, and to decrease religiosity, which in turn decreased church attendance. These two effects had a multiplicative interaction captured in the derived node labeled *Product*, since only those who attended church could partake of the church-provided social services, and the level of social services themselves was lower. We derived this multiplicative relationship from the regression equations in the paper. The combined effect of these two factors was to increase the cost of raising children, which decreased fertility. Berman et al. postulated that this was a better explanation for fertility decline in these countries than the standard explanation of rising participation of women in the workforce.

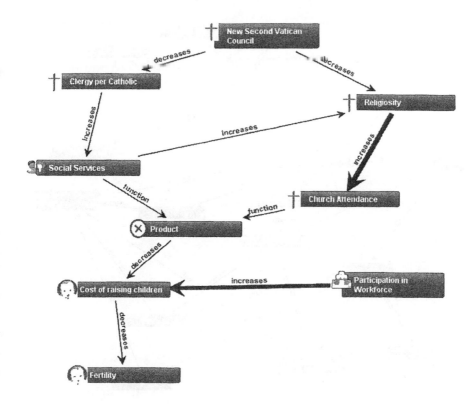

FIGURE 1.1 Fertility metamodel.

The second example comes from (Sambanis, 1999), which studies the factors leading to ethnic partition of countries and its effects. While the paper examined many possible causes and effects, it did not present a complete causal theory of its own, and our metamodel should not be taken as a representation of the author's causal theory, but rather as a possible model whose links can be subjected to validation.

The metamodel is shown in Figure 1.2. Some elements that were not present in the previous model include mutual exclusivity relationships (e.g., between Rebel Victory, Government Victory and Informal Truce), delays (e.g., between Partition and War Recurrence), and designation of no effect between two nodes (e.g., Partition and Postwar Democracy).

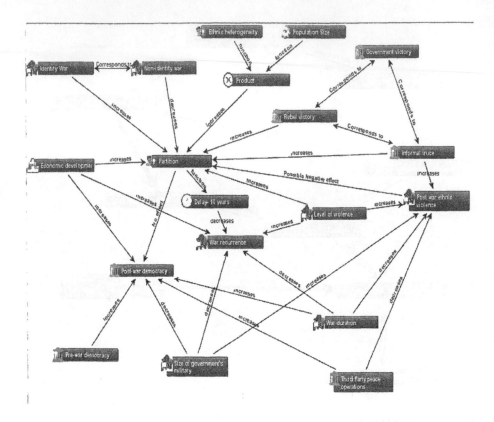

FIGURE 1.2 Ethnic partition metamodel.

CONCLUSION

While this is an ongoing research effort, we have made significant progress towards our objectives, restated here:

1. **It should be natural and intuitive for modelers without mathematical or computational training to use.** The language has been designed to be intuitive and graphical. We plan to work with social scientists to create metamodels to assess their ease of use.

2. **It should capture the important aspects of social science models.** We have constructed metamodels for a range of published social science models and found that our language was able to capture most nuances and details in these models.

3. **It should be independent of any specific computational modeling framework.** We have shown that our language can capture models created using linear regression, system dynamics, fuzzy sets and Bayesian networks, though the metamodels themselves are independent of all of these.

4. **It should lend itself to exploration and validation, including by specifying the key assumptions of the model so they can be independently validated.** Each component (i.e., link) of a metamodel is a separate claim that can be validated against data.

We have developed a validation analysis tool prototype that compares a dataset against the metamodel to provide deep validation and insights into problems with the model through a graphical front end. We are also developing tools that will allow the modeler to explore variations of metamodels to see how they fit the data, which can support the development and refinement of theoretical models, and tools to leverage the metamodel to support the verification of computational implementations of theoretical models.

ACKNOWLEDGEMENT

This work was performed under U.S. Navy contract number N000014-09-C-0463. The authors would like to thank Dr. Harold Hawkins of the Office of Naval Research for his assistance and guidance throughout this effort.

REFERENCES

Berman, E., Iannaccone, L. R., & Ragusa, G. (2007). From Empty Pews to Empty Cradles: Fertility Decline Among European Catholics. Unpublished Work.

Sambanis, N. (1999). *Ethnic Partition As a Solution to Ethnic War*. (Rep. No. Policy Research Working Paper 2208). Washington, DC: The World Bank.

CHAPTER 48

Cultural Intelligence Support for Military Operations

Amy Guthormsen, Ed MacKerrow, Ruth Morgart,Terence Merritt

Center for the Scientific Analysis of Emerging Threats
Los Alamos National Laboratory
Los Alamos, NM 87544

ABSTRACT

It has long been recognized that military success relies on knowledge of the enemy. In the context of standard warfare, adequate knowledge of the enemy may be gained by analyzing observable, measurable data. In the context of modern counterinsurgency operations and the global war on terror, the task of predicting the behavior of the enemy is vastly more complex and difficult. Without an understanding of the ways individuals in the host nation interpret and react to events, no amount of objective information can provide the insight required to accurately predict behavior. US military doctrine has begun to recognize the importance of the many ways that local culture can affect operation success. Increasingly military decision makers use cultural information in the service of operation planning, and troops are provided with pre-deployment cultural training. However, no amount of training can cover the breadth and depth of potentially useful cultural information, and no amount of careful planning can avoid the need to adapt as situations develop. Therefore, a critical challenge is to provide useful tools to US personnel in their efforts to collect, analyze, and utilize cultural information. Essential functions for cultural support tools include the following: 1) to narrow down a broad range of available data and focus the user's attention on context-relevant information, 2) to present cultural information in an easily understood form, 3) to prompt the user to seek relevant information in the environment, 4) to synthesize information, and 5) to predict outcomes based on possible courses of operation. In this paper, we begin by reviewing the ways in which military operations can benefit from cultural intelligence. We then discuss frameworks for analyzing cultural information in the context of a military operation. We conclude with a demonstration of our current efforts to develop a tool that meets the aforementioned functional challenges.

Keywords: Cultural intelligence, Software design

INTRODUCTION

The challenges facing the United States military have changed radically since the end of World War II, and even more so since the end of the cold war. Standard warfare among equivalent forces has given way to asymmetric warfare and counterinsurgency operations. At various points in recent history, voices have arisen to highlight the importance of gathering and understanding detailed information about the culture of both enemy groups and populations whom we seek to influence. Within the past few years, this theme has become a familiar refrain, as individuals from varied perspectives have all voiced a demand for a greater focus on cultural information. Respected military leaders such as General David Petraeus have garnered the attention of the military (Petraeus, David H., 2006). Messages carried by returning troops have provided specific examples to illustrate the demand(Montgomery McFate, 2005, p. 43). With the publication of the Army and Marine Corps Counterinsurgency Field Manual (FM 3-24) the call for cultural information has become part of military doctrine.

There remains a great distance between recognizing the importance of cultural information and achieving a state in which cultural information is routinely collected, analyzed, disseminated, and utilized at various levels throughout the military. Much of the writing on the topic of the military's relationship to cultural information focuses on what could be done but is not yet being done. The current distance between the ideal and the actual is understandable considering the vast amount of information that falls under the heading of cultural information and the infinite complexity of depth within any one topic. For example, Chapter 3 of FM 3-24, Intelligence in Counterinsurgency, covers such disparate topics as social structures, culture, language, sources of power and authority, and objective and motive identification. Each topic is rich and complex, rife with potential sources of data and possible ways of analyzing those data. To the extent that units succeed in acquiring data relevant to these topics, there remains the challenge of seeing that such data are available both to future individuals working in the same area as well as out-of-theater actors. Failure to achieve such bottom-up information transfer has been implicated as a major failure of the current intelligence community to support counterinsurgency operations(Flynn, Pottinger, & Batchelor, 2010).

FEATURES OF SOFTWARE TOOLS

The demands of the military for cultural information set the stage for the development of software tools designed to aid ground troops, intelligence personnel, and command level officers as they gather, analyze and utilize this information. Software tools are only useful to the extent they make work easier, and tools that merely veridicallly represent the complexity of the underlying information will undoubtedly overload the ability of the user to process the relevant information. Therefore, software tools must strike a balance between accurately representing the complexity of rich real-world information and providing a tool that is simple and elegant enough to provide value to its users. To support our own work on designing a tool to provide cultural intelligence support for in-theater end-users, we have reviewed the features that such a tool could potentially offer. We consider the following goals to be the essential ways that software tools can provide value to military users:

FOCUS USERS' ATTENTION ON CONTEXT-RELEVANT INFORMATION

Because the breadth and depth of cultural information is so vast, and because human beings are quite limited in the number of things we can think about at any moment, it is quite easy to see how a soldier or marine could encounter information overload. In a given situation, a deployed unit has a limited set of objectives. Not all cultural information will be relevant to these objectives. In order for a software tool to be helpful, it must provide a way to narrow down the range of data in view. The most helpful tools will be ones that accomplish this narrowing of information in a way that directly supports the goal of the user.

PRESENT CULTURAL INFORMATION IN AN UNDERSTANDABLE FORM

Cultural information can be complex, and often the sources from which it is drawn are academic, which can involve specialized vocabulary and subtleties that are not useful to the war-fighter. However, there is a significant danger in oversimplifying concepts, as this can result in a failure to communicate information that is useful. For example, a long explanation of the details of the Pashtunwali code of honor may garner little attention among soldiers and marines with limited resources. Conversely, a simplification such as "Pashtuns live in an honor-based society," may leave the user without any clear idea about what that means in terms of predicting the behavior of individuals in the environment(Grant, 2009). A good approach may be to highlight similarities and differences. Following this example, an entry about the importance of honor among Pashtuns could be:

> Maintaining the family's honor is one of the most important responsibilities carried by a Pashtun. Although the Pashtun sense of honor is similar to the American one in

terms of focusing on honesty and integrity, the focus of honor is the family rather than country or other organizations. It focuses heavily on the importance of maintaining a public image. Even an event that has no other impact, such as an adversary spreading negative rumors that all parties know to be false, must be revenged because it impacts the public perception of honor. Though a Pashtun tribe may not seek revenge immediately, affronts to honor are never forgotten, and revenge or repayment for the slight will be sought.

PROMPT USER TO SEEK RELEVANT INFORMATION IN THE ENVIRONMENT

Often the best sources of cultural intelligence are members of units who are actively interacting with local populations. Just as the volume of available cultural information can be overwhelming, so can the requirements associated with collecting such a great variety and volume of data. An important benefit of a software tool can be to provide troops with a way of structuring the data they gather, and a set of reminders about observable data to seek. For example, if a unit is making initial contact with a local population and is interested in evaluating the strength of Taliban influence in the area, a software tool could prompt them to seek information about such directly observable things as:

- Level of eye contact made by people encountered
- Color of beards of tribal elders
- Willingness of individuals to talk to American troops

If individuals in the population seem unwilling to talk or even to make eye contact with American troops, this behavior can be taken as a marker of Taliban influence, as fear of reprisal is a likely cause. Similarly, the observation of black-bearded tribal elders is an indication that the actual tribal elders (who would have grey or white beards) have been replaced by younger members of the Taliban.

SYNTHESIZE INFORMATION

If users are successful in gathering a great deal of useful information, we find ourselves once more in a situation of possible information overload. Merely being able to view that information may fail to provide the user with actionable analysis. A useful software tool will synthesize information so that general points can be easily viewed. If the user is interested in tracking the information on which these general points are based, the information should be readily available. The availability of underlying data allows the user to evaluate the accuracy and relevance of the conclusions based on what he or she knows about current circumstances.

It should be noted that synthesis is a much more complex task than the previous three goals. Even when it is clear that multiple pieces of information are tied to a general idea, it may not be clear how to weight them. Consider the example above, regarding assessing the influence of the Taliban on a local population. It may be

482

relatively easy to assess a situation in which the answers to all three questions all indicate Taliban activity. However, it may be less clear what to make of mixed information. What if the tribal elders, with grey beards, are willing to talk to the US forces, but the civilians routinely avoid eye contact? Is that a stronger or weaker indication of Taliban involvement than the reverse pattern? It is our opinion that it is better error on the side of simplicity, unless there is very strong data and theory to back up a more complex algorithm. In this case, we would suggest a three level generalization: 1) All data indicate low Taliban influence in the area, 2) Data regarding Taliban influence are mixed, and 3) All data indicate high Taliban influence.

PREDICT OUTCOMES BASED ON POSSIBLE COURSES OF ACTION

This feature is potentially highly useful, but also the most difficult to achieve and most subject to error. It is easy to imagine that a counterinsurgency-focused unit would be interested in assessing the possible outcomes of multiple courses of action. However, predictive models must always be based on a limited number of inputs, and it is quite likely that the actual outcome will be influenced by factors that are not part of the model. Assessing the danger of outside-model factors influencing the outcome relies on knowing the relative predictive power of the modeled and non-modeled factors. Even if this were knowable at a generalized level, with social and cultural factors it is often the case that interactions occur, in which some catalyzing agent changes the causal impact of the factors. Therefore, it is our position that predictive models should only be offered when:

- The model itself is very strong and well-supported by data
- Conditions of applicability are well-understood
- The user interface provides warnings about the potentially limited nature of the predictions' validity
- Users are encouraged/prompted to seek relevant information in the environment that may offer alternative ways of viewing potential outcomes

MULTI-OBJECTIVE ANALYSIS OF INDIGENOUS CULTURES (MOSAIC)

We now review our work on the development of a specific tool designed to assist both deployed units and strategic planners in their counterinsurgency work in Afghanistan. Although we can see many needs that a general tool could attempt to fill, our goal has been to focus on one specific aspect of cultural information that will be relevant to counterinsurgency units: tribal groups and the relationships between them.

Though Afghanistan is one of the most ethnically diverse places in the world, a unifying factor of its social organization is the tribal structure. Though tribe identity has become less important to some populations, primarily in the north, through displacement and urbanization, for the vast majority of Afghanistan's population, tribal identity is primary. This is especially true of the largest ethnic group, the Pashtuns. Pashtun tribal relationships are governed by an ancient code of honor known as Pashtunwali, which provides a source from which we can understand, monitor, and predict intertribal tensions.

In the following scenario, we consider the utility of the tool in light of the five features described above, in the context of a user seeking information about intertribal relationships in a specific area.

EXAMPLE USE CASE: USER SEEKS INFORMATION ABOUT TRIBAL RELATIONSHIPS IN MUSA QALA DISTRICT

Consider a user who is entering the Musa Qala district, located in Helmand province, Afghanistan. The user would consult MOSAIC to learn about the relationships between tribes in the area.

Step 1: User enters "Musa Qala"

MOSAIC responds by providing a list of possible locations that match this text string. This helps disambiguate the multiple possibilities that result from the fact that place names often appear in multiple places in Afghanistan. The user selects from this list the district Musa Qala.

MOSAIC displays a list of the tribal groups known to be present in this area. This list contains only one entry, the clan Alizai. This part of the example demonstrates **feature 1**: it narrows down the more than 4,000 tribal groups in Afghanistan, to one tribal group that predominates this particular district (see Figure 1). The display screen reminds the user that it may be informative to view the subgroups of tribes located in a particular area. This is because, within Pashtun culture, tribal tensions often exist between groups that are immediate subgroups of a common supergroup. This reminder demonstrates **feature 2** by providing a simple, understandable bit of information that is directly relevant to the user's goal.

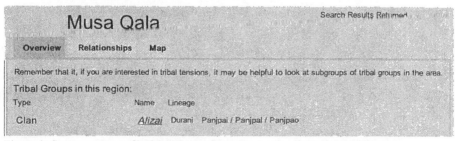

Figure 1: Screen capture of MOSAIC search results for the district Musa Qala

Step 2: User looks at subgroups of the Alizai

The user clicks on "Alizai" and is presented with basic information about this tribal group, including its subgroups. Recalling that it may be important to look at subgroups, the user clicks on the subgroup "Pirzai". Options for viewing intertribal relationships include viewing by:

- Hierarchical tensions: this returns all other groups whose immediate supergroup is the same as the currently selected group
- Recently selected tribes: allows the user to choose from a list of the tribes he or she has recently viewed
- Search by name or location

The user is primarily interested in hierarchical tensions and chooses the first option, which returns the same list of Alizai subtribes that originally led the user to view the Pirzai. Browsing these relationships shows some relationships with little data, but when the user clicks on Hasanzai, details about the history of a contentious relationship appear (see Figure 2) (Pratt, Phillips & Grant, 2008).

This information will be helpful to the user regardless if the intention is to 1) build a relationship with a single tribal group in the area, 2) encourage intertribal alliances in the area, 3) use intertribal rivalry to encourage a tribe to work against another tribe that is aligned with the Taliban.

By mousing over the data, the user is able to view the source of the information, the user who entered the data, the date of entry, and any comments left by either the original author or others who have viewed the data. The current user may also leave a comment. This allows the system to track not only raw data about intertribal relationships but also the subjective interpretations and commentary of various users.

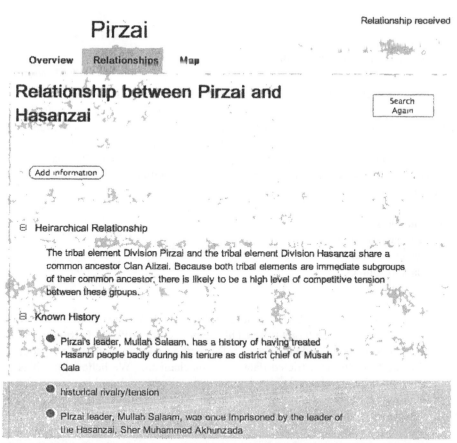

Figure 2: Screen capture of relationship data for relationship between Pirzai and Hasanzai

Step 3: User views a summary grid of relationships between Alizai subtribes

The following function is currently under development: The user may soon become aware that clicking on each pairwise relationship between the numerous subtribes in the area may be tedious and it will become difficult to track the details of the information that is revealed. To get an overview of the information available, the user selects "View Relationship Grid", and is given the option to select which of the currently listed tribes to include in the grid. The user selects all the Alizai subtribes, and is presented with a grid, which identifies the relationships for which available data indicate contention (red), cooperation (green), or mixed data (brown). Clicking on the colored square brings the user back to the screen that displays the details of the information about the relationship. This interaction demonstrates software **feature 4** – the individual pieces of data about the relationship are rolled up to a three-category set of possibilities, which are viewed through a simple and visually comprehensible interface. Yet the underlying, richer data are available at the click

486

of a mouse.

Step 4: User seeks and enters data

The user gains information about the relationship between the Pirzai and the Khalozai and wishes to enter that data. Currently, MOSAIC offers users the ability to add any data, which can be recorded along with commentary. We are currently working on developing an interface that will prompt the user to seek specific information about intertribal relationships by evaluating specific pairwise comparisons between the main tribe and the target tribe with regard to such features as relative:

- numbers
- military strength
- economic capacity
- land resources
- honor status

These comparisons can be evaluated either by direct observation or by interviewing members of the target tribe. This demonstrates software **feature 3**, the use of directed prompts to guide the user's information seeking. Note that this does not replace the flexible, unrestricted data entry mechanism. We believe that it is important to not allow guidance, which is intended to be helpful, to limit the user's vision or interest. The realities of the situation will guide the user to search for and record information in addition to that which the data entry rubric requests.

Predictions vs. Potentials

MOSAIC currently does not include any functions that demonstrate software **feature 5**, prediction of outcomes based on possible courses of action. As we noted above, we believe that such functions should only be included where the predictive model has very strong validity for the specific circumstances of its use. With our current focus on intertribal relationships, a question that a predictive model might try to answer would be, "Given this course of action, what are the likely effects on the intertribal relationships among groups in the area?" Two limitations drastically limit the extent to which accurate predictions can be made. First, in any real-world situation, it is certain that we would have access to only a subset of the existing information about the baseline sate of the intertribal relationship. US forces operating on foreign soil cannot be expected to have access to complete information about the social environment. With inherently incomplete data, efforts to make specific predictions are doomed to failure. Second, intertribal relationships are extremely complex and depend on many factors. Developing a model that would make accurate predictions in all situations would be impossible.

However, we believe it is possible to make progress by asking simpler questions. Instead of looking for all "likely effects" on intertribal relationships, we could set the goal of determining whether a course of action is likely to have an effect on the interests of either tribe. With basic input data about the interests of the local tribes, a user could answer a set of questions about the possible effects of the course of action, and MOSAIC could return information about whether the course of action preferentially benefits or harms the interests of some groups more than others. This situation could be flagged as a potential disruptor of the intertribal status quo. We are currently working on developing such a warning system.

CONCLUSIONS

As the US presence in Afghanistan increases in the near future, the engagement with local populations will grow, and with it will grow the volume of potentially useful cultural and social data about the people we are attempting to help. Correspondingly, the demand for actionable analysis of these data will grow. Software applications have the potential to make sure that the data collected are made available to the wide variety of personnel along the chain of command who need to make use of them. Such applications will only have the opportunity to fill this need if they provide value to the user. Our description of the development of the MOSAIC tool has illustrated the ways in which our development efforts place value to the user at the core of our design.

REFERENCES

Pratt, W., Phillips, D. & Flint, T. (2008). *Tribal Dynamics in Afghanistan: A Resource for Analysts*. Courage Services, Inc.

Flynn, M. T., Pottinger, M., & Batchelor, P. D. (2010, January). Fixing Intel: A Blueprint for Making Intelligence Relevant in Afghanistan. Center for a New American Security.

Grant, J. (2009). One Tribe at a Time. Nine Sisters Imports. Retrieved from blog.stevenpressfield.com

Montgomery McFate. (2005). The Military Utility of Understanding Adversary Culture. *Joint Forces Quarterly*, (38), 42-48.

Petraeus, David H. (2006). "Learning Counterinsurgency: Observations From Soldiering in Iraq *Military Review*, (Special Edition Counterinsurgency Reader,), 51.

U.S. Army Field Manual No. 3-24/Marine Corps Warfighting Publication No. 3-33.5, first issued on 15 December 2006.

Chapter 49

Politeness, Culture, Decision Making and Attitudes; Linking Brown and Levinson to Directive Compliance

Christopher A. Miller, Peggy Wu,
Vanessa Vakili, Tammy Ott

Smart Information Flow Technologies
U.S.A.

ABSTRACT

Language, culture and behavior come together in ways in which directives are issued, interpreted and acted on. We have been working with a functional and culturally universal theory of perceived politeness derived from Brown and Levinson
(1987). We have extended this theory toward a cognitive model of politeness effects on human decision making behaviors and attitudes. We describe these theories and then report an experiment we have conducted in which multi-cultural participants are given directives with varying politeness content in a testbed context
which allows us to control for relevant parameters from the models including power
and social distance of the directive giver and imposition of the directive. Results
clearly show significant impacts of politeness on a variety of directive compliance
behaviors. Our model proved accurate for predicting the relationship of social distance, politeness and directive compliance, but less so for power relationships.

Keywords: Culture, Politeness, Power Difference, Social Distance, Familiarity, Directive Compliance, Behavior, Attitude, Computational Modeling

INTRODUCTION

While "culture" is a deep, complex and multi-faceted construct, it is inevitably expressed in patterns of behavior, values and expectations characteristic of a group. In interpersonal interactions, such patterns are expressed in etiquette. In our sense, etiquette is precisely the "protocol" which members of a group—that is, a culture or subculture—both exhibit and expect and through which they interpret actions.

A large subclass of etiquette patterns are politeness protocols. We follow the work of Brown and Levinson (1987) in identifying culturally abstract, and therefore universal, functions for politeness—though these functions are expressed via culture-specific behaviors. Going beyond Brown and Levinson, however, we have postulated a model of the effects of, first, culture on perceived politeness and then, perceived politeness on decision making and compliance behaviors.

In prior work (cf. Miller, et al., 2007), we created a computational implementation of the Brown and Levinson model, and demonstrated its ability to make quantitative predictions of perceived politeness of an interaction given variations in the utterance and behaviors used and/or in contextual dimensions such as relative power, social distance and imposition. More recently, we have used a cognitive model of decision making to project the effects of perceived politeness on directive compliance behaviors such as compliance decisions, reaction time, and memory, as well as on subjective perceptions such as affect, trust and perceived workload. We call our linked model of perceived politeness and decision making CECAEDA for Computable Effects of Cultural Attributes and Etiquette on Directive Adherence.

In the remainder of this paper, we will briefly review the Brown and Levinson model of politeness and its functions, and then present our CECAEDA model deriving from it. Finally, we will review the results of recent experiments we have performed which support aspects of the CECAEDA model.

A MODEL OF POLITENESS FUNCTIONALITY

Brown and Levinson (1987) collected a large corpus of instances of politeness usage from multiple cultures and languages. From it, they developed what might be called a functional model of politeness. They proposed that the function of politeness behavior is to redress the *face threat* inherent in social interactions. In their model, all social actors are deemed to have "face" needs or wants (Goffman, 1967)—the desire to be accepted, to have one's views and desires approved of and to have freedom of action and freedom from imposition. But social interaction between individuals invariably threatens these desires. Even a simple greeting forces one to shift attention and thoughts from whatever path they were taking to ac-knowledge the greeting. When an interaction is made "baldly" with just its raw content, it is ambiguous as to whether the speaker believes that s/he has the right to demand that attention shift. Since we generally don't want to risk that ambiguity, we frequently make use of "redressive strategies" to mitigate or offset the face threat inherent in our interactions. These are the familiar verbal and non-verbal behaviors which we more frequently think of as "polite"—saying "please" and "thank you", using honorifics, apologizing, etc. The relationship of face threat (and its causes) and polite redressive behaviors is illustrated in Figure 1.

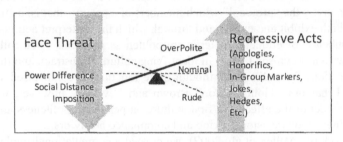

Figure 1. Conceptual depiction of Brown and Levinson's (1987) functional model of politeness.

How much face threat is present in an interaction is a critical determiner of how the interaction will be perceived—and of how much redress is required to offset it. Brown and Levinson propose three factors as influencing face threat:

- *Power Difference* that the Hearer has over the Speaker. A less powerful individual threatens face simply by addressing a more powerful one. Oth-er things being equal, if I ask a favor of a peer, I can use less redress than if of a boss or supervisor—because I have offered less face threat.
- *Social Distance* between the Hearer and Speaker. Social distance is rough-ly the inverse of familiarity. Familiar individuals (co-workers, family members, friends) are expected to address one another, thus familiarity re-duces face threat and social distance increases it. If I ask a friend for a fa-vor, I can use less redress than to a stranger.
- *Imposition* of the request or topic. Some topics are simply more imposing or face threatening than others. If I ask a small favor, I can use less redress than if I am asking for a large one.

These factors drive up face threat and require more redressive "value" (the value of the set of redressive acts used) to reduce it. Inherent in Brown and Levinson's work is the assumption, which we have made explicit and computational (Miller, et al, 2007), that using just enough redress to balance face threat yields an assessment of nominal politeness in the observer—an unremarkable, average, expected amount of politeness for the situation. The use of more politeness than the observer thought necessary yields a perception of over-politeness, while using less results in perceived rudeness. This explains two otherwise challenging observations:

1. That the same utterance or act may be perceived as polite, rude or nominal at different times. In fact, if differences in context include differences in power difference, social distance and/or imposition then, the model predicts, the perceived politeness will be different.
2. That the same utterance *in the same context* may be perceived as having different politeness by different individuals. If the individuals have differing interpretations about the power difference, social distance or imposition, or about the nature and value of the redressive behaviors, then their beliefs about the amount of politeness used would be expected to differ.

This latter phenomenon points to a strength of the Brown and Levinson model: by depicting the function of politeness in social interactions (and thanks to its derivation from a multi-lingual and multi-cultural corpus), it claims to be culturally universal. That is, while different cultures may use different behaviors to provide redress (e.g., taking off one's hat to show deference in the U.S., but removing one's sunglasses for the same purpose in Afghanistan) and may count different degrees of power difference, social distance and imposition in a context, all cultures behave similarly in using polite redressive behaviors to redress face threat and perceiving interactions as polite, rude or nominal on the basis of how nearly the redressive value used balances the face threat perceived.

In prior work (Miller, et al, 2007), we demonstrated the ability to use the Brown and Levinson theory as the basis for a computational model of perceived politeness. We have used that model in a language training game, demonstrating exponential scale up savings in software engineering costs. We have also shown that, for American English cultural examples, our knowledge representation is reliable across multiple users, and that the computational model produces results which correlate highly with those provided by naïve raters. Finally, we have also demonstrated the ability to "swap" knowledge bases of culture-specific rating values and provide both predictions of and explanations for alternate culture perceptions of interactions (in American English, Pasthun and Modern Standard Arabic).

POLITENESS AND DIRECTIVE COMPLIANCE

Building on the culturally universal nature of the model described above, we are seeking to construct a culturally universal model of the *effects* of perceived politeness on individual decision making and behavior. Whether politeness does impact decision making and behavior might be regarded as an open question, but we certainly behave as if it does—training our children to say "please" and "thank you", arguing over the seating arrangements at a diplomatic event, etc. Particularly with regards to *directives*—speech acts in which the speaker is directing the hearer to take some action—politeness would seem to likely to have some effect given the amount of effort and variation most languages and cultures place on directive uses. The "force" of a directive may range from begging through requesting to recommending or demanding all depending on the amount and type of politeness used.

We would like to project the effects of perceived politeness on directive compliance and associated behaviors and attitudes. Intuition and elements of the existing literature led us (via a method described in more detail in Miller & Smith, 2008) to the following predictions at the core of our Computable Effects of Cultural Attributes and Etiquette on Directive Adherence (CECAEDA) model (cf. Figure 2). Note that each prediction is made relative to the performance on that parameter under a situation in which nominal politeness is perceived.

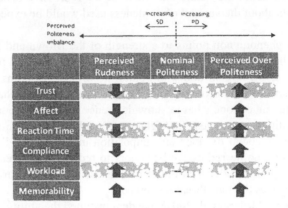

Figure 2. CECAEDA predictions based on politeness imbalance perceived by Observer.

- *Trust and Affect*: We hypothesize that, other factors being equal, both trust in a directive giver (DG) and affect about the situation will increase as the perceived politeness of the directive increases. Perceived rudeness will tend to decrease trust and affect. Relevant data for the relationship between trust and affect are summarized by Lee and See, (2004) and for the relationship between affect and performance by Norman, (2002).
- *Reaction Time*: Brown and Levinson themselves suggest (pp. 95-96) that reduced redress is permissible when action is both urgent and in the hearer's interest. Perceived rudeness may thus result in shorter reaction times because it suggests urgency, while politeness conveys reduced urgency.
- *Compliance*: We hypothesize a general increase in compliance with perceived politeness. This derives from the likely increase in trust and positive affect that comes with expected, pleasing interactions. Parasuraman and Miller, 2004 provides some data supporting the positive effects of politeness on compliance. Decisions about compliance are complex, though, and may be affected by ability and other factors.

- *Perceived Workload and Memorability*: A cognitive interpretation of the Brown and Levinson model suggests that deviation from expected (nominal) levels of politeness provokes increased reasoning about the interaction and its context. Thus, we would expect interactions perceived as very "off-nominal" (either very rude or very over-polite) to incur higher subjective workload as the hearer tries to decipher possible "hidden messages". Similarly, "memorability" (that is, memory for the interaction and its social context) might improve under off-nominal circumstances where the hearer spends additional time scrutinizing and evaluating assumptions.

A final element of the CECAEDA model is the relationship between culture and directive compliance. To some extent, we get this "for free" by using the culturally universal model of politeness perception from Brown and Levinson. Each of the hypothesized claims made above pertains to the politeness level that an observer *perceives* in an interaction—which is, itself, a function of the perceived imbalance between the perceived value of the redress used and the perceived face threat (in turn a function of perceived power difference, social distance and imposition). While each of these parameters can vary across cultures, insofar as we can predict what the face threat and the types and values of redress a given member of a culture will see in an interaction, we should be able to predict what level of imbalance will be perceived and therefore, what impacts on directive compliance will be expected.

TESTBED MANIPULATION OF POLITENESS

We have now completed a preliminary test of aspects of the CECAEDA hypotheses using a testbed that allows control of the timing, context and redressive value of directives, and the relationships of the directive givers (DGs) to the participant. Participants play the role of a dispatcher in a national park fire fighting scenario. As such, they have access to detailed information about the movements, destinations and schedules of various vehicles. Participants support 5 "field agents" working to fight the fire and these agents communicate (using textual messages) from time to time to ask for information that the participant has access to. It is the participant's job to respond to these requests as quickly and accurately as possible. Participants cannot alter vehicle movements and have no goal (and therefore, no imposition) other than the approximately constant one of answering directives.

The "directives" always request a piece of factual information the dispatcher can readily obtain from the displays of unfolding vehicle paths and schedules, and can be answered with one-word responses. The directive text itself is composed of a factual information request that is combined with a prefix drawn from one of three predefined sets: polite, nominal and rude text strings (as determined by our processing of the Brown and Levinson model for an American cultural perspective). Thus, for example, the information request "...the arrival time of UTRUCK 018?" can be combined with the polite prefix "Could you please let me know...", the nominal prefix "Tell me..." or the rude prefix "Stop being lazy and give me...".

Of the 5 DGs, 2 consistently use polite directives, 2 consistently use rude directives, and 1 consistently uses nominal directives. When directives are presented, an "incoming message" screen appears containing a 3x2 grid. Each DG is assigned one of the six squares in this grid and that DG's messages consistently appear in the same square. To help subjects uniquely identify each DG, we also place a unique icon for that DG in "their" square. Note that the use of text (instead of speech) and icons (instead of pictures or video) helps avoid providing information (such as DG gender, age or race) that might influence assumptions about power and social distance—even though this likely reduces the impact of politeness in the directives by reducing the 'channels' through which it is communicated.

In the two experiments we report here, we also varied the relationship of the participant to the DGs by means of a backstory and reinforced this through the icons used. In experiment 1, we varied power difference relationships by showing participants the organization hierarchy for the park system and telling them that, as a dispatcher, they occupied a middle level. Two of the DGs were "commanders" and were superior to dispatchers (and, hence, high power relative to the participant), while two were "rangers" and were inferior (low power), and one was a fellow dispatcher (nominal power difference). In this experiment, the unique icons for each DG were augmented with 1, 2 or 3 stars to indicate their "rank". In experiment 2, we varied the social distance relationship by team membership, telling participants that 2 of the DGs they would be interacting with were from "their" team (which they had been working with for a long time and got along well with—hence, low social distance/high familiarity), one was from "their" park but not on their team (nominal familiarity), and two were from another park they didn't get on well with (low familiarity). Icons for this experiment reflected team relationships by animal designation and background colors. Icon examples are included in Figure 3.

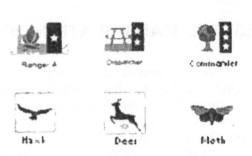

Figure 3. Sample icons used in Experiment 1 (top row) and Experiment 2 (bottom row).

One each of the "high" DGs in each experiment consistently used polite directives and the other used rude ones. The "low" DGs were similarly divided, while the nominal DG always used nominal politeness.

Subjects received 45 incoming message events during a 45 minute trial. 25 of these were single message events (only one message arriving)—for which we were primarily interested in subjects' response times. 20 were paired directive events (two messages arriving simultaneously)—for which we were primarily interested in compliance: which message the participant chose to respond to. In all cases, post test questionnaires asked subjects for ratings of the politeness of each DG, their liking (affect) and trust for each DG, and their perceived workload in interaction with each DG. A final set of memory questions assessed subjects' memory for which utterances were given by which DG.

Subjects (19 in experiment 1 and 20 in experiment 2) were selected from two Midwestern universities in the U.S. We recruited heavily from international students' organizations to increase the cultural diversity of our subject pool. Demographic questions about cultural affiliations were asked as part of a pre-test procedure. While U.S. citizens still accounted for the largest single block of our subjects, their proportion was lower than that of the university populations we drew from. Non-U.S. countries represented included India, Laos, Turkey, Germany, Japan, Korea, Egypt, China, Uganda, Mauritius, Peru, Bosnia/Herzegovina, and Taiwan.

The decision to make use of international students at American universities was a compromise. It allowed us easy access to a diverse population, while concurrently allowing us to conduct all experiments in English. On the other hand, these subjects were not completely representative of their respective cultures. Conducting these experiments in English, with students attending a U.S. university, therefore, represents a conservative test of cultural differences, since we would expect these factors to diminish culture-specific interpretations or decision making patterns due to their exposure to and daily need to interact in American culture.

EXPERIMENTAL RESULTS

First, results showed that participants saw DGs' politeness levels as we intended them, noticing and remembering which characters had been polite, nominal and rude ($M=$ 7.8, 6.2 and 3.1 respectively). DGs intended to be polite were rated as significantly more polite both in experiments: $F(1,21)=22.65$, $p<.001$; $F(1,19)=34.91, p<.001$.

Brown and Levinson predict that increased power of the speaker should increase perceived politeness of a constant utterance, while increased social distance should decrease it. This effect was born out for social distance ($F(1,19)=8.4$, $p<.01$) but not for power difference in our experiments—as shown in Figure 4.

Similar results were found for most subjective parameters. Polite DGs were rated significantly more likable than rude ones ($p<.001$ in both experiments), as more trustworthy in terms of both competence and advice given (p

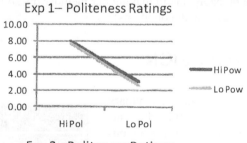

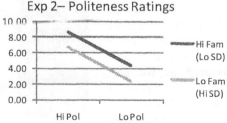

Figure 4. Ratings of DG Politeness.

<.001 in both experiments), and as producing less workload to interact with ($p <.05$ for experiment 1, non-significant but similar trend for experiment 2). Also, as for rated politeness above, predicted effects were found for social distance variations in experiment 2, but not for power difference variations in experiment 1. Decreased social distance (being more familiar) increased likability ($p <.05$), trust in advice ($p <.001$), and trust in competence ($p <.01$) in experiment 2, but there were no significant effects for power difference in experiment 1. There were no significant effects for perceived workload.

Compliance and reaction time results also followed a similar pattern: politeness had no effect in the presence of power variation, but did in conjunction with social distance. Compliance rates were significantly higher for high power DGs (as might be expected), but politeness made only a small difference in experiment 1-- subjects complied slightly less (~5%) with rude DGs, but this was not significant. Experiment 2, though, showed both significant politeness and social distance differences (both $p <.001$). Familiar (low social distance) DGs were complied with more than unfamiliar ones, but polite DGs were also complied with 26% more than rude ones.

No significant main effects or interactions were found for reaction time, though a trend seemed to exist. In both experiments 1 and 2, subjects were slower to respond to polite DGs than to rude ones—which was as we predicted—though this effect was small (~.4 sec). Similarly, high power DGs and familiar DGs were responded to about .4 sec faster than their counterparts. Interestingly, neutral DGs were always responded to more slowly than either polite or rude, high or low power, or high or low social distance ones. This effect ranged from ~.5 seconds to over 2 seconds in size.

Although participants could clearly remember which directive giver been more or less polite, no other memorability effects were found for either experiment— quite probably due to making our memory questions too difficult (performance was near chance in all conditions).

CONCLUSIONS

Figure 5 provides a summary of our results (in white arrows) versus our predictions (in black) about the effects of perceived politeness. In most cases, participants' compliance behaviors and attitudes occurred as we predicted. The one exception to this was workload, where we had predicted an increase in perceived workload under both polite and rude DG and

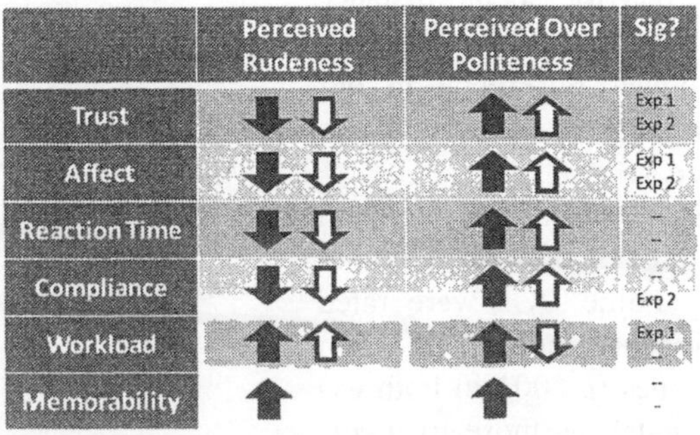

	Perceived Rudeness	Perceived Over Politeness	Sig?
Trust	⬇⬇	⬆⬆	Exp 1 / Exp 2
Affect	⬇⬇	⬆⬆	Exp 1 / Exp 2
Reaction Time	⬇⬇	⬆⬆	--
Compliance	⬇⬇	⬆⬆	Exp 2
Workload	⬆⬆	⬆⬇	Exp 1
Memorability	⬆	⬆	--

Figure 5. Results compared to CECAEDA predictions about effects of perceived politeness on directive compliance.

instead saw a reduction in workload with polite directives.

We are finding that the Brown and Levinson model provides a powerful framo work for understanding the culturally universal functions of politeness in social interactions Our attempts to extend it via the CECAEDA model seems to hold promise. The work reported here demonstrates clearly that perceived politeness *does* affect both whether a directive is complied with and how subjects feel about the person issuing it. Furthermore, social distance variations operated as predicted by Brown and Levinson and CECAEDA—with increased social distance operating to decrease perceived politeness and its effects (except for workload, which was increased, albeit non-significantly, as predicted). By contrast, we had predicted that power difference should enhance most politeness effects through making the same utterance from a powerful speaker seem more polite than when given by a weaker one. Since no such initial effect of power on perceived politeness was found, the lack of behavioral effects is, in a sense, confirmatory evidence for CECAEDA. Still, true confirmation must await a situation in which an effective manipulation of power difference does produce a perceived politeness variation—if that aspect of (our interpretation of) Brown and Levinson is to be supported and carried forward into a test of those aspects of the CECAEDA model.

As to why Social Distance might interact with politeness to affect directive compliance, while Power Differences seem largely independent of politeness—if these findings persist in future studies—we can offer only initial suggestions. We have, admittedly, proposed a simple model of the effects of politeness on decision making and compliance—not a model of decision making itself. One possibility which should be investigated is that other, independent and pre-existing attitudes, beliefs and goals may motivate compliance with power figures (especially among college students) to such an extent that the politeness or rudeness of those individuals makes little difference. Even so, while powerful DGs were complied with more than weak ones even if they were rude, their rudeness did have a negative impact on affect and trust for those working with them.

Since these experiments were done using a broad distribution of individuals from multiple cultures (albeit, all English speakers and attending American universities), we are at least tentatively able to conclude that the effects we observed are culturally universal. A more powerful line of future research, though, would test for systematic cultural variations in how different cultures perceive and afford the various dimensions of the Brown and Levinson model and seek to correlate these with systematic decision compliance differences. For example, perhaps high collectivist cultures (Hofstede, 2001) value social distance more than individualist cultures do—leading them to weight this factor more in politeness evaluations—with *predictable impacts on their compliance behaviors.* We have previously laid out an approach to modeling such impacts (Miller & Smith, 2008) but a thorough test of such hypotheses must await future research with larger populations of culturally diverse subjects in situ.

ACKNOWLEDGEMENTS

This work was supported by a Small Business Innovation Research grant (Contract # FA8650-06-C-6635) from the Air Force Research Laboratory. We thank Ms. Kellie Plummer and Dr. Rik Warren, our Technical Contract Monitors. We also acknowledge Dr. Curtis Hammond, Ms. Marie Kirsch, Dr. Michael Wade, and Mr. Harry Funk for their contributions.

REFERENCES

Brown, P. & Levinson, S. (1987). *Politeness: Some Universals in Language Usage*. Cambridge,UK; Cambridge Univ. Press.

Goffman, E. (1967). *Interactional Ritual*. Chicago: Aldine.

Hofstede, G. (2001). *Culture's consequences, 2nd Ed*. London: Sage Publications.

Lee, J. D. & See, K. A. (2004). Trust in computer technology: Designing for appropriate reliance. *Human Factors. 46* (1), 50-80.

Miller, C., Wu, P., Funk, H., Johnson, L. & Viljalmsson, H. (2007). A computational approach to etiquette and politeness: An "Etiquette Engine™" for cultural interaction training. In *Proceedings of BRIMS07* (pp. 189-198). Orlando, FL: Simulation Interoperability Standards Organization.

Miller, C. and Smith, K. (2008). Culture, Politeness and Directive Compliance: Does Saying "Please" Make a Difference? In *Proceedings of the NATO RTO Symposium HFM-142 on Adaptability in Coalition Teamwork*. Copenhagen, April 21-23, 2008.

Norman, D. (2002). Emotion & Design: Attractive things work better. *Interactions Magazine, 9* (4). 36-42.

Parasuraman, R. & Miller, C. (2004). Trust and Etiquette in High-Criticality Automated Systems. *Communications of the ACM, 47*(4), 51-55.

Chapter 50

Operator Trust in Human Socio-Cultural Behavior Models: The Design of a Tool for Reasoning about Information Propagation

Eric Carlson, Jonathan Pfautz, David Koelle

Charles River Analytics Inc.
Cambridge, MA 02138-4555, USA

ABSTRACT

Models of human socio-cultural behavior (HSCB) are often developed to aid decision makers or analysts by providing insight into current situations and/or by supporting simulations that reveal results of potential courses of action. However, in many cases, the operational users—the intended beneficiaries of sophisticated, complex HSCB models—exhibit a lack of trust in the models and the insights they might provide. This lack of trust can lead to a range of user behaviors from simply discounting models and model-based tools, to incorrectly interpreting model results, to over-reliance on models. In this paper, we discuss our experiences in the construction of tools to support reasoning about "influence operations," military operations that aim to shape human behavior to achieve a particular objective (e.g., encouraging all eligible citizens to vote in an upcoming election) and the mechanisms we designed to establish user trust.

Keywords: Trust in Automation, Cognitive Systems Engineering, Human Socio-Cultural Behavior Models, HSCB, Influence Operations, Information Propagation

INTRODUCTION

Models of human socio-cultural behavior (HSCB) are often developed to aid decision makers or analysts by providing insight into current situations and/or by supporting simulations that reveal results of potential courses of action. However, in many cases, the operational users—the intended beneficiaries of sophisticated, complex HSCB models—exhibit a lack of trust in the models and the insights they might provide. This lack of trust can lead to a range of user behaviors from simply discounting models and model-based tools, to incorrectly interpreting model results, to over-reliance on models. In prior work (Pfautz et al., 2009a; Pfautz et al., 2009b; Pfautz et al., 2009c), we have identified factors influencing trust in HSCB models, including, for example, the user's perceptions of the model's complexity, opacity, and pedigree/authorship.

In this paper, we discuss our experiences in the construction of tools to support reasoning about "influence operations," military operations that aim to shape human behavior to achieve an objective (e.g., encouraging all citizens to vote in an upcoming election). Here, we focus on one such tool, designed and developed to understand the impacts of information propagation or diffusion on the planning and execution of influence operations (e.g., word of mouth about the new website offering a free ice cream cone for every student with straight-As on their report card spread across the entire school by lunch time). The complexities inherent to how information propagates are clearly well suited to the benefits of behavior modeling, where many interacting factors could be represented in a model, and a simulation could support monitoring how those factors govern both individual and cross-group word-of-mouth behaviors. However, the targeted operational users of our tool require that we explicitly address the issue of operator-perceived trustworthiness.

BACKGROUND

As the role of automation increases, the models on which that automation is based becomes more sophisticated, and model-based reasoning is brought to bear on increasingly "fuzzy" domains, it becomes more difficult for the user of automated or partially automated systems to understand their functioning. This lack of understanding impedes the effectiveness of these systems as users either reject them through a lack of trust or become over-reliant on them through an overabundance of trust (Parasuraman, Sheridan, & Wickens, 2000). The study of trust in automation has grown of this phenomenon. See our companion paper, (Farry et al., 2010), for a thorough examination of this field.

Influence operations is one such domain where increasingly sophisticated models and model-based tools show promise in an area of growing importance. Globally, the current shift away from traditional warfare demands that military forces engage in an area that has not traditionally been their strength: shaping the behavior of the population without the use of (or threat of) military force (e.g., encouraging confidence in local government agencies, encouraging reporting of criminal activities). In influence operations, "operators" (those planning, analyzing, executing, and evaluating influence operations) need to identify messages (which contain arguments on one or more media channels) and/or actions that will encourage a population to act in a desired manner (a reasonable non-military analog is public service announcements—e.g., "Only you can prevent wildfires."). This requires those operators to understand the motivations, wants, desires, and needs of a population, and to consider social and cultural influences to determine how a message or action might be received (i.e., how a population perceives, interprets, understands, and decides to act in response).

Reasoning about influence is cognitively challenging and strongly structured, yet the state of the art is a whiteboard and a Microsoft Word document. Only now are more sophisticated tools becoming available via interest in human, socio-cultural behavior (HSCB) modeling. Still, operators are typically not a computer-savvy group; their interest skews heavily toward social science and away from technology. They also tend to be well informed and well read about their area of operations. Their background in social and behavioral sciences ranges from very little to specialists holding PhDs in fields such as anthropology, sociology, and psychology—though that level of education is less common. For data gathering and limited analysis, operators without as extensive an education may consult with cultural specialist, who are typically PhD-level social scientists and/or foreign nationals with first hand cultural knowledge. The nature of operators' work puts a high premium on rigor and accountability; they are required to provide justification for a message or action designed to shape behavior. That, plus the fact that they must receive approval within the military chain of command and military legal experts, requires that they be able to clearly communicate their reasoning. Within the community, there is a general acknowledgement—driven by the complexity of the domain and problem—that tools are needed to help with many aspects of influence operations. Because it is the operators' (and their organizations') ultimate responsibility to correctly assess and influence behavior, any analysis- or decision-support tools must be highly trustworthy. This leads to tools that enhance and extend the operators' expertise within their existing workflow and approval processes, rather than tools that attempt to replace the operators' expertise and judgment or introduce new and unvalidated processes.

The authors of this paper have amassed over five years of experience observing and analyzing influence operations using techniques from Cognitive Systems Engineering to identify work processes as well as individual cognitive challenges and socio-organizational complexities. These prior analyses of influence operations were conducted in the context of efforts to develop HSCB models, to apply HSCB models, and/or to develop tools to support the creation and use of HSCB models. We have previously reported on the results of these studies and on the tools we have designed as a result. In Pfautz et al. (2009c), we discussed methods to enable user-created and user-adaptable models—an approach suited to the most skeptical of users where specialized methods and interfaces allow the users themselves to build a representation of their reasoning about a particular situation. These flexible representations double as a kind of HSCB model that can be rapidly contextualized and applied perform limited forecasting about the situation. The resulting artifact of the users reasoning—the user's HSCB model—can be easily shared with others, which enables community-derived validation of users' reasoning and the model.

While user-created model methods are effective for this highly skeptical user group, they do not allow operators to exploit more sophisticated models capturing the expertise of social scientists. In the second publication stemming from our work on user trust in this domain, (Pfautz et al., 2009a), we reported on our creation of a meta-information taxonomy—along with enabling methods and interfaces for its application—to improve communication between modelers and users in such a way that users understand the operational limits of models sufficiently to apply them to their problems.

In this paper, we extend our prior work to define the meaning of "trust" as it relates to HSCB models and its impact on analysis and decision-making in the context of influence operations. We specifically examine considerations related to information propagation and course-of-action analysis, both tasks within the overarching process of conducting influence operations. Using the results of our analysis, we developed a number of designs aimed at generating and maintaining an appropriate level of operator trust. We worked within the paradigm of user-created HSCB models to address trust at a computational/algorithmic level, where models or model components need to be designed to accommodate an operator who may construct, compose, and/or adapt of human behavior models. Similarly, these designs interacted with associated user interface designs that would enable an operator not only to construct, compose, and/or adapt models, but also to understand, reference, select, and apply those models in particular situations. These designs resulted in the identification of necessary system infrastructure (e.g., the definition of meta-information (Pfautz et al., 2006; Pfautz et al., 2005) about models and model components that enable operational users to assess relevance, applicability, and validity). Based on these initial designs, we obtained initial operator feedback, which we are using to further refine our designs and gain additional insight into practical implications of perceived trust in HSCB models.

METHOD

The secure and highly distributed nature of operators' work makes naturalistic observation challenging. However, through the support of our sponsors and relationships with the community from prior involvement, we had access to operators in a number of contexts. First, we have conducted focused knowledge elicitation sessions with several highly experienced operators. Their collective experience encompasses planning, analysis, and execution in many operational settings. Furthermore, these experts are exceptional in their level of education, having a greater than normal knowledge of social science and related fields. Even with their advantageous social science backgrounds, we found these operators to be typical in their lack of comfort or trust when applying others' models. That is, they are much more comfortable employing theory based on their own background than they are applying and using the results of external models of reasoning. (In fact, we have observed that operators versed in social science can be even more resistant to the application of externally developed models, as these operators understand the models' limitations or may not trust their provenance. Less educated users tend either to completely discount model results or to over-trust models to their own detriment.) They are also typical operators in that they are highly knowledgeable about the political and social situation in their areas of operation.

To identify factors relevant to trust in models in this domain, we worked with these experienced operators to define an operational scenario representative of their requirements and working circumstances. We also worked with these experts to identify the particulars of their operational workflow. Once identified, we defined vignettes within this scenario that highlight particular instances where the operator reasons about the diffusion of information. These span planning, execution, product creation, and evaluation phases of operation. We then observed the operators as they proceeded through their processes within these vignettes by hand, using a whiteboard. From this, we were able to learn how operators currently reason about information flow through a population, and identify the visual and verbal formalisms they use to represent and classify this information flow.

In short, the problem has two parts: *representing the social and information environment* as it is, and *applying reasoning/analytics to forecast information propagation* of a message in that environment. Using tools from prior development in this and related domains, we were able to rapidly prototype a primitive capability to create a situation representation—i.e., a capability roughly aligned to operator needs in the first half of this process. We were then able to use this prototype to examine operators' analytical processes in detail.

Following further knowledge elicitation with this initial prototype, we have used it as a platform for experimentation and development, iteratively improving on its capabilities to interact with broader and more sophisticated parts of operators' process. As we have developed additional capabilities, we have continued to involve prospective users by testing capabilities and interfaces within enriched versions of the scenario identified in the original knowledge elicitation session. Our prototype is currently in its third iteration.

We have also included the input and feedback of a second, larger group of representative users in development of our prototype. We have had the opportunity to conduct limited cognitive task analysis in the course of actual operational planning and execution of influence operations, and while conducting formal evaluations of our tools. This has allowed us to verify the results of our knowledge elicitation sessions and to collect additional information on existing practices and procedures for, and challenges related to, forecasting information propagation. These opportunities have been especially enlightening for refining our use cases, identifying the data that will realistically be available in the field, and identifying elements of the doctrine that are routinely followed or ignored.

RESULTS

Experienced and educated operators apply social science theories under limited conditions to gain insight into what will happen in particular situations. That is, they use social science theories to inform their own pattern analysis and prediction for situations. In addition, they rely on analyses created by cultural consultants, so they are comfortable introducing and integrating information from third parties into their concept of a situation. Applying an external model to assess and simulate a situation involves those same factors: integration of external information and acceptance of others' reasoning. Therefore, we modeled our tool's capabilities on methods operators already use to apply social science theories and consultants' research to their own situation.

The highly educated operators that we studied are able to draw on their own social science backgrounds to apply a number of theories as appropriate. Though there are some individual differences based on their specific backgrounds, there is significant overlap in the theories they apply. Less educated operators, as well, tend to have a passing familiarity with the general concepts encapsulated in these same theories. There are a number of criteria that determine what theory an operator might be comfortable applying in a given situation.

The first, most rigorous criterion that these users consider to assess the validity of applying a theory (and, therefore, of a model built of external reasoning) is that the data used by a model must be valid. A tool to apply and provide products based on models must first establish that the data is accurate and that the uncertainty of that data is bounded. Communicating and establishing an acceptable level of uncertainty is especially important in this domain because ground truth is largely absent.

As in the user-created/user-adaptable work, one approach to establishing trust in the underlying data is to provide methods that allow users to create the pertinent situation representation themselves. This way, even though applied models are extrapolating from that data, users at least accept that the starting point was a state that accurately reflects their own understanding of the situation. Knowledge of the starting state of the situation representation also allows them to assess the results of the model accurately. The scope of the relevant situation or the heavy workload of the operator may render manual data entry unrealistic in many operational situations, though. In these cases, the data is either supplied by automated aids (e.g., text analytic tools that extract information from some corpus of source documents), from other operators who are tasked specifically to create a situation representation (e.g., planners who then hand their results off to the user responsible for execution), or researchers tasked to distill data from a wide range of sources. In this case, the provenance of data must be communicated effectively from those sources along with the data itself.

To support the communication of this meta-information in a situation representation, we defined an ontology of data types that captures the source and reliability of the data with granularity sufficient for both trust and the models that will be leveraging it to support analysis and simulation. Defining an ontology for this information serves multiple purposes. First, it represents this information in a structured fashion such that, within the tool, we are able to represent it in supporting visualizations designed to communicate it to bolster trust in the data. Figure 1.1 shows a visualization representing the sources of information backing the situation representation. This is an aggregate representation that allows the operator to easily assess bias and reliability of the representation at a glance. A compilation of specific sources referenced is displayed when the user moves the mouse over a set of sources.

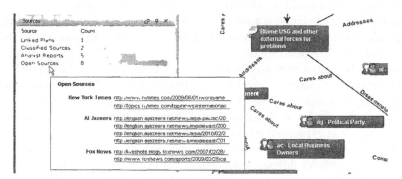

FIGURE 1.1 A visualization of the data sources used in the situation representation.

Second, using this ontology, we have been able to design supporting research tools that a researcher or automated collection agent can use to capture and communicate meta-information relevant to the operational user. This is an instantiation of the concepts discussed in Pfautz et al. (2009a). As in that paper, the underlying meta-information ontology helps to bridge the gap between the operational users and researchers who may not necessarily understand the principal considerations of the consumers of that information.

Once users trust that the underlying data is an accurate reflection of their situation, their next consideration is the applicability of a theory or model to their specific needs. We found that our experts apply theories that very narrowly provide the type of data they need to perform their task. In some ways, this is self-evident: of course operators do not want to use models that fail to tell them something they need to know. There are two important implications beyond this, though. First, complex models should not be applied generically for every situation. Instead, they are only applied at the explicit command of the operator. Additionally, leaving the user in control allows them to experiment to find what models are acceptable in which application and how far they can trust them. This also allows operators to use this tool for situation representation without the additional analysis being performed, when desired. Second, models should be presented to operators in terms of its utility to their specific goal, for a specific task. For example, our tool refers to a model applying Informational Cascades—a theory cited by one of our subject matter experts as useful in planning—as "Wisdom of Crowds." This is especially necessary for those users without our experts' educational background.

While a colloquial explanation may aid discovery of new and potentially relevant theories/models, the everyday use of a model is governed by trust, which we have found is partially established through communication of traditional measures of validity such as pedigree, community acceptance, or field of origin. These factors serve to communicate rigor as granted by a community of experts on the subject. In our tool, the most salient of these traditional measures are presented secondary to the application-focused name in a tooltip-style popup. Additional information in the form of a model dossier (based on another ontology we have defined) is available in a modal window that users can click through to.

Community based validation is another source of trust. As an example, consider the coordination between operators and researchers. In part, operators assess the utility of a researcher's product by the reputation of that researcher, both with that operator and in the community as a whole. A similar process can be applied to models. In our model description ontology, we have defined fields for ratings and comments to act as community-based validation. We hope to use this in future iterations as the basis for a distributed, collaborative component that allows operators to share their experience with various models.

In actually using a model, users form opinions of the applicability and trustworthiness of a model through independent verification of its functioning. In other words, the effect of the model must match the intuition of the user. To facilitate this validation, users must be able to observe a model's effect in detail. This is especially true in regards to one of the classical contradictions of modeling and simulation: any model that gives counter-intuitive results is discarded as being inaccurate, but a model that only tells you what you already know is useless. One method to overcome this is to enable the user to examine the functional subcomponents of the effect of a model. Trust in the basic building blocks of model behavior can then lead to trust and acceptance of the overall result.

In our tool, then, a model must be descriptive of its own mechanisms in addition to its results. Our tool makes this possible through several visualizations to allow the user to observe the theory "in action" as it is applied to the representation of the user's situation. These visualizations have several common characteristics. First, our tool emphasizes those specific elements of the situation representation on which the model is acting. For example, when applying a model based on informational cascades, because the simulation uses the heterogeneity of attributable previous actions that are observable by a specific group, we emphasize that variety of actions observable by a particular group. Next, rather than having model-derived changes exert an instantaneous effect on network information, changes in our tool are animated over time to highlight the change to entities and attributes involved. This draws users' attention to the area of the effect and helps to make the model's methods clear. For example, when a model changes a value of one entity based on values of another, the tool animates an effect on the link representing the relationship between those entities. This highlights the source, mechanism, and magnitude of a specific change, helping the user understand the results of the model. Captures from one such animation are shown in Figure 1.2.

In our tool, following application of a model, changes to the network are displayed explicitly alongside—but visually separate from—the starting state to allow the user to easily examine the effect of the model. Model-generated changes can be reverted at any time. This both helps gain the trust of the user and creates a transferrable justification of the conclusions drawn by user when using the tool. It also is useful in allowing the user to perform a "what is" versus "what if" (as in, "What if I take this action?") type of analysis, which is a consistent thread in our studies of user methods in this domain. This is especially important when applying theories with complex results of which users' may be skeptical.

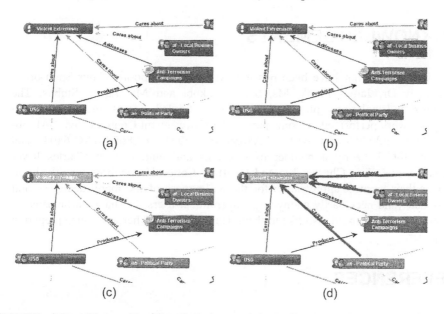

FIGURE 1.2   Steps of animation in model application: a) Initial situation representation; b) Emphasis of entities used as inputs by the model; c) Rendering of initial values affected by the applied model; d) Emphasis of relationships involved and of the resultant effect on the situation.

CONCLUSIONS & FUTURE WORK

Initial tests have affirmed that application of models in this domain and to this problem, in particular, does have great potential. Even more importantly, this need and potential is acknowledged by stakeholders within the domain. We have had initial success in building mechanisms and interfaces that encourage users to experiment with and use models and to assess their trustworthiness.

We are continuing to update this tool with explicit user involvement at each spiral. While our work to date has involved users with a high degree of expertise in influence operations and social science, we intend to begin formal tests in the near

508

future with a wider variety of users to ensure that the mechanisms we employ in this tool meet their needs. On a more technical note, in our work to date we have integrated models into this tool manually, filling in dossier information where needed and hand-tuning the visuals used to communicate models' process and results. There is great potential to apply this tool more generally as a platform for modeling and simulation in influence operations, and to generalize the model application mechanisms so that external models from nearly any source could be brought to bear on the influence operations. We are developing several other projects concurrently that might also contribute to this functionality.

ACKNOWLEDGEMENTS

This work would not have been possible without the support of our Sponsors—principally Dr. James Frank, Ms. Shana Yakobi and Mr. Jareen Stubbs. The concepts and solutions presented here were funded across multiple efforts, including work performed, in some part, under Government Contract Nos. N41756-09-C-4558, FA8650-04-C-6403, FA8650-08-C-6921, FA8650-04-C-6403, and N41756-10-C-3317 by a number of scientists and engineers at Charles River Analytics, including Chester Tse and Geoffrey Catto. The authors would like to express their greatest appreciation for the willingness of the many dedicated and passionate (and necessarily anonymous) operational users that have participated in our analyses. We would also like to thank Heidi Kador for her assistance preparing this paper.

REFERENCES

Farry, M., Pfautz, J., Carlson, E., & Koelle, D. (2010). Trust and Reliance in Human Socio-Cultural Behavior Models. In *Proceedings of 3rd International Conference on Applied Human Factors and Ergonomics*.

Parasuraman, R., Sheridan, T. B., & Wickens, C. D. (2000). A Model for Types and Levels of Human Interaction With Automation. *IEEE Transactions on Systems, Man, and Cybernetics, 30*(3), 286-297.

Pfautz, J., Carlson, E., Farry, M., & Koelle, D. (2009a). Enabling Operator/Analyst Trust in Complex Human Socio-Cultural Behavior Models. In *Proceedings of Human Behavior-Computational Intelligence Modeling Conference 2009*.

Pfautz, J., Carlson, E., Koelle, D., Potter, S., & Zacharias, G. (2009b). Using Meta-Information to Enable End-User Understanding and Application of HSCB Models. In *Proceedings of HSCB Focus 2010 Conference*.

Pfautz, J., Carlson, E., Koelle, D., & Roth, E. (2009c). User-Created and User-Adaptable Technosocial Modeling Methods. In *Proceedings of Technosocial Predictive Analytics: Papers From the 2009 AAAI Spring Symposium*.

Pfautz, J., Roth, E., Bisantz, A., Fouse, A., Madden, S., & Fichtl, T. (2005). The Impact of Meta-Information on Decision-Making in Intelligence Operations.

In *Proceedings of Human Factors and Ergonomics Society Annual Meeting*. Orlando, FL.

Pfautz, J., Roth, E., Bisantz, A., Llinas, J., & Fouse, A. (2006). The Role of Meta-Information In C2 Decision-Support Systems. In *Proceedings of In Proceedings of Command and Control Research and Technology Symposium*. San Diego, CA.

CHAPTER 51

Cultures and Networks in Security Communities

David L. Sallach

University of Chicago
Argonne National Laboratory

ABSTRACT

Security communities are ultimately based on shared cultures; however, cultural processes are rich and complex. Computational mechanisms representing culture are a vital component of models of historical processes. However, at present, computational models of cultural interaction are unrealistically simple. It is important to assess and integrate cultural descriptions from the social sciences, and identify computational representations relevant to social and policy processes. More specifically, this paper focuses on the cross-cultural modeling of security communities.

Keywords: culture, security, resources, rules, networks, communities, reification

INTRODUCTION

Communities that coordinate on security issues are based, in significant part, on the emergence and maintenance of shared cultures and (in a deeper sense) identities. This shared perspective allows them to develop a common stance relative to international threats, probable risks/dangers, and prospective conflicts. As a result, understanding the dynamics of interacting security communities requires comprehension of their cultural dynamics as well. This is true whether the security culture in question is within a single nation state (Jepperson, Wendt and Katzenstein 1996), or an alliance among multiple states (Adler and Barnett 1998).

In part because of the emergence and relative efficacy of security communities, there has been growing recognition of the relevance of culture to policy issues.

More recently, given advances in the computational representation, statistical analysis and graphical visualization of social networks, the relationship between cultural activities and network dynamics is becoming a significant research focus (Emirbayer & Goodwin 1994; Fuchs 2001).

Relatively little effort, however, has been invested in representing, or even understanding, what is known about cultural processes. On the contrary, computational cultural models have been reductively simple, frequently based on nothing more than strings of binary digits with each value representing a discrete cultural trait such as religion, language, identity, etc. (Axelrod 1997; Epstein & Axtell 1996; Lustick & Miodownik 2000), as if they are to be turned on or off like a switch. While such representations are easy to code and result in tractable implementations, they lack the complexities and subtleties of naturally occurring cultural dynamics, and the patterns they produce are subject to stylized effects.

The present paper reviews and synthesizes influential cultural and network theories, and considers their implications for the modeling of cross-cultural security communities.

CULTURE AS RESOURCES AND RULES

Rather than view culture as comprised of essentialist categories, contemporary sociological theory regards it as socially constructed (Geertz 1977). Cultures are the result of complex interactive processes and are, to a significant extent, shaped endogenously.

But what is it that is socially constructed? Giddens (1984) and Sewell (1992) consider culture to be constructed from rules[1] and resources, with the former providing the ideational source, and the latter the material basis of culture. Sewell emphasizes how the two intertwine: material resources (technology, artifacts, texts, etc.) enable capabilities, while schemata convey how such resources can be effectively employed in practice. The intertwined duality applies to the relationship between communication technologies and societal, institutional and organizational structures as well.

Whether defined as rules or schemata, the ideational aspect of culture is vague in this formulation, particularly if it is to provide the basis of a computational model. Prototype concepts (Neisser 1987; Rosch 1983) provide a more flexible, and non-essential (cf., Fuchs 2001), formalism. As concepts, they can represent rules, schemata, strategies, scenarios, and practices. As radial concepts, they distinguish core versus peripheral cases as well as the dimensions along which concept diversity emerges (Sallach 2006).

The idea that culture is composed of intertwined resources and concepts results in a highly decentralized definition. As Sewell (2005) cogently summarizes, modern cultures are contradictory, loosely integrated, contested, weakly bounded

---

[1] Sewell refers to 'schema' rather than 'rule' in order to avoid ambiguity associated with the latter.

and thinly coherent. Thus, in one phrase, he conveys the inadequacies of simple, discrete and/or static representations of culture. As a consequence of such traits, cultures are variegated. The described characteristics also identify social sources of cultural variation, fluctuation and evolution. The emergence of cultural coherence, when and to the extent that it is found, is a contingent local or regional accomplishment. To be of use, models must capture such forms of cultural variation and dynamics.

At the same time, cultures are also sometimes relatively static and unchanging. Historically, stable cultures sometimes persist for centuries. The paired patterns place a dual burden on prospective models. They must represent cultural processes to the point that fragmentation and coherence, and stability and change, are all capable of arising from the same process. Finally, of course, the diversity of actual cultural content imposes further demands.

To capture such richness and subtlety, cultural models will benefit from a shift to finer-grain representations. The first step in this process is to move down from the cultural level of analysis to social networks that create culture and, at the same time, are shaped and constrained by them.

CULTURE AS NETWORKS

A growing number of scholars have begun to focus upon the rich interrelationship between networks and culture (Breiger 2004:219). Perhaps the broadest view is provided by Fuchs (2001), who views culture as networks of communication and meaning, as a network of networks, and sometimes as a field.

> Assume that a "culture" is a recursive network of self-observations and distinctions from other cultures and non-cultures. Distinctions create boundaries of varying sharpness and permeability. They produce an inside and an outside, separating that which belongs to a culture from that which does not belong, does not yet belong, or belongs to a different culture.

This definition describes a holistic, self-organizing type of cultural network. Closer examination, however, reveals its sources in greater detail.

FORMS OF CULTURAL NETWORKS

Fuchs views cultural networks as having a radial structure, with strongly reinforced meanings at the core, and weak incoherent perspectives at the periphery. This framework appears to be consistent with Sewell's fragmented depiction of modern culture. Except, by also recognizing the power and coherence of network cores, Fuchs identifies a source, as well as a prospective mechanism, that can generate cultural stability as well.

Fuchs' core/periphery analysis identifies diverse effects based upon network *locations*. This generalization may be deepened in at least two ways. The first is to explore the fluidity of different types of networks. The second is to consider how networks are used strategically.

Regarding network fluidity, and its opposite pole, reification, Emirbayer and Goodwin (1994) cogently distinguish among three types of network structures: determinist, instrumentalist and constructivist. Network determinism either ignores or stipulates agent intent. Associated models manifest formal, reified relationships. Network instrumentalism considers only a limited, stylized agency such as that articulated in rational choice theories (cf., Elster 1986) and/or hard-coded rules. This type of structuralism appears appropriate for means-ends interaction, as found in some strategic networks (cf., Hay 1998). Network constructivism supports more expressive forms of interaction including robust action and identity evolution.

The key focus of the differences associated with the three forms of network structuralism is the extent to which the networks are reified (Thomason 1982), that is, that networks can effectively be viewed as nothing more than formal relations, a "mere configuration of structural locations" as Hay (1998) cogently summarizes. Cultural network mechanisms will benefit through incorporation of fluidity/reification as a variable network characteristic, and by theorizing its transitions.

Emirbayer and Goodwin ignore one critical aspect of fluid versus reified networks: this dimension is not only relevant to an analytical framework, it is also a characteristic recognized by social participants. Whether implicitly (as taken-for-granted structures) or explicitly (through reflection and/or explicit communication), social actors recognize the presence or absence of network fluidity. They respond accordingly, sometimes by reflexively opening reified networks to unanticipated, novel purposes (Padgett and Ansell 1993). In order to capture the non-linearity of cultural processes, it is important that social actors within computational models have the capability to reflect and act upon structures that appear fixed and, over time, to respond to previously problematic networks in strategic and ultimately routinized ways.

STRATEGIC NETWORKS

Hay (1998) extensively explores the ways in which networks constitute a strategic resource, and how this insight guides agent capabilities and the forms of network interaction. While within his framework, strategic networks can still be viewed as a narrowly instrumental resource, Hay also shows networks to be a more dynamic, endogenous and intentional focus of social action than is often recognized. In brief, Hay views networks as forming a strategic milieu and applies a theory of collective action to the social practice of networking.

From an agent modeling perspective, Hay would have social actors: 1) calculate how strategic resources will or will not be enhanced by network participation, 2) consider forming networks and/or recruiting participants as a means of advancing mutually reinforcing agendas, 3) shape network dynamics by influencing participant motivations and seek to transform enveloping contexts of the network, 4) learn from network failures, and realign network structures and/or practices in response to such failures, and 5) withdraw from or terminate the network when its usefulness is at an end. As such concrete options make clear, Hay sees networks as modes of coordination, produced and reproduced in and through action. In formulating strategies to advance their interests, such actors rely upon incomplete information in assessing configurations of constraint and opportunity, and more or less informed *projections* of the strategic motivations, intentions and likely actions of other significant players.

CULTURAL DISTINCTIONS

As Fuchs (2001) indicates, distinctions that specify cultural boundaries can be seen as unifying or divisive and, over time, can provide the basis for the emergence of an interior of a network (or family of networks), which distinguishes it from other networks, and is more or less coherent (Thagard 2000). The interactions of multiple cultural networks inevitably add to the complication of the process. The prospective dynamics of such interactions cannot be explored here, but they will need to be addressed in policy models of concrete problems.

Fuchs' emphasis on cultural markers and distinctions is consistent with Luhmann (2002:84), who writes:

> A system that is bound to use meaning as a medium constitutes an endless but complete world in which everything has meaning, in which everything gives many cues for subsequent operations and thereby sustains autopoiesis, the self-reproduction of the system out of its own products.

However, a somewhat awkward issue concerns whom it is that makes a cultural distinction. As considered in the next section, Fuchs' rejection of agency (such as that found in Hay's strategic-relational analysis) presents some difficulty in integrating the theoretical perspectives reviewed here.

NODES, OBSERVERS AND ACTORS

Fuchs (2001) views 'personhood' or 'agency' as arising exclusively from social attribution. Fuchs accepts that cultures, organizations, networks, groups and encounters make distinctions, and thus can serve as observers. However, at the same time, he considers it an expression of folk sociology (and essentialist) to assume that (other) individual humans do so. On the contrary, the only role for persons is as the final link in a causal chain, an effect rather than a cause. Networks are formed by relationships, and their nodes only serve as the passive products of those networks.

It is understandable that Fuchs prefers to avoid a methodological individualism grounded in common sense. However, interactionist sociologists such as Cooley and Goffman have articulated a social concept of self that is not an autonomous building block of social entities but, rather, a creation and residue of the larger social process within which it is embedded. It is also dynamic, as Rawls writes, "the social self needs to be continually achieved in and through interaction" (Rawls 1987:136). Further:

> Goffman does not begin with structures of consciousness [that] transcend the individual. He makes the social self dependent on encounters for its existence. The constants for Goffman are the constraints [that] the nature of the social self [places on] social encounters.

Thus, the social self can provide the micro phase of a dynamic cross-scale entanglement. Consistent with this view, what is needed is a conception of nodes in a social network as an interpretive substrate that can serve as the basis for collective actors.

An orientation field held by social actors interleaves prototype concepts with associated emotional valences (Sallach 2007; Sallach 2008b). Social actors exist at multiple scales, i.e., a corporation, army or state (as well as a human individual) can be regarded as a social actor. In the same way, an orientation field can be modeled in aggregate at multiple scales. A collective orientation can be more or less conscious, more or less shared, and more or less cohesive. A collective orientation can be expected to evolve over time, albeit sometimes abruptly.

Observers regularly attribute a particular orientation to a collectivity, whether with accuracy or in error. Indeed, some agents may attribute an orientation to a social actor (e.g., 'yuppies' or 'soccer moms') that others may not recognize as even existing. Clearly, when considering social actors, we have entered an amorphous domain. However, like corporations, armies, and states, most collectivities have an identifiable cohesiveness (whether as an integral actor or as an aggregation of actors with a common orientation) that makes social discourse conceptually coherent. The security communities considered below are examples of the latter.

NETWORKS AND FIELDS

Both the macro and micro levels can usefully be analyzed as fields. As a result, it is useful to more specifically consider the relationship between networks and fields at the meso level.

The representation of fluidity/reification as a variable characteristic of networks has previously been discussed. What has yet to be considered is how best to represent networks when they are in a fluidity mode. Moreover since, in any dense network topology, some will be in fluid and reflective mode, the field provides an effective representation of any extensive set of networks.

While network dynamics are inescapably shaped by their type, the broad patterns may give rise to field effects. Thus, field models have the potential to provide an integrative model of cultural dynamics.

SECURITY COMMUNITY DYNAMICS

Returning to security communities, let us now consider how cultural dynamics contributes to their co-evolution. While such dynamics are central to any cultural community, it is particularly relevant to security communities composed of cross-cultural alliances and interactions.

Each of the scholars discussed here has made a vital contribution to a concept of culture available to modelers. Sewell has dramatized the dynamic, sometimes chaotic, forms of modern culture. Fuchs explores the relationship between cultures and their constitutive networks, including the way in which core/periphery dynamics can account for both cultural stability and cultural fluidity.

Emirbayer and Goodwin clarify that, handy though it is, structural formalism is not a fixed characteristic of social networks, but a dynamic attribute, the sources of which must be studied in order to be understood. While they do not appear to recognize the *endogenous* communications and actions of embedded actors as the primary source of such transitions, their insights nonetheless invite such theoretical extensions. As previously noted, Fuchs also reveals an unexpected narrowness regarding fine-grain agency (and, therefore, reciprocal forms of cross-scale interaction) but, in both cases, explicit representation of social agency can overcome this weakness.

From a modeling perspective, this means designing agents capable of interacting with other social agents while maintaining an orientation toward multiple collectivities. This design provides a framework that allows the exploration of cross-cultural security communities.

Networks are often assessed in terms of homophily (cf., McPherson, Smith-Lovin and Cook 2001). However, to the extent that multiple, diverse networks are shaped by, and reciprocally constitute, an enveloping culture, they also define one basis for subcultural and cross-cultural processes. These variegated social relationships give rise to diverse social tensions and conflicts.

While the full scope of these tensions cannot be easily summarized, the challenge of organizing a cross-cultural security community can illustrate the advantages of using a network-based model of cultural interaction. Southeast Asia, for example, has been described as a "chaos of races and languages" (Hall 1964:5), yet it has given rise to the Association of Southeast Asian Nations (ASEAN), a security community arrived at through innovative forms of cross-cultural decision-making (Acharya 1998).

More specifically, based on principles such as respect for member independence and territorial integrity, pacific settlement of disputes, non-interference in each other's internal affairs, and multilateralism, ASEAN has been able to constitute a common cross-cultural security community (Jorgensen-Dahl 1982). In addition, ASEAN has established a fluid but coordinated security strategy, notwithstanding the fact that current and potential threats were differently viewed by member nations.

This is the focus of the interpretive agent (IA) research program (Sallach 2003), within which a first generation reference application has been constructed (Mellarkod and Sallach 2005; Ozik, North, Sallach, and Panici 2007:61-63; Ozik and Sallach 2006; Sallach and Mellarkod 2005), and a second is under development (Sallach 2008).

CONCLUSION

The present discussion reviews three ways that cultures may be represented as fields, as networks and as interactions. The three concepts are primarily scale-related corresponding to the macro, meso and micro levels, respectively. One must recognize that these concepts are from the perspective of an observer, in this case, an analyst. As a result, they are already proto-models awaiting implementation, but they cannot fully express lived sociality.

Each tier provides a useful view of cultural processes, but that view is necessarily partial and incomplete in at least two senses. First, no representation can fully capture the rich, complex and dynamic detail of lived reality.

The second source of incompleteness arises because the focus provided by each of the three tiers results in an obscuring of the other two levels of the same process. Fields represent large-scale processes of attraction and repulsion and ignore the orientational and strategic basis of interaction. At the same time, the macro effects of social and cultural fields serve as background tugs and undertows, experienced primarily as emotion and expressed through discursive and purposive action. The network tier simultaneously links cultural fields and oriented interaction through structures that simultaneously give shape to the field and channel the interaction.

All three levels are simultaneously at work in cultural processes. In fact, the three levels are themselves arbitrary abstractions that provide insight, but, at the same time, obscure a more fundamental process that is unitary, but not fully accessible.

When exogenous to the social process under study, and/or taken-for granted by participants, networks function as reified structures that are maximally aligned with formal models. But embedded participants are also able to approach structures reflexively, reconsidering prior events, relationships and commitments. When structures appear to be in flux, cultural dynamics will shift in complex and non-linear ways.

Field, network and interaction, all three levels, are necessary to understand cultural processes. However, it is not only the analyst who shifts between the three views, according to his or her focus of interest. The cultural participants also observe, reflect, shift focus and view the process in which they participate, and at multiple levels. Without representation of the endogeneity of cultural processes, effective modeling will remain elusive.

REFERENCES

Acharya, Amitav. (1998), "Collective identity and conflict management in Southeast Asia." Pp. 198-227 in E. Adler and M. Barnett, eds., *Security Communities*. New York: Cambridge University Press.

518

Adler, Emanuel and Barnett, Michael. (1998), "Security communities in theoretical perspective." In E. Adler and M. Barnett, eds., *Security Communities*. New York: Cambridge University Press.

Anderson, Benedict. (1991), *Imagined Communities*. London: Verso.

Axelrod, Robert. (1997), *The Complexity of Cooperation: Agent-Based Models of Competition and Collaboration*. Princeton, NJ: Princeton University Press.

Breiger, Ronald L. (2004), "The analysis of social networks." Pp. 519-520 in M. Hardy and A. Bryman, eds., *Handbook of Data Analysis*. London: Sage.

Elster, Jon, ed. (1986), *Rational Choice*. New York: New York University Press.

Emirbayer, Mustafa and Goodwin, Jeff. (1994), "Network analysis, culture and the problem of agency." *American Journal of Sociology* 99(5), 1411-1454.

Epstein, Joshua M. and Axtell, Robert. (1996), *Growing Artificial Societies: Social Science from the Bottom Up*. Cambridge, MA: MIT Press.

Fuchs, Stephan. (2001), *Against Essentialism: A Theory of Culture and Society*. Cambridge, MA: Harvard University Press.

Geertz, Clifford. (1977), *The Interpretation Of Cultures*. New York: Basic Books.

Giddens, Antony. (1984), *The Constitution of Society: Outline of a Theory of Structuration*. Berkeley, CA: University of California Press.

Hall, D.G.E. (1964), *History of Southeast Asia*. New York: Macmillan.

Hay, Colin. (1998), "The tangled webs we weave: The discourse, strategy and practice of networking." In W. J. M. Kickert, ed., *Comparing Policy Networks*. London: Sage.

Jepperson, Ronald L., Wendt, Alexander, and Katzenstein, Peter J. (1996), "Norms, identity and culture in national security." In P.J. Katzenstein, ed., *The Culture of National Security*. New York: Columbia University Press.

Jorgensen-Dahl, Arnfinn. (1982), *Regional Organization and Order in South-East Asia*. New York: St. Martin's.

Luhmann, Niklas. (2002), *Theories of Distinction*. Stanford, CA: Stanford University Press.

Lustick, Ian S. and Miodownik, Dan. (2000), "Deliberative democracy and public discourse: The agent-based argument repertoire model." *Complexity* 5(4), 13-30.

Mellarkod, Veena S. and Sallach, David L. (2005), "Interpretive heatbugs: Design and implementation." In *Proceedings of Agent 2005*. Chicago: Argonne National Laboratory.

McPherson, Miller, Smith-Lovin, Lynn and Cook, James M. (2001), "Birds of a feather: Homophily in social networks." *Annual Review of Sociology* 27, 415-444.

Neisser, Ulric. (1987), *Concepts and Conceptual Development: Ecological and Intellectual Factors in Categorization*. New York: Cambridge University Press.

Ozik, Jonathan, North, Michael J., Sallach, David L. and Panici, Joshua W. (2007), "Road map: Transforming and extending Repast with Groovy." In *Agent 2007: Complex Interaction and Social Emergence*. Argonne, IL: Argonne National Laboratory.

Ozik, Jonathan and Sallach, David L. (2006), "Voluntary exchange in interpretive heatbugs." In *Proceedings of Agent 2006*. Chicago: Argonne National Laboratory.

Padgett, John F. and Ansell, Christopher K. (1993), "Robust action and the rise of the Medici, 1400-1434." *American Journal of Sociology* 98(5), 1259-1319.

Rawls, Anne Warfield. (1987), "The interaction order *sui generis:* Goffman's contribution to social theory." *Sociological Theory* 5(2), 136-149.

Rosch, Eleanor. (1983), "Prototype classification and logical classification." Pp. 73-86 in E.K. Scholnick, ed., *New Trends in Conceptual Representation: Challenges to Piaget's Theory?* Hillsdale, NJ: Lawrence Erlbaum.

Sallach, David L. (2003), "Interpretive agents: Identifying principles, designing mechanisms." Pp. 345-353 in C. Macal, M. North and D. Sallach, eds., *Agent 2003: Challenges in Social Simulation*. Argonne: Argonne National Laboratory.

Sallach, David L. (2006), "Logics for situated action." Pp. 5-12 in S. Takahashi, D. Sallach and J. Rouchier, eds., *Proceedings of the World Congress on Social Simulation*. Kyoto, Japan.

Sallach, David L. (2007), "Orientation fields and collective coherence." In *Fourth Lake Arrowhead Conference on Human Complex Systems*. Lake Arrowhead, CA.

Sallach, David L. (2008a), "Interpretive agents and discourse-oriented games." In *Fourth Joint Japan-North America Mathematical Sociology Conference*. Redondo Beach, CA.

Sallach, David L. (2008b), "Modeling emotional dynamics: Currency versus field." *Rationality and Society* 20(10), 343-365.

Sallach, David L. and Mellarkod, Veena S. (2005), "Interpretive agents: A heatbug reference simulation." In *Proceedings of Agent 2005*. Chicago: Argonne National Laboratory.

Sewell, William H. Jr. (1992), "A theory of structure: Duality, agency and transformation." *American Journal of Sociology* 98(1), 1-29.

Sewell, William H. Jr. (2005), *Logics of History: Social Theory and Social Transformation*. Chicago: University of Chicago Press.

Thagard, Paul. (2000), *Coherence in Thought and Action*. Cambridge, MA: MIT Press.

Thomason, Burke C. (1982), *Making Sense of Reification: Alfred Schutz and Constructionist Theory*. Atlantic Highland, NJ: Humanities Press.

Chapter 52

Modeling the Reciprocal Relationship Between Personality and Culture

Sae Schatz, Denise Nicholson

Institute for Simulation & Training
University of Central Florida
Orlando, FL 32826, USA

ABSTRACT

The authors hypothesize that (individual) temperament, character, and behaviors are meaningfully correlated with (aggregate) beliefs and behaviors, and then to (emergent) cultural values. In this chapter, the authors describe an initiative intended to develop a theoretically driven model of these relationships through meta-analytic data collection and path modeling. This model will then be used to computationally explore and test the hypothesized relationships. The project's rationale and theoretical background are discussed.

Keywords: Culture, Personality, Temperament, HSCB, Modeling

INTRODUCTION

Culture and personality have a reciprocal relationship. On the one hand, culture helps build personality through the reinforcement and suppression of behaviors. On the other, individuals' personalities shape the culture, each contributing a thread to the overall fabric of a society. Thus, to understand culture, one must understand the complete system that comprises culture—that is, one must quantitatively articulate how a culture influences its people and, in turn, how the people shape a culture.

Evidence supporting the reciprocal relationship between culture and personality is readily available (e.g., Triandis & Suh, 2002). For instance, Hofstede and McCrae (2004) demonstrate a meaningful correlation between Hofstede's well-known cultural facets and the Five-Factor personality dimensions (i.e., Openness, Conscientiousness, Extroversion, Agreeableness, and Neuroticism). They conducted a statistical regression on data from 24 countries and found that the country of origin explained between 24–55% of the variance for each personality dimension. That is, culture predicted a substantial portion of variance in personality traits.

Next, consider a report by Kerr (2001) in which she discusses how the cultural judgment of various temperament characteristics affects individuals' self-images. She empirically demonstrates that the cultural perception of a temperament trait (in this case, shyness) can contribute to positive or negative self-assessments and to other downstream behaviors (e.g., later age of marriage). Further, Kerr subscribes to a theory in which shared cultural values shape the success of certain temperament traits within a society (i.e., the *goodness of fit* hypothesis). This theory suggests that cultural preferences can affect individuals' opportunities to reproduce and that over time certain genes (including those affecting behavior and temperament) will be more likely to be passed on to subsequent generations. That is, cultures can moderate the selection of certain temperament facets across generations.

Finally, McCrae (2009) offers a compelling discussion and some preliminary evidence suggesting a quantifiable link between personality facets, cultural ethos, and national character. He applied a trait-psychology approach to identifying ethos and achieved around 70% inter-rater reliability for the ethos ratings of different cultures. This figure is substantially greater than the typical 20–30% reliability found in most measures of national character, and McCrae's approach offers a tangible way to connect aggregate personality dimensions to the (often quite different) emergent properties of a culture.

These studies highlight the reciprocal nature between individuals and cultures. Each affects the other in a highly complex, but (the authors believe) ultimately predictable, manner. It is the authors' hypothesis that a model of the interrelations among individual, societal, and cultural attributes can be created, and that such a model will help reveal the contribution of specific variables on the overall system.

MODELING PERSONALITY AND CULTURE

In this section, the authors introduce the broad categories of variables that will be considered in the final model. These variables include individual attributes (i.e., temperament, character, and behaviors), aggregate attributes (i.e., group beliefs and group behaviors), and cultural values (i.e., emergent properties of culture).

INDIVIDUAL ATTRIBUTES

Traditionally, personality is divided into two major domains. *Temperament* corresponds with the genetic and biological bases of personality; thus, it refers to personality's "nature" component. It can be defined as "the automatic associative responses to emotional stimuli that determine habits and moods...those components of personality that are heritable, developmentally stable, emotion-based, or uninfluenced by sociocultural learning" (Cloninger, 1994: 64). *Character*, on the other hand, refers to the "nurture" aspect of personality. That is, it describes "the self-aware concepts that influence our voluntary intentions and attitudes...character is weakly heritable, but is moderately influenced by sociocultural learning" (Cloninger, 1994: 64–65). Character evolves in a nonlinear manner over a person's entire lifetime based upon his/her sociocultural education (Svrakic et al., 1996). As a general heuristic, about 50% of the variance in personality is attributed to heritable mechanisms (e.g., Turkheimer & Gottesman, 1991). This principle is now so well established within the behavioral genetics community that Turkheimer was compelled to name it a *law*, saying "The nature-nurture debate is over. The bottom line is that everything is heritable..." (Turkheimer, 2000: 160).

These attributes, created via combination of learned and inherited antecedents, comprise one's personality. However, personality can be further broken down into other dispositional variables, such as attitudes and intentions, and these finer traits ultimately influence how a person reacts to a given context, i.e. his/her overt behavior. By definition, personality is an enduring quality that can predict and explain human behavior. However, empirical support for behavioral consistency across contexts is slim. In order to gain greater predictive value, the intermediate steps between personality attributes and overt behaviors must be systematically articulated (Ajzen, 2005). For instance, the attitude–behavior displays a high correlation, and similarly intention and behavior share a high correlational value, up to around 90% (see Ajzen, 2005, page 100). Therefore, despite the inconsistency of behaviors, it appears that a relatively predictive model linking personality facets to overt behaviors may be possible, given a sufficient degree of granularity among the model's nodes.

AGGREGATE ATTRIBUTES

Reported or demonstrated individual behaviors are, of course, often aggregated in order to make group-level observations. This is also how the authors propose to proceed; i.e., by identifying societal-level mean values and standard deviations for the individual variables of interested mentioned above.

CULTURAL ATTRIBUTES

Finally, these aggregate values must be correlated with emergent cultural values. While the study of culture is often limited to describing the patterns of group behavior, these are not always synonymous with a culture's ethos or national character (Lee, McCauley, & Draguns, 1985). In other words, although traditional ethnographic observation of overt behaviors provides insight into cultural values, it does not necessarily correlate one-to-one with those cultural values. The facets that describe a culture are not always trait relevant (McCrae, 2009); the same shared belief may manifest differently among different cultures or may underlie behavior in non-obvious ways. Thus, it is critical to separately model and distinguish aggregate personalities and behaviors, from national character and cultural ethos. As McCrae (2009: 214) explains: "Aggregate personality is simply the mean profile of the citizens, whereas national character is the shared perception of the profile of the 'typical' citizen, a national stereotype;" further, one might think a culture's ethos "as the personality profile of the culture itself."

THEORIZED MODEL

Personality and culture clearly share a complex, reciprocal relationship. The authors hypothesize that a better understanding of cultural differences can be developed by investigating the correlational relationships, and theoretical causal links, between personal and cultural facets. Towards that end, the authors are engaged in a research initiative to develop a theoretically-driven model that links (individual) temperament, character, and behaviors to (aggregate) beliefs and behaviors, and then to (emergent) cultural values (see Figure 1.1).

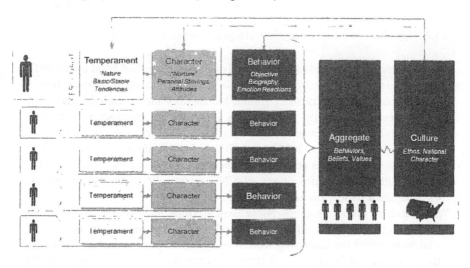

Figure 1.1 Simple model of the relationships among individuals, groups, and cultures

More specifically, the authors hypothesize that the probability for a person to demonstrate a certain behavior in a certain situation is partially dependent upon his/her personality. Further, a person's personality is constructed from genetic, biological, unique and shared (i.e., sociocultural) environmental inputs. Thus:

Behavior $p = f$ (Personality + Context)

Personality $p = f$ (Genes + Biology + Unique Environment
+ Shared Sociocultural Environment)

Second, a quantifiable relationship exists between the aggregate personality dimensions of a society (i.e., mean values and standard deviations) and that society's emergent cultural values. That is:

Cultural Value $p = f$ (Aggregate Personalities)

CONSIDER THIS EXAMPLE...

To help illustrate this approach and goals, consider the following example (adapted from McCrae, 2004). See Figure 1.2 for a visual diagram of the example.

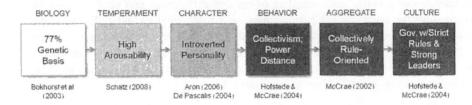

Figure 1.2. Hypothesized causal linkage between genetic foundations and cultural outcomes

Imagine a society of individuals who are highly arousable. As Bokhorst et al. (2003) demonstrated, arousability is strongly heritable and is set early in infancy; thus, a closed society could possibility share similar arousability facets. Individuals within this society would necessarily exhibit high introversion, because as numerous researchers have demonstrated, high arousability correlates strongly with introversion (e.g., De Pascalis, 2004; Aron, 2006). In fact, Eysenck (1967) defines introversion as lower stimulus response thresholds and higher cortical arousal (i.e., synonymous with common definitions of CNS arousability). Thus, our premise becomes: Imagine a society mainly comprised of introverts.

What kind of social structure might evolve from these introverts? McCrae (2004) suggests that there would be few natural leaders among the population, and those who were suited for leadership would easily rise to positions of authority and keep them. Given such an authoritarian government, and acceptance of the government from the introverted people, a rule-oriented society would likely emerge. As McCrae suggests, the society would then passively accept the leaders' dominance and dutifully follows his/her rules. McCrae's notions are based upon empirical evidence. Introversion substantially correlates with the cultural dimensions of collectivism and power distance ($r = 0.64$ and $r = -.57$, respectively; $n = 33$ countries; Hofstede & McCrae, 2004). Finally, a society emerges that possess a strict government and strong rulers, and people who tolerate (or even appreciate) this unambiguous structure.

Certainly, this illustration paints a simplified picture of the causal nature of temperament and culture; however, even this brief description generates many relevant questions. For instance, how resistant might this hypothetical society of introverts be to a new form of government? Which aspects of their culture may be changeable, and which cultural aspects might be most difficult to affect (because they are based upon foundational, genetic predilections)?

WHY MODEL PERSONALITY-AND-CULTURE?

To date, much mainstream cultural research has investigated the direct correlation between trait personality dimensions (e.g., Big Five) and culture. This trait personality approach attempts to define personality by subjectively creating descriptive categories of habitual patterns of behavior, thought, and emotion. Studying personality traits and culture has proven effective; however, like the discrepancies between the personality–behavior relationship, the (individual) personality–culture relationship cannot sufficiently support the emerging operational needs.

First, personality traits are highly intercorrelated. For instance, Musik (2007) demonstrated that the Big Five components correlate more than 90% ($r = .98–.99$) if the factors are constructed via oblique rather than orthogonal rotation. Others (e.g., Becker, 1999; Costa & McCrae, 1992a, 1992b, 1992c; Digman, 1997) have also reported correlations among the Big Five subscales even when orthogonal rotation is used. Obviously, this degree of overlap presents challenges for interpretation of the facets, and as Eysenck (1967, 1991, 1992) strongly argues, such substantial intercorrelations suggest that the most important factor (i.e., the higher-order factor) is being omitted by the trait-level analysis.

Second, personality traits may be difficult to reproduce. Personality traits, and their associated behavioral markers, are greatly influenced by context. As Tellegen (1991) so succinctly noted: Traits are not situation-free, instead a trait is a "disposition to exhibit reaction R under condition S" (17). Thus, concepts defined at the trait level may be elusive and difficult to reproduce from one study to another. However, higher-order concepts are more readily found across studies/populations (Eysenck, 1967).

Third, investigating temperament is critical for understanding the etiology of personality traits. In other words, understanding the genetic and neuro-biological foundations of personality can help contribute to an overall understanding of the causation of various behaviors. Evidence for this argument can be found from those factor analytic studies that failed to find meaningful results when they ignored the genetic components of personality (Cloninger, 1994).

In conclusion, the authors firmly agree that the trait psychology approach taken by many cultural psychologists has value; however, to be most meaningful, we argue that traits must be understood within a larger context. In other words, researchers in this area must integrate their understanding of trait psychology and culture with a wider understanding of foundational temperament facets and emergent cultural qualities, and they must articulate these relationships more clearly, at a sufficiently granular level, in order to better understand the complex, reciprocal relationship between personality and culture. The authors are beginning to embark on such an effort, titled "Modeling the Effects of Temperament on Society." Time will tell if this approach truly has merit, as the authors firmly predict that it does.

REFERENCES

Ajzen, I. (2005). *Attitudes, Personality and Behavior*. New York: McGraw-Hill.

Aron, E. N. (2006). The clinical implications of Jung's concept of sensitiveness. *Journal of Jungian Theory and Practice, 8*, 11–43.

Aron, E. N. & Aron, A. (1997). Sensory-processing sensitivity and its relation to introversion and emotionality. *Journal of Personality and Social Psychology, 73*(2), 345–68.

Becker, P. (1999). Beyond the Big Five. *Personality and Individual Differences, 26*, 511–530.

Bokhorst, C. L, Bakermans-Kranenburg, M. J., Fearon, R. M. P., van IJzendoorn, M. H., Fonagy, P. & Schuengel, C. (2003). The importance of shared environment in mother-infant attachment security: A behavioral genetic study. *Child Development, 74*(6), 1769–1782.

Buss, A. H. & Plomin, R. (1986). The EAS approach to temperament. In R. Plomin & Dunn, J. (Eds.), *The Study of temperament: Changes, continuities, and challenges* (pp. 67–79). Hillsdale, NJ: Erlbaum.

Carey, G., Goldsmith, H. H., Tellegen, A., & Gottesman, I. I. (1978) Genetics and personality inventories: The limits of replication with twin data. *Behavior Genetics, 8*, 299–313.

Cloninger, C. R. (1994). Temperament and personality. *Current opinion in neuroscience, 4*, 266–273.

Costa, P. T., &McCrae, R. R. (1992a). Four ways five factors are basic. *Personality and Individual Differences, 13*, 653–665.

Costa, P. T., & McCrae, R. R. (1992b). Reply to Eysenck. *Personality and Individual Differences, 13*, 861–865.

Costa, P. T., & McCrae, R. R. (1992c). *Revised NEO Personality Inventory (NEO-PI-R) and NEO Five-Factor Inventory (NEO-FFI) professional manual*. Odessa, FL: Psychological Assessment Resources.

De Pascalis, V. (2004). On the Psychophysiology of Extraversion. In M. Zuckerman & R. M. Stelmack (Eds.), *On the psychobiology of personality: essays in honor of Marvin Zuckerman* (pp. 295–327). San Diego: Elsevier.

Derryberry, D. & Rothbart, M. K. (1988). Arousal, affect, and attention as components of temperament. *Journal of Personality and Social Psychology, 55*, 958–966.

DeYoung, C. G., Peterson, J. B., & Higgins, D. M. (2001) Socially desirable responding is related to higher-order factors of the Big Five. Poster presented at the second annual meeting of the Society for Personality and Social Psychology. San Antonio, Texas.

Digman, J. M. (1997). Higher-order factors of the Big Five. *Journal of Personality and Social Psychology, 73*, 1246–1256.

Eysenck, H. J. (1967). *The biological basis of personality*. Springfield, IL: Charles C. Thomas.

Eysenck, H. J. (1991). Dimensions of personality: 16, 5, or 3?—Criteria for a taxonomic paradigm. *Personality and Individual Differences, 12*, 773–790.

Eysenck, H. J. (1992). A reply to Costa and McCrae: P or A and C—the role of theory. *Personality and Individual Differences, 13*, 867–868.

Eysenck, H. J. & Eysenck, M. W, (1985) *Personality and individual differences*. New York: Elsevier.

Goldsmith, H. H. & Campos, J. J. (1982). Toward a theory of infant temperament. In R. N. Emde & R. Harmon (Eds.), *The development of attachment and affiliative systems* (pp. 161–193). New York: Plenum.

Gray, J. A. (1964). Strength of the nervous system and levels of arousal: A reinterpretation. In J. A. Gray (Ed.), *Pavlov's typology* (pp. 289–366). Oxford, England: Pergamon.

Gray, J. A. (1981). A critique of Eysenck's theory of personality. In H. J. Eysenck (Ed.), *A model for personality* (pp. 246–276). New York: Springer.

Hofstede, G. & McCrae, R. R. (2004) Personality and culture revisited: Linking traits and dimensions of culture. *Cross-Cultural Research, 38* (1), 52–88.

Horn, J., Plomin, R. & Rosenman, R. (1976). Heritability of personality traits in adult male twins. *Behavioral Genetics, 6*, 17–30.

Jawer, M. (2005). Environmental sensitivity: A neurobiological phenomenon? *Seminars in Integrative Medicine, 3*, 104–109.

Kagan, J. (1994). *Galen's prophecy: Temperament in human nature*. New York: Basic Books.

Kagan, J., Reznick, J. S. & Snidman, N. (1986). Temperamental inhibition in early childhood. In R. Plomin & J. Dunn (Eds.), *The Study of Temperament: Changes, Continuities, and Challenges* (pp. 53–65). Hillsdale, NJ: Erlbaum.

Kerr (2001). Culture as a context for temperament: Suggestions from the life courses of shy Swedes and Americans. In Theodore D. Wachs, Geldolph A. Kohnstamm (Eds.), *Temperament in context* (pp. 139–152).

Kohn, P. M. (1983). Issues in the Measurement of Arousability. In J. Strelau & H. J. Eysenck (Eds.), *Personality dimensions and arousal*. New York: Plenum.

Lee, Y., McCauley, C. R., Draguns, J. (1985). Why study personality and culture? In By Yueh-Ting Lee, Clark R. McCauley, Juris G. Draguns (Eds.), *Personality and person perception across cultures* (pp. 3–22). Mahwah, NJ: Erlbaum.

Lerner, J. V. (1983). A goodness of fit model of the role of temperament in psychosocial adaptation in early adolescence. *Journal of Genetic Psychology, 143*, 149–157.

Levy, R. A. (1993). Ethnic and racial differences in response to medication: Preserving individualized therapy in managed pharmaceutical programmes. *Pharmaceutical Medicine, 7*, 139–165.

McCrae, R. R. (2002). NEO-PI-R data from 36 cultures: Further intercultural comparisons. In R. R. McCrae & J. Allik (Eds.), *The Five-Factor Model of personality across cultures* (pp. 105–125). New York: Kluwer Academic/Plenum Publishers.

McCrae, R. R. (2004) Human nature and culture: A trait perspective. *Journal of Research in Personality, 38*, 3–14

McCrae, R. R. (2009) Personality Profiles of Cultures: Patterns of Ethos. *European Journal of Personality, 23*, 205–227.

McNamara, F & Sullivan, C. E. (1999). Obstructive sleep apnea in infants and its management with nasal continuous positive airway pressure. *Chest, 116*: 10–16.

Mehrabian, A. (1995). Theory and evidence bearing on a scale of trait arousability. *Current Psychology: Developmental, Learning, Personality, Social, 14*, 3–28.

Mehrabian, A. (1996). Pleasure-arousal-dominance: A general framework for describing and measuring individual differences in temperament. *Current Psychology: Developmental, Learning, Personality, Social, 14*, 261–292.

Mehrabian, A. (1997). *Manual for the child trait arousability (converse of the stimulus screening) scale*. (Available from Albert Mehrabian, 1130 Alta Mesa Road, Monterey, CA 93940)

Musek, J. (2007). A general factor of personality: Evidence for the Big One in the five-factor model. *Journal of Research in Personality, 41*, 1213–1233

Rushton, J. P. (1985). Ethnic differences in temperament. In Y. Lee, C. R. McCauley, J. G. Draguns (Eds.), *Personality and person perception across cultures*, (pp. 45–64). Mahwah, NJ: Erlbaum.

Rushton, J. P. (1999). Ethnic Differences in Temperament. In By Yueh-Ting Lee, Clark R. McCauley, Juris G. Draguns (Eds.), *Personality and person perception across cultures* (pp. 45–64). Mahwah, NJ: Erlbaum.

Schatz, S. (2008). *Trait arousability and its impact on adaptive multimedia training.* Unpublished doctoral thesis, University of Central Florida.

Scherer, K. R., & Brosch, T. (2009). Culture-specific appraisal biases contribute to emotion dispositions. *European Journal of Personality, 23*, 265–288.

Svrakic, N. M., Svrakic, D. M., & Cloninger, C. R. (1996). A general quantitative theory of personality development: Fundamentals of a self-organizing psychobiological complex. *Developmental Psychopathology, 8* 247–272.

Tellegen, A. (1991). Personality traits: Issues of definition, evidence and assessment. In W. Grove & D. Cicchetti (Eds.), Thinking clearly about psychology: Essays in honor of Paul Everett Meehl (Vol. 2, pp. 10–35). Minneapolis: University of Minnesota Press.

Triandis, H. C. & Suh, E. M. (2002). Cultural influences on personality. *Annual Review of Psychology, 53*:133–60.

Turkheimer & Gottesman, 1991Turkheimer, E. (2000). Three laws of behavior genetics and what they mean. *Current Directions in Psychological Science, 9*, 160–164.

Vernon, P. E. (1982). *The abilities and achievements of Orientals in North America.* New York: Academic Press.

Wilson, J. Q. & Herrnstein, R. J. (1985). *Crime and human nature.* New York: Simon and Schuster.

Zonderman, A. B. (1982). Differential heritability and consistency: A reanalysis of the National Merit Scholarship Qualifying Test (NMSQT) California Psychological Inventory (CPI) data. *Behavior Genetics, 12*, 193–208.

CHAPTER 53

Auto-Diagnostic Adaptive Precision Training – Human Terrain (ADAPT-HT): A Conceptual Framework for Cross-cultural Skills Training

Kay Stanney[1], Christina Kokini[1], Sven Fuchs[1], Par Axelsson[1], Colleen Phillips[2]

[1]Design Interactive, Inc.
Oviedo, FL 32765, USA

[2]Applied Systems Intelligence, Inc.
Alpharetta, GA 30022, USA

ABSTRACT

This effort systematically examined relevant literature on cross-cultural competencies and focused on techniques for training tactical-level procedural skills, including cross-cultural communication, relationship building, and negotiation. A suite of measures based on trainee performance and neuro-physiological state was then established to determine a trainee's mastery level of cross-cultural skills along a continuum from unconscious incompetence (i.e., denial) to unconscious competence (i.e., integration). Taken together, these measures were used in the development of a conceptual framework - the Auto-Diagnostic Adaptive Precision Training – Human Terrain (ADAPT-HT) framework – which aims to provide effective cross-cultural training by tailoring adaptations and feedback to a trainee's mastery level. The ADAPT-HT framework aims to increase cross-cultural competency at the tactical level and assist decision makers in understanding and dealing with the 'human terrain,' and has applicability to the military, international business, medical training, and education domains.

Keywords: Sociocultural skills, communication skills, cultural training, cross-cultural, human behavior, serious game, cultural readiness

INTRODUCTION

Globalization is reshaping the world and as the world becomes ever-more interconnected, cross-cultural competency (3C) is rising in importance. Many are starting to assert that cultural intelligence will be a requisite competency to navigate the unique complexities of a global environment (Chin & Gaynier, 2006). Recognizing this, the military and many companies are developing 3C training initiatives. While there are several efforts aimed at providing training that imparts 3C declarative knowledge, there is little in the way of interactive trainers that consolidate and operationalize this knowledge. Social learning theory (Bandura, 1977; 1997) suggests that direct experience of events and consequences of one's actions are necessary precursors before behavior will be influenced. The objective of the current effort is to provide a framework that could be developed into an immersive trainer, which focuses on enhancing 3C procedural skills at the tactical level, specifically the ability of an individual to operationalize their cultural knowledge and exert positive influence in culturally complex environments.

THEORETICAL FOUNDATIONS OF ADAPT-HT

In terms of which 3C procedural skills to focus the ADAPT-HT framework on, the following skills have been found to most closely correlate with success when working across cultures: cross-cultural communication skills to avoid misunderstandings, interpersonal relationship building skills to work effectively with individuals from other cultures, and negotiation expertise to overcome differences and resolve conflicts (Mendenhall, Stevens, Bird, & Oddou, 2008). These skills can provide the foundation for ADAPT-HT's evaluative and diagnostic components to ensure effective training of cross-cultural procedural skills. Beyond cross-cultural procedural skills, two additional aspects that are critical for cross-cultural relations at the tactical level include: (1) cognitive orientation, and (2) traits and values (Mendenhall & Osland, 2002; Mendenhall et al., 2008). Cognitive orientation involves the mindset that is taken during interactions, of which cosmopolitanism (i.e., natural interest in and curiosity about different countries and cultures) is thought to be particularly important and a precursor to adaptation (i.e., incorporation of cultural awareness into daily operations and tactics). Furthermore, certain traits and values have been shown to have a positive impact on cross-cultural interpersonal interactions. Inquisitiveness (i.e., one's openness towards and active pursuit of understanding ideas, values, norms, situations, and behaviors that are new and different; willingness to seek to understand the underlying reasons for cultural differences) has been suggested as a precursor to exhibiting or improving competencies associated with relationship management and other cross-cultural competencies. Thus, for cross-cultural training to be effective, it is essential to understand the trainee's "cultural readiness," (i.e., cognitive orientation and traits and values) and how it may impact learning of communication, relationship building, and negotiation skills. Such "cultural readiness" to acquire cross-cultural competency could be assessed within the ADAPT-HT framework via probes (i.e., targeted but obscure questions) for both cosmopolitanism and inquisitiveness.

The aim of ADAPT-HT is to immerse trainees into culture-centric training experiences that "transform" (see Figure 1) them from unconscious incompetence (Howell's [1982] Level 1, of his levels of communication competency, where an individual does not recognize that he/she is miscommunicating or relating/ negotiating in a non-optimal way based on a given cultural context) to conscious incompetence (Howell's Level 2 – an individual recognizes the need to learn about other's culture, but is unable to correct or resolve culturally-related problems) to conscious competence (Howell's Level 3 – an individual believes that it is vital to consider cultural differences, and applies the cultural knowledge he/she has) and ultimately to unconscious competence/super-competence (Howell's Levels 4 and 5 – an individual can unconsciously adjust to cultural differences and incorporates "help me understand questions" into communications when cultural differences arise).

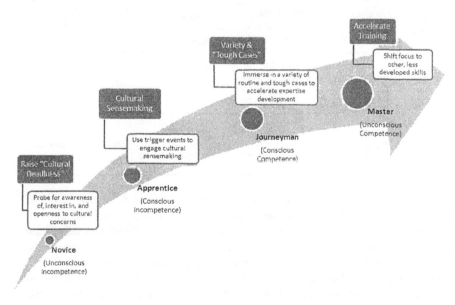

FIGURE 1. Supporting the transformation process in cross-cultural competency.

To support performance assessment, this multi-stage framework was further aligned with social learning theory (Bandura, 1997; 1977), where there are three components to the social learning process: attention, retention, and reproduction (see Table 1). Attention involves an individual becoming aware that his or her behaviors are different from the norms and values within a given cultural context (i.e., conscious incompetence; defense/minimization). Retention involves an individual storing and remembering learned skills and behaviors (i.e., conscious competence; acceptance). Reproduction involves internalized changes in behavior because an individual understands which behaviors and skills to execute or suppress in given situations (i.e., unconscious competence/super-competence; adaptation/integration). These three stages of social learning theory can be used to evaluate the effectiveness of cross-cultural training (Black & Mendenhall, 1990), and were thus integrated into ADAPT-HT. A fourth component was added to this theory, that of inattention, which would be associated with unconscious incompetence and denial.

Table 1 Measures of Transformation Process in Cross-cultural Competency

Trainee State	Training Progress	Social Learning Theory	Cross Cultural Skill Level	Workload	Engagement
Unconscious Incompetence	Novice - Denial	Inattention (few Areas of Interest [AoI] fixations)	No notice of cultural differences	Low	Low
Conscious Incompetence	Apprentice - Defense / Minimization	Attention (many fixations on AoI; imprecise scanpaths)	Aware of cultural differences	High	High
Conscious Competence	Journeyman- Acceptance	Retention (many fixations on AoI; more precise scanpaths)	Understand cultural differences	High	Low
Unconscious Competence	Master – Adaptation / Integration	Reproduction (many fixations on AoI; precise scanpaths)	Manipulate cultural differences	Low	Low

Performance assessment alone is an insufficient predictor of a learner's cultural competency, as much of culture is unobservable. Specifically, using the cultural iceberg concept, culture can be divided into two sections – above and below the water line (Peace Corps, 2000). Those aspects of culture above the water line pertain to observable behaviors. Under the water line contains aspects of culture that are unobservable (e.g. beliefs, values). These invisible aspects are thought to influence or cause the surface behaviors. While differences in the observable characteristics will be easy to identify and understand, the hidden aspects of cultural differences are much more difficult to discern. This is where neuro-physiological assessment can play a role. This dichotomy is further depicted in Table 1, where two stages are associated with incompetence (i.e., mostly incorrect responses to culturally-based performance measures), and two other stages are characterized by competence (i.e., mostly correct responses to culturally-based performance measures). Thus, transformation along these stages can only be unambiguously identified when consciousness is taken into account, which cannot be captured via observable behavior. Neuro-physiological measures, such as electro-encephalography (EEG), have been successfully used to identify the unobservable cognitive states of human users during information processing tasks. Specifically, indices have been developed and validated that indicate the cognitive workload and engagement associated with such tasks (Berka et al., 2007). EEG-engagement indices have been related to processes involving information-gathering, visual scanning, and sustained attention, whereas EEG-workload indices have shown increases with working memory load and with increasing difficulty level of mental arithmetic and other problem-solving tasks (Berka et al., 2007).

It is herein proposed that these EEG indices can be used in conjunction with behavioral metrics and performance indicators based on eye-tracking to unambiguously identify levels of cultural expertise (see Table 1). During novice performance (i.e., unconscious incompetence – denial - inattention stage of social learning theory), the trainee is not aware of the importance of the stimuli related to cultural competence. Thus, related cognitive processing (related to both perception and comprehension) is expected to be low. Also, few visual eye fixations on culturally relevant Areas of Interest (AoI) are expected at this stage because there is little to no focus on cultural cues. Once awareness of cultural issues is raised (transforming into conscious incompetence – defense/minimization - attention stage of social learning theory), mental effort is expected to increase on both EEG indices (i.e., workload and engagement) in an attempt to overcome the realized weaknesses and embrace the novel characteristics of the host culture. EEG engagement should be particularly high during this phase, given the novelty of the presented cultural aspects. Additionally, since awareness has been reached, a high level of visual fixations on AoI is expected at this stage and all subsequent stages; though scanpaths are expected to be imprecise at this stage, as means of differentiating between cues will not be well trained. When competency is first reached (i.e., conscious competence – acceptance - "retention" stage level of social learning theory), large amounts of active processing are still necessary, as the correct behaviors are known but not easily retrieved and implemented. While the novelty effect may have worn off at this time (resulting in decreased EEG engagement; Stevens, Galloway, & Berka, 2007), EEG workload is still expected to be high, due to the high mental workload requirements for accessing stored cultural rules from long term memory. Visual scanpaths are expected to become more precisely focused at this stage (e.g., QRS wave foveation length, visual fixation interpretation time, attention density, etc., Augustyniak & Tadeusiewicz, 2006). It is only when performance has reached automation (i.e., unconscious competence –

adaptation/integration - "reproduction" stage level of social learning theory), that conscious processing is expected to decrease as a result of easier memory retrieval processes and less executive function processing. At this point scanpaths should be very precise.

Identification of these expertise levels has important implications for implementing an adaptive training solution that considers trainee progress (i.e., transformation), and actively and effectively transitions trainees from one competency level to the next one until proficiency is reached (see Figure 1).

THE ADAPT-HT FRAMEWORK

ADAPT-HT is a conceptual framework that supports cross-cultural communication, relationship-building, and negotiation training through the measurement of trainee performance and cognitive state, the diagnosis of potential issues that may hinder training (e.g., issues with trainee state [i.e., performance and/or cognitive state] and cultural readiness [i.e., cognitive orientation and/or traits/values]), and then the adaptation of the training based on trainee state and cultural readiness. This framework has three main components: measurement, diagnosis, and adaptation.

ADAPT-HT MEASUREMENT COMPONENT

The ADAPT-HT framework contains a comprehensive measurement system that takes into consideration 1) trainee's "cultural readiness," 2) trainee's acquisition of cross-cultural procedural skills, and 3) trainee's neuro-physiological state.

Probes based on validated cultural questionnaires are used to assess the trainee's cognitive orientation and traits and values to ensure trainee "cultural readiness" to acquire cross-cultural procedural skills. Specifically, selected questions from the Cultural Competence Survey (Doorenbos, Schim, Benkert, & Borse, 2005; Schim, Doorenbos, Miller, & Benkert, 2003) and Cross-Cultural Adaptability Inventory (Davis & Finney, 2006) could be weaved into probes implemented as an interaction between a non-player character (NPC; e.g., Commander or Cultural Advisor), and the responses on these probes could be used to assess trainee's "cultural readiness." Example questions include:

- Have you found that race, gender, physical ability, age and sexual orientation are often good predictors of psychological or cultural traits? (Cosmopolitanism)
- Have you found it to be advantageous to form generalizations or stereotypes of groups of people?" (Inquisitiveness)

Once cultural readiness is achieved, trainee's acquisition of cross-cultural procedural skills is defined at the sub-skills level for communication, relationship-building, and negotiation skills. Specific performance measures for each procedural skill were obtained through a literature review and subject matter expert interviews and include proper verbal and non-verbal communication techniques, interactions in terms of eye contact, space criteria, hand-shake, conversation standards, etc.

Trainee's neuro-physiological state will be assessed using EEG to gauge levels of engagement and workload. Differences in cultural cue information-gathering, visual scanning, and sustained attention should be revealed in the EEG-engagement index, whereas differences in working memory load and problem-solving should be revealed in the EEG-workload index (Berka et al., 2007).

ADAPT-HT DIAGNOSIS COMPONENT

The neuro-physiological EEG measures, in conjunction with performance-based measures (both observable outcomes and eye tracking fixations/scanpaths) and cultural probes will be used to evaluate a trainee's transformation along the continuum from unconscious incompetence to unconscious competence (see Figure 1 and Table 1). This diagnosis is used to derive proper adaptations and feedback. If the trainee's performance is low (e.g., in a state of denial; low level of fixations on AoI), and both EEG engagement and workload are low, this indicates that the trainee is a "Novice" with respect to the cultural skill under consideration. In comparison, if the trainee's performance is low (e.g., defensiveness/minimization; knowledge of cultural factors but no integration into behavior) but visual fixations are on AoI but scanpaths are imprecise, and both EEG engagement and workload are high, this indicates that the trainee is an "Apprentice" with respect to the cultural skill under consideration. On the other hand, if the trainee's performance is high (e.g., acceptance; understand, apply, and comprehend cultural factors), there is a high level of fixations on the AoI and more precise scanpaths, and EEG workload is high, while engagement is low, this indicates that the trainee is a "Journeyman" with respect to the cultural skill under consideration. At the master level, culturally relevant behavior will be seamlessly integrated into one's conduct and thus EEG workload and engagement are expected to be low, while performance is high, with precise scanpaths and fixations on AOI.

ADAPT-HT ADAPTATION COMPONENT

Based on diagnosis of trainee skills acquisition and neuro-physiological state, several adaptations and feedback options are possible. Categorization of training progress provides an opportunity to optimally support the transition between each of the states depicted in Figure 1 by using different forms of adaptations.

Unconscious Incompetence - Focus: Raise Cultural Readiness

At the unconscious incompetence stage (i.e., denial during performance, low workload and engagement, and few fixations on AOI; see Table 1), skill training may not be effective, as the underlying motivation must first be established (i.e., "cultural readiness" must be achieved), so that the learner becomes aware of the existing shortcoming and their potential consequences in the real world. Thus, training strategies at this level should focus on conveying the relevance of cultural training by raising awareness of the importance of cultural concerns (see Figure 1). Probing questions could be posed to a trainee to gauge cultural readiness. If issues were identified, remediation would be instantiated (e.g., multimedia illustrations could be played that show bad consequences if improper cross-cultural skills were used), after which curiosity in cultural concerns could be increased through trigger conditions (Osland, Bird, & Gundersen, 2007) during training, including novelty (i.e., cultural incidents that come as a surprise, such as violating the personal space of a trainee) and discrepancy (i.e., situations that exhibit a contradictory nature and violate conceptualizations of expected cultural behavior). Trigger threshold characteristics (i.e., intensity, salience, persistence, accumulation, and timing [i.e., instantaneous or incremental]) as defined by Osland et al. (2007) could be manipulated until cultural readiness was achieved (i.e., EEG workload and engagement indices high, eye fixations on AOI, imprecise scanpaths; see Table 1).

Conscious Incompetence - Focus: Cultural Sensemaking

Once cultural awareness is achieved, procedural skills can be trained and cultural knowledge can be transferred until competency is reached. At this stage, a trainee could initially be presented with a number of worked examples (Kalyuga, Chandler, & Sweller, 2001), both successful and unsuccessful, with appropriate feedback, from which a trainee could learn appropriate cross-cultural behavior. After these worked examples, the trainee could be presented with trigger events (Osland et al., 2007) that should engage the trainee in cultural sensemaking (see Figure 1), which involves the trainee framing a situation when he/she identifies the cultural concerns being trained and then engaging in "indexing behavior," which involves noticing or attending to AOI that provide cultural cues about the situation. The trainee should then begin making attributions by a matching process in which cultural cues are linked to scenario events, inferences are drawn about the scenario, and then appropriate cultural scripts are chosen in response (e.g., choosing to engage indirect eye contact during negotiations). Deficiencies in responses would be addressed via adaptive scenarios/ feedback.

Conscious Competence - Focus: Variety of Tough Cases

To develop expertise, it is necessary to train beyond simple skill acquisition. Practicing skills from different angles and acquiring strategies for more efficient processing is necessary to reach a level of proficiency at which the processes related to each skill have become easily accessible and can be executed in an automated manner. Thus, to build adaptive expertise, a variety of scenarios – both routine and "tough cases" – would be presented to trainees from various viewpoints. Experts are said to learn more from their mistakes than from correct responses (Sonnentag,

2000), and thus "tough cases" that are meant to instigate errors would be used in order for trainees to learn how to handle complexity, overcome knowledge shields (i.e., rationalization of misunderstandings), and guide them in acquiring more adaptive knowledge that can be used to address the unexpected (i.e., tough cases) (Hoffman, Feltovich, Fiore, Klein, & Ziebell, 2009).

Unconscious Competence - Focus: Accelerate Training

Once trainees have reached a high level of competency for a particular set of procedural skills, training could be automatically refocused on other skill areas to ensure optimal use of training resources (see Figure 1). At this level refresher training may also occur to maintain proficiency.

CONCLUSIONS

Development of the ADAPT-HT framework has shown that multiple theoretical foundations exist that are compatible and can therefore be combined into a training framework for enhancing 3C procedural skills at the tactical level. A direct result of this is that static solutions are not practical for 3C training purposes, as training needs with respect to focus, content, and feedback change throughout the process. To effectively guide the trainee from novice (being unconscious about one's incompetence) to mastery (performing cultural tasks with unconscious competence), it is necessary to evaluate the trainee's progress along this route, so that training can be dynamically tailored to respective needs at any given progress stage.

It was further found that traditional performance measures alone are not sufficient to distinguish training progress stages to inform dynamic training adaptation. This is because 'cultural awareness' cannot be detected through observation of explicit behavior. Physiological sensors offer a practical solution here, as measures such as EEG and eye tracking provide a 'peek into the mind' of the trainee and enable the detection of the amount of consciousness employed during training. Thus, an optimal 3C training solution should combine explicit measures of competence with implicit indicators of consciousness, so that training can address trainee deficiencies in an organized manner (e.g., build consciousness before skills), content can be dynamically adapted to the current needs of the trainee, and appropriate type and level of feedback can be provided at any stage. Observing these findings in creating 3C training systems will ultimately result in improved training effectiveness and efficiency.

ACKNOWLEDGEMENTS

This material is based upon work supported in part by the Office of Naval Research (ONR) under SBIR contract N00014-09-M-0385. Any opinions, findings and conclusions or recommendations expressed in this material are those of the authors and do not necessarily reflect the views or the endorsement of ONR.

REFERENCES

Augustyniak, P. and Tadeusiewicz, R. (2006), "Assessment of electrocardiogram visual interpretation strategy based on scanpath analysis." *Phys. Measurement*, 27, 597-608.

Bandura, A. (1977), *Social Learning Theory*. Prentice-Hall, New Jersey.

Bandura, A. (1997), *Self-efficacy: The Exercise of Control*. WH Freeman & Co., New York.

Berka, C., Levendowski, D., Lumicao, M., Yau, A., Davis, G., and Zivkovic, V. (2007), "EEG correlates of task engagement and mental workload in vigilance, learning and memory tasks." *Aviation Space and Environmental Medicine*, 78(5, Section II, Suppl.).

Black, S., and Mendenhall, E. (1990), "Cross-cultural training effectiveness: A review and theoretical framework." *Academy of Management Review*, 15, 113–136.

Chin, C.O., and Gaynier, L.P. (2006), "Global leadership competence: A cultural intelligence perspective." Presented at the *2006 Midwest Business Administration Association Conference*; Accessed on February 14, 2010: http://www.csuohio.edu/sciences/dept/psychology/graduate/diversity/GlobalLeadership%2011206.pdf.

Davis, S.L., and Finney, S.J. (2006), "A factor analytic study of the cross-cultural adaptability inventory." *Educational & Psychological Measurement*, 66(2), 318-330.

Doorenbos, A.Z., Schim, S.M., Benkert, R., and Borse, N.N. (2005), "Psychometric evaluation of the Cultural Competence Assessment instrument among healthcare providers." *Nursing Research*, 54(5), 324-331.

Hoffman, R.R., Feltovich, P.J., Fiore, S.M., Klein, G., and Ziebell, D. (2009), "Accelerated learning?" *IEEE Intelligent Systems*, 24(2), 18-22.

Howell, W.S. (1982), *The Empathic Communicator*. Wadsworth, Bellmont, CA.

Kalyuga, S., Chandler, P., and Sweller, J. (2001), "Learner experience and efficiency of instructional guidance." *Educational Psychology*, 21, 5–23.

Mendenhall, M., and Osland, J. (2002), "Mapping the terrain of the global leadership construct." Paper presented at the *Academy of International Business*, San Juan, Puerto Rico, June 29th, 2002.

Mendenhall, M.E., Stevens, M.J., Bird, A., and Oddou, G.R. (2008), "Specification of the content domain of the Global Competencies Inventory (GCI)." The Kozai Working Paper Series, 1(1). Accessed on October 2, 2009: http://kozaigroup.com/PDFs/GCI-Technical-Report-Dec%202008-1.pdf.

Osland, J.S., Bird, A., and Gundersen, A. (2007, August), "Trigger events in intercultural sensemaking." Paper presented at the *Meeting of the Academy of Management, Philadelphia*. Accessed on November 20, 2009: http://business.umsl.edu/seminar_series/2009%20Spring%20Seminar%20Series%20in%20Business%20and%20Economics%20ppt%20files/Trigger%20Events.doc .

Peace Corps (2000), "Culture matters: The Peace Corps cross-cultural workbook." Peace Corp. ICE #T0087, Washington, D.C. Accessed on December 9, 2008: http://www.peacecorps.gov/wws/publications/culture/.

Schim, S.M., Doorenbos, A.Z., Miller, J., and Benkert, R. (2003), Development of a cultural competence assessment instrument." *Journal of Nursing Measurement*, 11(1), 29-40.

Sonnentag, S. (2000), "Excellent performance: The role of communication and cooperation processes." *Applied Psychology: An International Review*, 49(3), 483–497.

Stevens, R., Galloway, T., and Berka, C. (2007), "Allocation of time, EEG-engagement and EEG-workload resources as scientific problem solving skills are acquired in the classroom." *Proceedings of 3ʳᵈ Augmented Cognition International*, held in conjunction with HCI International 2007, Beijing China, July 22-27, 2007.

Chapter 54

Integrating Cross-Cultural Decision Making Skills into Military Training

W. Lewis Johnson, LeeEllen Friedland

Alelo Inc.
12910 Culver Bl., Suite J
Los Angeles, CA 90066 USA

ABSTRACT

Intercultural skills are increasingly recognized as critical to overseas military and humanitarian operations. As military operators engage in day-by-day and moment-by-moment decisions in the course of carrying out a mission, they need intercultural competence in order to make appropriate decisions and act on them in culturally appropriate ways. This paper describes an approach to intercultural skills training that focuses on the skills required by military operators in overseas operations. The overall capability is called Operational Language and Culture Training System (OLCTS) (pronounced "OLaCTS"). OLCTS provides anywhere, anytime language and culture training on a range of different training devices and platforms, including desktop, web-based, and hand-held. Immersive serious games provide an integrated learning environment in which trainees must make decisions about mission goals and logistics as well as cross-cultural and interpersonal interactions with socially intelligent virtual agents. The OLCTS capability is being rapidly transitioned into regular use by military service members in the United States as well as in allied countries.

Keywords: Cross-cultural decision making, intercultural communication, small unit training, immersive serious games

INTRODUCTION

Intercultural skills are increasingly recognized as critical to overseas military and humanitarian operations. For example, the US Defense Regional and Cultural Capabilities Assessment Working Group has identified the ability to integrate cultural knowledge and skills into mission execution as a critical cross-cultural competency for general purpose forces (McDonald, et al., 2008). As military operators engage in day-by-day and moment-by-moment decisions in the course of carrying out a mission, they need intercultural competence in order to make appropriate decisions and act on them in culturally appropriate ways.

This paper describes an approach to intercultural skills training that focuses on the skills required by military operators in overseas operations. The overall capability is called Operational Language and Culture Training System (OLCTS) (pronounced "OLaCTS"). OLCTS provides anywhere, anytime language and culture training on a range of different training devices and platforms, including desktop, web-based, and hand-held. It prepares trainees for the application of intercultural skills to simulated operational scenarios for civil affairs, humanitarian relief, and security cooperation missions. It provides training for the situations that trainees are likely to encounter as part of these missions, so that they can learn to take appropriate action in those situations. Immersive serious games provide an integrated learning environment in which trainees must make decisions about mission goals and logistics as well as cross-cultural and interpersonal interactions with "socially intelligent virtual agents" (Johnson & Valente, 2009). The OLCTS capability is being rapidly transitioned into regular use by military service members in the United States as well as in allied countries.

The paper is organized as follows. First it gives an overview of each of the components of the OLCTS training framework. This is followed by an example of how the US Marine Corps envisions employing OLCTS tools in cross-cultural skills training. The paper then outlines the methodology and technical approach used in OLCTS to develop and deliver training materials across the OLCTS training suite. The paper then summarizes current work in the Human Social Culture Behavior Modeling (HSCB) program to extend the capabilities of OLCTS and improve its ability to deliver high quality cross-cultural skills training.

THE OLCTS TRAINING FRAMEWORK

OLCTS includes immersive serious games in which trainees can learn cultural knowledge and skills, and then practice them in immersive simulations of intercultural exchanges. Figures 1 and 2 illustrate the different training platforms that are supported in the OLCTS training suite. These examples are taken from the OLCTS course for Dari language and Afghan culture. The image on the left of

Figure 1 is from one of the interactive lessons in the course, where trainees acquire basic knowledge and intercultural skills. The image on the right shows an immersive scenario in which the trainee's—or player's—character (on the left) is engaging in meeting with the elders in the village to discuss a reconstruction process. The non-player characters in the game are implemented using artificial intelligence and spoken dialog technologies, so that trainees can engage in conversations with the game characters and practice the intercultural skills they will need as part of their missions.

To meet the needs of advanced distributed learning (ADL), we have developed a Web-based delivery platform, called Wele, that is used to offer similar training over the Web. The basic version of Wele uses 2½ D animations instead of 3D animations, implemented in Flash, and menu-based interaction with animated characters. We also make available browser plug-ins and provide speech recognition and immersive 3D capability, to provide a highly immersive training experience similar to that of the PC games.

FIGURE 1. OLCTS PC-based training system

Although studies have shown that these PC-based immersive games can be very effective as intercultural skills training tools (MCCLL, 2008), they require trainees to dedicate training time in front of the computer, separate from their other training activities. Figure 2 shows additional training platforms that overcome these limitations. The left image shows an implementation of the interactive language and culture lessons on an Apple iPod Touch. Trainees can use this training device pretty much whenever and wherever they like, both prior to deployment and while in country. It does not yet have the speech recognition and 3D immersive capabilities of the PC-based trainers, but the added convenience and accessibility help to compensate for that. The screenshot on the right in Figure 2 is from our intercultural interaction plug-in for the VBS2 multiplayer simulation environment, which enables trainees to practice their intercultural skills as part of simulation-based training and mission rehearsal exercises.

 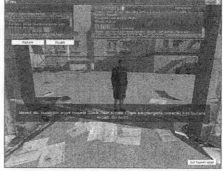

FIGURE 2. OLCTS handheld and multiplayer training platforms.

APPLICATIONS OF THE OLCTS TRAINING SUITE

The following example shows how one military service intends to use the OLCTS framework in training. It illustrates the need for anywhere, anytime language and culture training, and the most suitable conditions for that training to meet specific needs. This implementation calls for training that has the following characteristics: it should (1) prepare military service members to make decisions and take action in a mission context; (2) provide training and/or mission rehearsal opportunities that represent the specific types of situations that service members are likely to encounter in the field; and (3) be consistent across the range of training platforms that service members will employ.

TRAINING FOR US MARINE SMALL UNIT OPERATIONS IN AFGHANISTAN

US Marine Corps (USMC) small units in Afghanistan face a broad range of potential encounters with host nationals while performing non-kinetic missions. They must respond to changing circumstances as well as create and exploit opportunities to advance mission goals in environments where the lack of a shared language or cultural understanding can compound the challenges.

The complexity of the training need is illustrated by a "sea story" developed by the USMC Program Manager for Training Systems. In this sea story, which sketches out training needs in a platoon preparing to deploy to Afghanistan, a 2Lt is going to conduct a humanitarian assistance mission in an Afghan village that requires that he build rapport with local leaders and facilitate the work of NGOs in the area. In order to accomplish these tasks, the 2Lt would optimally require training in cross-cultural competence knowledge, skills, and personal characteristics (McDonald, et al., 2008); targeted cultural or regional knowledge, including etiquette and culturally appropriate practices; and, minimally, basic language skills, such as those required to greet people and show respect. Whether working with an interpreter or not, the 2Lt will need to make decisions and take action based on his assessment of explicit and implicit factors, including new and highly complex systems of beliefs and values among relevant Afghan communities.

In addition, the 2LT will need to bring similar training to bear in working with different sorts of organizations, including NGOs, which can bring into the mix such factors as organizational, international, or regional cultures; work processes and networks that cross traditional social boundaries, and well-established views about the relative merits of US military participation in civil affairs or humanitarian missions, While it might be tempting for trainees to assume that their counterparts in other organizations share the same goals, US military personnel can practice how to discover and assess other perspectives, and interact in appropriate ways for a given mission in a specific context, in the OLCTS immersive training environment

The sea story unfolds further as it focuses on the platoon and the different mission-related roles that will be played by squad members and squad leaders. While everyone completes pre-deployment training, the curriculum is tailored to the level, needs, and roles appropriate for different trainees. So squads train for junior-level roles in their mission scenarios and squad leaders do the NCO-level language and culture training. Finally, the sea story follows the Platoon Sergeant who creates new mission scenarios for training that incorporate location-specific cultural and human terrain information from the base culture and language center. This enables the platoon to become more familiar with the characteristics of their deployment location, pursue additional language and culture training targeted to their mission roles, and practice their new knowledge and skills in the immersive game environment that gives them feedback on their progress as they gain greater intercultural competence.

OLCTS DEVELOPMENT METHODOLOGY

As the above example illustrates, language and culture training content in OLCTS courses needs to be closely aligned with the operational needs of military service members, so that they can effectively apply their language and cultural skills in military decision making. Alelo develops course content using a Situated Culture Methodology that focuses training content on operational requirements in cultural context, and results in course content that is best suited for anywhere, anytime delivery.

SITUATED CULTURE METHODOLOGY

All OLCTS courses are designed with the same major goal: to teach effective interpersonal and intercultural communication via practical and task-based knowledge. OLCTS courses are developed using a dynamic methodology for identifying and teaching situated culture, that is, the cultural knowledge needed to successfully perform tasks or higher-level projects in a foreign country or unfamiliar sociocultural setting.

Trainees learn the cross-cultural competence required to successfully interact and communicate with people from different linguistic and cultural contexts – to quickly perceive the differences in a new cultural context, and to respond in a culturally appropriate way. Trainees learn to be more aware of cultural differences and cultural relativity, that is, they learn metacultural awareness. This improved metacultural awareness then becomes a tool kit that learners take with them to an international context and use to learn culturally appropriate and effective ways of speaking and behaving. The Situated Culture Methodology identifies the relevant contextual and sociocultural factors that interact with, and play a role in, a given situation at a given moment.

The methodology is broken down into three major areas of focus: context, sociocultural factors, and curriculum. The research and development process can be iterative, that is, research on sociocultural factors may feed back into work on context, which may then help determine the path of more work on sociocultural factors. All the research then feeds into curricula, which are meant to support trainees in reaching performance objectives and cross-cultural competence.

Micro-social factors are generally the most important factors for Alelo content development, because they play a role in face-to-face interactions, which are at the core of Alelo learning. Key among them is conversational culture, which encompasses the many parameters that come together to make a "normal" conversation, and ways of expressing politeness, distance or closeness, formality or informality.

Four types of subject matter experts, or SMEs are consulted in the course of cultural research and content development. Task SMEs, often provided by the client, are familiar with the client's project or mission, and possibly the specific geographic region as well. Culture SMEs are native to the region the project is focused on and preferably have some level of expertise in analyzing and explaining culture. Language SMEs have native or native-language competence in the language that is being taught as part of the project. Finally, academic SMEs provide high-level research perspective and advice to the project. Information from these SMEs is used to develop all major parts of OLCTS courses. Anthropologists and content developers also work on determining and documenting the sociocultural factors that will be relevant for trainees. These sociocultural factors come from all levels of social organization: macro-social, micro-social, and individual. Together they form the "cultural lens" through which trainees and the local citizens with whom they will interact view and interpret the world.

TECHNOLOGIES FOR CROSS-PLATFORM CONTENT DEVELOPMENT AND DELIVERY

OLCTS training content needs to be delivered and maintained on a range of different platforms. In the case of multi-platform delivery methods such as the Marine Corps training approach, this is required so that the content available on one training platform is consistent with that available on other platforms. This becomes especially challenging as training content is adapted for changing needs. The emphasis in OLCTS on mission-oriented training guarantees that training content will need to adapt over time as missions change. Additionally, OLCTS must support different combinations of training platforms and variations in technical feature sets. This variability is needed not just to support the range of needs and priorities across multinational coalition forces, but also the differences in training delivery strategies among military services and commands. For example, handheld delivery is currently a priority for the US Marine Corps, but is not yet a priority for the US Army. The features available in the Web-based components of OLCTS varies depending upon whether or not the military command receiving the training is willing to approve browser plug-ins with advanced technologies such as automated speech recognition. This has implications for the way content is authored as well as the way it is delivered.

The variety of training platforms used in OLCTS has made it necessary for us to develop authoring tools that make it possible to author content independent of the target platform (Johnson et al., 2008). The authoring tool portal, named Hilo, specifies the content of each lesson element and learning activity, independent of the screen dimensions and computing power of the device. Instead of adopting a What You See is What You Get (WYSIWYG) approach to content authoring, Hilo specifies the semantic components of each lesson element, their properties, and the relationships between them. For example, authoring a language lesson "page" in Hilo involves specifying the foreign language utterances to be covered, their translations, explanatory notes, and selecting appropriate voiceover recordings. This information is encoded in XML and stored in a content repository, and then used to generate presentations for the target device. The number of presentations, and the interactive features that they support, depend upon the capabilities of the device. For example, material that is presented in a single page in a Web browser is presented in series of pages on the iPod. If the target device has speech recognition capability enabled, activities tend to make heavy use of it, whereas if no speech recognition capability is available alternative interaction strategies that do not rely on speech recognition must be employed instead.

Although we do our best to generate equivalent realizations of content on all delivery platforms from the same content specifications, in practice it is simply not possible in all cases. In such cases authors have the option of authoring an alternative content element to use instead, or omit it from the platform. For example, scenario vignettes that utilize real-time 3D animation on PCs may appear as pre-rendered machinima instead. The author specifies for each content element the conditions under which it may be used. The same conditional rendering approach is employed to support customized courses for different user groups. This makes it possible to customize courses for different coalition partner countries or different military services.

FURTHER DEVELOPMENTS OF OLCTS UNDER THE HSCB PROGRAM

Under the auspices of the Human Social Culture Behavior (HSCB) program, Alelo has been undertaking research to develop improved models of sociocultural behavior and utilizing them to produce improved, more flexible training. The fruits of this research will be transitioned into future versions of the OLCTS training suite.

CULTURECOM

The CultureCom project is developing formal models of the cultural influences underlying dialog and utilizing them to increase the flexibility and realism of the behavior of non-player characters in training simulations. The work is being conducted in collaboration with Dr. Michael Agar of Ethnoworks and Prof. Jerry Hobbs of the University of Southern California. Cultural and linguistic anthropologists are developing validated sociocultural data sets for Afghanistan and other cultures of interest, consisting of annotated dialogs of cross-cultural interactions. Experts in artificial intelligence then use these data to develop logical models of sociocultural behavior based upon a formal ontology of microsocial concepts underlying interpersonal communication. The resulting models are validated against the original sociocultural data. We are adapting the agent behavior engines in VRP to use these models to drive agent behavior. The validated "swap-in" cultural models result in non-player characters whose behavior is culturally realistic and can be flexibly adjusted to varying levels of training and assessment difficulty.

SOCIALSIM-MR

The SocialSim-MR project has designed a novel hybrid virtual-constructive approach to simulation-based training. In collaboration with Prof. Barry Silverman of the University of Pennsylvania, we have designed and prototyped a cultural training system that enables trainees to practice their cultural skills in a simulated village, where the trainees' actions can produce effects that unfold over time. SocialSim-MR integrates with the VRP system so that trainees can interact with the host nationals in the village, just as they would in a basic VRP system. The next time trainees enter the simulated village, however, they will find that things have changed, partly as a result of actions they have taken and partly as a result of actions taken by insurgents and other simulated actors in the village. SocialSim-MR is designed to be able to model realistic missions such as the Zebat mission described above, which can only be achieved through a series of interactions and encounters with the host nationals in the village.

C-CORE

C-CORE is a framework for managing workflows in Cultural Content, Ontology, and Resource Engineering. C-CORE is an example of a Cultural Architecture Generator (CAG). It is a coherent suite of authoring tools for scenarios and computer-generated forces (CGFs) that results in highly authorable, culturally-aware communicative agents that are re-usable on a variety of serious game platforms, including the OLCTS platforms. The resulting CGFs will exhibit culturally appropriate behavior and track learner responses for After Action Review. To keep the scenarios up to date, C-CORE includes mechanisms for incorporating current sociocultural and human terrain data from a variety of sources, such as subject matter experts, HUMINT reports, and other media.

C-GAME

C-GAME, developed in coordination with C-CORE, is developing a framework for scenario-based training in operational cultural competence and other sociocultural skills relevant to military mission contexts. It will comprise a toolset for developing training scenarios, and a run-time execution engine that can be integrated with a variety of game-based training environments. The scenario construction toolset will enable military users to develop their own sociocultural training scenarios through a combination of libraries of reusable sociocultural models and authoring tools. It will include mechanisms for incorporating current sociocultural and human terrain data into the scenarios, to keep them up to date. It will also provide a way for trainers to define trainee performance standards. The run-time execution engine will incorporate instrumentation to compare trainee performance to the training objectives, and provide automated after-action review and remediation capabilities. The proposed framework will be designed to be platform independent, and at the same time integrate as seamlessly as possible with the user interface of specific game platforms, to provide trainees and trainers with a coherent, easy-to-use training environment.

CONCLUSION

OLCTS provides a flexible tool suite that is being used to integrate cross-cultural decision making skills into military training. The situated culture approach ensures that language and culture content is targeted to support critical mission goals and intercultural communication.

ACKNOWLEDGMENTS

Parts of this work have been funded by many sponsors, including the USMC, ONR, DARPA, CTTSO, and AFRL. Opinions expressed are those of the authors do not reflect official policy of the US Government or the British Government.

REFERENCES

Johnson, W.L., Valente, A., & Heuts, R. (2008), "Multi-platform delivery of game-based learning content." Proceedings of SALT 2008.

Johnson, W.L. & Valente, A. (2009), "Tactical Language and Culture Training Systems: Using AI to teach foreign languages and culture." *AI Magazine*, 30 (2), 72-83.

MCCLL (2008). "Tactical Iraqi Language and Culture Training System", *Marine Corps Center for Lessons Learned Newsletter*, 4 (8), 4.

McDonald, D. P., McGuire, G., Johnston, J., Selmeski, B., & Abbe (2008), "Developing and Managing Cross-Cultural Competence Within the Department Of Defense: Recommendations For Learning and Assessment." US Defense Regional and Cultural Capabilities Assessment Working Group.

References (Font: Times New Roman, size 10)

Courage, K.G., and Levin, M. (1968), *A freeway corridor surveillance information and control system*. Research Report No. 488-8, Texas Transportation Institute, College Station, Texas.

Daubechies, I. (1992), *Ten Lectures on Wavelets*. Society for Industrial and Applied Mathematics, Philadelphia, Pennsylvania.

Dennis, J.E., and Schnable, R.B. (1983), Numerical Methods for Unconstrained Optimization and Nonlinear Equations. Prentice Hall, New Jersey

Chapter 55

Modeling Cultural and Personality Biases in Decision-Making

Eva Hudlicka

Psychometrix Associates, Inc.
Blacksburg, VA 24060 USA

ABSTRACT

Cultural, personality, and affective biases in decision-making are well documented. In this chapter I describe an approach for modeling multiple decision biases due to trait (cultural and personality) and state (affective) influences, within the context of a symbolic agent architecture. The approach provides a uniform framework for modeling both *content* and *processing biases,* in terms of parameter vectors that control processing within the architecture modules. The associated simulation environment enables the modeling of a wide variety of decision-makers, in terms of distinct profiles, and the approach lends itself to exploring alternative mechanisms mediating decision biases.

Keywords: Decision biases; cultural, personality and affective biases; cognitive-affective architecture; decision bias mechanisms

INTRODUCTION

Cultural and personality biases in decision-making are well documented. Particularly dramatic cultural effects can be found in aviation, where a number of accidents have been attributed cultural factors, such as the adherence to strict hierarchical communication patterns (e.g., Korean Air Flight 801).

Increasing reliance on human-machine systems in decision-making environments, including aviation, healthcare, and the military, demands that decision-support systems be aware of cultural and personality biases. Decision-support systems in particular would benefit from an ability to recognize and react to cultural and personality decision-biases. This is the case for systems that support long-term collaborative efforts (e.g., Groupsystems' ThinkTank), but especially critical for systems designed to support decision-making in real-time, high-tempo and high-stress environments (e.g., space missions, aviation, healthcare, military).

An ability to model decision-biases in human behavior models embedded in decision-support systems would enhance their abilities to provide customized decision-support, for a range of decision-maker profiles. In addition, such models would enhance our understanding of these biases, both their mechanisms (Hudlicka, 2008), and their triggers and effects. Such improved understanding would in turn enable the development of training systems that would help decision-makers counteract the deleterious effects of biases, across a range of operational settings.

In this chapter I discuss an integrated approach to modeling a broad range of decision biases, within the context of a human behavior model, embedded in an agent architecture. The distinguishing feature of this modeling approach is the uniform parameter space it provides for representing the diverse influences of multiple types, and sources, of biases. The primary focus is on modeling cultural and personality biases, and biases due to transient affective states (e.g., anxiety and stress, boredom, frustration, and anger). Affective biases are included because emotions often act as the intervening variables for a number of decision biases. The modeling approach draws a distinction between *content-based* and *process-based* biases, in an effort to provide a categorization of biases that is based more on their mediating mechanisms, rather than their surface features.

The chapter is organized as follows. *First,* I provide background information about different types and sources of biases, emphasizing the role of emotion as the intervening variable. *Next,* I suggest that a categorization of biases into content-based and process-based is useful for both their modeling, and for elucidating their mechanisms. *Next,* I introduce the MAMID framework for modeling a broad range of decision-biases, and provide an illustrative example demonstrating its capabilities to model multiple biases. The chapter concludes with a summary and a brief discussion of the potential applications of the MAMID modeling framework.

DECISION BIASES

Research in judgment and decision-making has identified a number of *biases* in decision-making. Kahneman, Tversky and colleagues (Kahneman et al., 1982) identified a number of judgment and decision-making biases in individual decision-making, including the following: *availability bias* (the tendency to base decisions on the most readily available evidence, rather than the most appropriate evidence, hence the *primacy* and *recency* bias resulting from biased memory recall), *confirmation bias* (the tendency to prefer, or actively seek out, evidence that supports one's hypotheses, expectations and goals), and *framing effects* (the tendency to be influenced by, and respond differently to, the wording (as opposed to the substance) of a question or a decision).

Biases have been identified at various levels of processing complexity, ranging from biases on fundamental perceptual and cognitive processes (e.g., attention, memory), to biases on high-level cognitive processing (e.g., situation assessment, problem solving, decision-making). A number of individual differences contribute to decision-biases, including cultural and personality factors, and affective states. These are discussed below.

CULTURAL AND PERSONALITY DECISION BIASES

A number of personality trait sets have been proposed that aim to capture correlations between specific trait values and particular biases. These range from relatively low-level traits such as the Five Factor model (Openness, Conscientiousness, Extraversion, Agreeableness, Neuroticism) (Costa & McCrae, 1992), to more aggregated traits such as task-based vs. process-based styles of leadership. Several cultural trait sets have also been proposed, most notably Hofstede's Uncertainty Avoidance, Individualism-Collectivism, Power Distribution, Short vs. Long-term Orientation, and Femininity-Masculinity (1991).

Recent attempts to correlate the cultural factors with individual personality traits suggest that the observed behavioral regularities associated with high-level cultural factors such as Uncertainty Avoidance may eventually be explained in terms of individual personality traits, and even specific mechanisms; e.g., correlations have been identified between Uncertainty Avoidance and the Five Factor factor of neuroticism (Hofstede and McCrae, 2004).

Examples of personality-linked biases include: preference for self and affective-state stimuli associated with high-neuroticism and high-introversion, bias toward threat-cues associated with high-neuroticism, and sensitivity to reward and higher risk tolerance associated with high extraversion (Matthews et al., 2000). There are of course endless examples of cultural variability in beliefs, values and behavioral norms. More recently, differences in reasoning have also been identified (Matsumoto, 2001).

Cultural decision-biases are usually attributed to differences in knowledge, beliefs, values, attitudes, goals and behavioral norms. *Personality-linked biases* are also reflected in these types of differences, but are also evident in differences in processing, and differences in affective dynamics. For example, a decision-maker with high degree of extraversion and low degree of neuroticism may be biased toward the external environment, preferentially focusing on task cues and schemas. In contrast, a low-extraversion, high-neuroticism decision-maker may be biased toward self-cues and schemas, resulting in a focus on his own internal states. In high-stress situations, these differences will likely result in differences in decision outcomes. Whereas the extraverted/low-neuroticism decision-maker is likely to remain focused on the task, the introverted/high-neuroticism decision-maker is likely to focus on his internal state, resulting in task neglect. In addition, the associated negative emotions experienced by this decision-maker will exert further biasing effects on his decision-making, by influencing speed and capacity of attention and working memory, and further exacerbating the existing self-bias.

Personality and cultural traits represent long-lasting, permanent influences. In contrast to these *trait-based* effects on decision-making, psychologists and decision-scientists also study *state-based* effects. These are transient effects due to some temporary state of the decision-maker, most often states with a strong affective component (e.g., stress, boredom, surprise, frustration, anger, sadness), but also physical states (e.g., fatigue). Affective biases represent an important category of decision-biases and are discussed below.

AFFECTIVE DECISION BIASES

Influence of emotions on decision-making is well-documented, and emotions play an important role as the intervening variables of a wide range of decision biases. Both short-term *emotions* and longer-term *moods* influence decision-making. Some of the more robust findings include the effects of anxiety and fear, anger and frustration, and positive and negative mood. These effects are evident in both the fundamental cognitive processes (e.g., attention, memory), but also in the higher-level processes mediating decision-making and problem-solving. Examples of biases include: changes in attention capacity, speed and bias; changes in the speed and capacity of working memory; differential activation of specific perceptual and cognitive schemas that mediate the perception and processing of particular stimuli. Examples of specific biasing effects include: association between positive affect and lack of anchoring, overestimation of positive events, underestimation of negative events (Mellers et al., 1998); fear- and anxiety-linked attentional and perceptual bias toward the detection and processing of threatening stimuli (Williams et al., 1997); anxiety-linked focus on possible failure, and subsequent choice of protective behavior; anxiety-linked preference for analytical processing; mood-congruent bias in memory recall (Bower, 1981), and predictions of future positive or negative events (Matthews et al., 2000); and anger-linked increase in risk-tolerance, and attributions of hostility.

Recent research has explored the relationship between cultural traits and emotions, focusing in particular on emotion regulation. Matsumoto and colleagues (Matsumoto et al., 2005), have found correlations between the cultural trait of Uncertainty Avoidance and the individual's ability to control the generation and manifestations of emotions, primarily negative emotions, via a variety of regulatory coping strategies, such as re-appraisal or physiological methods (e.g., breathing, relaxation). Their findings indicate that low Uncertainty Avoidance (UA) correlates with more re-appraisal schemas for anxiety reduction, and more diverse coping strategies for emotion regulation, whereas high UA correlates with more rigid adherence to emotion display rules.

CONTENT AND PROCESS BIASES

The list of biases outlined above, along with the existing categorizations, provide a foundation for the development of computational models decision biases. However, the existing categories may not be the most useful ones for the development of computational models, or for understanding the mechanisms that mediate the broad range of biases outlined above.

For modeling purposes, it is more useful to divide decision biases into *content-based* and *process-based*. During the course of decision-making, both categories of biases may of course operate simultaneously. *Content-based* biases relate to differences in the specific content and organization of the knowledge schemas that mediate perception, and the cognitive processes involved in decision-making. These schemas represent long-term beliefs, values, goals, preferences, and attitudes of the decision-maker. The long-term influences of culture and personality are most easily represented in the specific contents and organization of these schemas. For example, a decision-maker coming from a high-collectivism culture (e.g., China) would be expected to have more memory schemas related to the well-being of the larger group (e.g., perceptual structures for assessing group state, and goals for group well-being). In contrast, a decision-maker from a high-individualism culture (e.g., US) would have more memory schemas devoted to perceptions and goals regarding his/her own well-being and goal priorities. Regarding the effects of personality traits: a high-neuroticism individual would have a larger proportion of schemas related to self- and threat-monitoring, as well as more differentiated set of schemas regarding the representation of self-relevant and threat-related states, in contrast to a low-neuroticism individual.

In contrast, *process-based biases* relate to the way stimuli are perceived and processed during decision-making, and the manner in which existing knowledge is used during decision-making. The primary sources of process-based biases are the current states of the decision-maker, most notably, the affective states. Effects of traits are also evident here however, since personality traits exert an influence on the dynamics of affective states, influencing their intensity and decay rates. Process-based biases are most readily implemented in terms of transient changes to the fundamental processes that mediate decision-making, altering the speed and capacity of attention and working memory, as well as the likelihood that particular content will be processed. For example, anxiety-linked threat bias is associated with faster processing of high-threat stimuli, subsequent bias toward selecting a high-threat interpretation, and a subsequent preference for self-preservation goals.

Categorizing decision-biases into content vs. process based implies distinct mediating mechanisms, and each category has distinct implications for modeling. Content-based biases are most appropriately modeled via differences in the structure, content and overall organization of the memory schemas mediating decision-making; that is, the long-term knowledge, within which stable and permanent differences in cultural and personality traits are reflected. In contrast, process-based biases are best represented by temporary alterations in the processes that mediate decision-making, and are most appropriately modeled via transient parameter-controlled variations in those processes. The MAMID framework and architecture for modeling decision-biases provides a means of modeling both categories of biases, as described below.

FRAMEWORK AND ARCHITECTURE FOR MODELING PROCESS AND CONTENT BIASES

The MAMID framework provides an approach for modeling a broad range of decision-biases, arising from both state and trait effects. The distinguishing feature of the MAMID approach is its *parameter space*, which provides a uniform means for representing the combined influences of a variety of state and trait effects. This in turn enables MAMID to model a broad range of *process-based decision-biases,* including cultural trait-linked biases (e.g., Uncertainty Avoidance-linked to preference for more higher-confidence cues), personality trait-linked biases (e.g., Neuroticism-linked self-focus and predisposition to negative emotions), and state-linked biases (e.g., anxiety-linked threat bias in attention and perception, anger-linked increased risk-tolerance). MAMID also supports representation of alternative knowledge schemas in the architecture's long-term memory, which represents the domain knowledge. Differences in the content and organization of these schemas, represented in terms of belief nets, are associated with differences in traits, both cultural and personality. This enables MAMID to model *content-based decision biases.* Below we describe the MAMID cognitive-affective architecture, and the approach that enables the modeling of multiple, interacting decision biases.

MAMID COGNITIVE-AFFECTIVE AGENT ARCHITECTURE

The MAMID architecture provides a computational modeling and simulation environment within which the effects of multiple traits and states on decision-making can be represented (Hudlicka, 1998; 2002; 2003; 2007a; 2008). Since emotion acts as an intervening variable in many biases, MAMID includes an explicit representation of emotions, their generation, and their effects on decision-making. MAMID has been instantiated within two domains, a peacekeeping scenario, where MAMID models individual commanders (Hudlicka, 2003), and a search-and-rescue game task, loosely based on the DDD task developed by Aptima, where MAMID models individual team members, as they search for a 'lost party', while encountering a series of obstacles and adverse events (Hudlicka, 2007b).

MAMID models recognition-primed decision-making, in terms of several modules which progressively map the incoming stimuli (cues) onto the outgoing behavior (actions), via a series of intermediate internal representational structures, termed *mental constructs* (e.g., situations, expectations, and goals) (figure 1). The architecture consists of the following modules: *Sensory Pre-processing*, translating the incoming raw data into high-level task-relevant perceptual cues; *Attention,* selecting a subset of cues for further processing; *Situation Assessment*, integrating individual cues into an integrated situation assessment; *Expectation Generation*, projecting the current situation onto possible future states; *Emotion Generation*, dynamically deriving the affective state from a combination of external and internal stimuli; *Goal Manager*, selecting the most relevant goal for achievement; and *Action Selection*, selecting the most suitable action for achieving the highest-priority goal within the current context.

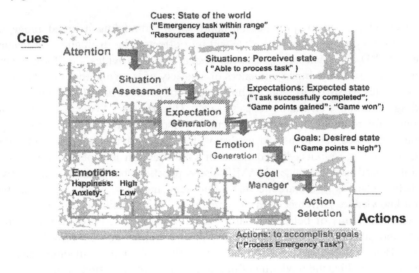

Figure 1. Schematic Illustration of the MAMID Cognitive-Affective Architecture

Each module has an associated long-term memory (LTM), consisting of either belief nets or rules, which represents the knowledge necessary to transform the incoming mental construct (e.g., cues for the "Situation Assessment" module) into the outgoing construct (e.g., situations for the "Situation Assessment" module). The currently-activated set of constructs within each module constitutes that module's working memory. Each module has parameters that determine the speed of processing and the capacity of the module's working memory. Parameters associated with the mental constructs determine the ranking of each construct, thereby influencing the likelihood of the construct's processing within a given simulation cycle. This enables the modeling of specific content biases, such as the bias towards processing threatening cues.

FRAMEWORK FOR MODELING MULTIPLE DECISION BIASES

The underlying assumption of the MAMID modeling approach is that the combined effects of a broad range of decision biases can be modeled *by varying the fundamental properties of the processes and structures* mediating decision-making

(Hudlicka, 1998; 2002). Examples of the properties necessary for *process-based biases* are the *speed* of the individual modules (e.g., fast or slow attention), and the capacities of the working memories associated with each module. Examples of these properties for the *content-based biases* are the structures available for storing long-term memory (LTM) schemas, their contents, and their overall organization.

The speed and capacities of the MAMID architecture modules are controlled by a series of parameters, whose values are derived from the decision-maker's state and trait profile, expressed as a vector of specific values for cultural and personality traits, and emotional states. The functions calculating the parameter values are weighted linear combinations of the specific factors influencing the associated process; e.g., value of the *rank parameter* of a mental construct (e.g., cue, situation) is a function of the following construct attributes: salience, confidence, threat level, decision-maker's anxiety level, individual history, neuroticism and uncertainty avoidance. The traits and states used to calculate a particular parameter, and their associated weights, are based on empirical data. Modeling different types of decision-makers is thus accomplished by changing the associated trait and state profiles, and mapping these onto the parameter vectors. Different values of the parameters cause different 'micro' variations in processing (e.g., number and types of cues processed by the Attention Module), which lead to 'macro' variations in observable behavior; e.g., high-anxious decision-maker misses a critical cue due to attentional narrowing and selects the wrong action. Figure 2 illustrates how MAMID models threat bias effects associated with neuroticism and anxiety. Figure 3 illustrates the changing affective profiles of low- and high-anxious decision-makers, within the search-and-rescue task, as a function of the situations encountered, and the resulting values of the working memory capacity parameter of one of the processing modules.

The MAMID parameter thus space provides a uniform means within which the combined influences of cultural traits, personality traits, and affective states can be represented. By providing the capabilities to represent a broad variety of traits and states, and integrate their biasing effects on perception, cognitive processing, and behavior, the MAMID modeling framework allows the integration of the cultural, personality and affective influences on decision-making.

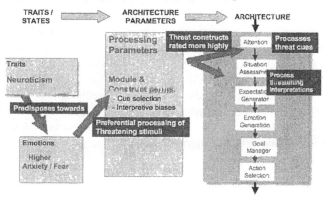

Figure 2: Modeling Threat Bias in Terms of MAMID's Parameter-Based Approach

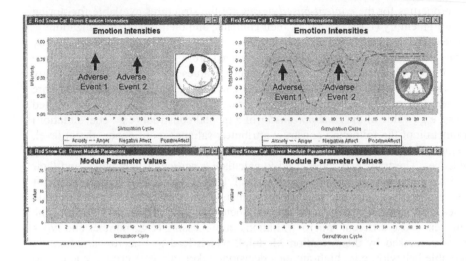

Figure 3: Variations in Two Distinct Decision-Makers' Affective Profiles (top), and Corresponding Module Parameter Values (bottom), During a Search-&-Rescue Task

CONCLUSIONS

This chapter discussed an approach for modeling decision biases in a symbolic agent architecture. The approach is based on the assumption that a broad range of interacting decision biases can be modeled in terms of parameters that manipulate properties of the fundamental processes mediating decision-making; e.g., speed, capacity, content bias. The approach is suitable for modeling traits (cultural and personality) and states (emotions), for both content and process biases. Its distinguishing feature is a parameter space which provides a uniform means of representing effects of multiple, interacting decision biases. By explicitly modeling individual cognitive processes mediating decision-making, and emotion generation and effects, the approach lends itself to modeling the mechanisms mediating decision-biases. In-depth understanding of these mechanisms would enable identification of attributes that contribute to decision errors, in individuals and teams. This would then enable the design of more effective human-machine systems in operational contexts, and more effective training systems.

REFERENCES

Bower, G.H. (1981). Mood and Memory. *American Psychologist*, 36, 129-148.

Costa, P. T, & McCrae, R. R. (1992). Four ways five factors are basic. *Personality and Individual Differences*, 13, 653-665.

Hofstede, G. (1991). *Cultures and Organizations: Software of the Mind*. NY: McGraw Hill.

Hudlicka, E. (2008). Modeling the Mechanisms of Emotion Effects on Cognition. In *Proc. of AAAI Fall Symposium on BICA*, TR FS-08-04, 82-86. Menlo Park, CA: AAAI Press.

Hudlicka, E. (2007a). Reasons for Emotions. In *Advances in Cognitive Models and Cognitive Architectures*, W. Gray, (ed.). NY: Oxford University Press.

Hudlicka, E. (2007b). *Application of Human Behavior Models to Risk-Analysis and Risk-Reduction in Human-System Design. TR 0507*. Blacksburg, VA: Psychometrix.

Hudlicka, E. (2003). Modeling Effects of Behavior Moderators on Performance. In *Proc. of 12th Conference on Behavior Representation in Modeling & Simulation*, Phoenix, AZ.

Hudlicka, E. (2002). This time with feeling: Integrated Model of Trait and State Effects on Cognition and Behavior. *Applied Artificial Intelligence*, 16:1-31. 2002.

Hudlicka, E. (1998). Modeling Emotion in Symbolic Cognitive Architectures. *AAAI Fall Symposium Series*, TR FS-98-03. Menlo Park, CA: AAAI Press.

Kahneman,D., Slovic,P. ,& Tversky,A. (1982).*Judgment under uncertainty*. Cambridge, UK.

Matthews,G., Derryberry,D.,& Siegle,G.J. (2000). Personality and emotion. In S.E. Hampson (Ed.), *Advances in personality psychology*. London: Routledge.

Matsumoto,D.,Yoo,S.H., & LeRoux, J.A. (2005). Emotional & Intercultural Communication. In *Handbook of Applied Linguistics*, H. Kotthoff & H.Spencer-Oatley, (eds.). Mouton.

Matsumoto,D., (2001). *The Handbook of Culture and Psychology*. NY: Oxford.

Mellers, B.A., Schwartz, A., Cooke, A.D.J. (1998). Judgment and Decision Making. *Annual Review of Psychology*, 49, 447-477.

Williams, J.M.G., Watts, F.N., MacLeod, C., & Mathews, A. (1997). *Cognitive Psychology and Emotional Disorders*. NY: John Wiley.

Chapter 56

Narrative Structure as a Cultural Variable in Modeling and Training Decision Making

Wayne Zachary[1], Lynn Miller[2], Stephen Read[3], Thomas Santarelli[1]

[1]CHI Systems Incorporated
1035 Virginia Drive, Suite 300
Ft. Washington, PA 19034

[2]Annenberg School of Communication/[3]Department of Psychology
University of Southern California
Los Angeles, CA 09989-0281

ABSTRACT

All cultures use narrative structures as a model for both making decisions and interpreting them. Current research suggests that narrative structures interact with individual motivations and personality factors to explain a large part of individual variability in decision making, with recourse to classical decision science constructs. Computational models and simulations based on narrative structure representations can be used train culturally adaptive decision making. Case studies from medicine and military negotiations illustrate the framework and its limitations.

Keywords: Narrative, Story spaces, motives, cognitive model, experiential learning

INTRODUCTION

There has been a growing agreement across a broad body of social science research over the last two decades that *situations* and *narratives* are critical constructs in understanding how people make sense of dynamic circumstances in a culture-specific way. The construct of 'situation' incorporates the perceived momentary configuration of actors and circumstances as an instance of a class or category, and the set of relationships that each actor has to the (salient) elements of the situation. Thus a group of people in room with one person doing the talking could be an instance of a 'telling a joke' situation, a 'listening to a lecture' situation, or even 'receiving battle orders before action' situation. A major way in which these situations differ is the understanding of how they are related to a chain of previous (or possible previous) situations and to a chain of expected (or plausible) future situations. Neither the participant nor a knowledgeable observer would expect that listening to a lecture would be followed by picking up armaments and moving to combat, or that listening to a lecture would be preceded by a situation in which a different person in the group be speaking in the same situation, as might be the case telling a joke. These abstracted and contingent chains of relationships among situations are *narrative structures* (or *story spaces*, the two terms are used interchangeably here).

Narrative structures represent an important form of knowledge in individual decision making. The narrative structures that an agent[1] posses can be both idiosyncratic or nomothetic. Idiosyncratic narrative structures represent abstractions of that specific agent's personal experience base of episodes involving interaction sequences having a common structure. These narrative structures are abstracted by the agent from its direct experiences and typically not widely shared in a community of agents except to the degree that they individual experiences are similarly structured by their physical or social environment. Nomothetic narrative structures, on the other hand, are based on culturally-prescribed or culturally sanctioned norms for behavior. They are therefore much more likely to be held across most or all of the agent's community. The nomothetic narrative structures are arguably more integrative from the perspective of social action, in that they provide the agent with a basis for interpreting the behavior of other agents, as well as for planning actions that are more likely to be understood and sanctioned by other agents.

In this paper, we briefly consider how narrative structures can be used as a basis for simulating interactive characters that provide a specific and recognizable cultural point of view. Such characters can then be used in experiential learning settings that teach others how to recognize and interact with individuals having these cultural points of view, and/or to learn to make and react to decisions based on those cultural perspectives. First, our approach to representing narrative structure is discussed, along with a computational framework for creating agents that reason and act based on their repertoire of culturally-situated narrative structures. Next, an experiential cultural learning framework involving those cultural agents is described, along with some examples of how it has been applied.

---

[1] Typically human, but also a simulated or synthetic character in a computer game or simulation.

REPRESENTING AND MODELING NARRATIVE STRUCTURES AND NARRATIVE REASONING

The knowledge organized into narrative structures provides broad and powerful constraints on cognition in general and on decision making in particular. First, the narrative structure discretizes the (continuous) space of action into an interconnected lattice of possible starting points, possible ending points, and intermediate situational states, each of which has some cultural significance, and each of which may provide branching points for future possible evolutions. In doing this, the narrative implicitly defines the limited classes of situations that it is important for the social agent to perceive and recognize, as that agent acts out or observes an instance of the narrative structure. Second, a narrative structure defines and constraints the culturally plausible behavioral options at each point in the structure. A given situation in the narrative structure will lead to either a *specific* subsequent situation as the story evolves or to a *set* of possible subsequent situations. The behavioral options that are associated with the range of possible next situations define a decision that is up to the agent, but one that is constrained by that set of behavioral options. There may be other conceivable behavioral options but the constraints of the narrative structure limit them to choices that are socially and culturally acceptable. Third, the narrative structure provides a common structure for talking and thinking about both the future and the past. When an sequence of action that is based on a narrative structure is underway, the sequence of situational states that have so-far occurred and the set of behavioral options that so-far been taken define a trace of the path that has occurred through the story space. This is the retrospective narrative of the past events, and it is connected seamlessly to the (prospective) portions of the story space that has yet to be traversed.

Finally, the narrative structure provides a basis for situated decision-making that is also socially situated, in that it relies on the sharing of knowledge about what defines situations and how those situations are linked into narrative structures. The sharing of narrative structures (and the situational definitions on which they rely) allows members of that population that shares them to interpret and understand the behavior of others and to make decisions about behavior that will make sense to others in the situation. In this sense, the sharing of narrative/story structures is a substantive part of the culture (or subculture) that binds this population together, and that make inter-cultural transactions difficult.

One question that arises from the above discussion is that of why, to the degree that the narrative structures are shared across agents, do people make different decisions in the same or highly similar situations? A general and powerful answer to the question is that individuals bring different underlying motives and personalities to these situations. Read and Miller (1991) showed that personality traits (McCrae & Costs, 1999), at the individual cognitive level, can be represented as configurations of motives, supported by situational plans and beliefs. Motives differ from the concept of goals, as widely used in cognitive science, in that motives are pervasive yet diffuse, persisting over very long periods of time yet are not satisfied by an specific individual action. Goals, on the other hand, are situational and are satisfied by some specific action, event, or configuration of actions and events. Thus, 'being elected mayor' is a goal that is a pursued through a protracted narrative of an election and satisfied by winning the election, while 'seeking power' is a motive that persists indefinitely and is not satisfied by any simple goal accomplishment.

Prior research by the authors here (Zachary, Le Mentec, Miller, Read, & Thomas-Meyers, 2005) into the role of narrative structures on decision making argued that the canonical decision points created by the branch points in narrative structures can differential affordances for pursuing different motives. Individuals would perceive those affordances and be likely to make decisions based on their individual long-term motivations. Continuing with the above example, the narrative structure shared by Americans as to how an election unfolds, individuals who are motivated to seek power (e.g., the "alphas") might decide to run for office, while others individuals that are motivated to seek social reinforcement might instead decide to align with some candidate as a supporter.

Modeling Narrative-based Decision-Making and Reasoning

We presume that the understanding of the motive affordances within a narrative structure is shared along with the rest of the structure. This enhanced narrative structure provides a possible explanation for the substantial variation that occurs as different individuals act out the same narrative. We note that this variation can be achieved and understood without any reference to classical decision science constructs such as optimality, subjective utility, or uncertainty. Clearly these constructs play a role in decision making, but it may well be case that the role may be secondary or even minor. In fact, it is possible to create a computational model that uses only this narrative-based theory of cultural decision making. A computational model of this type is introduced immediately below.

PAC (Personality-enabled Architecture for Cognition) is a social cognition architecture (Read et al, 2007) designed to predict social/interactional behavior and behavioral variability based only on the constructs of shared narrative structures, individual differences in motivations and personality, and motive affordances as described above. It was subsequently expanded to incorporate emotion (Read, Miller, Rosoff, et al., 2006) by incorporating the link between motives and emotions posited in Appraisal theory (e g , Roseman, 2001). PAC incorporates a reasoning cycle in which:

a) situations are recognized and understood, and

b) then matched against narrative structures to determine if the current situation advances a currently active narrative of may begin a new narrative;

c) when a situation representing a branch point in a narrative is reached,

d) the agent matches its individual set of motives and personality characteristics against the motive affordances of the current options, and

e) motive activation levels are adjusted based on a hierarchy of control processes (e.g., approach and avoidance systems).

f) Secondarily, an emotional response is generated through an appraisal process that relates the agent's situational understanding to the agents' currently activated motives.

g) The agent then decides on an action strategy, which is

h) implemented in a situationally appropriate manner, through a more detailed-level rational/planning system.

i) The action strategy's outcome will change the external world, possibly enough to create a new situation, in which case the cycle is repeated.

The PAC system was designed and implemented as a software engine to create and execute autonomous software agents; specifically agents that could drive the behavior of simulated human characters in simulated or game-based virtual environments. Figure 1 pictures this macro structure of PAC.

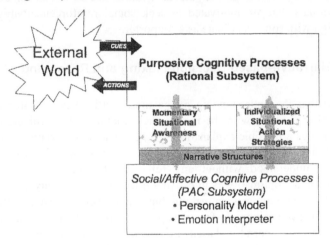

Figure 1. Macro-level PAC Agent Organization

PAC simulations are authored not by traditional knowledge engineering, such as creating rule bases, but by creating a base of narrative structures, including motive affordances. Individuals are created by assigning motives and personality parameters to them. Because all agents share the same narrative structure knowledge, any variation that arises from exercising the model does so solely from individual motives and personalities, and from any variability that exists in the outside environment.

APPLYING NARRATIVE MODELS TO LEARNING CULTURALLY-SITUATED DECISION MAKING

A common problem that people experience in interacting with people from other cultures or subcultures is finding a basis on which to interact. Even if language barriers do not exists or are ameliorated, there is a constant difficulty in making sense of what other people are doing and in having one's own actions be interpreted as intended. In short, the cultural narrative structures that the different parties are applying are not shared and do not 'synch up.' Simple decision like when to speak, what to talk about and how to say it, are misinterpreted and lead to is unsatisfactory and sometimes disquieting outcomes that produce anger and frustration. Learning the narratives that operate in a different culture, or even in a different subculture, can be problematic for this very reason. Learning by participating can become too difficult and even dangerous to try casually. At the same time, simply didactic instruction at the abstract level can also be incomplete, without the actual experience of participating in the narrative.

One solution to this is to allow the practice to take place in a virtual world, in which one learns to make decisions by practicing interacting in decision situations with synthetic characters represented by on-screen avatars. There are many advantages to this approach. If the transaction goes poorly, no harm can come either to the avatar or from the avatar. Moreover, the identical situation can be repeated without the avatar having any recollection of the prior repetition, if so desired. The characteristics and even the behavior of the avatar can also be controlled precisely, or bounded to conform to specific constraints, such as having specific personality traits or specific cultural beliefs. The learner should therefore be able to interact with such culturally-correct avatars in a virtual world, practicing and gaining mastery of cultural knowledge by acting out the specific cultural narratives while gaining realistic experience with and feedback from the avatars.

The authors are currently applying this framework to a range of training applications involving learning culturally-situated decision making skills. Two specific case examples are described below, each exemplifying a different approach to implementing the framework. Both of these approaches use a highly flexible tools set called Virtual Environment for Culturally Training to Operational Readiness® or VECTOR (Deaton et al, 2005). VECTOR supports development and delivery of interactive skills training via interaction with culturally-appropriate avatars in a virtual environment controlled by a commercial game engine. The avatar behavior can be controlled via a range of mechanisms, and the interaction in the virtual environment is integrated with learning support and various assessment and feedback mechanisms.

Explicit Narrative-based Scenarios for Culturally Adaptive Clinical Communications.

The first case example addresses helping American doctors and other clinicians adapt their clinical communication strategy to the subcultures of the various patients they encounter. The US is a large and highly diverse society, with many distinctive subcultures that contain their own beliefs and narrative structures about specific diseases, about dealing with doctors, and about dealing with adversity in general.

Clinicians generally spend much of their time interacting with patients verbally, but receive very little training in clinical communications and virtually none in deciding how to adapt these communications to different subcultural groups. As a result, of clinical interactions across subcultural lines can have poor outcomes that can lead directly or indirectly to poor clinical outcomes.

A training system called TEACH (Training to Enable/Achieve Culturally Sensitive Healthcare), was developed to enhance healthcare provider skills in delivering culturally adaptive care to African-American women with breast cancer (Santarelli, Maulitz, Zachary, Barnieu, and O'Connor, 2009). TEACH uses a population of virtual patients whose behavior incorporate different subcultural narrative structures involving medicine, adversity, and breast cancer. If the clinician can not recognize theses narratives structures and adapt to them, the resulting miscommunication can adversely affect the clinician's understanding of the patient as well as the patient's approach to treatment.

TEACH users (clinicians or medical students) interact with these synthetic patients at virtual clinical encounters representing different stages of the disease progression. In each encounter, the learner (clinician) has to recognize the narrative structures that the patient is using to structure the interaction, and adapt accordingly. In TEACH, the interaction between the learner and the patient avatar is guided by an analytical model of the subcultural narrative involved[2] for that specific patient and disease stage. The narrative is not modeled computationally, e.g., with PAC. Rather, the narrative structure is used to map out a scenario consisting of the space of transactions that delimit the possible paths through the story space defined by the narrative structure. For each possible path through the scenario, the narrative structure and analytical tools that make up the PAC model are used to define the emotions, motives, and decisions options that would characterize the patient. Animators then render the avatar accordingly for the interaction resulting from that path *into* the situation. The path forward in the scenario is determined by the clinician's decision on what to say and how to say it. For each situation in each 10-15 minute encounter, the patient avatar can be in no more than four different states (because of the structure of the scenario), so complexity is managed explicitly without compromising the effectiveness of the learning. The tradeoff of this enumerative approach with a model-based approach is discussed below.

Model-driven Cultural Avatars for Cross-Cultural Negotiations Training

The second case example addresses teaching American military personnel to learn how to carry out key non-kinetic tasks, such as negotiations, in Arabic cultures. The U.S. Military has recognized the importance of cultural and negotiations training and applied several training approaches, including live role-play exercises. While effective, live role-play suffers from problems of standardization (i.e., getting all the live actors to behave in consistent and standard way) and scalability (i.e.,

---

[2]The narrative structures are described in Santarelli, et al., 2009.

high cost and logistical complexities of using live actors). A VECTOR-based training system was created to demonstrate that the live role-play curriculum could be delivered through avatars in a virtual environment as well. The core problem in this example is teaching the learners not just how to recognize and participate in the target culture's narrative for negotiations, but to effectively use that narrative structure to achieve their own ends. Unlike the TEACH example described above, the VECTOR-Negotiations trainer utilized cultural avatars whose behavior was driven by a narrative based cultural model. Similar to the TEACH case, the narrative structure was used to develop a scenario that encompassed the story space of a canonical Iraqi Arabic negotiations narrative. The avatars for the key characters in the scenario were created and linked to a cognitive and emotion model would define their responses as the characters progressed through the scenario narrative and reacted to the soldier's (i.e., learner's) actions. In this case, the model was based on the much-earlier Ortony, Clore, and Collins (1988) Appraisal model of emotions integrated into a simple cognitive model of decision making.[3]

The VECTOR Iraqi was provided to the US Military Academy at West Point, where it was compared with the live role-play negotiations training that was being used there. Rinaldi and Rivera (2007) reported that trainees learning from this system were, by most measures, statistically indistinguishable from those learning via live role-play, and that on some measures the VECTOR-Negotiations learners performed marginally better. Although the study was limited by a small N, it did provide some validation for the idea that this framework for learning cultural decision-making worked as well as the only clear alternative, live role-play.

A comparison between the TEACH and VECTOR Iraqi case on the important dimension of model-based cultural avatar behavior versus enumerated cultural avatar behavior came to no firm conclusion, but pointed out the key dimensions of difference. The process of building the scenario for VECTOR Iraqi was simpler and faster. Without having to embed all the internal motive and affective states of the avatar into each possible evolution as was done in TEACH, less effort was needed to define the simple narrative-structure-based scenario. However, when the effort needed to build and debug the model-based behavioral engine was added back into the VECTOR Negotiations case, the time/effort differences largely disappeared. Both the model-based case and the enumerative case required comparable amounts of validation. However, when new changes to the scenario were made in the TEACH case, it was not necessary to re-validate all the existing states, as there were essentially hard-coded. This was not the case with the model-based version, as addition of new scenario variations required all model states (i.e., across all paths in the scenario) to be revalidated, because of the dynamic nature of the model's behavior. If, and as, these changes uncovered errors in the model-based code, corrections had to be made. Over time, the amount of this error correction decreased, suggesting that at some point the model would become stable enough

---

[3] This is because this model was built before PAC was completed. Effort is currently underway to integrate PAC into VECTOR Negotiations as a more-robust cultural decision and behavior engine.

that the testing could be largely automated, at which point the two approaches would be equivalent to maintain. Qualitatively, it can be concluded that more sophisticated narratives will ultimately be better addressed by model-based implementations, and simpler ones better addressed by an enumerative approach.

CONCLUSIONS

The research into narrative-based approaches to cultural decision training and aiding is on-going. Several implications and directions for future research are drawn below from the results to date.

The development of the PAC model creates an opportunity to experimentally assess and/or estimate the interactions among and the relative contributions of the various factors which the underlying theoretical model postulates as contributing to individual differences and variability in cultural decision making These factors include narrative structure , individual motives and their dynamics, personality factors such as the underlying motive control mechanisms (e.g., approach, avoid, focus), characteristics of the external situation, and the features of formally rational decision making such as estimation of uncertainty, optimization, subjective utility, etc. By collecting data that measure and control for these factors, parameters for each can be estimated by systematic variation of the PAC model. This would allow development of standardized or benchmark models and parameter values. Such benchmarks would provide contribution to the theory of cultural decision making and cognition, as well as simplify and add validity to the creation of model-driven avatars for virtual-environment based cultural decision training and simulation models.

Initial simulations with PAC and other model-generated avatars using VECTOR point out the need for more research on how incongruous cultural narratives interact. This occurs when agents have culturally different narratives for homologous tasks, such as purchasing in a marketplace. While the setting may have common constraints, different cultures may create substantially different narratives as how the interaction should unfold. The incongruities between these narrative structures may be unclear to the participants until problems have arisen. It would be very beneficial to have a means to understand and predict both where these incongruities would occur and what the consequences might be. This would be particularly useful in contexts where sudden and large scale cross-cultural interactions may occur, such as during disaster relief or sudden peacekeeping operations.

The results with TEACH, VECTOR Negotiations and other narrative-based trainers also points to the need for a way for learners to acquire narrative structures more quickly. While trial and error through simulated interactions can and will work, much greater efficiency could be generated by integrating methods for acquiring the structure abstractly and then mastering them through practice. Verbal methods such as the 'telling of war stories' and analogous cultural could be a useful model for this level of cultural learning.

ACKNOWLEDGEMENTS

The authors gratefully acknowledge support from the Air Force Research Laboratory and Office of Naval Research on the overall theory and on PAC, from the Army Research Institute for VECTOR, and from the National Cancer Institute and National Center for Minority Healthcare Disparities for TEACH.

REFERENCES

Deaton, J., Barba, C., Santarelli, T., Rosenzweig, L., Souders, V., McCollum, C., Seip, J., Knerr, B., & Singer, M. (2005) Virtual environment cultural training for operational readiness (VECTOR). *Jrnl. of Virtual Reality, 8*, 156-167

McCrae, R. R., & Costa, P. T., Jr. (1999). A Five-Factor Theory of Personality. In L. Pervin & O. John (Eds.). *Hndkbk. of Personality: Theory and Research* (2nd Ed). (pp. 139-153). NY:Guilford

Miller, L. C., & Read, S. J. (1991). On the coherence of mental models of persons and relation-ships: A knowledge structure approach. In G. Fletcher & F. Fincham (Eds.), *Cognition in Close Relationships*. (69-99). Hillsdale:Erlbaum

Ortony, A., Clore, G., & Collins, A. (1988). *The Cognitive Structure of Emotions.* Cambridge: Cambridge University Press

Read, S. J., Miller, L. C., Rosoff, A, Eilbert, J., Iordanov, V., Le Mentec, J-C., Zachary, W. (2006). Integrating Emotional Dynamics into the PAC Cognitive Architecture. *Proc. 14th Annual Conf. on Behavioral Representation in Modeling and Simulation, Baltimore, MD*

Read, S. J., Miller, L. C., Kostygina, A., Chopra, G., Christensen, J. L., Corsbie-Massay, C., Zachary, W., LeMentec, J. C., Iordanov, V., & Rosoff, A. (2007). The Personality-enabled architecture for Cognition. In A. Paiva & R. Picard (Eds.). *Affective Computing & Int. Interaction 2007.* Springer-Verlag.

Rinaldi, N. & Rivera, J. (2007). *Virtual Training Platform. Research Report For Pl490* Engineering Psychology, May 11, 2007. United States Military Academy, West Point, New York.

Roseman, J. J. (2001). A model of appraisal in the emotion system: Integrating theory, re-search, and applications. In K. R. Scherer, A. Schorr, & T. Johnstone (Eds.), *Appraisal processes in emotion: Theory, methods, research* (pp. 68-91). Oxford: Oxford University Press.

Santarelli, T., Maulitz, R., Zachary, W., Barnieu, J. and O'Connor, B. (2009) Training Healthcare Providers to Confront Diversity in Clinical Settings. In *Proc. of the Interservice/Industry Training, Systems & Education Conference (I/ITSEC) 2009.* [CDROM]. Arlington: Nat. Trng Systems Assoc.

Zachary, W., Le Mentec, J.-C.. Miller, L.C., Read, S. J., & Thomas-Meyers, G. (2005). Human behavioral representations with realistic personality and cultural characteristics. *Proc. Tenth International Command and Control Research and Technology Symposium,* Wash. DC.: DoD CCRP.

Chapter 57

AVATAR:
Developing a Military Cultural Role-Play Trainer

Daniel Barber, Sae Schatz, Denise Nicholson

Institute for Simulation & Training
University of Central Florida
Orlando, FL 32826, USA

ABSTRACT

The modern era of irregular warfare demands that today's military personnel have a sophisticated understanding of culture. Toward that end, new simulation-based training technologies are being developed to support Human, Social, Cultural Behavior (HSCB) instruction. This chapter describes one such system, called AVATAR, which supports tactical sociocultural training, cultural analysis, and knowledge elicitation. AVATAR employs a role-play approach that enables trainees to practice interacting with individuals from other cultures.

Keywords: Role-play, Simulation, Culture, HSCB

INTRODUCTION

In 2004, Major General Robert Scales (Retired) remarked that America is entering an "emerging era of culture-centric warfare" (p. 36), in which the human element underlies nearly all operations, from tactical to strategic duties. To be maximally effective in this environment, military personnel—even at relatively low ranks—must become familiar with their enemies', allies', and surrounding civilians' cultures.

With this statement from Major General Scales, as well as the inception of the *Human, Social Cultural, and Behavior (HSCB) Modeling Program* (Egeth, 2009), there is clearly a strong need for novel sociocultural technologies. Emerging virtual environments and simulation technologies can potentially support sociocultural training, forecasting, and experimentation (Losh, 2005); however, current systems suffer from several limitations. They are typically specific to one culture, with very little reconfigurability, and due to their lack of flexibility, they fail to readily support research into understanding culture for generalized data collection, knowledge elicitation, and modeling.

To help address these gaps, the authors are pursuing a research program that includes the development of a novel role-play culture trainer. This chapter outlines the development of the Automated Virtual Adaptive Training And Role-playing (AVATAR) trainer, a cultural role-playing system that uses a three-dimensional virtual environment, realistic human avatars, motion-capture technology, and behavior modeling to support tactical military culture training and foundational culture modeling research. In contrast to more procedural culture trainers, the AVATAR system is theorized to better engender development of adaptive expertise, help trainees generalize their instructional experiences, and be more rapidly responsive to evolving military requirements. Additionally, AVATAR is designed to support development of theoretically-based models for different culture by recording data generated by Subject Matter Experts (SME) through role-play. This supports investigation of specific cultural variables and their effects on outcome behaviors.

INSTRUCTIONAL ROLE-PLAY

Role-play, one of the most classic forms of simulation, is a well established teaching, training, and assessment technique (e.g., Van Ments, 1999). In its most basic form, instructional role-play only requires that participants use their imagination to act out a "role" within a given setting. Of course, contemporary role-playing can be additionally supported with instructional technology.

Role-play simulations, like all instructional simulations, involve situated learning and require participants to exercise their decision-making abilities to a greater or lesser extent. In addition to these general features, role-play simulations focus on the human element, highlighting the relative positions, motivations, beliefs, and attitudes of the individuals involved. In other words, role-play simulation "focuses attention on the interaction of people with one another" (Van Ments, 1999: 3-4).

VIRTUAL CLASSROOMS: STAR SIMULATOR

In an effort to improve retention of educators and teachers, members of the University of Central Florida's College of Education, along with researchers from the Media Convergence Laboratory (MCL) at the Institute for Simulation and

Training (IST), collaborated on a Small Business Innovative Research grant with Simiosys to develop a STAR Simulator (Dieker, Hynes, Stapleton, & Hughes, 2007). One of the goals of the STAR Simulator was to create a virtual classroom that better enabled teacher preparation for US urban public schools. More specifically, the STAR Classroom provided an immersive virtual environment capable of representing the various cultural and ethnic backgrounds a teacher may encounter in urban schools. This platform facilitated the instruction of the proven Haberman Educational Foundation's (HEF) STAR Teacher method. The developers believed this approach would improve learning outcomes for teachers-in-training, specifically regarding student interaction, and consequently, this would help increase teacher retention in urban settings.

The Virtual Classroom simulated a class of at least five students with various skin coloring, hair styles, and body shapes. The virtual students were capable of a range of facial expressions, including smiles, frowns, blinking, and mouth movements (i.e., talking). When not directly interacting with the participating teacher-in-training, each virtual student entered an idle state, where software controls its behaviors. However, once the teacher-in-training decides to interact with a particular student, that virtual student can be controlled by a hidden puppeteer, who manages the character's body and facial gestures, as well as dialog responses. So long as the puppeteer is a trained subject matter expert, this approach allows the interactions to be highly refined, natural, and accurate. Further, only one puppeteer was required to operate an entire virtual classroom.

Through a series of trials, the STAR Simulator researchers found that experienced urban school system teachers rated the behaviors of the virtual classroom as realistic, while those without experience thought the behaviors were extreme, suggesting they underestimated the actual challenge they would face (Dieker, et al. 2007).

AVATAR

The authors are currently building a system, called AVATAR, to support military sociocultural training and cultural analysis. AVATAR builds on the success of the STAR Simulator, extending the methods used to the cultural domain. Figure 1.1 shows a high-level depiction.

As shown in Figure 1.1, the AVATAR system includes a virtual environment with realistic three-dimensional cultural avatars. The avatars' behaviors are governed by the combination of a computational model (A) and a (human) expert role-player (B). The computational model contains theory-based human social cultural behavior (HSCB) data that can be easily modified to support different cultures. The computational model drives semi-autonomous agents' idle behaviors (i.e., when characters are not directly interacting with trainees), and it provides a framework within which to express agents' behaviors while the expert role-player puppeteers them.

Although, one or more cultural avatars can interact with a trainee, only one expert role-player is required. He/she interacts with trainees via through avatars he/she controls from the Role-Playing Station (C). The Role-Playing Station comprises a set of monitors that enable the expert role-player to observe the virtual environment and trainee reactions, a control station that allows the expert to execute subtle gestures such as facial expressions, and a low-cost tracking system that captures the experts' bodily movements. The expert role-player only controls one avatar at a time, but he/she can switch control among the different agents within the scenario. While the expert role-player controls a specific avatar, he/she is be able to express body and facial gestures, as well as respond appropriately to communication and gestures performed by the trainee. These behaviors are then displayed on the trainee's screen, in the Interaction Station (D).

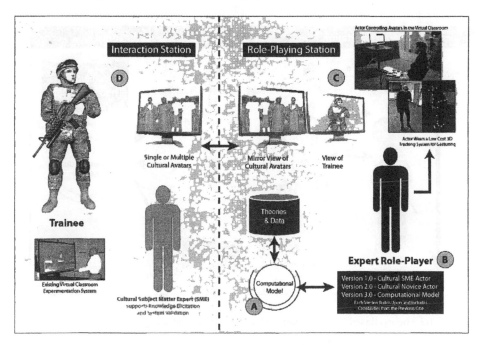

FIGURE 1.1 AVATAR Cultural Role-Playing System Design

The AVATAR system is designed to be low-cost, portable, and support distributed simulation. In addition to this paper, the authors anticipate presenting a live demonstration of the prototype system at this conference's exhibit hall. The initial version of AVATAR leverages a low cost tracking system that is able to identify the location of Infrared (IR) reflective markers on a puppeteer's joints, head and face. With these tracked markers, the puppeteer is capable of controlling arm and hand gestures, body posture, and facial expressions of the virtual characters, producing natural and realistic movements. An added benefit of this feature is a dual ability to record custom animations and gestures for future use by other software programs which will control the cultural avatars (e.g., during idle states). Using these additional pre-recorded animations, the system can generate realistic behaviors if a tracking system is not available, or feasible, through the use of alternative interfaces (e.g. keyboard, mouse). Another feature being developed to support portability is network connectivity. Using a network connection, a puppeteer or computational model can control the actions of the virtual characters without needing to be co-located near the user of the system.

In addition to an immersive training experience for users of AVATAR, a long term goal is the development of theory-based computational models capable of representing different cultures. To achieve this goal, a multi-stage (Figure 1.2) approach is used which takes advantage of data collection and recording capabilities built into the system.

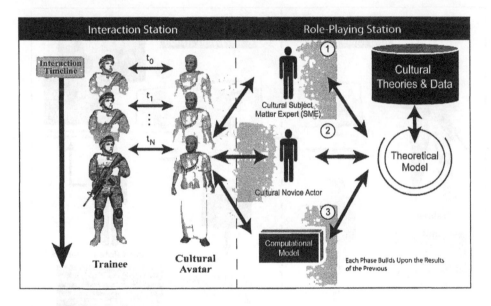

FIGURE 1.2 Theory-Based Computational Model Development using AVATAR

DEVELOPING THEORY-BASED COMPUTATIONAL MODELS

Before a complete computational model can be implemented, a theory-based model must be developed first, (1). In this first phase, a subject matter expert (SME) for the culture under investigation plays the role of the puppeteer in the virtual environment. Through a series of trials with different individuals representing the trainee, rules of behavior and other observations are collected for different scenarios. With the help of the SME and other social scientists, a model is developed from the collected data which represents the desired beliefs, desires, and attitudes of the virtual characters for the culture under investigation.

Upon the completion of this first phase, the initial model must be further vetted for completeness. During the second phase (2), a cultural-novice replaces the SME as the puppeteer for the virtual characters, and the SME becomes a third party observer to a new round of interactions. Using the model and rules previously developed during phase-one, the cultural-novice interacts with a new set of trainees to replicate the interactions of the first phase. During these trials, gaps in the model are discovered by the observing SME, allowing for further model refinement and generalization.

Using the refined model from the second phase, computer scientists and engineers will be able to develop an implementation of a computational model (3). This computational model will supplant the need for a live actor within AVATAR. Using the same process from the second phase, this computational model will be validated for the culture and scenarios under investigation by expert-observers.

CONCLUSION

The system presented in this paper, AVATAR, is being constructed to support HSCB research and training using proven methods. Building off of the lessons learned from the Virtual Classroom, this project extends to the cultural domain with an interactive immersive virtual environment for role-play, letting trainees perform naturally within realistic scenarios for different cultures. In addition to Warfighter support through immersive training, a multi-stage approach to the development of computational models based on social science-theory is presented. This approach to model development is unique in that it integrates Subject Matter Experts, and ongoing validation, with literature lessons-learned from the social sciences. It is the authors' belief that this combination of training, expert input, and theory-based modeling makes AVATAR and ideal platform for HSCB related research.

REFERENCES

Dieker, L., Hynes, M., Stapleton, C., Hughes, C. (2007). Virtual Classrooms: STAR Simulator. *Proceedings of the New Learning Technology SALT.*

Egeth, J. (Ed.). (2009). *Human Social Culture Behavior Modeling Program, Issue 2* (ADA496310). Arlington, VA: Strategic Analysis, Inc.

Losh, E. (2005). In Country with Tactical Iraqi: Trust, Identity, and Language Learning in a Military Video Game. *Proceedings of the 6th Digital Arts and Culture Conference.*

Scales, R. H. (2004). Culture-centric warfare. *Proceedings of the United States Naval Institute, 130*(10), 32-36.

Van Ments, M. (1999). *The effective use of role-play: Practical techniques for improving learning* (Second Edition). London: Kogan Page Limited.

Chapter 58

Culture and Escalation of Commitment

Hsuchi Ting[1], Michele, J. Gelfand[1], Lisa M. Leslie[2]

[1]Department of Psychology
University of Maryland
College Park, MD 20742-4411, USA

[2]Carlson School of Management
University of Minnesota
Minneapolis, MN 55455-9940, USA

ABSTRACT

Escalation of commitment describes individuals' tendencies to persist in a chosen course of action and self-justification is the primary antecedent for this behavior. We posit that escalation tendency should differ between people who emphasize individualism versus collectivism and advance the notion that individualists feel the need to justify their decisions only when their self-efficacy is in doubt whereas collectivists feel the need to justify their decisions only when they can be evaluated by others. We present a theoretical perspective to show that the escalation tendency depends on the individual's culture orientation and the availability of public appraisal.

Keywords Escalation of commitment, Culture, Collectivism, Individualism

INTRODUCTION

Escalation of commitment is the phenomenon where individuals are unable to extricate themselves from a failing course of action in which they have already committed to and failed to achieve their goal. The traditional account of escalation of commitment assumes that sunk cost is the primary antecedent for escalation of commitment (Staw, 1981). Such behavior is considered economically irrational because decisions should be made based upon the expected benefits and costs of each incremental investment, not on how much has been invested in the past.

Various theories exist to explain the robust link between sunk costs and escalation of commitment, and the most prominent theory is the self-justification need (Brockner, 1992). The accounts rely on cognitive dissonance theory in that abandoning past investments would force individuals to admit that they made their decisions in error, thus creating a psychological feeling of discomfort. According to this account, to reduce this feeling, the person unconsciously justifies or rationalizes the prior decision and in essence concludes it was justified. Therefore, when a decision-maker has chosen a course of action that later becomes undesirable, the decision-maker will defend the prior decision by continuing to invest in the same course instead of withdrawing from it, resulting in escalation of commitment.

In the classic escalation experiment demonstrating these notions (Staw, 1981), participants were asked to invest in either a consumer or industrial division within a company. After making the initial decision to invest, participants were told that their decision was made in error and their prior investment had been lost. When given another chance to make new investment, most participants chose to pour the new investment into the same division they had invested earlier. Researchers attributed this pattern of escalation to self-justification because pouring new resources to another division would be construed as admitting that the initial investment was made in error (Staw, 1981). Additional experiments in which the initial investment were manipulated to be made either by the participants themselves or by a fictional character played a crucial role in escalation. Staw (1981) found that participants were more likely to escalate when they were responsible for making the initial investment as compared to the condition in which the fictional decision-player was responsible for the initial investment.

ESCALATION OF COMMITMENT IN THE CROSS-CULTURAL CONTEXT

Although research on escalation of commitment has yielded important theory and practical implications, it has nevertheless largely been tested in the West, largely with Caucasian samples. To date, there has been little research on this bias in other cultures. One recent paper that has examined escalation of commitment that we are aware of found that Asians tended to escalate more than Westerners because of higher propensity to take risk (Keil et al., 2000). Unfortunately, there are several problems with this interpretation. First, the precise probability that the course of action can succeed is unknown, therefore, participants in the study made decisions under uncertainty, not under risk (Knight, 1921). Second, the relationship between the propensity to take risk and tendency to escalate is at best correlational as the study did not manipulate the riskiness of investing in company projects. Finally, if escalation of commitment could potentially lead to success, then the escalation behavior is economically rational, not irrational.

We forward an alternative framework to explain conditions under which decision-makers from different cultural backgrounds would *irrationally* escalate. Given that what causes self-justification is different across cultures ((Hoshino-Browne et al., 2005; Kitayama Snibbe, Markus, & Suzuki, 2004) have found, we expect that different conditions activate escalation for different cultural groups. In the case of individuals high on individualism, the need for self-justification may be based on the person's own judgment about him- or herself. In contrast, individuals high on collectivism will only feel the need to justify their own decisions when they know that their performance will be appraised by others. That is, individuals with collectivistic mindsets are motivated to adjust to and fit in with the expectations of socially meaningful others (Morling Kitayama, & Miyamoto, 2002). As a consequence, approval and acceptance by others figure prominently in the definition of the collectivistic self. In sum, both individualists and collectivists have interests in maintaining self-efficacy. Based on this logic, collectivistic individuals would escalate to avoid doubts about self-efficacy if the performance could be appraised by others. Individuals with the individualistic mindset would escalate in a failing course of action in order to avoid doubts about self-efficacy and should be unaffected by the presence of appraisers. The hypothesis is depicted in Figure 1.1. In this presentation, we will present two studies that examined these notions.

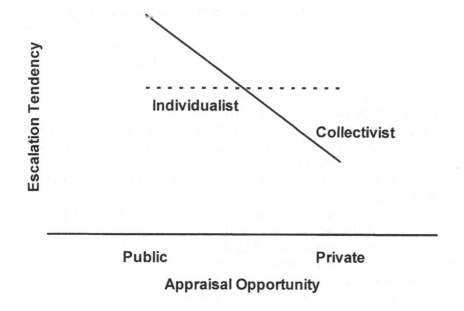

FIGURE 1.1   Tendency to irrationally escalate as a function of appraisal opportunity and cultural mindsets.

CONLUSION

Our theoretical framework departs from earlier effort in that we examine how one defines self influences the decision processes behind escalation of commitment. The current research incorporates advances in cross-cultural research to address a well-known decision-making bias, namely the escalation of commitment, and results suggest that the concepts of the self can have a profound impact on decision strategies. Most important, research on conflict resolution should consider the roles appraisers and culture mindsets might play when trying to persuade decision-makers to de-escalate.

REFERENCES

Brockner, J. (1992). "The escalation of commitment to a failing course of action: Toward theoretical process." *Academy of Management Review*,

17(1), 39-61.

Hoshino-Browne, E., Zanna, A.S., Spencer, S.J., Zanna, M.P., Kitayama, S., & Lackenbauer, S. (1997), "On the cultural guises of cognitive dissonance" The case of Easterners and Westerners." *Journal of Personality and Social Psychology*, 89(3), 294-310.

Keil, M., Tan, B.C.Y., Wei, K., Saarinen, T., Tuunainen, V., Wassenaar, A. (2000). "A cross-cultural study on escalation of commitment behavior in software projects." *MIS Quarterly*, 24(2), 299-325.

Kitayama, S., Snibbe, A.C., Markus, H.R., & Suzuki, T. (2004). "Is there any "free choice? Self and dissonance in two cultures." *Psychological Science*, 15(8), 527-533.

Knight, F. (1921/2005). *Risk, Uncertainty, and Profit*. Cosimo Classics, New York

Morling, B., Kitayama, S., & Miyamoto, Y. (2002). "Cultural practices emphasize influence in the United States and adjustment in Japan." *Personality and Social Psychology Bulletin*, 28(3), 311-323.

Staw, B. (1981). "The escalation of commitment to a course of action." *Academy of Management Review*, 6(4), 577-587.

Staw, B., Barsade, SG., Koput, K.W. (1995). "Escalation at the credit window: A longitudinal study of bank executives' recognition and write-off of problem loans." *Journal of Applied Psychology*, 82(1), 130-142.

CHAPTER 59

Cultures of Fate: Implications for Risk-Taking

Michele J. Gelfand[1], C. Ashley Fulmer[1], Arie W. Kruglanski[1], Abdel-Hamid Abdel-Latif[2], Hilal Khashan[3], Hussein Shabka[2], Mansoor Moaddel[4]

[1]Department of Psychology
University of Maryland
College Park, MD 20742-4411, USA

[2]Sociology, Anthropology, Psychology, and Egyptology Department
The American University in Cairo
New Cairo, 11835, Egypt

[3]Department of Political Studies and Public Administration
The American University of Beirut
Beirut, 1107 2020, Lebanon

[4]Sociology, Anthropology, and Criminology Department
Eastern Michigan University
Ypsilanti, MI 48197-2207, USA

ABSTRACT

Fatalism denotes perceptions that whatever happens must happen (Bernstein, 1992, p.5). Such a belief arises when individuals, in the face of low personal control, seek order and structure from an external source. It is therefore surprising that fatalism should lead to a high level of risk-taking—behaviors that further lower one's control over the environment. While some studies have examined the link between fatalism and risk-taking within a country, little research has focused on this phenomenon cross-culturally or experimentally. To investigate this relationship both at the cultural and individual levels, our research project conducted a multicultural comparison of fatalism and risky behaviors and an experimental study in which we primed individuals' fatalistic beliefs and observed their risk-taking preference. A discussion of these results will be presented during the conference meeting.

Keywords: Fatalism, Culture, Risk-Taking

INTRODUCTION

Che sera, sera / What will be, shall be.
    —Christopher Marlowe, *the Tragic History of Doctor Faustus* (Scene I).

It is a universal human tendency to prefer order and structure over randomness and chaos (Kay, Gaucher, Napier, Callan, & Laurin, 2008; Kruglanski, 1989; Kruglanski & Webster, 1996). As individuals do not always have control over their life events, however, belief in an alternative source of control can often take place. In particular, people with a sense of low personal control may endorse external forms of order, such as fate. This belief that one's destiny is predetermined, or that events in life are meant to happen, is likely present across cultures (Norenzayan & Lee, in press).

Acevedo (2005) defined fatalism as an attitude that is "associated with feelings of personal helplessness and vulnerability, resulting from an overwhelming external force that has a total control over personal action" (p. 78). It denotes the belief that whatever happens must happen (Bernstein, 1992, p.5). Fatalism has been found to be a valid cultural dimension by which countries vary in their degrees of endorsement (Aycan et al., 2000; Leung & Bond, 2004), and has been proposed to be related to excessive external controls, such as harsh economic environment and extreme government regulation, which decrease individuals' perceptions of personal control (Moaddel & Karabenick, 2008).

CROSS-CULTURAL EVIDENCE OF FATALISM AND RISK-TAKING

As people subscribe to fate to gain a sense of control over their environment, it may seem counterintuitive that fatalism should lead to a high level of risk-taking. Can fatalistic belief prompt individuals to engage in risky behaviors that increase randomness and further lower their own control? Surprisingly, a number of studies have demonstrated the positive link between fatalism and risky behaviors. For example, compared to individuals low on fatalism, individuals high on fatalism have been found to be less likely to engage in safe sex practices (Hardeman, Pierro, & Mannetti, 1997; Kalichman, Kelly, Morgan, & Rompa, 1997), screen for cancer (Power & Finnie, 2003; Powe & Johnson, 1995), prepare for natural disasters such as earthquakes and tornadoes (McClure, Allen, & Walkey, 2001; Sims & Baumann, 1972; Turner, Nigg, & Paz, 1986), and use seat belts (Colón, 1992; Connell, 1969).

The current research project sought to examine the implications of fatalism—at both cultural and individual levels of analysis—for risk-taking behavior and preference. We expected that a reduction in individuals' perception of personal control could lower their sense of responsibility and consequently raise their propensity toward risk-taking. Based on this consideration, we predicted that high fatalistic beliefs would lead to greater risk-taking preference and behavior related to safety and health than low fatalistic beliefs, and that this relationship would be evident at both the cultural level and the individual level.

To test this hypothesis, we first conducted a cross-cultural correlation study that examined national fatalism scores and indices of risky behaviors. Specifically, we examined whether fatalistic belief is positively related to the likelihood of obesity, tobacco use in adults, motor vehicle-related deaths, as well as mortality due to cardiovascular diseases and injuries. Second, we conducted a priming study with a student sample that focuses on experimentally primed fatalism and risk-taking preference, using the domain specific risk-taking scale (DOSPERT; Weber, Blais, & Betz, 2002). Again, we focused on the area of safety and health, including "Drinking heavily at a social function," "Engaging in unprotected sex," "Driving a car without wearing a seat belt," "Riding a motorcycle without a helmet," "Sunbathing without sunscreen," "Walking home alone at night in an unsafe are of town," and "Driving while taking medication that may make you drowsy." We will present the results for these two studies during the conference meeting.

CONCLUSION

Fatalism has been shown to be universal but vary in its degrees across cultures (Aycan et al., 2000; Leung & Bond, 2004; Norenzayan & Lee, in press). We proposed that such a belief lowers people's sense of responsibility and increases the likelihood that they will engage in risky behaviors. To test this hypothesis, we conducted two studies to examine this phenomenon: a multicultural comparison of fatalism and risky behaviors and an experimental study of primed fatalism and risk-taking preference. Altogether, our research highlights the theoretical and practical implications for cultures of fate. There is clearly a need for cross-cultural decision-making research to further understand the complex interplay among culture, risk, and fate.

ACKNOWLEDGEMENT

This research is based upon work supported by the U. S. Army Research Laboratory and the U. S. Army Research Office under grant number W911NF-08-1-0144.

REFERENCES

Acevedo, G. (2005). Turning anomie on its head: Fatalism as Durkheim's concealed and multidimensional alienation theory. *Sociological Theory, 23,* 75-85.

Aycan, Z., Kanungo, R. N., Mendonca, M., Yu, K., Deller, J., Stahl, G., et al. (2000). Impact of culture on human resource management practices: A ten country comparison. *Applied Psychology: An International Review, 49*(1), 192-220.

Berstein, M. H. (1992). *Fatalism.* Lincoln: University of Nebraska Press.

Colón, I. (1992). Race, belief in destiny, and seat belt usage: A pilot study. *American Journal of Public Health, 82,* 875-877.

Council, F. M. (1969). *Seat belts: A follow-up study of their use under normal driving conditions.* Chapel Hill, NC: University of North Carolina Highway Safety Research Center.

Hardeman, W., Pierro, A., & Mannetti, L. (1997). Determinants of intentions to practice safe sex among 16-25 year-olds. *Journal of Community and Applied Social Psychology, 7,* 345-360.

Kalichman, S. C., Kelly, J. A., Morgan, M., & Rompa, D. (1997). Fatalism, current life satisfaction, and risk for HIV infection among gay and bisexual men. *Journal of Consulting and Clinical Psychology, 65,* 542–546.

Kay, A. C., Gaucher, D., Napier, J. L., Callan, M. J., & Laurin, K. (2008). God and the Government: Testing a compensatory control mechanism for the support of external systems. *Journal of Personality and Social Psychology, 95,* 18-35.

Kruglanski, A. W. (1989). *Lay epistemics and human knowledge: Cognitive and motivational bases.* New York: Plenum Press.

Kruglanski, A. W., & Webster, D. M. (1996). Motivated closing of the mind: "Seizing" and "freezing." *Psychological Review, 103,* 263-283.

Leung, K., & Bond, M. H. (2004). Social Axioms: A model for social beliefs in multi-cultural perspective. *Advances in Experimental Social Psychology, 36,* 119-197.

McClure, J., Allen, M. W., Walkey, F. (2001). Countering fatalism: Causal information in news reports affects judgments about earthquake damage. *Basic and Applied Social Psychology, 23,* 109-121.

Moaddel, M., & Karabenick, S.A. (2008). Religious fundamentalism among young Muslims in Egypt and Saudi Arabia. *Social Forces, 86*(4), 1675-1710.

Norenzayan, A., & Lee, A. (in press). It was meant to happen: Explaining cultural variations in fate attributions. *Journal of Personality and Social Psychology.*

Powe, B. D., & Finnie, R. (2003). Cancer fatalism: The state of the science. *Cancer Nursing, 26*(6), 454-467.

Powe, B. D., & Johnson, A. (1995). Fatalism as a barrier to cancer screening among African-Americans: Philosophical perspectives. *Journal of Religion and Health, 34*(2), 119-125.

Sims, J. H. & Baumann, D. (1972). The tornado threat: Coping styles of North and South. *Science, 176(4042),* 1386-1392.

Turner, R. H., Nigg, J. M., & Paz, D. (1986). *Waiting for disaster: Earthquake watch in California.* Berkeley: University of California Press.

Weber, E. U., Blais, A.-R., & Betz, N. E. (2002). A domain-specific risk-attitude scale: Measuring risk perceptions and risk behaviors. *Journal of Behavioral Decision Making, 15*, 263-290.

CHAPTER 60

Simulating the Afghanistan-Pakistan Opium Supply Chain

Jennifer H. Watkins, Edward P. MacKerrow, Terence Merritt

Center for the Scientific Analysis of Emerging Threats
Los Alamos National Laboratory
Los Alamos, NM

ABSTRACT

This paper outlines an opium supply chain using the Hilmand province of Afghanistan as exemplar. The opium supply chain model follows the transformation of opium poppy seed through cultivation and chemical alteration to brown heroin base. The purpose of modeling and simulating the Afghanistan-Pakistan opium supply chain is to discover and test strategies that will disrupt this criminal enterprise.

Keywords: opium cultivation, heroin, drug smuggling, credit system

INTRODUCTION

The rampant cultivation of opium poppy impedes the diversification of the Afghan economy and associated rural skill building. Heroin strains health and law enforcement in the countries where the drug is used. In between the cultivation of opium poppy and the use of heroin, the supply chain requires drug trafficking and money laundering. The money generated through these illegal activities can be used to disrupt the ISAF (International Security Assistance Force) war efforts in Afghanistan and to fund terrorism. In short, it is imperative that the flow of opium-based drug money be disrupted.

In a situation as complex as the Afghan opium supply chain, any interdiction measures are bound to result in unintended consequences if the wider system is not taken into account. One means to model the local and global effects of an interdiction strategy is through an agent-based simulation. The purpose of developing an agent-based simulation of the opium supply chain is to provide a useful predictive tool. The primary goal is to develop models of adaptive decision-making in illicit cross-border supply chains and illicit economies. These models will be used in computer simulations to anticipate and predict the interactions within and between economic agents based on local incentives, constraints, and decision-making processes. The challenge of this research is to simulate how the decisions of these agents will adapt and evolve under different economic, political, and counter-terrorism scenarios. This modeling effort has begun with a focus on the foremost opium-producing region in the world, the Hilmand province of Afghanistan.

HILMAND CASE STUDY

The Hilmand (also spelled Helmand) province of Afghanistan produces 59% of Afghanistan's opium (United Nations Office on Drugs and Crime [UNODC], *Survey*, 2009). This province is an optimal climatic zone for many crops, although the soil requires high levels of fertilizer (Shairzai, Farouq & Scott, 1975). The southern portion of the province receives consistent irrigation from the Hilmand River regulated through the Kajaki dam and canals. Each fall, the farmers decide which crops to plant. In 2007, they planted the highest acreage of poppy yet recorded in Hilmand with 102,000 ha (UNODC, *Assessment*, 2009). However, the acreage dedicated to poppy has declined since then to 70,000 ha in 2009 due to production in excessive of world demand (UNODC, *Summary*, 2009).

Once a farmer decides to plant poppy, the opium supply chain is initiated. Opium tar is found in the seedpod of the poppy plant. To harvest the tar, vertical slits are made in the pod to allow the tar to ooze out. Each poppy seedpod requires successively deeper lances as the harvest continues, until no more tar is available. This labor-intensive process often demands that the farmer hire additional labor. The demand for labor has produced a class of itinerant harvesters who begin harvesting in Hilmand's lower elevations and travel to higher elevations as the crop ripens. Harvesters may be farmers, soldiers, or students from universities or madrassas who use the opportunity to earn money (UNODC, *Access to Labour*, 1999).

Traders purchase the raw or dried opium tar from farmers and their laborers who are paid in kind. Many transitory traders were originally livestock dealers. They moved into the opium trade when the devastating drought of 2000 killed most livestock across Afghanistan (Pain, 2006). Farm-gate traders offer the service of transporting the opium to other buyers where the taxes incurred through road travel as well as the risks of seizure or theft deter the farmer. Traders maintain a tightly knit network formed through kinship and ethnic ties, that follows traditional trade routes, and is organized with satellite phones (Pain, 2006).

Traders facilitate the movement of opium tar to processors. The crude heroin labs in Hilmand require nothing more than metal kettles to boil the opium along with precursor chemicals. In Marjah, a district of Hilmand, where in February of 2010 there was estimated to be hundreds of processing labs, the labs serve as a profitable alternative to growing opium. Processors earn around $10 a day compared with $5 in a cash-for-work program (Tassal, 2010). Usually, the output of a heroin-processing lab is brown heroin base. This form of heroin is smokable, but must be further refined and mixed with cutting agents to produce the injectable white heroin desired in the West. This degree of refinement is beyond the capability of most labs in Afghanistan, thus it is brown heroin that is smuggled out of the country (UNODC, *Survey*, 2009).

Although, due to space considerations, this paper will not detail the distribution portion of the supply chain, suffice it to say that Hilmand is situated ideally for smuggling. The province borders Pakistan's Balochistan region to the south and has a traditional trade route with Iran to the west. Because Hilmand lacks a strong governing presence, this border economy has a history of smuggling goods. Those who were historically gun smugglers have moved into the profitable opium trade. Unlike almost all other provinces in Afghanistan, Hilmand plays a key role in every step of the supply chain from cultivation to processing to trafficking.

THE OPIUM SUPPLY CHAIN

Supply chain modeling is a technique typically employed to improve the efficiency of logistical and business processes. However, the purpose of modeling the opium trade as a supply chain is to sustainably disrupt, not improve, the process. As in any supply chain modeling, the focus of the present model is on understanding product, financial, and information flows. Both the product flow, the transformation of poppy seed to heroin, and the financial flow, the movement of money through the system, is briefly described in this section.

The depiction of the information flow in the opium supply chain is reserved for the next section. As the goal of this model is to anticipate and predict agent behavior, information flow is the primary phenomenon of interest. Information flow is modeled as the timing and execution of agent decisions. The agents involved in the opium trade, like other illegal trades, must adapt quickly to law enforcement pressures (Kenney, 2007). Thus, the challenge is to model agent decision-making so as to capture adaptation when condition change.

Product Flow. The Hilmand opium supply chain begins with the planting of opium poppy and ends with the trafficking of heroin across the borders. Along this chain, poppy seed is transformed to opium tar, the tar to morphine base, and the morphine to heroin. The product conversions are tracked in Table 1.

TABLE 1. The process that transforms poppy seeds into heroin including delays and conversion amounts The Hilmand data (derived from UNODC, *Survey*, 2009) shows that around 70,000 ha of land dedicated to poppy produced 584,000 kg of heroin worth around US$1.3 billion.

	GROW	HARVEST	REFINE	PROCESS
Products:	*papaver somniferum*	opium tar poppy seeds poppy oil leaves (fodder)	morphine base	heroin base
Inputs:	fertilizer		water, kettles, burlap, ammonium chloride	acetic anhydride, slaked lime, activated charcoal, sodium carbonate
Conversions:	0.45 kg seeds/ 0.40 ha (1 lb / 1 acre)	60,000-120,000 poppy plants / ha 120,000-275,000 opium-producing pods	10-100 mg opium/pod average: 80 mg opium/ pod 8-15kg dried opium/ ha	0.45 kg morphine / 0.31 kg heroin
Delay:	120 days	7 days	several days	1 day
Hilmand 2009:	69,833 ha	58.5 kg/ha 4 million kg opium	7:1 opium to morphine 584,000 kg morphine	584,000 kg heroin $1,284,800,000

Financial Flow. Tracking the value chain of an illegal resource is difficult. It would be ideal to track the farm-gate prices that farmers receive through the prices that intermediary traders receive up to the final sale price of heroin as it crosses the Afghanistan border. There are additional financial considerations beyond the price at which the opium products are bought and sold. Transit and protection costs are an important component of each intermediary leg. The Taliban escorts shipments along the smuggling routes. Each laboratory will pay between $590 and $1180 per month to the Taliban in escort costs (Tassal, 2010). Figure 1 presents a simplified accounting of costs to produce heroin based on data for the Hilmand province.

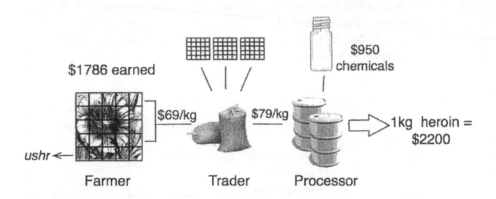

FIGURE 1. A simplification of the financial flow in Hilmand's opium supply chain. Farmers owe 1/5 of their harvest in *ushr* (an Islamic tax), the remaining harvest can be sold to a trader at the farm-gate. Given the average farm-gate price of $69/kg dried opium, an average Afghani opium-growing household earned $1786 in 2009 from the harvest. Traders engage in arbitrage, earning money by exploiting price differences between the farm-gate and bazaars. Acetic anhydride, an essential precursor chemical for heroin refinement, is a controlled substance that must be smuggled into Afghanistan significantly increasing its cost. Seven kg of opium plus $950 worth of precursor chemicals are required to produce 1 kg of heroin. In Hilmand, this amount of heroin is worth $2200. Across Afghanistan, the heroin to opium price ratio is 26:1. All data derived from (UNODC, *Survey*, 2009) and apply to Hilmand province in 2009 unless otherwise stated. Prices are in US dollars.

Supply chain models are often simulated using a system dynamics methodology. Here the factors of interest include inventories, desired inventories, number of items in transit, desired number of items in transit, current demand, and expected demand. The focus on stocks and flows glosses over the complicated decision processes used by those involved in the supply chain. In fact, some supply chain simulations exogenously include the supply chain actors' real-time decisions in a game-like format, rather than model those decisions in computer code (e.g., see the Beer Distribution Game, Sterman, 2000). In order to focus on the decisions of the opium supply chain actors, an agent-based approach is used instead. The next section describes the models of the decision processes of the agents.

AGENT DECISION MODELS

To model agent decisions, an understanding of the options available is necessary. These options are both physically and culturally prescribed. For example, in the tribal agricultural societies, there are multiple ways to excuse debt besides currency or collateral. For example, farmers have been known to marry off young daughters to excuse a debt (Mansfield, 2006). The triggers that lead to a decision point are also both physically and culturally defined. For example, it is a combination of traditional growing seasons with the current and predicted weather that determine when a farmer plants his crops. Finally, the inputs that are taken into consideration are crucial to an accurate depiction of agent behavior.

The following sub-sections outline a sampling of the roles included in the model. For each role, a description and a representative decision are outlined. The agents in the opium supply chain may adopt one or more roles. Each role encompasses a set of decisions the agent will make given a trigger. Each decision has a number of potential options and the option chosen is dependent upon the state of one or more inputs at the time of the decision.

THE FARMER DECISION MODEL

The farmer role encompasses a few positions in the Afghan agricultural scene. We can describe these positions based on the amount of land owned or the arrangement under which land is temporarily obtained for farming. Large landowners own more than 100 jeribs of land. A jerib is a traditional unit of land measurement in Afghanistan equivalent to 1/5 of a hectare. Medium landowners own 10 to 100 jeribs, while small landowners own 1 to 10 jeribs. These are owner-cultivators. Additionally, landowners may allow sharecroppers or *bazgan* to farm some of their land in exchange for a portion of the final harvest. Landowners may also lease farmland to *Ijara* or tenant farmers for a period of years in exchange for a fixed amount of money or kilograms of crop per jerib.

The choice of which crops to grow is a complicated and important one for the farmer and his family. Drought, disease, anticipated crop prices, soil fertility, land available, laws and security, debt, and credit availability are all components of the choice. Crop choices can be segregated into four kinds: cereals grown mostly for domestic use, industrial crops for selling, horticulture used mostly for selling, and forage crops for animal fodder and soil health. For modeling purposes, these crops choices are represented by wheat, poppy or cotton, onion, and clover, respectively.

Decision: Should I Farm?

Trigger. The decision to farm occurs in the autumn. In Hilmand, the chosen crops of wheat, onion, poppy, and clover can be planted starting in September. Thus, the decision to plant must occur by September for farmer agents located in Hilmand.

Inputs. The decision to farm is based on land availability, soil health, water availability, and credit to debt ratios. If a farmer cannot gain access to land on which to plant, he cannot plant. Otherwise, he may choose not to plant if the soil is unfit, water is unavailable, or if there are better opportunities to provide for his family.

Options. This decision is restricted to two options, to farm or not to farm. If the farmer chooses not to farm, that ends all other farmer role decisions for the agent until the "should I farm?" decision is triggered next year.

For the opium supply chain, the most important farmer decision is that of deciding which crop to plant. One important interdiction avenue would be to change farmers' valuations of each crop such that they simply choose not to plant poppy at all. For example, in 2008, Afghanistan wheat prices sharply increased rendering poppy a less desirable crop.

THE LABORER DECISION MODEL

Laborers are often farmers whose farms are not yet or have already been harvested. Thus, one agent can manage both a farmer and a laborer role. Most often laborers travel in kin-based groups to a bazaar where the local farms are in need of harvest. The experienced harvesters train and vouch for their inexperienced relatives. Farmers hire the laborers with a stringent oral contract that sets forth the duration of the harvest and the amount to be paid. The terms of the contract vary with the type of crop that will be harvested and the amount of labor available.

Decision: Sell opium now or later?

Trigger. After the harvest, the farmer usually transports the laborers back to the local bazaar. The laborers have been paid in a portion of the opium harvest and have the opportunity to sell that opium immediately at the bazaar.

Inputs. The decision to sell the opium is based on the immediate needs of the laborer and the current versus expected price of the opium. Usually, when the laborers arrive at the bazaar all local farms have been recently harvested, so the bazaar is flooded with opium and the price is low. The laborers would earn more by retaining the opium for sale at a later date, but this may be impossible if there are pressing needs for money or other goods.

Options. This decision is restricted to two options, to sell immediately at the bazaar or to maintain the opium in stock. The clearing of opium from stock is handled by a separate agent decision.

THE TRADER DECISION MODEL

The role of the trader is to buy farm crops, especially poppy, from the farmer at the farm-gate or bazaar. If the traders meet the farmer at the farm-gate, they save the farmer from the dangerous and expensive task of transporting the crops to market.

Traders can be classified according to trading volume, seasonality, stock management, and geographical reach (Pain, 2006). The risks of trading include the up to 30% daily opium price fluctuations and the risk of theft and seizures. Additionally, inexperienced traders run the risk of buying adulterated product (Pain, 2006).

A trader may offer credit to the farmers in exchange for a fixed amount of a crop at the harvest, an advanced payment known as *salaam*. Opium is the preferred crop. A trader offers a cash advance that is half the current market price of opium. Given that advance payments are usually required during the opium off-season, the price of opium is high. The payments are repaid in opium immediately following the harvest, when opium prices are at their lowest.

Decision: Should I offer credit?

Trigger. Farmers most often require a *salaam* payment in the autumn prior to planting their crops or during the lean winter months. A farmer seeking advanced payment is the triggering event.

Inputs. The decision to offer credit is a classic risk calculation based on the likelihood the loan will be repaid. The best credit risks are those that own land, so that they may mortgage or sell all or a portion of the land to repay the debt. Farmers may also have to agree to sell household commodities to the trader for a portion of their market value should they default (Mansfield, 2006). For those without land, the threat of eradication or a drastic opium price decrease must be calculated.

Options. The trader is able to offer credit in any amount up to the full amount requested based on the determination of the amount the farmer will be able to repay should the opium harvest not go as planned.

Debt and access to informal credit is an important driver of opium cultivation. The Taliban ban on opium cultivation in 2001 left many farmers in debt. The price of opium soared because supply was low, leaving those with outstanding debt facing, in essence, a 1500% interest rate (Mansfield, 2006). The crop with the best return for repaying the debt was opium and many lenders refused to offer further credit for any crop other than opium poppy.

CONCLUSION: EVALUATING STRATEGIES

Once the agent decision models are entered into computer code and combined with a product flow and financial flow framework, the simulation can be populated with initial conditions. From this base simulation, intervention strategies can be tested and evaluated. One of the benefits of the agent-based approach is the ability to identify unintended consequences. For example, previously, the favored solution to

Afghanistan's opium problem was the eradication of the poppy crop. This straightforward solution has the benefit of stopping the supply chain when it is easiest to catch---when the poppy is rooted to the ground. However, there are a number of unforeseen and unfavorable results of the eradication policy.

When poppy crops are eradicated, the farmer is often left in debt, as he borrowed money against the expected harvest. To get out of debt, the farmer's best option is often to cultivate even more poppy next season, as this is the crop with the highest return. Additionally, uneven eradication policies favoring those who live near roads and markets and those who cannot afford to bribe eradication teams results in further mistrust of the government and lends the Taliban, who may offer protection against eradication, a stronger hold.

The second unforeseen result of the eradication policy is that eradication increases worldwide heroin prices resulting in higher profits for those who sell the drug. Morphine base, a pre-cursor of heroin, stores well and is stockpiled and sold when the price increases.

Armed with the knowledge of how eradication, or any intervention, affects farmers, laborers, traders, and drug processors, decision-makers can better implement strategies to reduce the flow of money to the Taliban. Agent-based modeling provides an opportunity for the modeler to take an in-depth look at the problem and allows future users the chance to try out interventions in a consequence free environment.

REFERENCES

Kenney, M. (2007), *From Pablo to Osama: Trafficking and Terrorist Networks, Government Bureaucracies, and Competitive Adaptation.* Penn State Press, University Park, PA.

Mansfield, D. (2006), *Exploring the 'Shades of Grey': An Assessment of the Factors Influencing Decisions to Cultivate Opium Poppy in 2005/06.* Afghan Drugs Inter Departmental Unit of the UK Government. Available at: http://www.fco.gov.uk/resources/en/pdf/pdf15/fco_driverscultivationopiumrpt06

Pain, A. (2006), *Opium Trading Systems in Hilmand and Ghor.* Afghanistan Research and Evaluation Unit (AREU). Available at: http://www.areu.org.af/index2.php?option=com_docman&task=doc_view&gid=353&Itemid=99999999

Shairzai F., G. Farouq and R. Scott (1975), *Farm Economic Survey of the Hilmand Valley*, Kabul: USAID/DP.

Sterman, J. D. (2000), *Business Dynamics: Systems Thinking and Modeling in a Complex World.* McGraw-Hill: Boston.

Tassal, A. A. (2010), *Afghanistan Heroin Menace Grows In Hilmand.* Eurasia Review. Available at: http://www.eurasiareview.com/2010/02/31619-afghanistan-heroin-menace-grows.html

United Nations Office on Drugs and Crime. (2009), *Afghanistan Opium Survey 2009*. Available at: http://www.unodc.org/unodc/en/drugs/afghan-opium-survey.html

United Nations Office on Drugs and Crime. (2009), *Afghanistan Opium Survey 2009: Summary Findings*. Available at: http://www.unodc.org/documents/crop-monitoring/Afghanistan/Afghanistan_opium_survey_2009_summary.pdf

United Nations Office on Drugs and Crime. (2009), *Afghanistan Opium Winter Assessment*. Available at: http://www.unodc.org/documents/crop-monitoring/ORA_report_2009.pdf

United Nations Office on Drugs and Crime. (1999), *Access to Labour: The Role of Opium in the Livelihood Strategies of Itinerant Harvesters*. Available at: http://www.unodc.org/pakistan/en/report_1999-06-30_1.html

Cross-Cultural Decision Making Strategies: Contrasts Between Cherokee Native American and Australian College Students

Jeff King[1], Kitty Tempel[1], Juris G. Draguns[2]

[1]Center for Cross-Cultural Research
Western Washington University
Bellingham, WA, 98225, USA

[2] The Pennsylvania State University
Department of Psychology
University Park, PA, 16802, USA

Abstract

Previous research has indicated that decision making strategies differ between individualistic and collectivist societies (Sanz de Acedo Lizarraga, Sanz de Acedo Baquedano, Oliver, & Closas, 2009). The Decision Making Questionnaire (Mann, 1977) and its revised version, the Melbourne Decision Making Questionnaire (Mann, Burnett, Radford, & Ford, 1997), label certain strategies as positive and others as negative. As a cross-cultural measure, the importance of labeling certain strategies must be examined. This study utilized the DMQ across Native American and Australian college student samples and found significant differences in decision making strategies. This study suggests that collectivist-individualist culture account for these differences and that values assigned to decision making strategies must be re-evaluated.

Keywords: Decision making, Native American, Cross-cultural, Assessment

Introduction

In the economics of today's world, global interactions are becoming more expansive and necessary. How we communicate with each other and make decisions across cultures is crucial to the success of organizations (Sue, 2008). In the world of cross-cultural decision-making, our choices and behaviors are shaped by the cultural values within that environment (Yi & Park, 2003). Western European societies manifest primarily an individualistic orientation while many non-western societies prefer a more collectivist approach (Mann, Burnett, Radford, & Ford, 1997; Sanz de Acedo Lizarraga, Sanz de Acedo Baquendano, Oliver, & Closas, 2009; Yi & Park, 2003). These two very different perspectives provide distinct frameworks by which individuals' function and process information. Furthermore, it is through this framework that information is interpreted and acted upon; in short, it is how decisions are made (Mann, Radford, Burnett, Ford, Bond, Leung, Nakamura, Vaughan, & Yang, 1998).

History of Decision Making Theory

Simon (1960) first described the organizational decision-making process in three phases. The first was the scanning of the environment for the conditions that called for the decision; the second was the creation, evaluation, and organization of the potential paths of action. The third was making the choice from the options available. Each of these phases was incredibly complex and each were decisions in and of themselves. Simon (1960) goes on to say that within this structure is the traditional decision-making technique of habit and memory. Because of the complexity of the information process that leads us to a decision, heuristics takes us to our quickest choice (Simon, 1960; Zimbardo, 2008). Since the environment is constantly changing in the decision process-thereby complicating the organization of the information and therefore the choice-one tends to revert to the quickest and the most familiar "rule of thumb" course of action. In today's world, one must ask, are these strategies universal, or do they fall exclusively within the framework of Western individuals, organizations and societies?

Although Herbert Simon in 1960 wrote from a distinctly Western position supporting the organizational perspective of that era, his framework is nonetheless a generic model of decision-making. Within his traditional decision-making technique, one may implant any culture into habit and memory, and information is all encompassing, in any environment. Mann, et al also mention fundamental to every culture is "fulfillment of human needs, protection of the individual, promoting group survival, and maintenance of community norms and standards" (1998, p. 326). Janis and Mann (1977) also describe universal psychological forces that motivate the individual into decision-making and advocate that patterns for managing difficult choices are common to all people. Moreover, they add that we are all plagued by differing degrees of "conflict, doubts and worry, struggling with incongruous longing, antipathies, and loyalties, and seek relief by procrastination, rationalization, or denying responsibilities to one's own choices (Janis & Mann, 1977, p. 15)". It is also universal that we all have coping mechanisms that we fall back on when being confronted by difficult decisions, but, it is ultimately *the way* that we cope and *how* we come to a decision that is different and separates Western and non-western cultures (Janis & Mann, 1977).

Cross-Cultural Decision Making

In the cross-cultural decision-making environment the division lies in the Western world's individualistic perception of the self and the non-western world's view of the collectivist consciousness (Yi & Park, 2003). Countries such as America, Canada, Western Europe, and Australia are considered Western and individualistic in orientation (Yi & Park, 2003; Sue, 2008; Mann et al, 1997). While China, Japan, Taiwan, Venzuela, India, and all non-European tribal cultures are considered collectivistic cultures (Guss, 2004; Yi & Park, 2003; Mann et al 1998). Triandis (1995) wrote that the individualist begins his life in the collective, with his family and slowly begins to separate in varying degrees as the societal perspective begins to have stronger influence on the individual. He becomes more autonomous and detached from his family, the group. The social behavior is based upon interpersonal connections and the goals and needs usually have precedence over the needs of the collective. He further adds that there is minimal detachment from the family in the collectivist's world. That the individual believes himself to be one part of a whole, that personal goals and needs are subservient to the desires of the collective. The behaviors that manifest are a result of complying with the obligations, duties, and norms put forth by the group. It is the well-being of the collective that supersedes the individual's chase to freedom and to be free from the group. He does mention though "that no society is purely individualist or collectivist" (Triandis, 1995, p. 27).

The two elements of the cross-cultural decision-making process involves cultural values and the various dynamics of that particular culture which is interpreted and applied to the decision making process. Yi & Park (2003) discuss in their paper the importance of child development and the upbringing of independent, self-directed, self-reliant, and very competitive individuals. This is usually the case for American children. Asian children, Japan, Korea, China, are not as competitive and are brought up to focus much more on harmony, interconnectedness, level of cooperation, level of helpfulness in the family, and ultimately how helpful and how much they bring to the group and society. This interpretation is vital as the child grows up and interacts in society. It is the child's understanding of his cultural values that determine the behavior and the manner of communication with outside groups, it is with these values that he will conduct himself, interact, transmit information, work, and make choices in life.

In Yi & Park's (2003) study of college students in five countries, five decision-making styles were established: cooperative, collaborative, avoidant, competitive, and dominant. The results were not as expected as the abrupt element of sub-culture emerged in the study. For example, Koreans although considered a collectivist culture, show strong tendencies toward individualism and had the highest scores for dominant decision-making, Americans scored second in this category. Koreans also scored highest in the competitive decision-making along with the Japanese, Canadians and Americans scoring the lowest, which was a strong contraction to their hypothesis. The Chinese showed the highest avoidant type of decision-making with the Koreans, Canadians, and Americans coming in second. Their study showed many inconsistencies in attitudes of the groups and mentions that variable factors such as the subculture should be controlled in further studies.

Mann's Decision Making Questionnaire

Guss (2004) used Mann's DMQI & II self-reports of American, Australian, and New Zealand students for the individualistic cultures and Japan, Hong Kong, and Taiwan for the collectivist. The Asians scored higher on the buck-passing, avoiding, and hypervigilance than the Westerners. Americans and New Zealanders showed a much higher confidence level than the Asians. This study portrays a link between culture and self esteem which one may interpret as a positive or negative style. In the same paper Germans and Indians participated in a computer program of a business simulation in Malaysia. There were no differences found in the strategic and tactical errors, as well as no differences in the overall decision-making behavior. Concluding differences showed up in the general success of the company. The Germans used more expansive and risky behaviors, the company was more successful, they acted typical of individualist cultures, and the Indians were described as having more defensive and incremental behaviors, valuing security, avoiding risk, typical of collectivist cultures. Since this measure is continuing to be used as an indicator of decision making styles, it's efficacy for assessing American Indian decision making was examined.

American Indians as a Collectivist Society

American Indian tribes typically are collectivist societies (King, 2009). This paper assesses the decision-making strategies of Cherokee Indian college students in eastern Oklahoma, United States and Flinders University students in Australia, using Mann's (1982) Decision Making Questionnaires (DMQI and DMQII). These questionnaires are used around the world to evaluate the decision-making strategies of diverse cultures. DMQI and II are based on Janis and Mann's (1977) conflict theory of decision-making. DMQI measures self-esteem as a decision maker and DMQII measures different styles for decision-making.

Methods

Subjects. Sixty-two Cherokee American Indian college students, 40 female and 22 male, at Flaming Rainbow University in eastern Oklahoma were administered the Decision Making Questionnaires (DMQI & DMQII) among other measures. They received five dollars for their participation. Also used in this study were the DMQI & II scores of Flinders University students in Australia collected by Mann (1982).

Measures

Mann's Decision Making Questionnaires (DMQI and DMQII) were used to assess decision making. Mann (1982) developed the Decision Making Questionnaires (DMQI and DMQII) and is currently using this measure to assess decision making strategies among various cultures worldwide. DMQI and DMQII are based on Janis and Mann's (1977) conflict theory of decision making.

DMQI measures self esteem as a decision maker and DMQII measures different styles for decision making. The DMQ contains subscales: (1) self esteem, (2) vigilance, (3) hypervigilance, (4) defensive avoidance, (5) rationalization, (6) buck-passing, and (7) procrastination.

Results

DMQ subscale scores were initially tested for differences across gender, due to the ratio of females to males. Cherokee sample non-significant differences were found. DMQ subscale scores were then compared across groups. T-test analyses revealed significant differences between groups on all seven subscales. The Cherokee students used fewer of the positive decision making strategies (self-esteem and vigilance) and more of the negative decision making strategies (hypervigilance, defensive avoidance, rationalization, buck passing, and procrastination) than the Australian students (see Table 1).

Table 1 Decision-Making Strategy Subscales

	Australians N=262		Cherokees N= 62		
DMQ Subscales	Mean	S.D.	Mean	S.D.	Probability
Self esteem	8.42	2.16	6.42	2.13	p<.001
Vigilance	9.42	2.31	7.62	2.43	p<.001
Hypervigilance	3.52	1.99	5.23	2.05	p<.001
Defensive Avoidance	3.18	2.28	4.77	2.42	p<.001
Rationalization	3.61	2.05	4.84	2.07	p<.01
Buck Passing	3.20	2.27	4.25	2.64	p<.05
Procrastination	2.92	2.16	4.15	2.25	p<.001

Discussion

Clearly, the Cherokees differed from the Australian students in decision making strategy idealization. At first glance it appears that the Indians used fewer positive and more negative decision making strategies than the Australians. However, American Indians in general employ a worldview which emphasizes the group over the individual (King, 2009). In this case one would expect the Cherokee students not to choose decision strategies which emphasized the self over others (e. g., self esteem decision making) but rather would choose strategies which include others in the decision making process (e.g., buck passing) or which do not put the self ahead of others. Further, American Indians traditionally have not been time oriented (e.g., "Indian time" is a term used among American Indians which has to do with not arriving on time) but rather event oriented. This orientation may influence their decision strategies as well (e. g., procrastination). Given these factors, it would be unwarranted to state that the Cherokee sample were poorer decision makers.

Within their culture, the style of decision making employed would be more appropriate than the decision strategies utilized by the Australian students. Procrastination as a decision making strategy would not have the negative connotation. Rather, it would be seen as an effective group-based decision of waiting until everyone's voice was heard. Blame Avoidance would not be seen as such, but rather as a strategy that shared the decision with the group. Buck Passing would be similar. It is not passing the responsibility for decision making on to others, but rather taking the responsibility to share the decision making with the group. In a collectivist society, such as our Cherokee sample, these "negative" decision making strategies were actually positive strategies that served to maintain group cohesiveness.

Thus, while Mann's subscales seem to suggest that there are certain decision making strategies that are better than others (e.g., vigilance versus procrastination), it appears that the specific labeling of these strategies do not apply cross-culturally. These strategies take on different meanings when viewed within a specific cultural context. This has been demonstrated by this study. It seems that collective societies (such as the Cherokee sample) employ more decision strategies that would be labeled by western societies as "ineffective" and do not use decision strategies that western societies would label as "healthy." Yet, when viewed in the cross-cultural context of societal structures of individualist and collectivist, these strategies carry different meanings and would be considered very "positive." Thus, it is imperative that we closely examine how we label decision making strategies as we employ these measures cross-culturally.

Summary

As we move further into a globalized world, the way we communicate with each other and make decisions across cultures is crucial to the success of organizations and to cross-cultural understanding. In the world of cross-cultural decision-making, our choices and behaviors are shaped by the cultural values within that environment. This study has demonstrated significant differences in decision making strategies between Australian (Individualist) culture and Cherokee (Collectivist) culture. It has further demonstrated that ethnocentric labeling of certain decision making strategies can be misleading and misinterpreted without a consideration of the cultural context in which these take place. Researchers need to be extremely careful when assigning positive or negative characteristics to certain decision making strategies and when assessing these across cultures.

References

References

Guss, C. D., (2004), "Decision-making in individualistic and collectivistic cultures." In *Online Readings in Psychology and Culture* (Unit 4, Chapter 3), (http://orpc.iaccp.org) International Association for Cross-Cultural Psychology.

Janis, I. L. & Mann, L. (1977), *Decision making: A psychological analysis of conflict, choice and commitment.* New York: Free Press.

King, J. (2009), "Psychotherapy Within An American Indian Perspective." In, M. Gallardo, & B. McNeill (Eds.), *Casebook For Multicultural Counseling.* Lawrence Erlbaum Associates, Inc.: Mahwah, New Jersey.

Mann, L., Burnett, P., Radford, M., & Ford, S. (1997), The Melbourne Decision Making Questionnaire: An instrument for measuring patterns for coping with decisional conflict. *Journal of Behavioral Decision Making, 10,* 1–19.

Mann, L., Radford, M., Burnett, P., Ford, S., Bond, M., Leung, K., Nakamura, H., Vaughan, G., & Yang, K. S. (1998), "Cross-cultural differences in self-reported decision-making style and confidence." *International Journal of Psychology, 33,* 325 – 335.

Sanz de Acedo Lizarraga, M. L., Sanz de Acedo Baquedano, M. T., & Cardelle-

Elawar, M. (2007), "Factors that affect decision-making: gender and age differences." *International Journal of Psychology and Psychological Therapy. 7(3)*, 381 – 391.

Simon, H. A. (1960), *The new science of management decision*. New York: Harper and Row.

Sue, D.W. (2008), "Multicultural organizational consultation: A social justice perspective." *Consulting Psychology Journal: Practice and Research,* 60(2), 157-169.

Triandis, H. C., (1995), *Individualism and Collectivism*. Boulder, CO: Westview Press.

Yi, J. S., Park, S. (2003), "Cross-cultural differences in decision-making styles: a Study of college students in five countries." *Social Behavior and Personality,* 31(1), 35-48.

Zimbardo, P. & Boyd, J. (2008). *The time paradox*. New York: Free Press.

Chapter 62

Developing a Multidisciplinary Ontology: A Case Illustration from ICST's Research on Competitive Adaptation in Terrorist Networks

Kurt H. Braddock[1], John Horgan[1], Michael Kenney[2], Kathleen Carley[3]

[1]The International Center for the Study of Terrorism
Penn State University-University Park
130 Moore Building
University Park, PA 16803, USA

[2]Penn State University-Harrisburg
W160G Olmsted Building
Middletown, PA 17057, USA

[3]Center for Computational Analysis of
Social and Organizational Systems
Carnegie Mellon University
5000 Forbes Avenue
1325 Wean Hall
Pittsburgh, PA 15213, USA

ABSTRACT

The execution of a project involving researchers from multiple disciplines is often beset with conceptual difficulties, theoretical differences, and downright misunderstandings. Perhaps the most pressing of these challenges is the fundamental disagreement across key terms and concepts. Because operational clarity is imperative for the successful execution of a multidisciplinary project (particularly in the context of a developing HSCB 'science'), a mechanism for standardizing terminology is paramount. In a project investigating the adaptive practices of terrorist organizations, this presentation describes the progress, challenges and emerging issues from an effort to develop an ontology for use in interdisciplinary, multi-method collaborative research. Through the effective development of this ontology, psychologists, political scientists, communication scholars, and computer scientists were able to develop a common understanding of terms and concepts that had previously held different meanings in their respective disciplines. This presentation not only details the challenges associated with human, social, cultural, or behavioral research that employs multidisciplinary research teams, but provides suggestions for overcoming associated difficulties.

Keywords: Competitive adaptation, terrorist networks, ontology, terrorism, competition, multidisciplinary research

INTRODUCTION

There is a large and growing recognition within among academics and counterterrorism officials that the study terrorism suffers a general lack of primary source field research. This can be explained by the hesitance with which researchers' approach the prospect of directly interacting with active and former terrorists in the field. As a consequence, the majority of our understanding of terrorist activity and counterterrorist response are based on popular news media reports and secondary sources. This has led to a systematic bias in data analyses within terrorism studies, leading to the development and proposal of infeasible, impractical, or ill-informed policy recommendations.

To redress this shortcoming, researchers from the International Center for the Study of Terrorism (ICST) at Penn State and the Center for Computational Analysis of Social and Organizational Systems (CASOS) at Carnegie Mellon University (CMU) have undertaken a project intended to combine the strengths of first-hand ethnographic field research with the semantic and behavior modeling capabilities of computer scientists. Experienced terrorism researchers from Penn State will interview militants, their supporters, and government officials across three terrorist movements (the Provisional Irish Republican Army in Northern Ireland, the Revolutionary Armed Forces of Colombia, and al-Muhajiroun in the United Kingdom and Lebanon). In concert, computer scientists at Carnegie Mellon will analyze the interview data using structured qualitative data analysis tools and social network and agent-based modeling software.

Through interviews at Penn State and computer modeling at Carnegie Mellon, this project aims to provide an organizational framework to understand how terrorist networks and coutnerterterrorism agencies learn from each other in adaptive systems. This project employs draws from a number of perspectives, including organization theory, social psychology, network analysis, and agent-based modeling. Ultimately, this project is designed to (a) assist in making policy recommendations, (b) evaluate the potential effect of certain interventions, and (c) predict future developments in terrorist activity in response to counterterrorist behavior.

Currently, Penn State researchers are in the process of collecting interview data from individuals involved with the groups mentioned above. Thus far, our data is populated with news sources and interview transcripts concerning individuals related to al-Muhajroun. In the near future, ethnographic fieldwork conducted by Penn State researchers will yield analogous data for the Provisional Irish Republican Army and the Revolutionary Armed Forces of Colombia. Prior to this stage of the project, however, there were initial steps that were taken to facilitate cooperative research among the social scientist and computational modelers involved with the project.

MULTIDISCIPLINARY RESEARCH

One of the greatest strengths of this project is its emphasis on multidisciplinary cooperation for the sake of advancing understanding of complex adaptive systems. With individuals from clinical psychology, organizational psychology, political science, communication science, and computational modeling, this project is the product of several academic perspectives on the problem of political violence.

Although the multidisciplinary nature of the research team affords the project a number of benefits, there were also initial problems associated with researchers' differing research perspectives and practices. The most notable of the miscommunications and misunderstandings experienced on this project came in the form of a knowledge gap between the social scientists at Penn State and the computer scientists at Carnegie Mellon. Of course, as social scientists and computational modelers, Penn State and Carnegie Mellon researchers had different respective methodological strengths and weaknesses. Penn State researchers, for example, possess knowledge regarding the specific terrorist groups being investigated. Methodologically speaking, for this project, Penn State specializes in obtaining interview and open source data regarding these groups. In contrast, Carnegie Mellon specializes in modeling the interview and open source data that Penn State collects. For Penn State and Carnegie Mellon researchers to understand each other's methods, constant contact between the two branches of the research team is paramount.

In addition, meetings between personnel at Penn State and Carnegie Mellon illustrated that there were significant miscommunications regarding what would be needed to move the project forward in a meaningful way. Central to this problem was an initial lack of a shared language with which the social scientists and computational modelers could communicate with one another. Given the importance of tailoring the interview and open source data to match Carnegie Mellon's data analysis capabilities for the execution of this project, it became imperative for the terrorism specialists at Penn State and the computational modelers at Carnegie Mellon to develop a shared understanding regarding the terminology that would be used in the context of the project. Quite simply, before we could collect and analyze data, we had to ensure that other researchers on the project understood what we were talking about. As such, one of the fundamental deliverables associated with this project is the development and utilization of a shared ontological reference guide to be utilized by all personnel on the project.

THE NEED FOR A SHARED ONTOLOGY

As a consequence of this project's focus, there are a number of terms that required exposition prior to moving forward with data collection and analysis. Because of this project's emphasis on learning, competition, and adaptation, there was a great deal of attention paid to explicating these and related terms. Unfortunately, researchers from Penn State and Carnegie Mellon had decidedly different understandings about how several key terms should be characterized.

This section will detail a select few of these respective understandings. Although the number of terms on which there were conceptual disagreements and misunderstandings is too numerous to document here, concepts that are considered central to the theoretical and methodological underpinnings of the project are mentioned below.

Agents, Knowledge, and Competitive Adaptation

Some of the central goals of this line of research are to identify (a) how agents act in a given system, (b) how individuals and organizations acquire knowledge relevant to their goals, and (c) how individuals and organizations "compete" with and adapt to those entities that seek to undermine their goals. These concepts are closely related and may easily be confounded. For this project, however, they are conceptually distinct and should be understood as such. The first step to differentiating these issues for the sake of objective analysis was to develop consensual definitions for terms central to the project's goals.

Agents

Because of the limited understanding of computational modeling on Penn State's end of the project, the social scientists knew little about what constituted *agents* in that context. It was generally assumed that because we are interested in the adaptive practices of terrorists in response to counterterrorist practices, the network models developed by Carnegie Mellon would characterize only terrorists and counterterrorists as agents. Although this was true to an extent, the computational modelers at Carnegie Mellon understood agents to be a more comprehensive concept than the one assumed by researchers at Penn State.

For CASOS, agents are entities that possess the ability to make decisions and act in pursuit of a particular goal. Agents are typically identified as being human actors, but other entities that are cognizant and capable of making decisions could also be considered as such. Penn State agreed with the former part of this assertion, but the latter part seemed confusing.

To reconcile this misunderstanding, Penn State and Carnegie Mellon agreed to define agents as *only* being human actors for the purposes of this project. Although Carnegie Mellon *generally* characterizes agents as being human or non-human actors, the implications for only allowing humans to be agents are minimal, provided the computational modelers are aware of this distinction. Although one conceptualization of the term (i.e. only humans) was adopted at the expense of the other (i.e. humans or non-humans), communication between Penn State and Carnegie Mellon allowed the computational modelers to adapt their analysis techniques to accommodate the definitions proposed by the social scientists.

Knowledge

Most members of the research team understood *knowledge* as being a nebulous concept that refers to essentially "what one knows." This general conceptualization, however, did not assist in making scientific claims about the adaptive practices of terrorist organizations and their adversaries. There were several on the research team, however, that had more scientific understandings of what constitutes knowledge. For the Carnegie Mellon computational modelers, knowledge refers to coherent bits of information that may prompt an individual or organization to act in a particular way. For the organizational psychologist at Penn State, knowledge can refer to either an abstract, technical understanding of a specific mechanism (which is typically learned through study of documents or through formalized instruction) or an intuitive, practical understanding of how to perform a particular act (which is typically learned through performing the particular activity).

Although the respective Carnegie Mellon and Penn State conceptualizations of knowledge were not diametrically opposed, they were characterized by different emphases. Further, although Penn State's understanding of knowledge was perfectly reasonable, because knowledge is a specific facet of one of Carnegie Mellon's modeling methods, a common understanding had to be reached to facilitate data preparation and analysis on CASOS' end.

Ultimately, a comprehensive definition was agreed upon that incorporated both emphases listed above. For Penn State researchers, it was important to note that terrorists acquire knowledge in a variety of ways. For Carnegie Mellon researchers, it was imperative that knowledge, regardless of how it is obtained, be represented as a means by which agents (i.e. terrorists and their enemies for this project) take action.

Competitive Adaptation

Contrary to the knowledge gap associated with conceptualizing "agents," for *competitive adaptation*, the Penn State social scientists had a more comprehensive initial understanding than the computational modelers from Carnegie Mellon. For CASOS personnel, competitive adaptation represented a methodological problem. Because there are any number of ways that an individual or organization can alter their behavior, and modeling different behaviors requires conceptually distinguishing them, it became important to discriminate different types of behavior that could qualify as competitive adaptation. To accommodate this need, ICST researchers determined competitive adaptation to be either small-scale tactical changes or large-scale, organization-wide strategic changes in response to an enemy's behavior. By developing this broad conceptualization, ICST and CASOS mutually agreed upon types of behavior that could be characterized as competitive adaptation, thus facilitating data analysis for CASOS.

THE BENEFITS OF A SHARED ONTOLOGY

The above-referenced terms represent only a small sample of the concepts that required explication for the sake of data analysis and moving the project forward. Still, although it is certainly convenient for all researchers on a project to be "on the same page," there are *tangible* benefits for developing a shared language with which to conduct research across academic disciplines.

Most notably, the development and use of a shared ontology has been the facilitation of understanding and communication regarding the output produced by CMU. As mentioned above, the social scientists at Penn State specialize in collecting interview and open source data. The complex research questions presented here, however, require data analysis methodologies beyond the scope of ICST's expertise. Although it was initially attractive to send raw data to Carnegie Mellon with expectation that they would analyze the data and send us perfectly organized and comprehensible output with which we could draw conclusions, it quickly became clear that such a practice would not be possible.

Because of the sophistication of Carnegie Mellon's modeling capabilities, the Penn State social scientists would have been unable to understand or interpret the output being presented without some measure of consensus regarding the terminology being used or the nature of the input being fed into CMU's modeling tools. By mutually developing a shared language with which Penn State and Carnegie Mellon researchers could identify different types of data, the research team avoided splitting the research team into those who could understand the output (i.e. computational modelers) and those who could not (i.e. social scientists). Instead, the shared ontology has allowed the research team to go over output as a whole, facilitating informed data-specific questions from Penn State personnel for the Carnegie Mellon modelers.

The Benefits of a Shared Ontology: Moving Forward

Although the ability for social scientists to understand computational modelers' language (and vice versa) has been paramount for this project thus far, the significance of a shared ontology becomes even more pronounced as the project moves forward into the next phase. This project is in its early stages, as Penn State researchers are still in the process of collecting open source data regarding the terrorist groups mentioned above. Soon, however, scholars from Penn State will be going into the field to gather interview data from former terrorists and counter-terror practitioners related to those groups. One iteration of questionnaire data (transcripts of interviews with al-Muhajiroun members) has been sent to Carnegie Mellon and is beginning to be analyzed. Through this first example, researchers on both sides are beginning to understand how the developed ontology can inform questions to be used during interviews for the other terrorist groups.

Because the social scientists and computational modelers have been able to come to consensus on terms and concepts related to competitive adaptation, the interview protocols to be utilized in the field can be developed to get at these mutually-understood concepts. For example, because we, as a research team, have developed a mutual understanding for what constitutes "knowledge," the Penn State researchers going into the field can develop interview questions that reflect this mutual understanding. If we collectively understand knowledge to be actionable information that can be obtained through formal training or informal experience, questions can be developed that ask about what terrorists or counter-terrorist agents have learned through formal or informal means, respectively. Analogous questions can be developed for any number of other concepts that we, as a research team, deem appropriate.

Because we expect the interview data to prove more conceptually rich and useful than the open source data in demonstrating terrorist and counter-terrorist tendencies with respect to adaptation, the development of interview protocols that reflect relevant (and mutually understood) concepts will be extremely important moving forward.

CONCLUSIONS

Although this project is in its early stages, and will certainly evolve as more data is collected and analyzed, it may be useful to look back and reflect on the exercise of developing a shared ontology among researchers. Despite being an arduous task at times, it has become clear that the development of a shared language has facilitated communication and understanding among personnel from different academic backgrounds. That said, there are several lessons that can be drawn from ICST's experience with ontology building in a multidisciplinary research context.

First, it is imperative that researchers from different disciplines meet as regularly as possible to clarify outstanding concerns and confusions on a research team. Without regular meetings between Penn State and Carnegie Mellon personnel, the conceptual nuances of the data to be collected could not have been comprehensively fleshed out for this project. As the second phase of the project approaches, meetings between social scientists and computational modelers grows increasingly important, as it must be determined exactly what questions would be most useful to include in interview protocols to get at the concepts described above. Because the Carnegie Mellon personnel are responsible for structuring the data prior to analysis, it is equally important for them to understand how relevant terms are conceptualized as it is for the Penn State personnel.

Second, for all its strengths, multidisciplinary research also has a number of practical problems. Most notable of these problems, of course, is the aforementioned knowledge gap experienced for researchers from different scholastic backgrounds. Although individuals' academic deficiencies can be overcome, doing so requires a special attention to instructing research personnel about the major tenets of each academic perspective to be tapped in the project. In the context of this project, for example, because knowledge of a special form of social network analysis called dynamic network analysis (DNA) is imperative, Penn State social scientists unfamiliar with this term and concept travelled to Carnegie Mellon to receive training from the computational modelers that were to be working on the project. Although the social scientists' knowledge of DNA could not come close to matching that of the computational modelers, simply explaining the basic tenets of the methodology facilitated understanding of how data will be analyzed among the Penn State contingent of researchers.

Finally, for those projects with aspects that require collaboration (such as the ontology described here), an online mechanism that allows for remote access may prove useful. Because the ontology was (and is) a living document, open for editing by any of the research team members from Penn State or Carnegie Mellon, it was necessary to have it readily available for all team members. To this end, we developed the ontology as a "Google Doc." A Google Doc operates like a Microsoft Word document, but rather than saving and sending different versions of a document to all team members (which invites confusion about the most recent and/or comprehensive version of the document), a document is hosted online and team members can remotely access it for editing at any time. Although the Google Doc function may not be the most appropriate in all cases, for those components of a research project that require input from multiple individuals, an online openly-editable version of that component facilitates remote collaboration.

Despite the concerns expressed above, multidisciplinary terrorism work remains a valuable, but greatly underrepresented category of research. ICST's endeavors in investigating competitive adaptation in terrorist networks have revealed the extent to which talented researchers eager to study terrorist practices are present within academia. Bringing together multiple disciplines to address hypotheses and research questions associated with political violence is only the first step. Once brought together, to fully utilize the respective knowledge bases possessed by research personnel, a common language must be developed. Only through a means by which academics from different disciplines can communicate will multidisciplinary research realize its fullest potential in the context of any research project.

Chapter 63

Adversarial Behavior in Complex Adaptive Systems: An Overview of ICST's Research on Competitive Adaptation in Terrorist Networks

John Horgan, Michael Kenney, Peter Vining

International Center for the Study of Terrorism
130 Moore Building
Pennsylvania State University
University Park, PA 16802, USA

ABSTRACT

There is widespread recognition among scholars and practitioners that the counterterrorism literature suffers from a lack of primary-source field research This shortcoming is largely due to a failure to integrate ethnographic research into modeling efforts, as well as a failure more broadly to appreciate the significance of ethnographically valid data in human, social, cultural, and behavioral studies in a systematic investigation of adversarial behavior. The project briefly outlined in this preliminary paper seeks to redress this deficiency by combining the strengths of ethnographic field research (collected by social scientists at Penn State) with the sophisticated modeling capabilities of computer scientists (at Carnegie Mellon University). Specifically, we are analyzing data from interview transcripts, news reports, and other open sources concerning the radical Islamic group al-Muhajiroun. Using competitive adaptation as a comparative organizational framework, this project focuses on the process by which adversaries learn from each other in complex adaptive systems and tailor their activities to achieve their organizational goals in light of their opponents' actions. Ultimately, we will develop a meso-level model of terrorist networks that combines insights from organizational theory, psychology, network analysis, and computational modeling. This model will assist counterterrorism practitioners in their decision-making regarding the impact of specific interventions.

Keywords: Terrorism, Competitive Adaptation, Organizational Learning

INTRODUCTION

Growing numbers of scholars and practitioners recognize the value of mixed-methods and interdisciplinary approaches to studying militant organizations that commit acts of terrorism. Moreover, there is growing recognition that the terrorism and counterterrorism literatures suffer from a lack of primary-source field research. Much of our knowledge and understanding of militant movements comes solely from news reports and other secondary sources, contributing to a systematic bias in available data. As a result of this bias, vague and impractical policy recommendations are often generated from results that do not correlate with the reality of militant behaviors. The project that we discuss in this paper has begun to address these issues in addition to providing an interdisciplinary framework from which to study the behavior of militant groups that use terrorism. Building on the concept of competitive adaptation, our research team[1] is investigating how militant groups learn and adapt when interacting with governments, civilians and other militant groups. We are combining the analytical richness of ethnographic research with sophisticated modeling techniques in order to provide a meso-level model of militant networks that function in complex-adaptive systems. More specifically, we are analyzing data from interview transcripts in addition to news reports, public statements and other open sources in order to study competitive adaptation in three militant groups across four countries. This paper discusses our project's theoretical approach and initial focus on the radical Islamic group al-Muhajiroun in the United Kingdom. First, we present our theoretical framework of competitive adaptation and how it assists us in the investigation of our research questions. Second, we discuss the development of our data collection procedures, methods of analysis and how our mixed-methods approach is expected to generate new and interesting findings. Third and finally, we conclude with a discussion of challenges we have addressed thus far, lessons learned and how the project will evolve over the course of the next few years.

---

[1] Lead Principal Investigator – Dr. John Horgan, Penn State; Co-Lead PI – Dr. Michael Kenney, Penn State Harrisburg; Co-Investigator – Professor Kathleen Carley, Carnegie Mellon University; Co-Investigator – Dr. Mia Bloom, Penn State.

Organizational Learning and Competitive Adaptation in Militant Groups

Terrorism researchers have found that pathological characterizations of those who participate in terrorism are inconsistent with empirical realities. Work by Horgan (2008), Taylor and Horgan (2006), Sageman (2004) and Kimhi & Even (2004) demonstrates that those who commit acts of terrorism tend not to be psychologically abnormal or disturbed individuals as they are often popularly characterized. To the contrary, these scholars have found that profiling terrorists is extremely difficult and often unproductive. In fact, individuals who commit acts of terrorism and the groups that they belong to often reflect a broad diversity of backgrounds. From these findings, scholars reason that terrorist organizations themselves may not be fundamentally different in structure and membership from other organizations, such as governments, political parties or corporations. Terrorist groups often exhibit attributes of more conventional organizations, such as merit-based hierarchal structures, public relations and media branches, or even the use of non-violent political activism and social engagement. The groups Hamas and Hezbollah, for example, are both known to engage in terrorism while also pursuing roles of governance and provisioning social welfare to supporters. Some scholars have thus sought to study terrorist groups through the complex lens of organizational theory, using a broad array of methodologies such as ethnographic research, rational choice modeling and social network analysis (Abrahms 2008, 2006; Asal & Rethemeyer, 2008; Freedman, 2007; Helfstein, 2009).

A review of the terrorism literature has indicated to us that organizational approaches can offer guidance when investigating the behaviors of militant groups. Organizational perspectives of terrorism are broadly distinguished in Crenshaw's (1987) discussion of why traditional and instrumentalist approaches to the study of terrorism cannot account for many unsolved puzzles regarding militant behavior. Instrumentalist, or strategic approaches to the study of terrorism focus on how the use of terrorism can potentially coerce target states into making policy concessions during the strategic interactions between militant groups and states (Crenshaw, 1981; Dershowitz, 2002; Lake 2002; Pape, 2003; Brym and Araj, 2006; Freedman, 2007). Whereas these approaches have furthered our understanding of how terrorism may be logically employed within to context of a dyadic conflict, it has become clear that there is also strong merit in attempting to dissect militant organizations themselves in order to understand the many instances in which a group's use of terrorism does not appear to fit within strategic models. Specifically, Crenshaw (1987) suggests that the use of terrorism itself may offer distinct internal benefits to the organizations that employ it as they pursue group survival and cohesiveness, *in addition* to the achievement of stated political goals. Examples of recent work building on these organizational perspectives of terrorism include Abrahms (2008; 2006) and Freedman (2007). Both of these scholars observe that in reality, the use of terrorism often appears at face value to be counterproductive to achieving stated organizational goals. Therefore, the relationship between a militant group's use of terrorism and the achievement of organizational goals is likely to be more complex than often conceptualized by strategic approaches. Moreover, it becomes clear that explaining militant behavior requires an in-depth examination of militant groups themselves. Specific questions about militant organizations that our research is addressing include:

- What are the social network properties of the relationships within the militant organization as well as between one militant organization and another?
- What is the cultural significance of familial kinship and friendship networks in relation to all stages of organizational involvement (e.g. pre-socialization and cultural risk precursors, indoctrination, recruitment, initial involvement, sustained involvement, engagement in violent activity, disillusionment, disengagement, and de-radicalization)?
- How have individual and organization profiles evolved from the terrorist movement's inception to today?
- What are the social network properties of the relationships within the militant organization as well as between one militant organization and another?
- How does loyalty and group power work in the organization, and what role is played by ideology in this (both in terms of ideological content, and ideology as a process)?
- What factors influence the nature of the relationship between the organization and the community it claims to represent, and the organization's use of violence to achieve its aims?
- How do militant organizations successfully rejuvenate themselves in the wake of external successes against them?

While a wealth of knowledge is available to describe organizational processes, learning and decision-making in conventional political entities (see Breslauer & Tetlock, 1991; Eden, 2004; Etheridge, 1985; Goldgeier, 1994; Haas, 1990; Haas, 1992; Hall, 1993; Heclo, 1974; Khong, 1992; Nye, 1987; Reiter, 1996; Sagan, 1993; or Weir & Skocpol, 1985 for just a few prominent examples from this expansive literature), far fewer studies examine learning and decision-making in nonconventional, covert, illegal or violent groups such as militant organizations. Nevertheless, some recent work has offered useful insights into how militant organizations learn and adapt within the adversarial environments that they operate. A team of RAND researchers, led by Jackson (2005), examines organizational learning in several militant groups, including the Provisional Irish Republican Army, Aum Shinrikyo, Jemaah Islamiyah, Hizballah, and the radical environmentalist movement. Jackson models organizational learning as a four part process involving the acquisition, interpretation, distribution and storage of information and knowledge. This knowledge may be explicitly expressed in knowledge artifacts and organizational routines, or tacitly expressed in the skills and experience of organization members. A separate study by Hamm (2007; 2005) draws on court documents contained in the *American Terrorism Study* database and the criminological literature on social learning to explore how some political extremists acquire the skills to perform their violent tradecraft. Leweling & Nissen (2007) provide conceptualizations of militant groups as industries that attempt to perform the "work" of terrorism while interacting and competing with similarly-conceptualized counterterrorism and governance industries. While these studies offer insights into how numerous militant groups train their members and develop certain technological innovations, they do not systematically examine the *internal* processes of group learning and interpretation, as experienced by militants themselves. Moreover, these studies also do not take into account the broader competitive environments in which militant groups operate.

Drawing on organizational and complexity theory (see, for example Fiol & Lyles, 1985; Dodgeson, 1993; Argyris & Schon, 1978; Argyris & Schon, 1996; Hedberg, 1981; Tetlock, 1991; Levitt & March, 1988; Epstein & Axtell, 1996; Axelrod, 1997; Axelrod & Cohen, 1999; Carley & Prietula, 1994; Carley, 2002; Holland, 1995; Jervis, 1997) in addition to fieldwork in Columbia, Kenney (2007) describes how organizational knowledge is leveraged by competing groups that interact in complex adaptive environments. Kenney dubs this process competitive adaptation, which explains how organizational learning occurs within an environment that is typically (though not always) characterized by hostility and multiple actors pursuing diametrically opposing goals. Often, organizations that exhibit learning and adaptation in competitive environments will vie for a common, finite resource (such as the support of a population or control of territory). Thus, competitive adaptation may be described as the tit-for-tat process of organizational learning that is associated with the pressures of a typically zero-sum contest between organizations, *in addition* to the pursuit of broader organizational goals as they are discussed by conventional organizational theorists. Competitive adaptation is the framework from which we approach our study of militant groups and generate falsifiable hypotheses that are consistent with our research questions, as follows:

- Participants in terrorist movements are organized into overlapping social and organizational networks
- Social and organizational network properties, such as centralization, hierarchy, and density of network ties, vary across different terrorist movements
- Social networks are a primary mechanism by which participants in terrorist groups and organizational networks share information about their activities and government counterterrorism efforts
- Terrorist networks "learn" when their participants receive information about their activities, process this information through knowledge-based artifacts, and apply the information to their practices and activities
- Networks with fewer management levels and decentralized decision-making will process information and make decisions more quickly than those with greater management levels and centralized decision-making.

- Networks with decentralized and dispersed authority structures will be more susceptible to fragmented, localized learning than organizations with centralized, tightly coordinated administrative structures.
- Terrorist networks, with their flat decision making hierarchies and informal rules of engagement, adapt quickly in response to external pressures

Methodology and Data Collection

A major limitation to the study of organizational learning and competitive adaptation in militant groups is the lack of available primary data. Jackson (2005: 200) points out this problem in his concluding remarks, noting that "[t]he lack of data on group decision making and the factors that drive groups' interpretation of their own successes and learning opportunities is a major handicap to complete understanding of learning in these organizations. Additional study and exploration of specific elements of organizational learning in these and other terrorist groups is therefore warranted." Our own research is opening this "black box" of terrorist groups by capturing them in their own words, in addition to leveraging available primary and secondary data sources. Specifically, we combine ethnographic data, collected from primary sources in the field, with computational modeling to augment our understanding of how participants in terrorist networks acquire knowledge and adapt their operations in response to counterterrorism agencies with which they share complex adaptive systems. We are thus using an interdisciplinary and mixed methodology in addressing our research questions.

The first step in the progression of our research was the development of an interdisciplinary ontology in order to arrive at mutual understandings of the important terms and concepts associated with competitive adaptation. This ontology was developed through meetings between our ethnographic research teams, led by Drs. John Horgan and Michael Kenney at Penn State University and Penn State Harrisburg, respectively, and our computational modeling team, led by Dr. Kathleen Carley at Carnegie Mellon University. The development of our ontology was a major undertaking and is described further in a forthcoming paper.

Following the development of our ontology, the research team began to collect primary and secondary data related to the first militant group we chose to study: al-Muhajiroun (henceforth referred to as AM) in the United Kingdom. Recent work by Pantucci (2010:226) claims that AM and the related organization Supporters of Shariah "have been the connective thread through most Islamist terrorist plots that have emanated from the United Kingdom." Founded in 1996 by Omar Bakri Mohammed and officially disbanded in 2004, AM members continue to operate in the United Kingdom under several splinter organizations that we seek to include in our study. AM was appropriate as a first choice for developing our data collection procedures because the group is relatively accessible and members of our own research team were able to access transcripts of interviews with AM members that were relevant to the analysis. Data collection for AM thus consisted of these interview transcripts, in addition to:

- Publically available interview transcripts, autobiographical sources and documentary accounts of AM members
- Press releases, website op-eds, magazine articles and other materials written and published by AM members
- Newspaper articles, blog postings and other media accounts of AM
- Peer-reviewed articles, books and other major publications about AM

From evaluation of these primary and secondary sources, our research team was able to construct lists of known AM members with their associated roles in the organization, AM front and splinter organizations (of which there were many) and timelines of AM activities and events, in addition to a "blue team" timeline documenting activities of the UK government that are related to AM. These data were extracted and prepared in consultation with our computational modeling team at Carnegie Mellon University for the purpose of ensuring that the requirements of both social and computer scientists were met for the overall analysis. The data are currently being coded using the program NVivo for qualitative analysis and were also provided to the computational modeling team for:

- Analysis of semantic networks and meta-network variables from texts using the text analysis program AutoMap
- Analysis of how cellular and distributed network structures within AM and between AM and other militant groups change over time using the dynamic network analysis (DNA) software ORA
- Identification and analysis of critical network nodes according to who they know, what they know, and what they are doing using the program Organizational Risk Analyzer.

In addition to these analyses, we are working with our partners at CMU in order to develop common procedures for the documentation and storage of primary source data in digital format as well as determining what data should be collected in upcoming field research. We are thus integrating ethnographic data and computational modeling from the outset of the project. Regarding the collection of new data on AM and the other militant groups we seek to study, our researchers are developing questions and procedures for conducting semi-structured interviews in the field. These interviews will seek to generate codeable data about information-gathering and analysis within groups as well as organizational adaptability. Factors to be addressed include the role of economic resources, group survival, laws, rules and regulations, organizational structure and bureaucracy, organizational culture, knowledge-based artifacts, the education and training of members, personal lifestyles and characteristics as well as the rotation of members. We will thus generate rich ethnographic data from militant groups themselves that will be subject to qualitative and quantitative analyses.

Challenges, Lessons Learned and Next Steps

Several challenges have presented themselves over the course of the project which we are currently addressing. One of these challenges is associated with the interdisciplinary, mixed-methods approach that we have chosen to adopt. Our work has required reviewing literature in multiple fields and has required researchers to explore concepts, ideas, literature and knowledge in fields outside of their specialties in order to establish baselines of common understanding. To address these challenges, the research team has implemented several measures to ensure that the project retains this interdisciplinary character throughout its lifespan. The construction of our ontology was a major first step in achieving this goal. Moreover, our ethnographic researchers at Penn State have received training in the use of network analysis software that our computational modeling partners at Carnegie Mellon will be drawing on to analyze our data. As previously discussed, we have consulted our partnering institutions in data collection and formatting procedures and will be cooperating to develop a codebook which will be used for our interview transcripts. Among the technological tools that we have found to be helpful in collaboration thus far include Google Documents, which permits multiple users to edit and comment on documents in real time.

Another major challenge to our research is associated with our gathering of data. Gathering rich, primary source information about militant organizations from militants themselves requires a careful approach which balances researcher's questions with realities in the field. While it has been shown that interviewing militants can be done safely, ethically and does produce valid, policy-relevant data, several important concerns affect the sorts of questions which should be asked and the extent to which detail can realistically be pursued. Many of these concerns originate in the ongoing conflicts between militant groups and states, which will likely cause an active militant to limit the type and extent of information they are willing to provide. Moreover, it is typical for active members of political organizations to respond to interviewer questions with ideology and propaganda. While such responses are inherently valuable to researchers, they may prevent our own researchers from gathering the process-related information they are seeking. Nevertheless, the questions that are asked and the way in which they are asked will often determine the type of answer that is provided. Horgan (2008) finds, for example, that asking militants "how" they became involved in terrorism often results in very different responses than those provided when asking "why" they became involved. Therefore, we are addressing the difficulties associated with interviewing militants by carefully developing a rubric of questions to be covered during interviews, while leaving the interviews themselves semi-structured in order to maximize the relevant information provided. Moreover, the primary investigator is providing all members of the field research team with advanced training in interview techniques meant to address these issues.

As our computational modeling team continues to analyze the data we have provided to them on al-Muhajiroun, our field researchers will in 2010 begin to collect new primary source data on AM in the United Kingdom and Lebanon, as well as the Irish Republican Army (IRA) in Northern Ireland and the Revolutionary Armed Forces of Columbia (FARC) in Columbia. Using our theoretical framework of competitive adaptation, our research team will analyze this data with the objective of understanding how many of the internal processes affecting militant organizations affect their behavior, their use of terrorism and how they learn and adapt. The resulting model of competitive adaptation will contribute an evidence-base to inform decision-making and counter-terrorism training in addition to evaluating the impact of specific policy interventions and policy forecasting.

REFERENCES

Abrahms, Max, 2006. "Why Terrorism Does Not Work." *International Security* 31(2): 42-78.

Abrahms, Max, 2008. "What Terrorists Really Want: Terrorist Motives and Counterterrorism Strategy." *International Security* 32(4): 78-105.

Asal, Victor and R. Karl Rethemeyer, 2008. "The Nature of the Beast: Organizational Structures and the Lethality of Terrorist Attacks." *Journal of Politics* 70(2): 437-449.

Axelrod, Robert, 1997. *The Complexity of Cooperation: Agent-Based Models of Competition and Cooperation*. Princeton: Princeton University Press.

Axelrod, Robert and Michael D. Cohen, 1999. *Harnessing Complexity: Organizational Implications for a Scientific Frontier*. New York: Free Press.

Breslauer, George and Philip Tetlock (eds), 1991. *Learning in U.S. and Soviet Foreign Policy*. Boulder, CO: Westview Press.

Brym, Robert J. and Bader Araj, 2006. Suicide Bombing as Strategy and Interaction: The Case of the Second Intifada. *Social Forces* 84(4): 1969-1986.

Carley, Kathleen and Michael Prietula, eds, 1994. *Computational Organization Theory*. Hillsdale: Lawrence Erlbaum.

Carley, Kathleen, 2002. "Intra-organizational Computation and Complexity," in The Blackwell *Companion to Organizations*, edited by Joel A.C. Baum. Malden, Mass.: Blackwell: 208-232.

Crenshaw, Martha, 1981. "The Causes of Terrorism." *Comparative Politics* 13(4): 379-399.

Crenshaw, Martha, 1987. "Theories of Terrorism: Instrumental and Organizational Approaches" in *Inside Terrorist Organizations* ed. David C. Rappoport, 2001. London: Cass.

Cyert, Richard and James March, 1963. *A Behavioral Theory of the Firm*. Englewood Cliffs, New Jersey: Prentice-Hall.

Dershowitz, Alan, 2002. *Why Terrorism Works: Understanding the Threat, Responding to Challenge*. New Haven: Yale University Press.

Eden, Lynn, 2004. *Whole World on Fire: Organizations, Knowledge, & Nuclear Weapons Devastation*. Ithaca, NY: Cornell University Press.

Epstein, Joshua M. and Robert Axtell, 1996. *Growing Artificial Societies: Social Science from the Bottom Up*. Washington, D. C.: Brookings Institution Press.

Etheridge, Lloyd, 1985. *Can Governments Learn?* New York: Pergamon.

Freedman, Lawrence, 2007. "Terrorism as a Strategy." *Government and Opposition* 42(3): 314-339.

Goldgeier, James, 1994. *Leadership Style and Soviet Foreign Policy: Stalin, Krushchev, Brezhnev, Gorachev.* Baltimore, MD: John Hopkins University Press.

Haas Ernst, 1990. When *Knowledge is Power: Three Models of Change in International Organizations* Berkeley: University of California Press.

Haas, Peter (ed), 1992. "Knowledge, Power, and International Policy Coordination," special issue of *International Organization* 46 (1): 1-35.

Hall, Peter, 1993. "Policy Paradigms, Social Learning, and the State." Comparative Politics 25: 275-296.

Hamm, Mark, 2005. "Crimes Committed by Terrorist Groups: Theory, Research and Prevention" U.S. Department of Justice.

Hamm, Mark, 2007. *Terrorism as Crime: From Oklahoma City to Al-Qaeda and Beyond.* New York: New York University Press.

Helfstein, Scott, 2009. "Governance of Terror: New Institutionalism and the Evolution of Terrorist Organizations." *Public Administration Review* 69(4): 727-749.

Heclo, Hugh, 1974. Modern *Social Politics in Britain and Sweden: From Relief to Income Maintenance.* New Haven, Yale University Press.

Holland, John, 1995. *Hidden Order.* Cambridge: Perseus Books.

Horgan, John, 2008. "From Profiles to Pathways and from Roots to Routes: Perspectives from Psychology on Radicalization into Terrorism." ANNALS, *American Association of Political and Social Science* 618: 80-94.

Jackson, Brian A. et. al., 2005. *Aptitude for Destruction Volume 1: Organizational Learning in Terrorist Groups and its Implication for Combating Terrorism.* Santa Monica, CA: RAND.

Jackson, Brian A. et. al., 2005. *Aptitude for Destruction Volume 2: Case Studies of Organizational Learning in Five Terrorist Groups.* Santa Monica, CA: RAND.

Jervis, Robert, 1997. *System Effects: Complexity in Political and Social Life.* Princeton: Princeton University Press.

Kenney, Michael, 2007. *From Pablo to Osama: Trafficking and Terrorist Networks, Government Bureaucracies, and Competitive Adaptation.* University Park, PA: Penn State Press.

Khong, Yuen Foong, 1992. *Analogies at War.* Princeton: Princeton University Press.

Kimhi, Shaul and Shemaul Even, 2004. "Who Are the Palestinian Suicide Bombers?" *Terrorism and Political Violence* 16(4): 815-840.

Lake, David A., 2002. "Rational Extremism: Understanding Terrorism in the Twenty First Century." *Dialogue, International Organization*, Spring: 15-29.

Levitt, B. and J. March, 1988. "Organizational Learning," *Annual Review of Sociology* 14(3): 319-340.

Leweling, Tara and Mark E. Nissen, 2007. "Defining and exploring the terrorism field: Toward an intertheoretic, agent-based approach." *Technological Forecasting and Social Change* 74: 165-192.

Nye, Joseph, 1987. "Nuclear Learning and U.S.-Soviet Security Regimes." *International Organization* 41 (3): 371-402.

Pantucci, Raffaello, 2010. "The Tottenham Ayatollah and the Hook-Handed Cleric: An Examination of All their *Jihadi* Children." *Studies in Conflict & Terrorism* 33: 226-235.

Pape, Robert A., 2003. "The Strategic Logic of Suicide Terrorism." *American Political Science Review* 93(3): 1-19.

Reiter, Dan, 1996. *Crucible of Beliefs: Learning, Alliances, and World Wars.* Ithaca, N.Y.: Cornell University Press.

Sagan, Scott, 1993. *The Limits of Safety: Organizations, Accidents, and Nuclear Weapons.* Princeton: Princeton University Press.

Sageman, Marc, 2008. *Understanding Terror Networks.* Philadelphia: University of Pennsylvania Press.

Taylor, Max and John Horgan, 2006. "A Conceptual Framework for Addressing Psychological Process in the Development of the Terrorist." *Terrorism and Political Violence* 18: 1-17.

Weir, Margaret and Theda Skocpol, 1985. "State Structures and the Possibilities for 'Keynesian' Responses to the Great Depression in Sweden, Britain and the United States," in *Bringing the State Back In*, edited by Evans et al. Cambridge: Cambridge University Press.

Chapter 64

Analyzing ABC of Disaster Experience Using Text Mining

Halimahtun M. Khalid[1], Jenthi K. Radha[1], Xi Yang[2], Martin G. Helander[2]

[1]Damai Sciences Sdn Bhd
59200 Kuala Lumpur, Malaysia

[2]Nanyang Technological University
Singapore 639798, Singapore

ABSTRACT

This paper presents an analysis of A̲ffect, B̲ehavior and C̲ognition (ABC) of people who have experienced disasters, namely, natural, human induced, and pandemic. The goal is to understand their attitudes in disaster risk situations. We used secondary data from personal stories, research and media articles as corpus, sourced from digital libraries and the Web. The textual data was analyzed using Latent Semantic Analysis and Leximancer, to explore their capability in ABC mining. The classification of ABC was performed manually by two classifiers; there was substantial agreement between them. Univariate Analysis of Variance (ANOVA) was performed on the ABC data. The results showed that both text mining tools were similar in identifying ABC words. However, there were highly significant effects of disaster and corpus types on the frequency of ABCs. The findings were used to refine the citarasa model.

Keywords: ABC, Disaster risks, Text mining

INTRODUCTION

The ABC model by Albert Ellis (Wirga & Bernardi, 2008), explains the relations between the environment situation and emotional and behavioral reactions. *Affect* refers to human emotions or instinct such as anger, happiness, sadness; it also represents sensory (physical) experiences of feelings. As a prime determinant of subjective well-being, affect plays a central role in many activities. Generally, people are imprecise about their feelings as they are confused between affect and cognition. Therefore, they often rely on feelings for doing or not doing something (Berger, 2002). *Behaviors* are overt, observable responses and actions. They are measurable, and therefore more easily identified than affect or cognition (Berger, 2002). Some behaviors are much more difficult to develop or change than others, especially behaviors that are derived from cultural values (Oltedal et al., 2004). Once identified, it becomes clearer what needs to be done and what is within a person's reach. *Cognition* encompasses human beliefs, values, decision-making, and perceptions of self, others, and the world.

People experience risk in two ways: risk as analysis, and risk as feelings (Slovic et al,, 2004). Risk as analysis requires the use of logic, reasoning, and scientific deliberation. Risk as affect brings about fast, instinctive, and intuitive reactions to danger. These approaches of perceiving risk can be used to formulate an experiential system and an analytic system. The experiential system is affective and intuitive. It is fast, mostly automatic, and not very accessible to conscious awareness. The analytic system supports analysis using algorithms and rules, including formal logic, and risk assessment. Although analysis is important in some decision-making process, dependence on affect and emotion is quicker and easier. This is because affective reactions are immediate and automatic, and guide information processing and judgment (Loewenstein et al., 2001).

To understand disaster risk attitudes, we used the citarasa model (Khalid, 2007). *Citarasa* originates from Sanskrit but is widely used in the Malay world, to mean "intent" (cita) and "feelings" (rasa). Citarasa refers to individual's emotional intent, and it is used in this research to operationalize the measurement of risk attitudes. So, citarasa encompasses ABC and other system variables associated with disaster experience.

CITARASA MODELING

To analyze ABC of disaster experience, we adapted the citarasa model (Khalid, 2007) as illustrated in Figure 1.

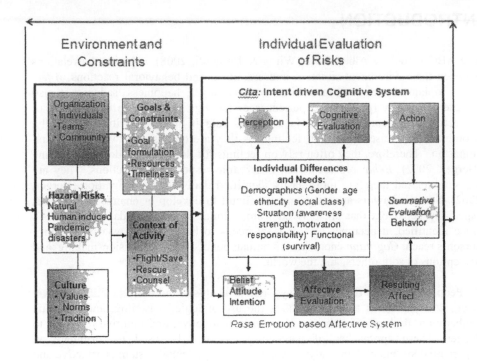

Figure 1. Citarasa model for disaster risk assessment

There are two subsystems: Environment and Constraints, and Individual Evaluation of Risks.

- **Environment and Constraints.** This subsystem documents information derived from sources that are associated with hazardous risks, namely: Organization of Teams, Context of Activity, and Culture. The goals and constraints could motivate or inhibit individuals from achieving their goals (e.g. rescue). The goals depend on the context of activity. For example, if the individual is the victim, the goal might be to survive; if a paramedic, the goal would be to aid victims. Constraints in the form of resources such as manpower and time affect an individual's goals. The nature of a disaster could also trigger organizational efforts driven by individuals, teams, or community. The cultural values, norms and traditions of individuals, teams or communities, play an important role in the organizational effort.

- **Individual Evaluation of Risks.** This subsystem describes the individual's evaluative process, resulting from the cross-functioning of affect and cognitive systems.

  Cognitive System. The intent driven cognitive system follows the human information processing model. Perception of risks triggers cognition and memory recall which lead to evaluation and action (Sjöberg, 1996).

  Affective System. The affective system generates a variety of emotions based on belief, attitude and intention. A belief that an earthquake was caused by upset

anocotoro would affoot attitudo and bohavior; pooplo may want to cvacuate, and yet they feel that they betrayed their forefathers and culture. The final decision for action then is negotiated between the affective and cognitive system.

Individual Differences. Individual differences may affect a person's perception and belief towards a risk. Gender differences also influence how risk may be represented. For example, Israeli women felt lesser control than men in the wake of a terrorist attack (Shiloh et al., 2007).

RESEARCH GOAL

The purpose of this research was to identify ABC of disaster risks in three categories of disasters: Natural disasters (e.g., tsunamis, earthquakes, hurricane, floods, fires), Human-induced disasters (e.g., terrorist attacks, major industrial accidents, vehicle accidents, technological disasters such as nuclear failures), and Pandemic disasters (e.g., swine flu, Spanish flu, Asian flu, SARS). In this paper we will report on one type of disaster in each category, namely: tsunami (natural), terrorism (human induced), and swine flu (pandemic).

We used two types of text mining tools: Latent Semantic Analysis (LSA) and Leximancer (LXM). *Latent Semantic Analysis* (LSA) is a method for capturing and analyzing the similarity of words and text passages by statistical computations applied to a large corpus of text. *Leximancer* (LXM) is a data-mining tool that can be used to analyze the content of one or more collections of text documents. The extracted information was displayed by means of a conceptual map that provided an overview or map of the text analyzed. This map represented the main themes and concepts in the text and how they were related.

METHOD

RESEARCH DESIGN

The investigation involved collection of secondary textual data on disaster experiences, gathered from digital libraries and Internet sources. The data comprised narratives of personal stories, research studies and media reports. The text was transformed into a standard format for content analysis by text mining.

HYPOTHESES AND VARIABLES

We tested three hypotheses. The first hypothesis, H1, assumed that there is a difference between the text mining tools in identifying words including ABCs. The underlying rationale is that if the tools have the same functionality, they should be able to mine the same words. However, there may be a difference in the frequency of words due to the algorithm being used in the mining process.

In the second hypothesis, H2, we predicted an effect of disaster type in the type of ABC mined. The reason is that the different disasters may have a different impact on attitudes due to the emotional intent involved. For example, a pandemic may be viewed less traumatic than a terrorist attack or a tsunami.

Thirdly, we postulated that the type of corpus might influence the type of ABC. This is because personal stories may have fewer expressions of ABC than research investigations, which focus on the purpose of the research.

The independent variables were: tool type, disaster type and corpus type, while the dependent variable was frequency of ABCs.

CORPUS CONTENT AND SIZE

The corpus size is 208,545 words, and the breakdown by disaster and corpus type is shown in Table 1.

Table 1: Corpus size by disaster and source type

Disaster type	Corpus type	Corpus of words
Tsunami	Research/media reports	52,694
	Personal stories/interview chats	39,931
Terrorism	Research/media reports	63,317
	Personal stories/interview chats	16,485
Swine flu	Research/media reports	19,022
	Personal stories/interview chats	17,096

PROCEDURE

The process is summarized below:

1. The corpus specimens were gathered from digital libraries and Internet sources;
2. Each specimen was screened to ensure that they contained ABC;
3. The text was converted to .doc format and transferred into a specimen template;
4. The number of words for each corpus specimen was counted;
5. Frequency analysis was performed using LSA and LXM;
6. The identified concepts were transferred into Excel spreadsheets;
7. ABCs were identified manually using two research assistants. They evaluated each concept and classified it as either A, B, or C. The LSA parts-of-speech, thesaurus, and corpus were used as references during classification;
8. The number of ABCs identified was investigated using an inter-classifier reliability test. To determine agreement, Cohen's Kappa value was used;
9. ABCs were tabulated for statistical analysis using univariate ANOVA to test the hypotheses. The Statistical Package for Social Sciences (SPSS) Version 15 was used to perform the analysis.

Text Analysis Technique

LSA involved four steps: First, essential words from the corpus were extracted. To form the word-by-document matrix, the corpus was subjected to two text processing stages: the corpus was first assigned to parts-of-speech. The natural language processor POS Tagger from Stanford University was used (Toutanova, et al., 2006). Then content-bearing words were extracted from the corpus, including nouns, verbs and adjectives.

Second, a word-by-document matrix and log-entropy matrix were created. The former counted the number of occurrences of each essential word in each utterance. The word-by-document matrix may be transformed to a log-entropy form, which expresses both a word's importance in the particular passage and the amount of information the word carried in the passage. Once a word-by-document matrix has been created, the number of occurrences of each word was calculated by adding each row in the matrix. Third, Singular Value Decomposition (SVD) was applied to construct a k-reduced matrix. MATLAB® was used to implement SVD and to calculate the k-reduced approximate matrix. Finally, semantic coherence was measured. MATLAB® was utilized again to compute correlation between any words or passages.

For LXM analysis, the same data set was used and preprocessed to convert the raw documents into a format in order to identify sentences and paragraphs. Important concepts were automatically identified from the text. To obtain meaningful and rich concepts, the identification was set to 500 concepts. The concepts were transferred into an Excel spreadsheet for classification into ABCs.

RESULTS

INTER-CLASSIFIER RELIABILITY

The classifications of the two classifiers were compared, and Cohen's Kappa was calculated as shown in Table 2.

Table 2. Classification of ABCs by two classifiers

		Classifier 2				Total
		None	Affect	Behavior	Cognition	None
Classifier 1	None	0	7	38	2	47
	Affect	18	111	9	5	143
	Behavior	15	1	362	4	382
	Cognition	5	1	11	57	74
Total		38	120	420	68	646
Measure of Agreement : Kappa value						.678

Table 2 showed the agreements and disagreements between classifiers. There were agreements between the two classifiers for Affective (N = 111), Behavioral (N = 362) words, and Cognitive (N = 57) words. The value of Cohen's Kappa is 0.678, p<0.0001, indicating that there was substantial agreement between the classifiers in classifying ABCs. The averages of the identified ABCs were used to test the following hypotheses.

HYPOTHESES TESTING

H1 - *There is a difference between the text mining tools in identifying words including ABCs.*
The ANOVA revealed F (1, 216)=0.06, n.s. Therefore, the hypothesis was rejected. Both tools identified similar words of Affect = 42, Behavior = 52, Cognition = 14.

H2 – *There is a difference between disasters in the type of ABC mined.*
Table 3 shows that there is a significant effect of disaster type, F(2, 530) =6.95, p< 0.001. The hypothesis was supported. ABC was also highly significant, F(2, 530)=13.47, p<0.0001. But the interaction Disaster * ABC was not significant.

Table 3. Effects of disaster and ABC on frequency counts

Source	Sum of Squares	df	Mean Square	F	Sig. level
DISASTER	3218.316	2	1609.158	6.948	.001
ABC	6236.987	2	3118.494	13.465	.0001
DISASTER * ABC	1758.420	4	439.605	1.898	.109
Error	120661.087	521	231.595		
Total	243706.000	530			

Figure 2 illustrates that there are more Affective and Cognitive words in terrorism, while tsunami generated more Behavioral type words.

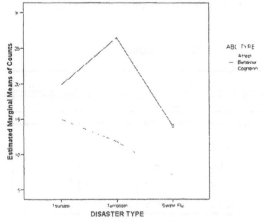

Figure 2. Average values of ABC for disaster types

H3 – There is a difference between the sources of corpus in the type of ABC mined
The results are shown in Table 4, where F (1, 530)= 45.43, p<0.0001, with research articles producing more ABCs than personal stories. This confirmed the hypothesis. ABC was also significant, F(2,530)=3.76, p<0.05. The interaction between ABC and corpus type was highly significant, F(2,530)=7.01, p<0.001.

Table 4. Effects of corpus type and ABC on frequency counts

Source	Sum of Squares	df	Mean Square	F	Sig. level
CORPUS	9484.666	1	9484.666	45.429	.000
ABC	1570.086	2	785.043	3.760	.024
CORPUS * ABC	2926.847	2	1463.423	7.009	.001
Error	109400.066	524	208.779		
Total	243706.000	530			

In Figure 3, the increase in affective words in research articles relative to personal stories was greater than for behavior and cognition.

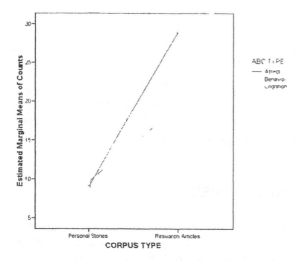

Figure 3. Average values of ABC for corpus types

From Figure 3, it is clear that there are more affective words in the research corpus, as compared to behavior and cognition words. Personal stories, however, contained relatively fewer ABC words. The difference was greatest for the affective words, which were much less frequent in the personal stories than in the research articles.

632

DISCUSSION

COMPARISONS BETWEEN LSA AND LXM

From the analyses, LSA took about 2 hours to process a data set (e.g., 40,000 words), while LXM took about 20 to 30 minutes. LSA misclassified some words because it separated the words according to parts-of-speech, whereas LXM omitted words that were non-lexical and weak semantics in order to identify meaningful concepts. LSA generated correlations between target words and other words, while LXM indicated the likelihood of occurrence between the target concept and other concepts. LXM produced thematic concepts maps that can be analyzed further at the node level to develop a semantic framework or ontology. Figure 4 illustrates the concept maps for terrorism and tsunami from personal stories.

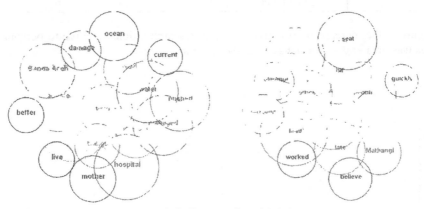

Figure 4. Concept maps – tsunami (left), terrorism (right)

ABC, DISASTER AND CORPUS TYPES

ABCs are important factors in understanding the effect of disasters, be they natural, human-induced or pandemic disasters. In this study large corpus of text describing the disasters were used to distinguish if there were differences in ABC for the different disaster types. Human induced disasters, such as terrorism, produced the greatest affect and greatest cognitive activity. Tsunami was second in affect and cognition, and swine flu was third, which may due to the fairly mild effect during the last year. But both terrorism and tsunami resulted in greater behavioral actions. The ABC of natural disasters is relatively smaller than for terrorism, although the cognitive activities are of similar size. This is not unexpected since there is much an individual can do to escape a tsunami or similar disaster; thereby, emotional intent in natural disaster is both affective and cognitive. It was assumed that personal stories would be a good source of ABCs, but it appeared that research corpus generated more affective words than behavior and cognition as the investigations may be directed at understanding emotions associated with disaster risks.

CONCLUSION

On the basis of its greater usability and added functionality, Leximancer was the preferred tool for future validation of Citarasa descriptors comprising ABCs. Due to traumatic experience, human induced disasters such as terrorism generated greater emotional impact than pandemic such as swine flu. Emotional intent in natural disaster, such as tsunami, is both affective and cognitive. The use of both types of corpus is important in developing comprehensive citarasa ontology and for further refinement of the citarasa model.

ACKNOWLEDGMENT

The authors acknowledge the support from the US Air Force, Grant No: FA2386-09-1-4009.

REFERENCES

Berger, P.L. (2002). The Cultural Dynamics of Globalization. In: P.L. Berger & S.P. Huntington (Eds.), *Many Globalizations: Cultural Diversity in the Contemporary World*. New York: Oxford University Press.

Khalid, H.M. (2007). CATER system for efficient mass customization of vehicles. In: *Proceedings of WWCS 2007*, KTH, Stockholm, Sweden.

Landauer, T. K., Foltz, P. W., and Laham, D. (1998). An Introduction to Latent Semantic Analysis. *Discourse Process, 25*(2&3), 259-284.

Latent Semantic Analysis, Website: http://lsa.colorado.edu/

Leximancer, Website: http://www.leximancer.com/

Loewenstein, G. F., Weber, E. U., Hsee, C. K., and Welch, N., (2001). Risk as Feelings. *Psychological Bulletin*, 127(2), 267-286.

Oltedal, S., Moen, B., Klempe, H., and Rundmo, T. (2004). *Explaining Risk Perception. An evaluation of cultural theory*. Norway: Rotunde Publication.

Shiloh, S., Guvenc, G., and Onkal, D. (2007). Cognitive and emotional representations of terror attacks: A cross-cultural exploration. *Risk Analysis*, 27(2), 397-409.

Sjöberg, L. (1996). A discussion of the limitations of the psychometric and Cultural Theory approaches to risk perception. *Radiation Protection Dosimetry*, 68, 219-225.

Slovic, P. (1987). Perception of risk. *Science, 236*, 280-285.

SPSS, Website: http://www.spss.com/

Wirga, M., and Bernardi, M. D., (2008). The ABCs of Cognition, Emotion, and Action. http://www.arcobem.com/publications/ABC%20of%20Cognition.html

Printed and bound by CPI Group (UK) Ltd, Croydon, CR0 4YY

21/10/2024

01777107-0019